THE HISTORY OF BIOLOGY

CHAPTERS IN
MODERN BIOLOGY
AND
BIOMETRICS

By Raymond Pearl:
THE BIOLOGY OF POPULATION GROWTH
ALCOHOL AND LONGEVITY
THE RATE OF LIVING

By Julian Huxley:
ESSAYS OF A BIOLOGIST
ESSAYS IN POPULAR SCIENCE

By William Morton Wheeler:
FOIBLES OF INSECTS AND MEN

GIRALAMO FABRIZIO

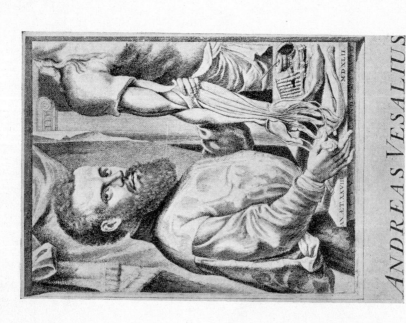

ANDREAS VESALIUS

By ERIK NORDENSKIÖLD

TRANSLATED FROM THE SWEDISH BY
LEONARD BUCKNALL EYRE

THE
HISTORY
OF
BIOLOGY

A SURVEY

TUDOR PUBLISHING CO.
MCMXLII   New York

Originally issued as

BIOLOGINS HISTORIA
in three volumes
1920–24
STOCKHOLM, BJÖRCK & BÖRJESSON

MANUFACTURED IN THE UNITED STATES
BY MONTAUK BOOK MFG. CO., INC.
NEW YORK CITY

FOREWORD

THIS work, which is here presented in the English language, is based on a course of lectures given at the University of Helsingfors, Finland, during the academic year 1916–17. It is the author's intention to present a picture of the development of biological science throughout the ages, viewed in conjunction with the general cultural development of mankind. Regarded thus as a link in the general history of culture, the problems of biology will, it is hoped, prove of interest not only to young university students, for whom this book is primarily intended, but also to a still wider public. With regard to modern times, for obvious reasons it has only been possible in such a brief history as this to give a very summary account of recent developments. A more thorough knowledge of the results of specialized biological research will be gained by reference to the literary works of professional biologists, which often contain a historical survey by way of introduction. On the other hand, the theoretical principles on which research work has been carried out have been discussed here in greater detail, both for the reason that records of them are not so easily accessible and on account of the influence they have exerted upon culture in general. In accordance with this principle a number of typical representatives of each trend of thought have been selected for inclusion and their work described, while no attempt has been made to present a complete record of all personalities that have figured in the biological world. In this, as in other historical *works, the selection has of course been made by a process of elimination, which to a certain extent was bound to be subjective; especially in a work dealing mainly with a general historical development it has been necessary to exclude the names of a great many brilliant specialists, in spite of the fact that their work may be of lasting value, while other personalities, perhaps in themselves of less importance, have been mentioned on account of the part they have played in the general cultural development of their period. For the same reason representatives of scientific progress in the various civilized countries of the world have been included, as far as space has allowed, in order to present as comprehensive an idea as possible of the progress of science and the contributions that different peoples have made thereto.*

For their assistance in preparing the English edition I take this opportunity of recording my thanks to Mr. Leonard Bucknall Eyre, B.A. Cantab., of Stockholm, who has translated the book from the Swedish, and to Mr. Alfred A. Knopf, who has promoted its publication.

Stockholm, November 1927   THE AUTHOR

CONTENTS

PART ONE

BIOLOGY IN CLASSICAL ANTIQUITY, THE MIDDLE AGES, AND THE RENAISSANCE

*

THE HISTORY OF BIOLOGY DURING THE RENAISSANCE

*

PART TWO

BIOLOGY IN THE SEVENTEENTH AND EIGHTEENTH
CENTURIES

CONTENTS

*

PART THREE

MODERN BIOLOGY

BIOLOGY DURING THE FIRST HALF OF THE
NINETEENTH CENTURY

*

FROM DARWIN TO OUR OWN DAY

LIST OF ILLUSTRATIONS

LIST OF ILLUSTRATIONS

PART ONE

BIOLOGY IN CLASSICAL ANTIQUITY,
THE MIDDLE AGES,
AND THE RENAISSANCE

CHAPTER I

THE DEVELOPMENT OF BIOLOGY AMONGST THE PRIMITIVE PEOPLES AND THE CIVILIZED NATIONS OF THE EAST

Primitive man's speculations upon life

THE EARLIEST FOUNDATION of all our natural scientific knowledge is to be sought in the observations of nature collected in the course of thousands of years by prehistoric peoples who had reached a primitive stage of civilization. This empirical folk-knowledge, which the student of folk-lore in our own day investigates from a historical and national-psychological point of view, has not only been the starting-point for all scientific thought, but has also, right up to the most recent times, to a certain extent influenced scientific research itself; increased its store of facts with material for observation and even now and then given rise to problems which science has debated. Primitive man's speculations upon life have naturally been influenced by his mode of life in various climates and under varying conditions. Common to them all, however, would appear to have been the fact that the first thing that has induced man to reflect upon life has been its cessation: death. And to the aborigines what we call a natural death is actually the most wonderful; that a man should fall in a fight against wild animals or his enemies is all part of the order of the day, but that the powers of a sound and healthy man should suddenly and without reason begin to fail and life to cease with or without the accompaniment of pain — that is a thing one finds it hard to acquiesce in. And the thing becomes all the more remarkable for the fact that again and again at night the departed one appears in dreams to those who have survived him. These dreams have given rise to a belief in ghosts, spectres, and spiritual powers of various kinds, both friendly and evil, and this belief has in its turn called forth measures with a view to deriving advantage from the well-disposed and avoiding the snares of the wicked. Thus measures of many and various kinds were adopted in regard to the bodies of the dead, which were either cremated or otherwise destroyed in order to render it impossible for them to return amongst the living, or else, on the other hand, they were elaborately cared for by the preservation of the skeleton or by embalming, which was intended to make the dead well-disposed towards their survivors. From these manipulations arose the first knowledge of the anatomy of the human body, while observations of the actual course of death created certain physiological ideas. Men learnt

3

to observe the heart-beat and to connect life with its continuance or cessation, and thus the heart itself was regarded as the organ of life. Breathing was also observed to be an essential condition of life, and in particular the deep expiration which indeed so often attends the actual moment of death gave rise to the idea of life as having something of the nature of air, being dependent upon the respiratory organs and leaving the body through them. In mediæval church paintings this belief reappears in a particularly naïve manner: the soul of the dying is seen to leave the body in the form of a little child creeping out through the mouth. Likewise the words of the biblical story of the creation to the effect that God breathed into man's nostrils the breath of life testifies to the same kind of idea. And so there arose, as a further development of these ideas, the belief that the breath or spirit lives when the body dies. The contrast between body and spirit which is an outcome of the ideas described above is included in the speculations of the earliest natural philosophers as a fundamental principle.

Relationship to animals

HOWEVER, in the mind of primitive man these lines of thought, proceeding from the contrast between life and death, are crossed by others, which have their origin in his relationship to the rest of the world of living creatures. The great wild beasts — bears, lions, elephants — were difficult to overcome; it was often necessary to try, as far as was possible, to make friends with them. Other beasts were regarded with terror for their night-roving habits and horrible cries, such as hyenas and owls; while some possessed otherwise enviable natural gifts — the fox his cunning, the deer his swiftness, etc. It is out of all this that we must explain the origin of the mass of animal superstition that has filled the life of both wild and civilized peoples. As forms in which this superstition has developed may be mentioned totemism, or the custom existing among certain wild peoples of adopting animals as a kind of guardian spirit and family symbol, as well as the belief in and worship of holy animals, which, even amongst highly civilized peoples, such as the Egyptians and the Romans, have played such an important part in life. This animal superstition has naturally contributed towards increasing the interest in and knowledge of animals, both as regards the habits of life of those which were worshipped as gods, and the anatomy of those which were offered in sacrifice and were most minutely examined with a view to divining portents for the future from their internal structure.

Primitive surgery and medicine

FINALLY, a third extremely important source of biological knowledge has been medical science. Primitive surgery, which originated in attempts to cure various bodily injuries, must of course eventually lead to a certain amount of knowledge of the anatomy of the human body, a knowledge which was increased by the process of comparison with the experience gained from the slaughter of wild and tame animals. As to the natural diseases, the same holds good for these as what has just been mentioned in regard to death; for lack of ability to explain them naturally, people took refuge in a belief in supernatural causes. The belief in enchantments of various kinds which arose therefrom and which has been maintained even amongst civilized peoples for a surprisingly long time, fills one of the darkest chapters in the history of civilization. Disasters of supernatural origin of course demanded corresponding remedies, and consequently the earliest practice of medical science among all races of mankind has been that of magic: they sought to remove the evil by setting sorcery against sorcery. However, the regular course of certain processes of disease could not fail to be observed and conclusions drawn therefrom as to the functions of the body in sickness and health. By a comparison of these observations a number of primitive ideas were acquired on physiology and pathology. Hand in hand with this was evolved the theory of pharmacology, based on experiments — originally for the most part for magical purposes — with plants which experience proved to be poisonous or otherwise capable of affecting the life-process. Through observations of this kind the knowledge of life was still further enhanced. It was not given, however, to just anyone to acquire all this knowledge, the origins and development of which have been described above. The supernatural and mysterious elements in them made them a privilege for certain qualified persons: magicians, sorcerers, sacrificial priests. Among these classes of people they were shared and handed down as professional secrets, until in course of time a division of them took place — the magical and ritual customs became the professional sphere of the priests, while the amassed knowledge of nature, released from the obstructive bonds of magic, was developed by independent inquirers into a free sphere of learning. The people amongst whom this independent natural science first arose were the Greeks. But long before Greek culture appears in history, the people of the East had already bequeathed historical evidences of their civilization, and these deserve all the more to be carefully examined for such contributions to biological knowledge as they may have to show, seeing that the whole of Greek culture was so highly influenced by the oriental.

Babylonian science

THE earliest home of human civilization is now generally supposed to have been Babylon, and a high standard of culture was maintained there under the dominion of various types of peoples up to the latter part of the Middle Ages. The "oriental wisdom" which has played such an important part in the mystical literature of all times also originates from there through a more or less varying number of intermediate stages. Actually, the mystical and the magical have from the earliest times played a predominant role in that country's learning, undoubtedly owing to the fact that all knowledge was nurtured and developed by a powerful priesthood. The conception of nature was influenced thereby: the early knowledge of astronomy was placed at the service of mystical powers, as were also mathematics and medicine. The latter science, however, in certain respects made no small progress. The knowledge of anatomy was considerable; preserved clay-models of certain of the viscera of the body prove this and give evidence that the dissection of corpses must have taken place in spite of the horror which Orientals have always felt for the dead and their spirits. It is clear from preserved writings on medicine that the heart was regarded as the organ of intelligence, and the liver as that of the blood-circulation; the blood was divided into "light" and "dark" blood — arterial and venous. The knowledge of higher animal forms was, as extant lists of nomenclature go to prove, quite considerable, and kings and princes kept rare live animals in their gardens. Even animal-doctors are mentioned in preserved inscriptions.

Egyptian medicine and natural knowledge

AGAIN, in the other oldest civilized country of the West, Egypt, there was developed at an early period an art of healing which was based not merely upon superstition, but also upon actual observations. The early perfected religious practice of preserving dead bodies from putrefaction by conserving the skeleton and, later, by embalming offered an opportunity of acquiring anatomical knowledge which proved of great benefit to medical science. The sacred animals were likewise studied with minute care, and writings have been discovered giving in detail the history of the development of the sacred scarab, and even the metamorphosis of the frog and the fly. The parasitic worms that so infected Egypt were also objects of investigation and speculation.

Israelitic conception of nature

WITH regard, finally, to the Israelitic people, their cultural contribution has been in a sphere entirely different from the natural-scientific; namely, the ethical-religious. Their material, and thereby also their scientific, culture was borrowed from the earlier developed and powerful neighbouring peoples and may therefore be passed over here. Nevertheless the Israelitic conception of nature as preserved in the Old Testament, has, owing to religious causes, right up to our own day had a deeply significant influence. The part played by the six days' creation as a co-determining factor even in purely scientific explanations of the world is too well known to need close examination. Likewise the ordinances of the Mosaic law regarding clean and unclean animals have had their great importance for the conceptions of nature held by the Christian peoples, while even the well-known problem of the ruminant hare is still today a subject of lively discussion in certain circles. And undoubtedly, even in the far distant future, the religious·

dogmatic currents of thought which have always had, and indeed always will have, a powerful influence on the development of human culture will receive guidance from this quarter.

Hindu and Chinese science

THE civilized peoples of eastern Asia, the Hindus and Chinese, have likewise contributed very little of importance to the development of the science of biology. Hindu science, indeed, especially in the sphere of mathematics, reached a high standard, and the tendency to employ figures even in the other branches of learning which this people cultivated is unmistakable. Thus a Hindu work on medicine states that the human body has seven skins, 300 bones, 107 joints, 900 tendons, 700 blood-vessels, and 500 nerves. But they had very primitive ideas as to the functions of these organs, and similarly the various fluids and kinds of air which provide for the body's renewal are of interest to them more from the numerical than from the functional point of view. Chinese culture, again, has essentially occupied itself with ethical and social problems. Chinese medicine has on the whole advanced little beyond that of primitive peoples, although certain isolated instances of progress achieved — for example, smallpox inoculation — might perhaps be traced back to the experiences of this people. Even pure zoology has on the whole made no advance; as early as about a thousand years before Christ mention is made of an imperial zoological garden, but the thorough study of the causal connexion in living nature did not come within the sphere of Chinese interest.

CHAPTER II

THE EARLIEST GREEK NATURAL PHILOSOPHY

The Greeks: creators of natural science

IF THE BABYLONIANS and Egyptians thus succeeded in collecting quite a considerable mass of individual facts of science, it was nevertheless left to the Greek nation to deduce from these facts a consistently realized conception of nature — not free from mystical and magical influences, it is true, but still striving more and more after a natural explanation of the laws of existence. There has been much speculation as to why it should be amongst just this people, who were not only few in number, but were also politically divided, that such a splendid development of human thought should have taken place. The deepest cause is surely to be sought in the much discussed, yet fundamentally so inexplicable national character, in the spiritual and cultural disposition of the people. It may, at any rate, be worth while briefly considering its manifestations in the social sphere, in order to gain some idea of the external conditions of development under which free thought was here able to expand.

The people of Greece, as is well known, never achieved political unity; it remained divided into a number of small communities independent of one another, consisting usually of a city with its surrounding country district. Trade and shipping rather than agriculture were the people's main source of income. Over-population gave rise to splendid colonizing activity along the coasts of the Mediterranean; the colonies, which from the very beginning were made independent of the mother city, adopted the latter's institutions. A strong national feeling prevailed everywhere and was maintained by law and custom. Outside the boundaries of his own town the Greek was a foreigner without rights, without the possibility of acquiring civic privileges elsewhere, and with no prospect of winning the consolations of religion. Religion was in fact as localized as the communities themselves; every town had its own gods, which could be worshipped only by its citizens and within its boundaries. Such a local form of religion was naturally primitive and remained so even at the time when Greek culture was at its zenith. It was just on account of this lack of a more highly developed religion, however, that free thought was able to develop as it did. Here there was no priesthood, as there was in Babylon, Egypt, and India, to reserve to itself alone the right to the higher learning and to ensure that its results

8

did not conflict with the ancient religious usages. The religious persecutions that occasionally took place in Greece against thinkers, as, for instance, against Socrates, were rather the work of the mob than of the defenders of religion and were therefore of a purely incidental and transitory nature. On the other hand, we find that many of Greece's oldest philosophers were priests or at any rate the sons of priests. And just as religion in ancient Greece was primitive, so also were the moral ideas: provided the citizen obeyed the ancient laws of the State, he need not worry much about what further duties were owed to his nearest and to himself. Thought was thus at liberty to turn to external nature and devote itself to speculations on how things arose and why the world and the living creatures in it were formed just as they were. The oldest Greek thinkers were therefore natural philosophers, while it was not till later that the ethical problems — which, for instance, among the thinkers of the Jewish people, the prophets, had from the very beginning dominated the soul — through Socrates found a place in Greek thought and finally, in late classical times, entirely supplanted the interest in nature and its phenomena.

These, mankind's earliest natural philosophers, went about their work under conditions which in most respects were utterly primitive. The general education amongst their neighbours was extremely limited and far from widespread — in fact, throughout the whole of the classical period of Greek culture it was confined to a very few. The public instruction provided by the State for the benefit of its citizens was of the simplest kind; in Athens in the time of Pericles, when the greatest philosophers and poets of Greece were assembled there, the citizens had to learn in the State schools only the simplest rudiments of reading, writing, and arithmetic, besides music and gymnastics, which were necessary for military service, while it is said that at the same period in the more conservative country of Sparta the majority of the people were illiterate. Anything that the private individual wished to study beyond that, he had to find out for himself as best he could. Nor were there in ancient times any private professional teachers. If a person of studious mind happened to belong to a family connected with the priesthood, its traditional learning was naturally at his disposal as a foundation; for the rest he had to rely upon whatever knowledge he could acquire in his own city from foreign travellers and such of his countrymen as had travelled abroad, unless he himself was rich enough to travel and visit learned men in their own homes. Fortunately hospitality in Greece in ancient times knew no bounds; in actual fact it took the place of learned schools and universities and even of books and writings. For if the knowledge of writing was rare, this was to a great extent due to the difficulty of obtaining writing-materials. The Egyptians had discovered a cheap material in their papyrus, the Chaldees another in their clay tablets, but the ancient Greeks had nothing but metal tablets and animal hides, both of which were expensive to get and inconvenient to preserve. The learned therefore had to express their opinions in short and weighty compositions, preferably in the form of verse, so that they could be easily learnt by heart. Thus learning became the asset of a privileged few; they had to be wealthy in order to be able to undertake the journeys that were essential for the acquiring of knowledge, and of high standing in order to be able, both at home and abroad, to gain access to the masters who were primed with the wisdom of the period. But in point of fact the scholars of those days were highly respected: the various states summoned them to be lawgivers and rulers, paid the expenses of their costly journeys, and gave them financial assistance when they ruined themselves over their research work. On the other hand, they were often persecuted by hostile political factions and were sometimes condemned to end their days in exile.

The earliest scientists of Greece: the Ionian philosophers

THE earliest of these Greek natural philosophers, the so-called Ionic philosophers, all lived in, or at any rate originated from, the colonies which the Ionic tribes of Greece founded on the coast of Asia Minor. Through trading with the Orient these cities rapidly grew wealthy, and through contact with the more highly cultivated peoples of the East there arose a keen desire for knowledge, and means for satisfying it were obtainable. Chaldean and Egyptian travellers were able to tell of the great learning of their priests and physicians; journeys to the East gave the ambitious Ionians opportunities of acquiring at least something of that secret knowledge. And on this foundation they themselves built further. — Nature presented a great number of phenomena in constant variation, and herein it was proved that certain phenomena always stood in a certain regular relation to one another. A common primary cause of the variations of existence had to be discovered — a common element out of which everything originated. What was this primary element and how have things originated from it? These two questions occupied the minds of the Ionian thinkers. Nature, the Greek $\phi\acute{\upsilon}\sigma\iota\varsigma$, became the one great problem, and these ancient philosophers who studied the problem of nature were therefore called physicists, a name which later on was reserved for those who carried out research in a limited sphere of natural science. The investigations carried out by these ancient physicists, however, led them just as often into the realms of metaphysics, and it is just that lack of insight into the insuperable bounds of natural science that gave to their speculations that vague and fantastic character which is so conspicuous in them.

As one of the earliest of the natural philosophers in Greece is mentioned THALES of Miletus. Even in ancient times very little was known of his life and activities. The very epoch in which he and his immediate successors lived has been so variously stated that the dates differ by centuries. It is most generally assumed, however, that he lived between 650 and 580 B.C. Historians agree in declaring that he left no writings; perhaps he was not even able to write. He was probably of Phœnician origin — some assert that he immigrated from Phœnicia. At any rate it is clear that he had educated himself by travelling and studying in the East. He was very rich and of high standing and collected around him a number of disciples. Of his philosophy it is mentioned that he regarded water as the cause of all things. The earth floated like a disk on a vast sea which surrounded it on all sides. The details of his philosophy are unknown, but the assumption mentioned above is to a certain extent reminiscent of the story of the creation in Genesis, with its definite assertion of "waters which were under the firmament" and "waters which were above the firmament." That we are here dealing with a theory of oriental origin seems beyond all doubt. That Thales was the pioneer of the Greek natural philosophy is undeniable; he is unanimously acclaimed as such by the thinkers of antiquity. The very name of philosopher and philosophy probably originates from him. Once asked whether he was a wise man (σόφος in Greek), he modestly replied that he could not call himself one; he was merely a lover of wisdom (φιλόσοφος).

A younger fellow-countryman of Thales, and in all probability a disciple of his, was ANAXIMANDER, who lived approximately between the years 611 and 546 B.C. Concerning his life and personality about as little is known as of that of Thales. On the other hand, it is known that he described the results of his scientific researches in a poem On Nature (περὶ φυσεῶς), which is quoted by several later philosophers. Even Aristotle declares that he had read it, but it seems to have been lost as early as the later classical period; the people of antiquity had not such great respect for "classical" authors as we have in our time. Through quotations and references in the writings of later authors, however, it is possible to form some idea of this, the first work on natural science ever written. Just as for Thales, the most important question for him is: What is the material cause of the universe? As mentioned above, Thales held that water was the causal principle; Anaximander conceives it to be "apeiron" (ἄπειρον), from which he supposed the things on earth to develop themselves and into which he supposed them to return. What he actually meant by this "apeiron" it is difficult to say, but the word probably means "the quality-less, the indeterminate." Out of this primordial cause have arisen heat and cold, from these water, and from that again earth, air, and fire, which last surrounds the atmosphere and is radiated through the stars. The earth came into being through a kind of condensation of water; it was originally composed of pristine mud and then became solid and floats as such on the water, in form like a spherical segment — that is, very much like a loaf. He is said to have

designed a map of the world and even to have made a celestial globe of spherical form, with the earth suspended in the centre of the circular vault. Living
beings he conceives as having evolved through a kind of primordial procreation in the mud which formerly covered the earth. Thus, first there
arose animals and plants, and then human beings, who, originally formed
like fishes, lived in the water, but afterwards cast off their fish-skin, went
up on dry land and thenceforth lived there. We see, then, that Anaximander
produced a complete theory of evolution, childishly clumsy, it is true, but
interesting for the audacity with which he deduced his conclusions from
his premises. Nor is he afraid of letting the world-process continue into infinity; the present universe has been preceded by others, which were evolved
out of the primordial element and returned once more to it, and so it will
always continue. We should not go too far, however, in making a comparison
between the ancient Ionian's theory of creation and the evolution theory of
our own day. An attempt has been made to prove him a predecessor to Darwin on the ground of his above-mentioned conceptions of the origin of
man. This is an entirely unhistorical view of the matter, although an easily
accountable one; highly debatable theories have always sought for direct
predecessors as far back in time as possible. Anaximander's theory of the
origin of man is in reality most reminiscent of his fellow-countrymen's
legends of autochthonism — stories of how men were born of the earth they
lived on, which was one of the very popular myths of these periods of migration of peoples, whereby they sought to justify their title to the country
they possessed on, as it were, semi-natural, semi-divine grounds. But if we
must exercise caution in gauging the speculations of Anaximander by modern
standards, they at any rate are worthy of our high admiration. Natural research in our own day endeavours to discover an explanation, based on a natural connexion of causes, of the origin of things and of the variations they
present, and the philosopher who was the first to realize the necessity of
such a natural explanation and who worked it out, although incompletely,
must be regarded in all ages as one of the pioneers of human thought. The
religious-poetical myths of the creation, which up to that time had had to
serve amongst the Greeks, as in the East, for an explanation of the cosmos,
became from this time part of the sphere of poetry and the life of religious
faith, while scientific research went on building upon the foundations laid
down by Anaximander.

Among his contemporaries Anaximander enjoyed a high reputation, and
several of his disciples are mentioned in history. In his native city his work
was carried on by ANAXIMENES, who chose air for his principle and considered
that this not only enveloped the world, but also penetrated living beings
and represented their life-principle. Shortly after his death the city of Miletus
was ravaged by the Persians and razed to the ground (494 B.C.), and with

that the city of Thales and Anaximander disappears for ever from cultural history. Their theories, however, had been widely dispersed, and when the Asiatic Greeks lost their cultural supremacy, philosophers and philosophic schools were already to be found scattered throughout the world of Greek culture.

Thus DIOGENES OF APOLLONIA, in Crete, is regarded as one of the Ionian school of philosophy. He lived in the first half of the fifth century and is not to be confused with the more famous Cynic Diogenes, who lived in the time of Alexander the Great. His explanation of the universe is based on An-aximenes' theory of air as the primary matter. Out of the air are formed all other elements in the world through a process of condensation. He conceives life to consist of warm air moving like currents through the veins and thus preserving the strength in the body. Diogenes has described the ramifications of the venous system in man, or rather in mammals, and this description is still partially extant — the earliest anatomical work known. For the rest, we know of Diogenes that he conceived living beings to have been produced out of the earth through the influence of solar heat — that is to say, a fur-ther development of Anaximander's theory. Also, he believed that the embryo in the uterus was developed by the warmth of the mother out of the semen of the father. His embryological statements must therefore, like his anatomical ideas, have been based on dissection. A contemporary of his was HIPPO, who is also said to have engaged in embryological research. Unfortunately we know very little about him; we are not even certain of his birthplace, which some say was the Isle of Samos and others Rhegium, in the south of Italy. His reputation as a philosopher seems to have been in-ferior to his fame as a naturalist, which is largely the reason for his having been almost forgotten. He is said to have maintained Thales' theory of water as the origin of matter.

This survey of the old Ionic natural philosophy shows that there was a serious attempt made to discover a natural connexion in the events of earth, in the existence, origin, and decay of matter. However, partly through the adoption and development of its ideas and partly through fresh influences from the East, there grew up side by side with it other lines of thought, hav-ing in some ways a deeper vision of the phenomena of life, but at the same time also a tendency to mysticism and fanciful ideas which had been foreign to the Ionian philosophers. Although it was not particularly interested in biological research, it is necessary here to mention the Pythagorean philoso-phy, owing to the important part it plays in the history of culture in general. Its founder, PYTHAGORAS, is one of the most extraordinary figures in cultural history. Scientist, religious prophet, and statesman, mathematician, and mys-tic all in one, he has become in the tradition of posterity a purely legendary figure. Born in Samos off the coast of Asia Minor, he travelled widely in the

East — how far is not known with any certainty — and afterwards taught in his native island, but on account of political disturbances he was forced to migrate to the Greek colony of Croton in the south of Italy. There he carried on as a research-worker and religious and social reformer until his death, probably about the year 500 B.C. His dates are in any case highly uncertain and much disputed. There is no doubt that much of the wisdom he taught emanated from the East; his famous theory of the wandering of the soul, for instance, already existed in India long before his time, and the geometrical theorem named after him had already been proved by Indian mathematicians long before he lived. His cosmological theories, however, are extremely interesting. He conceived fire to be the origin of matter. It should be borne in mind that to the people of antiquity and to many succeeding generations fire was not a chemical process, but an element, like air, water, and earth. Pythagoras believed that all things originated in a primordial fire forming the centre of the cosmos. Around this primordial fire revolve all the celestial bodies, the earth, the planets, and the sun. The shape of the celestial bodies is spherical and their orbits circular. This cosmology, as compared with that of the Ionians, represented an immense advance. Through him the fact of the globular shape of the celestial bodies was introduced into science, although it took a thousand years to penetrate the consciousness of the world in general. Still more remarkable was his theory of the earth as a moving, revolving body. This theory the people of antiquity found it impossible to accept;[1] Copernicus was the first to take up the theory anew and was actually accused by his opponents of Pythagoreanism. In the sphere of mathematics Pythagoras was also a pioneer; he discovered the regularity of number-series and was led by his speculations in mathematics to propound the fanciful and mystical theory that numbers govern matter, and even that numbers are the principle of matter. Further, he even included tonal harmony in music in this mystical system of thought; his theories on the "harmony of the spheres" are as well known by name as they are difficult to understand in substance.

Pythagoras' influence on scientific development was very great and was also considerable in the political life of his day. His disciples founded a strict order, or kind of sect, which worshipped in Pythagoras a divinely inspired prophet. Persecuted during the most brilliant period of Greek democracy for their pronounced aristocratism, they were finally dispersed, but their teachings experienced a revival in late classical times.

The Greeks of the West

WITH Pythagoras the nationality of western Greece assumes a place in scientific history. Southern Italy as far as Naples, as well as Sicily, had been

---

[1] Aristarchus of Samos taught, it is true, at a far later period that the sun is the centre around which the earth revolves, but this theory soon fell into oblivion.

colonized by the Greeks at an early period, and by the extermination or the assimilation of the aboriginals a homogeneous Greek nationality had grown up here, split up into small states, as in the mother country, and living, if possible, under still worse conditions of political unrest. But their intellectual culture rivalled that of the Asiatic cities. A peculiarity of the western Greek philosophers was that a number of them were, like Pythagoras, imaginative prophets and imperious statesmen, and at the same time keen research-workers, while they founded schools of a far stricter order than their Ionian predecessors. One philosophic school of this kind was the Eleatic school, called after the city of Elea, in southern Italy. There came to this city at the end of the sixth century B.C. a man whose name was XENOPH-ANES, born at Colophon in Asia Minor and a disciple of Anaximander. Disturbances in his native city had driven him into exile and he had wandered far and wide in the greatest poverty, supporting himself by reciting his own poems in the towns he visited. Finally in Elea he found a place of refuge and died there about 490 at a very advanced age. The results of his scientific researches he has described in a poem, similar to his master Anaximander's treatise *On Nature*. Some fragments of this poem are still extant. In spite of the extraordinary audacity of the ideas which it contained, it won its author a great reputation and a large number of disciples. He based his ideas on Anaximander's theory of the origin of the world through the condensation of water and primordial mud, and he developed it still further. Of interest in this connexion is his pointing out fossilized marine animals high up in the mountains, which he declared to be a proof that the mountains were at one time under water. These ideas were neglected, mainly owing to the fact that Aristotle and his disciples regarded fossilization as one of the *"lusus naturæ."* It was not until the Renaissance that Xenophanes' more correct views once more came into their own. But speculations as to the origin of the world drove Xenophanes further and further over to purely theological problems. He became a keen and eloquent opponent of his fellow-countrymen's belief in a plurality of gods, which he despised on account of their purely human limitations; horses and oxen, he declared, would, if they thought as men, imagine gods in the form of horses and oxen. On the other hand, he for his part maintained the eternity and unfathomableness of divinity, and consistently therewith the eternity, unity, and immutability of the world in which we live. This did not prevent him, however, from embracing Anaximander's theory of alternating evolution and annihilation of the earth and all that lives on it. But it was the theory of immutability that his disciples further developed. The most famous of these, PARMENIDES of Elea, vigorously maintains the unity of being. The world is conceivable, he declared, only if we disregard the variations and changes and seek the immutable. In this connexion he warns us against relying upon the senses, whose judgment is

misleading and from which he appeals to reason. His abstract conception is
also seen in the antitheses by which he expresses existence: hot as opposed to
cold, light as opposed to darkness, the *"ent"* (being) as opposed to the *"non-ent"* (not-being). In regard to man, he conceived the soul, which to him was
the same thing as life, as hot and the body as cold. For the rest, he accepted
Anaximander's theory of the origin of living creatures and the doctrine of
the Pythagoreans as to the globular form of the celestial bodies. His greatest service to mankind lies in his insistence upon logical consistency in
thinking; in this he far outstripped the Ionian philosophers and strongly
influenced the thinkers of the ages that followed. The later Eleatics finally
pursued the theory of immutability to sheer absurdity and thereby rendered it
untenable. The Eleatic Zeno, for instance, denied all change and even motion.

Of far greater importance for the development of biology than the
Eleatics, however, was another western Greek philosopher, EMPEDOCLES,
of Acragas, in Sicily. His period of activity is generally placed in the middle
of the fifth century. Around his personality and way of life there has grown
up, as around Pythagoras, a number of legendary tales, which prove, if
nothing else, that he must have very greatly impressed both his contemporaries and posterity. And this seems to have been very much his own intention. He boasts about himself in the writings, fragments of which have been
preserved to us, concerning his own supernatural gifts; he claims that he has
power to heal the sick and cure the infirmity of old age, raise the dead,
change the direction of the wind, and bring rain and sunshine upon the earth.
And he delights in being acclaimed; adorned with chaplets and flowers, he
goes in procession into the city that besought his help and is hailed by the
inhabitants almost with the reverence due to a divinity. In our days this
would, of course, be characterized as shameful humbug, but in early times
it was apparently not so. Undoubtedly Empedocles himself believed in his
miraculous powers, and the taste for pageantry he shared with his own
countrymen. History also relates a number of serviceable acts he performed;
for instance, he improved the hot and unhealthy climate of his native city by
making a breach in the mountain wall which shut out the cool north wind;
he rid a neighbouring town of malaria by arranging for the draining of the
district. He was, besides, a leading politician; although descended from a
distinguished family, he was a keen democrat; he overthrew the oligarchies
in his native city and set up a popular government; the honour of kingship,
which was offered to him, he declined. However, his enemies prevailed over
him and he had to flee in exile to Greece, where he died. Shortly after his
death Acragas fell into the hands of the Carthaginians and was razed to the
ground. True, the city again flourished in the time of the Romans under the
name of Agrigentum, now Girgenti, but the part it played in the history of
culture was at an end.

The four elements

As a philosopher Empedocles bases his theories on Parmenides. The world is uniform, immutable. Nevertheless changes do take place, but these must be explained by movements in existing matter and by the alternating commixture and dissolution of its component parts. As the fundamental material causes Empedocles postulates fire, air, water, and earth — in other words, the four elements or roots. The theory of these elements, which has been maintained, one is almost tempted to say, up to the present day, originates, as was universally acknowledged by antiquity, from Empedocles. And if he asked what it was that produced the motions in the four elements which caused the changes in their constitution, he would answer: love and hate; love acts as attraction, hate as repulsion. Through their alternating predominance are produced all the changes in nature. Love, when it ousts hate, gives rise to new worlds; when hate predominates, they are again dissolved. The world was formed by parts of the four elements becoming united in a work; through the same kind of motion the water elements were flung out of the originally humid earth-mass, so that the latter became dry and habitable. If fresh beings arise or such beings as have existed perish, it is not to be concluded from this that something can arise out of nothing or become nothing; for whence could anything new come to that aggregate of reality that exists, or whither could anything already existing disappear? No, the cosmic matter was and remains the same, only its component parts are mixed in different ways through love and hate alternately predominating. Living creatures he conceives as having arisen out of the earth; first, plants, whose life he compares in detail with that of the animals; their nourishment is procured through pores in the stem and leaves, and their germination is comparable to the reproduction of animals. Animate organisms have likewise originally sprung from the earth; first arose individual limbs and later, through the powers of attraction of love, these were conjoined; from them then arose animals themselves. But this development did not proceed undisturbed, for as the conjoining of the separate parts took place by accident, it was entirely a matter of chance whether beings capable of life or malformed monsters arose. Mankind also originated in a similar way; through the co-operation of the subterranean fire there were cast up out of the interior of the earth shapeless lumps which formed themselves into limbs, from the union of which man developed. Men, who are of a warmer temperament, came into being in a southern climate, while women, the more cold-blooded sex, were created in a more northerly climate. In reproduction the embryo received some parts of the body from the father's and the others from the mother's seed. Growth in childhood is due to increase of warmth in the body, and the infirmity of age to its diminution. Respiration he believed to be effected not only through the windpipe, but also through the pores of the skin; as the blood is conveyed alternately to and from the skin, the air is inhaled in conjunction with it. The perceptions of the senses are due to the objects that are perceived or sensed giving off fine particles, which unite themselves to the corresponding components in the organs of sense. Thus the various elements are apprehended by corresponding elements in the organs of sense: water by water, air by air, etc. Tones are created by the air brought into movement forcing its way into the auditory duct of the ear as into a trumpet. Empedocles is said to have been the first to describe the labyrinth of the ear. The eye he compares to a lamp; light is distinguished by the eye's fire-components, darkness by its water-particles. Even in the operation of thinking he sees a purely corporeal function; the blood, in which all the elements are most minutely commingled, is the seat of intelligence, but in this other parts of the body can also co-operate; elsewhere in the operation of thinking the four elements in the thinker and the thought seek one another. The more finely and evenly the elements are mixed in a man, the better does he think; if the correct mixture is confined to particular parts of the body, then these parts are more highly developed than the others. — In curious contrast to this materialistic theory of sensation are several utterances of Empedocles in which, like Parmenides, he warns us that the evidence of the senses cannot be implicitly trusted; they can deceive, while the reasoned thought is infallible.

We cannot here enter into a discussion of Empedocles' religious theories. Like Pythagoras, he believed in the migration of the soul and likewise forbade his disciples to kill and eat animals. Even certain plants were in this respect sacred. For the rest, it is difficult to reconcile his mystical religious pronouncements and his miracle-working activities with his natural philosophy, which so strictly emphasized the doctrine of causality. From a closer knowledge of the conditions under which he lived we might well have been able to explain the riddle; now he stands as a unique, strange phenomenon in the history of biological research.

Empedocles' scientific influence

In this sphere he undoubtedly deserves a high place. In particular his speculations on the constitution of matter and its changes are worthy of attention. While his predecessors and even his successors for long ages afterwards had no other natural grounds of explanation to offer in regard to these phenomena than motion in space, Empedocles comes forward with a kind of doctrine of affinity, crude and clumsy, it is true, but nevertheless containing within it the germ of a number of ideas which it was only possible for far later ages to think out. And the fact that these ideas were not adopted by his successors, that antiquity failed to produce any science of chemistry, does not detract from their interest. His physiological speculations, naïve though they are, also give evidence of his keen powers of observation and combination. In his curious theory of the creation of living organisms from scattered parts brought together by chance an attempt has been made to see a kind of primitive theory of selection. That is, of course, an exaggeration; his anthropogeny gives even clearer evidence than Anaximander's of being derived from the old legends of autochthones.

With Empedocles the western Greek school made their final and most important contribution towards the history of biology. A couple of hundred years later this people saw the birth of their greatest philosophical genius, the physicist Archimedes, who lived to see the destruction of the liberty of western Greece. He was indeed a specialist in his own sphere, which has no place in this work. We return therefore to Asia Minor, whence sprang the whole of Greek natural philosophy and where the Ionian succession was still being carried on with achievements of great importance for its further development.

From its point of departure in the Ionian philosophy HERACLEITUS of Ephesus (about 510–450) developed an entirely new idea of nature. He was one of the greatest philosophers of antiquity, called "the dark," on account of both his obscure style of writing and his gloomy view of life. He came from a distinguished priestly family, but resigned an eminent government post to devote himself entirely to philosophy. His life, too, however, was disturbed by political revolutions. As a thinker he seems to have been essentially auto-didactic, and though his system is to a certain extent based on the Ionian philosophy, mainly on that of Anaximander, it nevertheless bears an entirely original stamp. In contrast to the Eleatics' assertion of the immutability of the universe, Heracleitus sums up his teaching in the proposition that everything is mutable, that mutability is the essence of existence. "Struggle is life" is a saying that comes from him; "All is flux" is another. He regards fire as the causal principle of the cosmos. Everything has arisen from a primordial fire, to which everything returns, for worlds arise and perish alternately. Heracleitus saw divinity in the primordial fire. Fire is also the soul of man; fire is inhaled in breathing, and its cessation is therefore identical with death. Disease arises mostly through water, the enemy of fire, predominating in the body; intemperance in drinking clouds the soul, since the wine makes the soul humid; "The driest soul is the wisest," he expressly states in his writings. His special biological investigations have not been preserved. He is said to have dissected animals, but it is not known whether he drew any fresh conclusions therefrom. The service he rendered to science lies in his general view of existence; in his constant insistence on the modification of the principle, and at the same time on incontrovertible natural law governing the universe — these two factors being the essence of existence. In this he exerted great influence upon the natural philosophy of succeeding ages, particularly upon Plato and, through him, upon Aristotle.

Contemporaneous with Heracleitus, however, there appears another line of thought which is far more concerned with a close study of nature and makes it the basis of the entire cosmic system — the atomic theory. The founder of this philosophy is said to have been LEUCIPPUS, a thinker of whom we know nothing except that he was the teacher of DEMOCRITUS, one of the foremost natural research-workers and natural philosophers of all time. Democritus was born at Abdera, a Greek colony on the Thracian coast. In his time Abdera was a rich and powerful city, but in the course of subsequent centuries it declined and its citizens became notorious for their stupidity, which gave rise to the epithet "Abderitic" as a universal expression for extreme foolishness. His period of activity is, like that of most of the early Greek philosophers, not known for certain, but it is generally assumed that he was born between the years 470 and 460 B.C. and died at a very advanced age — in fact, not far short of a hundred years. From his father, one of the richest and most eminent citizens of his native town, he received a large inheritance, which he is said to have spent entirely on long journeys undertaken for the purpose of acquiring knowledge from various countries and peoples. On his return home he was supported by a brother until his fellow-countrymen, proud of his scientific fame, granted him a pension sufficient for all his needs. Known for his mild and friendly disposition and surrounded by admiring disciples, he grew old in peace and died without suffering from ill health. He was known as one of the most productive scientific authors of antiquity, and his writings seem to have embraced many and varied subjects. Except for a few fragments, however, they are entirely lost and indeed seem to have already been so as early as the late classical period. Through other ancient authors, however, particularly Aristotle, who always mentions him with respect and seems to a large extent to have made use of his learning, we are able to form a fairly good idea of his scientific point of view.

Materialistic theory of the universe

HIS teacher, Leucippus, seems to have based philosophy on Parmenides' theory of the immutability of matter, and to have come across this in the paradoxical form that it assumed among the later Eleatics. In order to preserve the elements of truth in this theory and at the same time to make possible the changes which are incontestably observable in the world, Leucippus conceived the universe as composed of a quantity of particles moving in empty space. Democritus adopted this theory and developed it further. He thus became the founder of the atomic theory, one of the most fruitful ideas in natural science. And on this atomic theory he, as no one else has done either before or since, based the whole of his theory of existence, both spiritual and material. No thinker of antiquity ever produced such a consistent and materialistic theory of the nature of the cosmos; none has advanced further in the endeavour to satisfy the demand — upon which Anaximander had already insisted — for a natural explanation of the origin of matter.

From the fragments of his writings a number of general principles have been gathered which are characteristic of his point of view; some of them sound surprisingly like modern natural science.

Out of nothing comes nothing; nothing which is can be reduced to nothing. All change is merely an aggregation or separation of parts.

Nothing happens by chance or intention, everything through cause and of necessity.

There is nothing but the atoms and space; all else is an impression of the senses.

The atoms are infinite in number and shape. Their movement is eternal; in endless space they impinge upon one another, thereby producing vortices out of which worlds arise, only again to perish. All the qualities of matter are due to the shape, size, number, and motion of the atoms of which it is composed. The atoms possess no other qualities than those mentioned above. The soul consists of the finest and most mobile atoms; they fill the body and give it life; if they leave the body, death ensues. Fire likewise consists of small mobile atoms. On the whole, Democritus seems to have shared Heracleitus' idea of the mutual affinity of the soul and fire. The stars he considered to be like the earth, but owing to their rapid motion they were fiery bodies. He seems also to have observed the mountains in the moon.

Biological knowledge of Democritus

DEMOCRITUS has achieved important work in the science of biology and he is here, as in many other directions, the finest of Aristotle's predecessors; how much of his learning his successor borrowed we do not know, but there are grounds for supposing that it was far more than posterity has ever guessed. Without doubt he performed dissections of both the higher and the lower animals and was the first to differentiate between them according to the quality of the blood. The distinction between sanguiferous animals (vertebrates) and bloodless animals, which was the principle of classification adopted by Aristotle, originates from Democritus. In contrast to Aristotle he believed that even the minutest animals possess perfected organs, although, owing to their transparency, they are invisible to the human eye. In the embryonic development the external organs arise first and the internal afterwards. Many of Democritus' ideas we know only through Aristotle's polemics against them, and in this respect modern research not seldom proves Democritus right. For instance, he considers that the spider's web is produced from inside its body while Aristotle maintained that it is cast-off skin. The sterility of the mule he seeks to explain by assuming a contraction of its uterus. The construction and functions of the human body, however, were naturally the main object of his studies. He conceives man to be a world in miniature, a microcosm in which every kind of atom is represented. He regards the brain as the organ of thought, the heart as that of courage, and the liver as that of sensuality. His estimation of the brain places Democritus once more in front of Aristotle, who believed the brain to serve only the purpose of cooling the blood. Democritus considered life and the soul to be one and the same thing, and the latter he believed, as already mentioned, to consist of fire atoms which, owing to their lightness and mobility, are constantly being given off by the body. Through inhalation the body receives a fresh supply of them; if respiration ceases, then life departs from the body. Sleep and asphyxia he declared to be also due to a loss of soul-atoms, but on a smaller scale. Hydrophobia in dogs and human beings he considered to be caused by inflammation of the nerves. Epidemics he believed to be the result of atoms falling upon the earth out of other celestial bodies. Sensation is due to the movement of atoms which emanate from the objects perceived. In connexion with Democritus' materialistic idea of the soul he believed in spiritual beings and revelations, a belief which other thinkers who regarded the soul as matter — as, for instance, Swedenborg — have shared with him. On the other hand, he denied the divine beings of popular belief, without, however, like Xenophanes, substituting any unified and eternal divine power in their stead. Necessity, which, according to his views, governed the universe, was purely impersonal.

Democritus on the whole represents the climax of the endeavour of Greek philosophy to arrive at an explanation of existence based on a natural connexion of causes, an endeavour which, with Anaximander as its instigator, gave rise to a long series of heterogeneous explanations of the cosmos, of which only the most important can be considered here. In several fundamental respects — for instance, in the strict theory of causation, the atomic theory, and, in connexion therewith, the principle of motion, the emphasizing of the importance of the brain for the function of thinking, the insistence upon the complicated organism of the lower animals — Democritus achieves results similar to those that have been attained by natural research in our own day, although his speculations were in detail often very primitive, even when compared with the achievements of philosophers of later antiquity. The promising idea was not pursued, however, by succeeding generations; for certain reasons which will be more clearly explained when accounting for Aristotle's theoretical views, shortly after the age of Democritus the Greek natural philosophy started to work out a method of explaining the cosmic process entirely different from that indicated by the atomic theory. Democritus thus represents the close of the first and the purely natural scientific period of Greek philosophy. The results achieved by research during that period have been mentioned above; what it failed to achieve may also be briefly described here. The most serious defect from which it suffered was undoubtedly the lack of material for investigation, which rendered it difficult to follow up the principle of causation which had already been determined. It should be remembered that antiquity knew nothing whatever of chemistry, so that the ideas of chemical association and affinity were entirely lacking as a basis for the changes in nature, and for these ideas the vague and dogmatic theory of vortical motion was of course a very poor substitute. Generally speaking, there did not exist in antiquity the ideas of force and energy in the modern sense, but philosophers had to be content with motion as an explanation of all changes. And, finally, they drew no definite line between objective facts and subjective opinions; they had no idea whatever of hypothesis. The whole of Democritus' cosmology was purely dogmatic and was condemned to give way to other similarly dogmatic explanations of the universe, which, though more attractive to his contemporaries, nevertheless proved fatal to the future development of natural science.

Reaction against natural philosophy

THE first sign of the coming reaction is given by the philosophy of ANAXAGORAS. This thinker was a contemporary of Democritus, probably born somewhat earlier than the latter, at Clazomenæ in Asia Minor. His activities, however, were pursued in Athens in the time of Pericles — thus in the community which proved to be the first great power of Greek nationality. This is the first time in Greek scientific history that the town is mentioned which was afterwards to become for nearly a thousand years the centre of thought of classical antiquity. Athens was a city with an extremely mixed population and an equally diverse intellectual life: side by side with the most daring novelties in the region of thought dwelt crass superstition and fanatical intolerance — to which latter Anaxagoras fell a victim. Accused of "godlessness," he was first cast into prison and then had to flee from Athens. Nevertheless, his philosophy, as compared with that of Democritus, must be regarded as idealistic. He conceived that the driving force in the universe is what he calls the cosmic reason or the cosmic soul; a kind of spiritual power to which he ascribes unity, omnipotence, and omniscience. This same power is part of all living creatures and in them represents life itself. Matter he believed to consist of an endless number of primary elements — that is, in the same sense as that in which the old Ionians conceived this idea. He does not appear to have gone in for biological research, nor do his natural-scientific views in general show any advance over those of Democritus. But he was not without influence upon the succeeding ages and is therefore worthy of mention.

The Sophists

OF far greater influence on the general course of development, however, was a school of philosophers which appeared contemporaneously with him and which began to lead Greek thought along entirely different channels. These were the famous Sophists, whose founder and leading personality was

PROTAGORAS, likewise a contemporary of Democritus. Protagoras introduced scepticism and subjectivism into Greek thought: "Man is the measure of all things" is one of his fundamental principles; "Contrary assertions are equally true" is another. To such a thinker any consideration of nature was foreign; what he desired to teach was an art of living, such as would free mankind from the fetters which traditional ideas in the sphere of religion and morality had imposed. Highly acclaimed by the younger generation, bitterly hated by the representatives of the past, not least because, in contrast to the earlier philosophers, he taught for payment, which was at that time regarded as equivalent to usury, Protagoras starts an entirely new era in the Greek view of mankind, in relation both to his fellow-creatures and to nature. But before going on to describe this new tendency we must make a brief survey of Greek medical science, at that period a specialized science which had achieved results of lasting value to general biological development.

CHAPTER III

THE EARLIER PHASE OF GREEK MEDICAL SCIENCE AND
ITS SIGNIFICANCE FOR THE DEVELOPMENT OF BIOLOGY

Magical beginnings of Greek medicine

THE EARLIEST BEGINNINGS of the science of medicine among the Greeks were as always in primitive medical practice, based on magical religion. Æsculapius, the god of the healing art, had a numerous priesthood, in which secret knowledge of the forces of nature and their use for the curing of disease was jealously guarded and handed down from generation to generation. Pilgrimages were made to the temples of Æsculapius by crowds of sick persons — both those in real and those in imaginary ill health — to be cured of their maladies; and the latter class, the hypochondriacs, were not slow to spread abroad and confirm stories of the most wonderful miracles of healing performed there. Immense hospitals were erected in the neighbourhood of these temples for the benefit of those who required lengthy treatment, and the very necessity of having to watch over and follow the course of these patients' diseases must naturally have created a large supply of purely empirical observations of immense value to the sacerdotal miracle-workers, who were thus enabled to estimate the result of their methods of treatment. In time there grew up, as a result of the widespread adoption of these empirical observations and methods, a class of purely secular healers, no longer directly associated with the temples of Æsculapius, who nevertheless, in order to denote the origin of their art and to take advantage of the confidence inspired by religious belief, called themselves Asclepiads — that is, descendants of Æsculapius. As other professions did at that time, they formed a private guild, the members of which taught their art preferably only to sons and kinsmen. Outsiders too were able by paying large sums to obtain an insight into the secrets of the profession, while the children of the family of Asclepiads always had the right to receive free instruction from any of their father's colleagues. The wording of the oath which the young physician had to swear before being allowed to begin the practice of his profession is still extant. By this oath he pledged himself to help teachers and his professional colleagues, give free instruction to their sons, share with them any fresh experiences and discoveries, to the best of his ability heal the sick, and refrain from mixing poisons and producing abortions.

25

Earliest medical writings

THE oldest known medical writings date from a period when medical science was still materially, if not also in its ideals, dependent on the temples of Æsculapius. There were in particular three famous shrines of the god from which profane medical science was thus derived, and these were situated on the islands of Rhodes, Cnidus, and Cos, all near the coast of Asia Minor. The situation of these places shows, as did also the first nurseries of philosophy, that oriental influence had been at work, and this influence did in fact prove of incalculable importance for Greek medical science; particularly in Egypt the art of healing had from the very earliest times been highly developed, and in historical times, too, was highly reputed even amongst foreign peoples. Of the Greek temple schools mentioned, that on the Isle of Rhodes was the earliest, that at Cos the most famous. The Coan school is principally indebted for its fame to the family of Asclepiads originating from there, which gave the world one of the greatest pioneers of medical science known to history — namely, HIPPOCRATES.

History has preserved the memory of seven Greek physicians called Hippocrates; the one here in question is generally spoken of as Hippocrates the Second or the Great. He is believed to have lived between the years 460 and 377 B.C. — that is to say, at the time of Democritus. He was born in the Isle of Cos of the family of the Asclepiads which for several generations had been attached to the temple of Æsculapius on the island. He received his medical education from his father, Heracleides, who, however, evidently died when his son was still a youth. The young Hippocrates then betook himself to Athens, where he studied philosophy with the Sophist Gorgias of Leontini, and afterwards made several journeys in the Balkan Peninsula and Asia Minor, eventually settling down in Thessaly, where he established a large practice and finally died in the city of Larissa. His sons and grandsons were likewise doctors whose reputation was high, though not comparable to his own.

Hippocrates' fame as a pioneer of medical science is based chiefly on his medical authorship. Even in ancient times, however, there prevailed some uncertainty as to whether the writings that go under his name are genuinely his, and in our own time only a few treatises out of what is called the Hippocratic Collection can be accepted as coming from his own hand; the rest are supposed to be partly the work of his school, partly Asclepiad writings dating from before his time, partly, and mostly, essays by considerably younger authors. Hippocrates' own main treatise, *Airs, Waters, and Places*, contains extraordinarily brilliant observations on climatical and geophysical conditions and their influence on mankind in sickness and health, in their material and spiritual aspects. The treatises in the Hippocratic Collection dealing with anatomy and physiology, which are thus of interest for the history of biological research, are all believed to be of later date than Hippocrates and are at any rate influenced by his views. They give evidence of a close study of anatomy and physiology with the help of dissections and vivisections. From the enunciation of the anatomy of the human body, however, it is clear that it is not based on dissections of the dead body, but that the results of experiments carried out on the dead bodies of animals have been applied to the human body without further evidence. To dissect the human body, to violate the dead bodies of human beings, was from the very earliest times considered a dangerous procedure and on that account was forbidden; the reason for this was undoubtedly the universal fear of ghosts, which has already been touched upon. It is certain, at all events, that, except in a few particularly unprejudiced cultural epochs, it has always been difficult, even in more modern times, to procure dead bodies for purposes of scientific investigation and to obtain permission to utilize them. The first part of the human body to be studied with any great care was the bone-construction, which could be examined in the skeletons of long-decomposed and consequently less dangerous individuals, while the ancient custom of preserving corpses by the preservation of the skeleton — a practice known in most countries from prehistoric times — must have given cause for observing and getting to know the various bones of the body. The musculature also, at any rate its external layers, which it was possible to study in living human beings in wrestling and athletics, has been comparatively well understood, while the internal organs — the digestive, respiratory, and circulatory systems — remained longest shrouded in obscurity, as regards both their construction and their functions. These facts have also had an influence on surgery; while fractures and sprains were carefully studied and cleverly treated even in Hippocrates' time, the art of arresting hæmorrhage was extremely primitive; from antiquity to the Middle Ages there existed no better method than cauterizing with red-hot iron, and it was not until the sixteenth century that people learnt how to apply ligatures when operating.

Hippocratic physiology: the four temperaments

THE Hippocratic treatises assume with Empedocles and his successors that the human body is composed of the four elements: fire, air, water, and earth. To these elements correspond four "juices" in the body: blood, phlegm, yellow bile, and black bile. Of these the yellow bile is produced in the liver, and the black bile in the spleen. Proof of the existence of these juices was found in the condition of the blood on coagulation, when, as is known, its component parts become separated; in the undermost, black part of the clot was recognized the black bile, in its uppermost, red part the blood; the yellow bile was seen in serum, and the phlegm in the fibrin.[1] The condition

---

[1] See Fåhraeus: *The Suspension Stability of the Blood* (Stockholm, 1921).

of the body was due to the existence and commingling of these four primary elements; if they existed in the proper proportions, health was the result; if the harmony between them was disturbed, sickness followed.[2] For the rest, Hippocrates and his successors seem to have shared the conception which originated in Heracleitus that the soul, the life-principle, consists of fire or — in later writings — of substance akin to fire, called pneuma (breath).

In regard to human anatomy, osteology was, as has been mentioned, comparatively carefully studied. The skull in particular was radically investigated and a large number of the names of its bones and sutures are derived from these works. The bones of the face were also minutely studied. The knowledge of the backbone was more defective, while, on the other hand, the bones of the extremities were well described. Of the muscles several, particularly the muscles of the extremities, are correctly described, though the actual muscle substance is not accurately distinguished from a number of other internal organs. The separate parts of the digestive canal are named, but their connexion and function were extremely vaguely known. The various sections of the intestinal tube are given different names and characteristics in the different writings on the subject. The liver is an organ which particularly interested the people of antiquity, as also the spleen, but there were very confused ideas of their respective functions. A good deal was known about the glands, especially the lymphatic glands; the pancreas, on the other hand, was unknown. The function of the glands was believed to be to segregate water from the body. Of the respiratory organs the larynx and the trachea in particular were completely described, while the lungs were treated only summarily. Respiration is believed to serve the purpose of cooling the heart. Again, in another treatise it is asserted that the inhaled air spreads to the various parts of the body, to the brain, the body cavity, and the arteries. The enunciation of the circulatory system is, as mentioned above, vague. The different cavities of the heart are fairly accurately described, but here exists already the delusion which it has since been so difficult to eradicate that the left cavity of the heart does not contain blood, but some kind of airy substance, which is proved by reference to slaughtered animals, in which the arterial blood is drained off by severing the jugular veins! The apprehension of the venous system is far vaguer than that of the heart, and, moreover, the description of the former in the different Hippocratic treatises is highly contradictory. The right side of the heart

---

[2] The doctrine of the four temperaments, which counts for something even in our own day, is based originally on the theory of these four juices and their distribution in the body; in sanguine people the predominating factor is the blood, in phlegmatic people the phlegm, in choleric people the yellow and in melancholy people the black bile. Melancholy, owing to its origin in the black bile, is called spleen or atrabiliousness.

is the starting-point of the circulation of the blood; the blood flows to it from the rest of the body, and there obtains warm temperature from the left cavity — that is to say, the blood coming from the body is "cold." The left heart-cavity gets warmth from the air through the pulmonary veins. The arteries likewise contain this warm kind of air, called pneuma, which maintains vital action, and they disperse it throughout the body. It is truly remarkable that, in spite of observations made from innumerable dissections and vivisections performed by various research-workers, this primitive butcher's experience of the emptiness of the left heart-chamber and of the arteries should have been maintained up to the final era of the science of antiquity.[3] The warmed-up blood is forced from the right cavity of the heart out into the body. In contrast to these primitive conceptions, however, certain of the Hippocratics have had some idea, though a vague one, of the movement of the blood as an actual circulation.

Of the nervous system the Hippocratic school had, if possible, a still vaguer idea than of the circulation of the blood. The brain was believed to be a gland which segregates water and possesses the function of cooling the blood and collecting mucus out of the body. This mucus is then segregated, together with the water, by a catarrhal affection through the nose. Certain later Hippocratics, however, probably influenced by Democritus, have a more accurate view of the functions of the brain, believing it to be the centre of thought, feeling, and motion. The nerves are invariably confused with the tendons, sometimes also with the veins, and for this reason all ideas of the functions of the nervous system are already ruled out. Certain of the most important cerebral nerves are, however, described and named. The construction of the eye was fairly thoroughly studied; its membranes and fluids, as well as the pupil, were known, but the lens was unknown. Sight was produced as a reflection of the object seen on the pupil. Of the ear the bony labyrinth, the auditory canal, and the tympanum were known. The urogenital apparatus is described in its main features; regarding fertilization there existed then, as indeed throughout antiquity, extraordinarily fantastic ideas.

With zoology proper the Hippocratics naturally had little cause to concern themselves. Nevertheless, there exists a treatise *On Diet*, in which there are enumerated fifty-two different edible animals, arranged on a certain definite system; first quadrupeds, tame and wild, birds, fish of several kinds, including coast-fish, mud-fish, river-fish, mussels, and crayfish. This so-called Coan animal system has the advantage of differentiating between various categories of living creatures — a first primitive attempt at proper systematization.

---

[3] This idea possessed, it is true, the sanctity of religion; in sacrificial animals the arteries were naturally empty, and the interior parts of these animals were examined with a view to basing on them prophecies of the future. To deny the results of such examinations would of course have involved wounding time-sanctioned religious susceptibilities.

CHAPTER IV

THE END OF NATURAL-PHILOSOPHICAL SPECULATION. THE PREDECESSORS OF ARISTOTLE

The results of natural philosophy

AN ENTIRE ERA in the history of biology closes with the philosophers that have here just been characterized — Democritus, Hippocrates, and his school — an era which may properly be called the era of natural-philosophical speculation. The results achieved by their researches cannot be regarded as anything but magnificent; for the first time in the history of humanity to have built up a real natural science is an achievement worthy of the highest admiration, however modest the results may have been in certain details. This research work had so far been directed by three illustrious representatives, Anaximander, Empedocles, and Democritus. All of them sought an explanation of existence as a natural course of events; in this direction Democritus proceeded as far as human thought has at any period proved capable of going. Nevertheless, even in his own lifetime the revolution was being prepared which was shortly to lead Greek thought in entirely different directions. The first ideas on which this change was based originated, as hinted above, from the school of the Sophists. The new principle that they taught was subjectivity: "Man is the measure of all things." To this really true assertion the ancient "physicists" could make no objection; indeed, their cosmic explanations were as numerous as themselves, and each one of them could only declare dogmatically that his own views were the true ones and thereupon produce a number of more or less illogical arguments in support of them. The claim of the Sophists as to man's being the measure of all things thus for the time did away with all objective explanations of natural phenomena, for what was the use of disputing about matters which all viewed from different standpoints if all could be equally right? Sophistry itself, however, when consistently applied, led to pure nihilism, both intellectual, in that the object was, by means of ingenious turns of phrase ("sophisms"), to prove absolutely anything, and moral, in that all generally accepted sound traditions were held in contempt as laying a restraint upon the individual's freedom of action. If all scientific thought were not to be destroyed altogether, its preservation must be sought by turning the whole trend of thinking into quite a different direction. And this was found by maintaining that human thought, however much it may be the measure of all things, such as they appear here on earth, is nevertheless itself subject to eternal laws, more infallible than those which the old physicists saw in existence. The men who thus saved the Greek philosophy from degenerating into empty rhetoric and worthless quibbles were two Athenians, SOCRATES and his disciple Plato. Socrates worked exclusively in the ethical sphere; in this he sought for standards binding for all and emanating, not from ancient tradition, but from the conscience of the private individual. "Anyone can become virtuous if only he accepts a knowledge of virtue"; that is his principal doctrine, and this knowledge he for his own part derived from a divine voice in himself, which he desired also to awaken in his fellow human beings. Nature did not interest him in the least; the streets of Athens were his haunt, he said, and neither trees nor stones had anything to teach him.

PLATO, a disciple of Socrates, generalized the doctrine of standards in the ethical sphere which he learnt from his master, so that it was applied to embrace the whole of the intellectual life of man. Born at Athens in 429 of a distinguished family, he attached himself to Socrates at the age of about twenty. Upon his master's death he left Athens and made extensive journeys, afterwards returning to Athens, and there he established a school or college known as the "Academy," which survived long after his death. He died in 347. Like Pythagoras, he was a clever mathematician and, also like him, combined an inclination for the conclusive deductions of mathematics with a strong attraction for the mystic. In the dialogue *Timæus*, in which he propounds his theory of the origin of the universe, the functions of the human body, and the relation of man to nature, he has evolved a history of creation, poetically very fine, but at the same time purely mystical, testifying, it is true, to his high ethical aims, but of no greater value as natural science than any of the ancient popular cosmogonical myths. The world was created by an eternal and perfect god, and therefore there can be no question of an endless number of worlds, as Anaximander and Democritus made out, but only one, and this single world must have received the most perfect of all shapes, the sphere. He accepted Democritus' atomic theory to the extent that he believed matter to be composed of particles, though these again are not of endless variety, but are five in number, corresponding to the five regular polygons of the geometry of space (Plato was one of the founders of this science and one of the foremost geometricians of all time) — so that each element has its atomic form: fire the pyramid, the earth the cube, the air the octahedron, and water the icosahedron; the dodecahedron represents the heavens. Now, this geometrical atomic form is in itself no more dogmatic than the entire atomic theory of Democritus, but it forms the starting-point for a process of natural speculation which is diametrically opposed to it. For while Democritus takes as his starting-point the atoms and the matter formed by

them as being the only really existing element, Plato finds true reality in
the world of abstract thought and maintains that what is perceptible by
the senses is an imperfect image of the eternal ideal, the divine intelligence, conceivable only by abstract thinking. The nearer things are to
the divinity, the more perfect and animate do they become. Thus the
stars of heaven possess a higher animate life than man and are created in
greater likeness to the supreme intelligence. Of man the first thing to be
created was the head, which Plato, like Democritus, regarded as the organ
of the animate soul; the head has very nearly the spherical form corresponding to the ideal, and trunk and limbs are created to save the head the
trouble of rolling along the ground. In the trunk dwells a lower, mortal
soul, whose best part, the heart, the organ of courage, is separated by the
diaphragm from the organ of animal desires, the digestive apparatus, the
most vital part of which, the liver, however, has the merit of producing
dreams during sleep, from which the future can be predicted. The plants on
our earth are created to provide man with food; animals, again, are sprung
from men whose souls have degenerated and have consequently been given
an inferior dwelling; first, women have come into being out of cowardly
men's souls, then birds and quadrupeds out of such human beings as have
neglected their intelligence, and, finally, the most worthless souls have been
placed in aquatic animals, "which may not even breathe pure air." This
theory of the migration of the soul is clearly reminiscent of that of Pythagoras; from him, too, no doubt originates Plato's mystical theory of number,
the details of which do not belong to our present subject.

Systematization of thought

IT may seem unnecessary to dwell so long on these fantasies. They are, however, well worth noting, for their originator has exercised a rare and radical influence on human culture in its entirety and even, as we shall find later
on, no small influence upon the development of biological science. His philosophical speculations related above have in this respect had but little significance. Plato's greatest contribution has been made in the sphere of the
purely ideal intellectual life. The spirit of man is, he said, bound by laws far
more abiding than those that may be deduced from natural phenomena. To
these inflexible standards for intellectual activity he is led by speculations
in the ethical sphere, in which, following Socrates, he found a definite parallel between human actions and their consequences. Having advanced thus
far, he devoted himself to developing this theory of the conditions, bound
by immutable law, of the world of ideas, in which, as mentioned above, he
saw the true existence, of which the phenomena visible in natural life are
mere images. Everything on this earth, then, has its eternal idea as its prototype; every individual horse is an imperfect image of the idea horse, which
is eternal and perfect. Through this reasoning Plato came to be the founder

of the system of ideas, which has played an important part in biology; out of the idea horse in contradistinction to the individual has arisen the notion of species, and gradually likewise all higher systematical categories. It is easy to realize the immense influence this has had on biological research. A mass of detailed ideas in systematization—for instance, the dichotomic classification tables of genera and species — come from the Platonic school. And, mathematician as he was, Plato further endeavoured to make his inferences as conclusive as possible; the listener and the reader should realize that the result of the investigation must be so and could not be otherwise. Thus for the first time in the history of science he not only made assertions, but also submitted proofs of them, based, it is true, upon abstract reasoning, as was the entire world of thought which he built up, but at least formally convincing.

Theory of ideas

BUT if Plato, in laying the foundation of biological systematization, made a powerful contribution to the progress of biology, nevertheless his activities proved in other respects unfortunate for it. The enunciation of the world of ideas as the true essence of being led to the underestimation of nature and of the senses by which man observes nature. Plato realized the relativity and limitation of observation through the senses,[1] but not the arbitrariness to which abstract thinking may lead if it is not controlled by observations. In the *Timæus* he expressly states that no true knowledge is to be acquired through the observations of the senses, but only a pleasure to the eye suitable for a diversion — a statement which has been repeated after him by innumerable idealistic philosophers, both major and minor. Plato, the creator and perhaps also, so far, the greatest upholder of idealistic philosophy, is likewise responsible for the contempt with which this trend of thought, which otherwise deserves such high praise for contributing towards the development of human thought, regarded natural philosophy. History indeed shows that the more the idealistic philosophy governed man's desire for knowledge, the greater became his indifference to the study of nature. In a far greater degree than the traditional religions, idealistic philosophy has been the antipodes of natural science.

On the whole, Plato's disciples followed in their master's footsteps. Of the exact sciences mathematics interested them most and they worked at it with great energy; otherwise, like many other schools of thought emanating from the Socratic circle, they mostly occupied themselves with ethical questions. Even in Plato's lifetime, however, one of his disciples in the Academy had begun to discard his ideas in a number of essentials and started to guide philosophy in a new direction. This man was Aristotle, the greatest biologist of antiquity and one of the most many-sided natural philosophers of all time.

---

[1] As a matter of fact, Democritus had already realized this, but was unable to develop the thought further.

CHAPTER V

ARISTOTLE

ARISTOTLE was born in 384 B.C. at Stagira, a small Greek colony on the Macedonian coast. His father, Nicomachus, belonged to an old family of the Asclepiads and was, like several of his ancestors, body-physician to the Macedonian royal family. His predecessors among the Greek philosophers had lived amongst the mobile and restless communities of the city republics, while Aristotle spent his childhood at a royal court, and a semi-barbarous one at that. This fact undoubtedly put its stamp on his personality and way of thought; he became in every respect an upholder of authority and conservatism. At an early age he lost his father, and his mother, Phæstias, retired to her native city and brought up her children there. Aristotle received his earliest education, in accordance with the ancient Asclepiad tradition, from his father's colleagues, who initiated him into the biological and medical learning of their profession. It was necessary, however, for a properly educated physician to receive also philosophical instruction, and for this purpose Aristotle was sent at the age of eighteen to Plato's Academy at Athens. There he remained for twenty years, was initiated into the teachings of his master, wrote his first treatise, and already at that period began to oppose his master's authority, which the latter is said to have observed with no little displeasure. After Plato's death he was passed over at the election of a successor as head of the Academy, in spite of his already established reputation, and retired to Asia Minor, where he settled down at the court of the Persian vassal-prince Hermeias of Atarneus, who gave him his niece in marriage. Some years later, however, Hermeias was deposed under a revolution, and Aristotle had to flee to the country of his birth, Macedonia. There he was charged with the task of educating the heir to the throne, Alexander, the future conqueror of the world, and he held this post for three years (338–335). What influence the master exercised on the pupil it is of course difficult to decide; the relations between them, however, were on the whole good, though Alexander's increasingly despotic character and barbaric outbursts of passion must have offended the cultured, self-controlled Aristotle. His profession as teacher at any rate brought Aristotle illustrious honours and made him a wealthy man, able to choose his place of abode and his sphere of activity. He then moved back to Athens and lived there under the protection of Alexander, highly esteemed by a constantly

34

increasing throng of pupils for the space of twelve years. During this period he displayed indefatigable activity. He was granted the right to use for educational purposes a temple dedicated to Apollo Lycæus, after whom the place was called the Lyceum, the archetype of learned educational institutions throughout the world. Here every morning Aristotle gave scientific lectures to his chosen pupils, often old and highly reputed men of science, who collaborated with him; and, further, every evening he held more popular courses for younger collegiates. Moreover, he found time to write an incredible amount on very different subjects: logic, metaphysics, art, politics, psychology, and biology. This extraordinary activity testifies to his inexhaustible energy and splendid powers of organization. It is obvious that his disciples had to carry out the rough work. Aristotle kept aloof from public life; indeed, he was a foreigner in Athens. He was a conservative and a monarchist, however, and when after Alexander's death Athens rebelled against the Macedonian supremacy, his position became dangerous. For lack of other means of calumniating him he was accused, like Socrates, of "godlessness." In order, as he himself said, to save the Athenians from committing a fresh crime against philosophy he fled to the island of Eubœa; there he died shortly afterwards, in the year 322. In external appearance he is said to have been of small stature and corpulent; his carriage was proud, his manners arrogant and sarcastic, his dress and way of living courtly, refined, and elegant. These latter characteristics brought him personal enemies, who sought to blacken his character. It is not possible, however, to bring any serious accusations against him as a private person. It is true he appropriated with considerable lack of bias the results of the work of earlier philosophers, but the ideas of literary copyright were not so strict as they are now. On the other hand, he treated different thinkers with true humanity; his polemics, when he went in for them, were always courteous and his arguments founded on facts. Towards his family, his friends and pupils, and even his slaves he was affectionate and considerate.

Aristotle's sphere of activity was, as mentioned above, extraordinarily extensive, and equally universal has been his influence during these thousands of years in such widely separated spheres as biology, metaphysics, statesmanship, and art. In the present work it is of course possible to deal at any length only with his biological work; besides this, however, we must touch upon his general ideas so far as they affect his views on life in nature. And as varied as his own interests have been the judgments passed on him by others; he has been by turns elevated to the skies and dragged in the mud. On the whole the biologists have been the most loyal; up to recent times, and not least in our own day, he has had devoted admirers, and his works have been remarkably free from the bitter criticisms which biologists of more recent times (Linnæus, for example) have passed on his predecessors. The philosophers proper have judged him far more sternly; they have realized far more clearly the deficiencies in his system of thought, and the weaknesses inherent in its structure. The idealistic philosophers have accused him of vitiating Plato's theory of ideas, while adherents of critical philosophy have criticized his dogmatic ideas of the universe. His extraordinary influence both on his contemporaries and on posterity, however, no one can deny.

Form the true reality of matter

As a thinker Aristotle bases his system on Plato. According to Plato the ideas of eternity are existing realities, of which the things of our earth are an imperfect image. Aristotle adopted the theory of ideas, but sought to overcome the difficulty arising out of the questions how ideas are really related to things and how they influence them, by placing ideas not outside things as something independent and apart from their existence, but in things themselves. And he regarded the form of everything as its idea, as its true reality. Form is the thing's reality, matter is a potentiality, to which form gives reality. The bronze of which the statue is made is a potentiality; the form which the sculptor gives it makes of the statue a reality. This method of observation, derived from human life, Aristotle applies with inflexible consistency to the whole of nature, both animate and inanimate. Thus the seed is a potentiality out of which the germinating plant develops reality. The same is true of the egg and the embryo in relation to the creature which develops therefrom. Consequently every lower stage of development is a potentiality in relation to the higher stage of development which represents its full realization. Thus we get a whole series of stages of development, beginning with completely formless matter, which is an exclusive potentiality without any reality at all, through inanimate nature, in which matter is stronger than form, to the animate, in which form governs matter. Form in living creatures is the soul, and the more highly developed it is, the more does it control the corporeal matter. Plants have a lower kind of soul, which only lives, but does not feel; animals possess a higher, sensitive soul, and, finally, man has conscious reason. The means whereby form gives expression to its dominance over matter is motion; all that occurs in the universe is motion, and the more form-perfect the motion is, the higher the development it represents. Our earth, therefore, with its manifold irregularities, is a lower form of existence than the heavens, whose celestial bodies possess the most perfect motion, circular motion. The heavenly bodies keep their position and motion owing to their being enclosed in transparent spheres, one outside the other. Thus they represent to the mind of Aristotle, as to that of Plato, a higher form of existence than the earth with its creatures, including man, and outside the outermost celestial sphere is the world of form free from all matter, the highest intelligent existence, God, the fundamental origin of all motion. Since, then, existence has its origin in a supreme intelligence, it is natural that everything that happens has an intellectual cause, and every-thing exists to serve a given purpose. This purpose is, above all, the develop-ment of a higher form, the striving towards a higher intellectual existence. Natural necessity and its cause, chance, have, it is true, their part to play on the earth, but only as an attendant of incomplete matter; in the heavenly spheres nothing happens by chance, but all is intellectual, while on the earth the higher intelligence is victorious over its lower opponents. But con-sequently terrestrial life, including man, is also governed by the higher intelligence of the heavenly bodies, which get their impulses direct from God.

The first evolutionist

THUS Aristotle makes his biological theories a link in the general cosmogony which he built up on the fundamental principle of the domination of form—that is, of the spirit over matter, and of motion as the origin of all things. The whole world of this thought-structure is as foreign to our modern ideas as it could possibly be; Democritus' atomic theory is far nearer our own notions. Nevertheless, Aristotle's theory implies an absolute advance in the sphere of biology. Here we find enunciated for the first time a really complete theory of evolution. To the old natural philosophers, Democritus among them, existence was a casual change of different forms. Again, Aristotle saw a consistent evolution from lower to higher forms of being, and although it is based on purely metaphysical speculation, this idea has proved for all time a fertile one in the biological sphere, for the very reason that it is here in agreement with actual fact. Quite in accordance with this his fundamental principle, Aristotle also made special investigation into the development of animals from the egg and embryo to the perfect state, and in this sphere he has made his most important contribution to biological research. But other-wise his philosophical and educational activities embraced the whole of biology, as it was known at that time, as well as all natural phenomena in general. Of his purely biological works the following are extant: ten books *On the History of Animals*, of which, however, three are considered spurious; four books *On the Parts of Animals;* five books *On the Reproduction of Animals;* and three books *On the Soul*. In these treatises he has collected all contempo-rary knowledge of animal life, not only his own and his pupils' personal observations, but also all the knowledge that his extensive collections of books could impart regarding the observations of the earlier philosophers. All this material he worked up with a view to including it in his general cosmic theory. It has been said that "never before or since has a scheme been so completely carried out with a view to incorporating biology in one com-mon science, while at the same time by personal observations and literary notes systematically building it up into one unit out of the phenomena" (Burckhardt). This is true, and the reason for it is to be found in that unique genius for the purely formal side of thinking which belongs to Aristotle alone; he is in fact the founder of formal logic and in that sphere his laws hold good to this day. He is, therefore, also the originator of biological classification, not only because he determined those categories in which human thought has since primitive times sought to arrange natural objects, and because he subordinated these to the general laws of thought which he created, but also in that he sought to interpret and combine into a law-bound whole those phenomena which accompany life in all its forms. By this work he has paved the way, as no one else has done, for the further development of biology as a science based on fixed principles. And his influence has extended to our own day; many of the biologists of our own time have revived certain expressions out of his terminology and have been influenced more or less directly by his ideas, particularly in regard to evolution. But his great gift for form has also its darker side. He who saw in form the true content of existence could not imagine a world to be other than finite — spherical, for the sphere is the most perfect form — and as he could not visualize an infinite world, he could not imagine infinite potentialities of knowledge; on the contrary, he expressly declared that his own system, complete as it was, would make it possible to solve all problems. But for that reason he takes up all questions for discussion, even such as in our time would be received with the old proverb "A fool can ask more than seven wise men can answer." Often when reading his writings one can fancy that one hears an inquisitive pupil offering objections which the master takes up with undisturbed calm and answers with unerring assurance in accordance with the principles of his system; as, for instance, why men, and not women, become bald; why the sow produces many pigs while the cow bears only one calf, etc. The result is that, while the reader can sometimes trace in his writings the pen of a biologist with almost a modern view of life's phenomena, on the next page he may receive the impression of a master of scholastic disputation from a mediæval university. Many of the irregularities may certainly be due to his treatises not having been carefully planned out; a number of them are probably notes of lectures published by his pupils, and as such are perhaps in places based on misapprehensions of the meaning of the lectures.

To give a complete description of Aristotle's biological works would be a voluminous task and would result in a wearisome mass of detail. In order, however, to give some idea of the peculiarities of his work, a brief account must here be given of his position as regards the various main sections of biology.

Aristotle's classification of animals

IN regard to the classification of animals, Aristotle, as was hinted above, has made an essentially important contribution to the subject by differentiating between and analysing and characterizing, from different points of view, a number of systematic categories. "Animals may be characterized according to their way of living, their actions, their habits, and their bodily parts." According to the three first-named principles they are divided into land-animals and aquatic animals; certain of the latter live entirely in the water — the fishes; others live most of their time there, but breathe and breed outside it — otters, beavers, crocodiles. The water-animals can partly swim, partly creep, and in part they are adherent. The land-animals have similar characteristics as regards habitat and way of living, feeding, habits, and character. The most important bases of classification, however, are the parts of animals' bodies, both external and internal: motive organs, respiration, organs of sense, blood-circulation. By combining various qualities the groups are defined and characterized. These groups are variously extensive: "Many animals allow of association into large divisions, such as birds, fishes, and whales," and, further, ink-fish, shell-fish or mussels, and crayfish. Others are more difficult to classify, such as the quadrupeds, which may certainly be classified as oviparous and viviparous, but amongst these it is not possible to make subdivisions, the animals having to be characterized each separately. The categories which Aristotle thus established he never summarized; his tabulated and generally recognized "system" has been extracted from his writings by others and need not be repeated here, all the more so as it is reproduced in different ways by different authors. Otherwise, his systematic categories are only two in number, the *genos* and the *eidos*, the latter corresponding to the individual animal form — horse, dog, lion — the former to all combinations of a higher degree. That is really the reason why his system cannot be compared with the Linnæan, with its manifold categories, though it by no means detracts from its pioneer importance for all time.

His knowledge of forms

In connexion with the system of Aristotle a few words may be said about his knowledge of form — about the material out of which he built up his system. In his writings have been recognized about 520 of the species which present-day zoology has classified. These forms all belong to Greece and its seas, and it seems that marine fauna interested him more almost than land fauna; fishes, molluscs, and crustaceans are better represented in his works than land-animals — in sharp contrast to what was afterwards the case with Linnæus. Exotic animals Aristotle knows only from the descriptions of others; it has often been stated that Alexander used to send him material for investigation from the countries he conquered, but this can hardly be true. The crocodile, for instance, he describes in the exact words of Herodotus and he accepts as true without further comment the latter's statement that the upper jaw of that creature is jointed on the lower jaw — which he would never have said had he seen a crocodile. Still more remarkable is the fact that he apparently never saw, or at any rate never carefully examined, a lion; in fact, he says of this animal that it has no cervical vertebræ, but instead has a single conjoining bone. In another place he declares of a lion's bones that they are so hard that they give off sparks when struck, like flint, and that they are "said" to have no medullary cavity. How much Aristotle in general borrowed from the wisdom of his predecessors it is impossible to determine, as their writings have been lost and he never quotes others except with polemical intent. But examples such as those cited above undoubtedly testify to a quite uncritical exploitation of foreign sources, and it is manifest that the value of Aristotle lies not so much in the facts he established as in the systematic working-up of the scientific material he had at his disposal. And this systematizing work of his was of course as comprehensive as was possible with the means available at the time. It is not only the outward form of animals and their existence that interests him; he studies the migrations of birds and the wanderings of fishes and tries to discover the causes that underlie these habits; he critically examines the outward manifestations of animal intelligence, and everywhere he closely compares different forms of life.

He chiefly occupies himself, however, with the anatomical and morphological structure of animals, as well as with their reproduction and evolution. These two spheres of investigation he has elaborated with the utmost care, and here we find he stands out most prominently as the founder of comparative natural philosophy. His treatises on human anatomy have been lost, but his great descriptive work on animals is extant and deals mainly with anatomical questions. Here he at once, in the very beginning of the work, lays down that anatomical research should be comparative; the less known should be studied by comparison with the better known, and since the structure of the human body is best known, that should be the point of departure. Following this method, he goes through the parts of the body, the external as well as the internal. He bases his general anatomical ideas on the same principles as Empedocles and the Hippocratic writings, and certainly also as Democritus. With the first-mentioned he holds that all beings are composed of four elements, while for him, too, the contraries hot and cold are of fundamental importance. His description of human anatomy, which forms the first book of his work on animal life, is of unequal value and, as regards its details, undoubtedly very much dependent upon the statements of others; but his manner of summarizing his description is excellent. As to his description of the internal anatomy, he frankly acknowledges that it is founded upon conclusions drawn from the dissection of animals, and, broadly speaking, it is not very different from the Hippocratic writings referred to above. Thus the heart is to Aristotle the organ of the soul and the intelligence, the brain serves the purpose of producing mucus and cooling the blood; in this respect his ideas are inferior to those of both Democritus

and Plato. The circulatory system is described in the same way as in Hippocrates. The various parts of the digestive canal are described in some detail,
but as regards the physiology of the digestive process he has extremely
primitive and vague ideas, which is not to be wondered at, seeing that
the science of chemistry did not yet exist. "Cooking" plays the essential
part in his physiology. The food is "cooked" in the intestinal tube; the
heart pulsates through the regular "ebullition" of the blood. With regard to
the nervous system his notions are equally vague; as is indicated above, the
brain is cold, the spinal marrow is hot; nerves and tendons are confused. The
aural cochlea is described, as are also the membranes of the eye; the moisture
of the eye is believed to receive the visual impressions. It is curious to note
the amount of popular superstition that is accepted, as, for instance, predic-
tions read from the lines of the hand, or the idea that flat-footed people are
of a treacherous disposition. In the comparative anatomy and morphology
of animals Aristotle shows his many-sided interest in all kinds of life-forms
and his immense power of combining observations of various qualities with
striking characteristics. "Four-footed beasts which produce their young
alive have hair; four-footed beasts that lay eggs have scales." "A single-
hoofed animal with two horns I have never seen. . . . No animal has at the
same time tusks and horns." He also makes many sound observations regarding birds and reptiles, as, for example, in reference to the outer structure
of the sensory organs. A distinction is made between whales and fishes, and
the gills forming the breathing-apparatus of fish are described with emphasis
on the difference between osseans and sharks. Of the lower animals — the
bloodless, as Aristotle calls them, after the example of Democritus — the
ink-fish in particular is minutely described, many carefully observed details
being given. Crayfish, too, and insects are cleverly described in part, though
with some inaccurate details.

Reproduction of animals asexual and sexual

In his work on the reproduction of animals Aristotle differentiates between
animals which reproduce themselves by sexual means, by asexual means,
and by spontaneous generation. The latter occurs in a number of lower
animals which are produced out of putrefying substances; among these are
specially mentioned certain insects, such as fleas, mosquitoes, and day-flies
(other insects, such as grasshoppers, wasps, and flies, have sexual reproduction). Among the shell-fishes some are produced by asexual means through
bud-formation, others through self-generation. The possibility of this latter
method is explained by the fact that the whole of nature is full of life-spirit
or "soul"; this, under certain circumstances, gives form to the inanimate
matter and so gives rise to new beings. Sexual reproduction, again, is due to
the occurrence of male and female individuals. Of these two the male — or
more properly the man, for Aristotle takes as his starting-point here, as

always, man — represents the more complete, "warmer" element, and the female, the woman, the more incomplete, the "colder" element. The masculine represents, above all, form, motion, activity; the feminine is matter, the passive, and consequently potentiality, which achieves reality through form. It is expressly asserted that the earth represents in the universe the womanly and maternal, the sun the manly — that is to say, the ancient view held by most natural religions. The male sex-product, the seed, is a product of the blood, which through complete "cooking" receives the purest and most form-creating qualities. The woman's sex-product is the menstrual blood, which is an undeveloped sperm — "half cooked," because the woman is weaker, "colder" than man, and has not the power to cook her product completely. In impregnation the man contributes to the future child form, motion, soul; the woman matter, body; his contribution is compared with the work of a carpenter, hers with the timber of which things are made. Holding this view of reproduction, Aristotle, when making his thorough investigation into the question of heredity, can of course only involve himself still deeper in abstract speculations. Against the opinion of earlier philosophers that the seed is derived from all parts of the body and therefore gives rise to similar individuals as issue, he asserts that on the contrary the seed goes to the different parts of the body, through which process a remainder is left over for the next generation, "as with a portrait-painter, a certain amount of colour is left over similar to that which was used for the portrait." If the man's form-building power is sufficiently strong, the child will be a boy; otherwise a girl; for that reason very young and very old fathers have mostly girl-children. The cold north wind also favours the birth of girls, for warmth is strength. The explanation of why children resemble partly their parents and partly their ancestors is very complicated and ingenious, for all its abstractness, but it would take too long to examine it here. Of far greater value than these metaphysical speculations, at any rate, are Aristotle's observations on the reproduction of animals. He draws up a scale in which the animals are placed according to their development and points out that those animals are highest which have a warm and moist, and not an earthy, nature. For all animals with lungs are warmer than those without lungs, and of those provided with lungs, again, the warmest are those which have not tough, spongy, and anæmic, but soft and sanguineous lungs. The most perfect animals, those which possess most warmth and moisture, whose young are born alive and immediately start growing, are the mammals; those which possess moisture, but less warmth lay eggs which afterwards develop inside the female animal and are born alive: sharks. Warm and dry animals lay "complete" eggs, such as birds and reptiles; cold and earthy animals lay "incomplete" eggs, such as osseans, frogs, and ink-fish; and finally the lowest animals of all — that is, of those that propagate in a sexual way — breed worms which give rise to eggs, as, for instance, insects, whose pupæ Aristotle regarded as eggs, whereas the true insect eggs were unknown to him. His descriptions of animal development contain a mass of extraordinarily sound details; as always, it is marine animals that interest him most. His account of the breeding of sharks is especially well known, while the pairing and growth of ink-fish are also described with a thorough knowledge of his subject. Embryology, too, he discusses in great detail, chiefly the development of the hen's egg, which Hippocrates had also studied; in particular the evolution of the heart and the first blood-vessels is carefully explained. Various speculations on colour-variations, change of teeth, and other problems of development complete Aristotle's work on the reproduction of animals, which, more than any other of his biological works, testifies to both his greatness and his limitations.

His evolution theory really dogmatic

ARISTOTLE's great contribution to the development of biological science lies, as has already been pointed out, not so much in the sphere of discoveries as in the thought-system, embracing all the phenomena of life, which he created and consistently worked out in all its details. The finest merit of this system of thought lies in the fact of its being based on an evolution subject to rigid laws and proceeding from the lower to the higher. But as this theory of evolution is, as has been shown above, primarily based on a predominant guiding intelligence, it acquires a dogmatic arbitrariness; the subjection to law is not an act of nature itself, but rather a product of divine wisdom, or, in other words, human speculation. This could, then, it is true, solve with abstract catchwords all the problems against which Democritus' atomic theory was powerless, but any such method of solution failed to stimulate thought to continued search; on the contrary, it induced a feeling of self-complacent satisfaction with the limited cosmogony produced out of unreal systems of thought. Thus it came about that Aristotle, the founder of systematic biology, became at the same time the father of the scholastic philosophy of the Middle Ages; that the man who was the first to introduce and logically to apply to the conception of the entire universe a theory of evolution from the lower to the higher appeared fifteen centuries later as the founder of a system of stagnation and obedience to authority. What the whole of this long period lacked was a conception of nature which would have associated Aristotle's theory of subjection to law with Democritus' theory of the dominance of necessity in nature. Subjection to law caused by a personal, guiding will, and necessity with pure chance as its driving force: the choice lay between these two alternatives until, through Galileo and Newton, the impersonal, law-bound force, operating by natural necessity, was made the basis on which to interpret the course of events in the universe. For the fact that such a long time should have elapsed before this conception of nature as held today could make itself realized, Aristotle and the system which he created, unexcelled in its perfection of form as it is, are mainly responsible. Under such circumstances it is not to be wondered at that the biology of antiquity, in spite of its splendid achievement, never succeeded in advancing beyond Aristotle's conception of the phenomena of life.

But if Aristotle thus represents, in most fields of science, the highest that the culture of antiquity could attain, beyond which no further development took place, this is not merely due to the influence of his own personality and system. At the time of his death the Greeks had already seen their best days. The civic spirit which, in spite of sordid party strife and sanguinary border-feuds, had borne forward the petty Greek states both before and in the throes of the Persian wars, disappeared as soon as the states lost their independence, first through the hegemonies which the Athenians and the Spartans in turn exercised over them, and later on through the Macedonian and the Roman conquests. And with the sense of patriotism disappeared also the intellectual power and will to act. The semi-oriental monarchies into which the empire of Alexander became split up were certainly often governed by enlightened princes who generously patronized the sciences, but their lavish pensions had nevertheless to be purchased with obsequious flattery, and the proud self-respect which induced Empedocles to refuse a royal crown, and Heracleitus to decline the office of high-priest, existed no longer. The great systems of thought which were created by the noblest spirits of later antiquity were in fact essentially founded upon ethical aims; they were intended to reinforce the individual in the struggle against the ever-increasing difficulties which life in those days presented. The exact sciences, again, were divided up more and more into special spheres and the research work carried out during the succeeding centuries gave substantial results, until here, too, the spiritual weariness from which that epoch suffered claimed its due.

In these circumstances it may seem suitable in the following chapter to pay special regard to the attempts at a general explanation of natural phenomena which were made after Aristotle, after which we shall view the results achieved by the biology of antiquity as a special line of research.

CHAPTER VI

NATURAL-PHILOSOPHICAL SYSTEMS AFTER ARISTOTLE

Aristotle's followers

WHEN ARISTOTLE FLED from Athens he left his school in the hands of THEOPHRASTUS, who had been his faithful friend and follower ever since his student days with Plato. Though ten years younger than his master, Theophrastus was an old man when he assumed the leadership of the Lyceum, but he lived much longer and for more than thirty years presided with honour over the education of the pupils. Already under Aristotle he had paid special attention to the study of botany and he continued to work in this science in the spirit of his master. His two treatises on plants are to botany what Aristotle's works were to zoology. Furthermore, there is extant a "history of physics" by him, which has always been the main source of our knowledge of the ideas of the ancient natural philosophers. He also wrote a zoological work, which has been lost, but on the whole it seems to have contained nothing essentially new that is not found in Aristotle.

On the other hand, Theophrastus' successor, STRATO, developed Aristotle's theory along entirely fresh lines. Unfortunately the present age is acquainted with his point of view only through the references of other authors, but from these it is clear that he was a truly independent thinker. Born at Lampsacus in Asia Minor, he became a disciple of Theophrastus at an early age, and after the latter's death held his professorship for eighteen years. His numerous writings, now lost, dealt particularly with problems of natural science and procured him the title of "the physicist." In contrast to Aristotle he denied the existence of a dominant intelligence outside the universe; he imagined that the forces that govern the course of events dwell in things themselves and operate by natural necessity. Further, the soul of man he believed to be a force inhabiting the body, expressing itself as motion, and having the brain for its organ. On the other hand, he seems to have attacked Democritus' theory of the atoms and infinite space and considered the whole world to be finite.

Strato's successors appear to have been men of little importance, and although Aristotle's school survived down to the sixth century after Christ, it nevertheless ceased to act as a guiding light in science; its teachers and pupils became for the most part involved in specialized investigations into grammar, literature, and ethics, and the keen interest in the natural sciences which the founder of the school had evoked disappeared entirely from their circle. Instead, philosophical speculations on nature were undertaken by another school of thinkers, who revived the atomic theory of Democritus, which was thus given a fresh lease of life and survived not only the classical period, but through the Middle Ages until the Renaissance. This new line of thought was directed by EPICURUS the Athenian, one of the most discussed philosophical personalities of antiquity. He lived between the years 342 and 271 B.C., and in his native town founded a school whose members lived quietly and carried out joint researches under their master's guidance. As already mentioned, his theoretical standpoint was Democritus' atomic theory, which he adopted without really developing it further. In direct opposition to Aristotle he taught that universal space is infinite, that bodies are composed of particles indivisible in themselves, whose motions are the cause of everything that happens and through whose alternate association and dissolution worlds arise and perish. Even the soul of man consists of atoms and is thus a purely corporeal organ. There is no universal intelligence, but all things happen through natural causes. What these causes are Epicurus did not bother much about. In fact, he considered it hardly worth while trying to find out the secrets of nature; thus it might well be, he expressly assures us, that the moon borrows light from the sun, but it might equally well be self-illuminating. The main thing was that one assume a natural explanation of the world and the universe; it mattered little what this explanation turned out to be like in detail if only man rid himself of the superstition which always accompanies a belief in supernatural powers. Epicurus' system, in fact, was expressly based on the idea of creating by the aid of philosophy a pleasant existence, and man attained this best by being an opportunist in all the main problems of life. That on account of this he gained numerous followers is generally acknowledged, but also the principle of a cosmic conception which lays most stress on the exclusion of all supernatural elements, and in doing so does not bother much about the difficulties arising out of questions of detail, has always had, and will indeed always have, keen adherents. It is no matter for surprise, however, that the followers of Epicurus, with such a view of the functions of scientific research, could never claim any really great natural philosopher of antiquity amongst their number. Epicureanism survived as a mode of living which invited people during depressing and hopeless times to seek life's happiness and goal in pleasure, spiritual as well as material. To Epicurus himself and to his friends pleasure was essentially spiritual; their material needs were extremely modest. But matters became worse when his doctrine was brought to Rome. In that world-capital it degenerated into an unbridled worship of pleasure, particularly under the Empire; the fact that Nero and his friends called themselves Epicureans was not calculated to heighten the school's reputation. Even in Rome, however, it gained adherents of a nobler character. One of these, Lucretius, deserves further mention, especially as his enunciation of the atomic theory is the most detailed of its kind that has been handed down to us from antiquity, and as such it is also of interest on the grounds of the biological particulars which it contains.

Titus Lucretius Carus was probably born in 99 and died in 55 b.c. He belonged to a famous patrician family and seems to have been acquainted with several well-known persons among his contemporaries, but although the epoch in which he lived — he was contemporary with Cæsar and Cicero — is without doubt the best known in antiquity from a historical point of view, nothing is known of his life with any certainty. He seems to have kept aloof from the political struggles of his time and devoted himself entirely to philosophical and literary study. An early father of the Christian Church declares that he died by his own hand; the statement may indeed be true, for in the deeply unhappy age in which he lived, this desperate way out of life was, as is well known, resorted to by many. It was not until after his death that his great work *On the Nature of Things* was published, in which he recorded the results of his philosophical speculations. Following the example of the earlier Greek philosophers, particularly of Empedocles, whom he greatly admired, he has clothed his thoughts in verse form; he is the last of the ancient philosophers to do so. The wealth of imagination and the high inspiration which fill his poem have given him a place amongst the greatest poets of antiquity, but he is of great interest also as a philosopher. The most striking feature of his poem is his passionate love of truth and his absolute conviction that thought ultimately will succeed in penetrating the true nature of things. To the glorious mission of philosophy to seek after truth he opposes the grim picture of darkness and superstition called forth by traditional theism, a miserable state from which he hopes free-thought will save humanity. The cosmic explanation which he accepts as the only possible right one is the atomic theory, in the form in which Epicurus expounded it; Democritus is mentioned only in passing. It can hardly be said that he contributed anything towards the development of the general principle of that theory, but at any rate his conception differs from that of which an account has been given above in connexion with his predecessors. The worlds are infinite in number, formed of atoms moving in empty space. Their motion is due to gravity and consequently represents a constant descent; the fact, however, that they strike against one another, as presupposed in the atomic theory, is due to their fall's being for internal reasons not quite perpendicular, but deflecting to one side.

Lucretius' soul theory

The most interesting of all, however, is Lucretius' attempt to apply the atomic theory in detail to the phenomena of the senses and the processes of the soul. Following Aristotle, with whose theory of the finality of all things he nevertheless sharply disagrees, he assumes three kinds of soul, *animus*, *mens*, and *anima*[1] — that is, the spirit, the understanding, and the soul or life-principle. He does not, however, consistently differentiate between these three categories, but discusses them mostly as one single idea. In quality the soul is material, an organ, like the rest, formed of extremely small atoms distributed throughout the body, very mobile, and therefore easily dispersed. These atoms are of three determinable kinds: warmth, air, and "aura," a more rarified kind of air corresponding to the "pneuma" of Hippocrates. But besides the three categories of atoms just named, the soul contains still a fourth component, which has no name, but which forms the real percipient, the consciousness in the soul, and whose atoms are the smallest and most mobile. They give impulse to the other soul-atoms and thereby indirectly to the movements of the body. These component parts of the soul, being variously commingled, produce the varying soul-characteristics in different individuals and make themselves felt in different degrees in the same individual in different states of mind; heat in anger, cold air in terror, etc. The soul, which in life is contained by the body, as a vessel contains whatever is kept therein, dissolves at death into the simplest component parts and is annihilated. The immortality of the soul as maintained by Plato and his disciples is attacked by Lucretius with passionate intensity. Again and again he seeks to prove that this, combined with a belief in gods, is the cause of all human miseries. Sense-perceptions are, according to Lucretius, due to things' giving off from their surface a kind of light particles which, formed like the things themselves, float about in space and influence the organs of sense. As a proof that such images are given off he cites, *inter alia*, the change of skin in snakes and insects. All sensations — sight, hearing, smell, and taste — are thus excited by different atoms which affect the organs of the body. The ideas arising herefrom are caused by a mass of still more subtle images of things, floating about in space even after the things themselves have disappeared. Thus one sees in imagination images of individuals long since dead, and owing to the images' sometimes coalescing, one receives impressions of creatures which have never existed in reality; for instance, through the coalescing of a horse and a human image is produced the mythical centaur. Through such images still remaining in the soul dreams arise. Primitive as these sensory physiological speculations are, they are nevertheless accompanied by a number of extremely striking observations regarding different kinds of sensations. In particular Lucretius discussed in detail sensations in the sphere of sexual life, which he describes with a curious mixture of

---

[1] These names, which Lucretius undoubtedly himself invented on the Greek model, are again found in the psychological terminology of the Middle Ages, and even in Swedenborg.

minute observations from a natural-scientific point of view and poetic inspiration.

Lucretius' influence upon posterity has been both lasting and important. It is undoubtedly due mostly to him that atomism survived throughout the Middle Ages, although in obscurity, owing to the hostility of the Church.[2] During the Renaissance he was held in high estimation; the greatest thinker of that epoch, Giordano Bruno, was strongly influenced by him and in imitation of him wrote several of his scientific works in verse form, and even the free-thinkers of the eighteenth century studied him closely. Yet it can hardly be said that he advanced the natural sciences. He has not succeeded in improving upon the atomic theory as created by Democritus; such progress as Aristotle made in the sphere of biological development he rejected with the theory of finality upon which it rested, but without succeeding in substituting any better subjection to law. On the whole, atomism became a theory that brought no benefit to natural research until in the beginning of the nineteenth century it was, through Dalton, adopted in chemical research and, thanks chiefly to Berzelius, became the most fruitful working hypothesis of that science. Since then it has been one of the most important foundations on which our idea of nature, both inorganic and organic, is based; but the universal application which the ancient atomists ascribe to it it has not received; no true natural philosopher of today hopes to be able to explain the phenomena of animate life with its aid, although in quasi-scientific popular literature attempts have been made to do so.

---

[2] In his *Divina Commedia* Dante relates that in hell there were several thousand "Epicureans," among them many of the most eminent men of his own time.

CHAPTER VII

The anatomists of Alexandria

I F, THEN, the ancient conception of nature failed to advance beyond the point to which Aristotle brought it, nevertheless there developed after his time and on the foundations laid by him a specialized form of biological research which during the following centuries produced rare and abundant harvests. The centre of natural research during this period, not only in the biological, but in other branches, was Alexandria, the purely Greek capital of Egypt. Under the patronage of the refined and generous kings of the Ptolemaic dynasty there was here established an institute of scientific research the like of which the ancient world never saw before or afterwards. Even the founder of the dynasty, Ptolemy I (died in 283 B.C.), was a highly cultured man who collected books and was himself an author. His son Ptolemy II founded the Museum of Alexandria (*mouseion* — a temple of the goddesses of song and wisdom, the Muses), an institution where scholars from every country received lodging and maintenance and substantial assistance for the furthering of their research work. It was conducted on the lines of an academy with the chief librarian as chairman; the highest authority was exercised by the high-priest of the Muses, who was religious head of the college. All the branches of science known to classical antiquity were studied here; the science of biology was chiefly pursued in connexion with medicine, like anatomy and physiology.

It has been mentioned above that medical science was from early times highly developed in Egypt, inasmuch as the custom of embalming bodies in that country contributed towards increasing the knowledge of human anatomy. Herein, then, lay certain preconditions for the stimulus given to the study of anatomy in Alexandria; substantial grants of money and literary and material aid from the princes rapidly advanced the development of the school of medicine. Two teachers of more than usually prominent gifts, Herophilus and Erasistratus, finally brought to the school a reputation such as no other attained in classical times. We know little of the personal history of these men, and not much about the general ideas of nature which they embraced. It is assumed, however, that they were influenced by the scepticism of Pyrrho, who apparently found most of his adherents amongst the Alexandrian physicians. PYRRHO of Elis (376–288 B.C.) taught that no knowledge of things is really possible; man can know nothing and prove nothing, not even the impossibility of knowledge or justification for doubt. Such a fundamental principle naturally precluded any theoretical conception of nature, whether Democritean or Aristotelean, but it was just this very circumstance which drove a philosopher seeking after knowledge to become all the more deeply engrossed in specialized science and its practical application. It was thus exclusively through detailed anatomical research that the Alexandrian medical school advanced the science of biology.

HEROPHILUS was a native of Chalcedon in Asia Minor, studied in the Asclepiad schools in Cos and Cnidus, and afterwards worked as a teacher and researcher in Alexandria. The dates of his birth and death are unknown, but his activities fall within the decades about the year 300 B.C. His writings, too, are lost, except for a few fragments; their contents are known to us only through the references of other authors. That Herophilus was one of the most prominent anatomists of antiquity is, however, universally acknowledged, both by classical and by modern authors. His fame is based on the numerous discoveries he made, particularly in human anatomy. Every part of the human body was investigated by him, and what more than anything else attracted the attention of his contemporaries was the fact that he employed human bodies for the purposes of investigation. Sceptics as they were, he and his pupils despised the traditional fear of dissecting human bodies, and the enlightened Ptolemaic rulers placed material at their disposal. It is even declared that Herophilus took advantage of opportunities offered to him to carry out investigations on living human beings — criminals condemned to death, whose internal organs he studied in a living state.[1] Among the organs which he described in detail may specially be mentioned the brain; he discovered and gave an account of its membranes and its venous blood sinuses, which still bear his name, the torcular Herophili. Moreover, he studied the ventricles of the brain, being particularly interested in the fourth, which he regarded as the organ of the soul. He gave similar close study to the eye, its membranes, film, and retina. He also described the alimentary canal; the name "doudenum" for its upper section comes from him. The liver he carefully studied with regard to the variations in its shape in different individuals. The circulatory system he also made the subject of close investigation; he compared the walls of the arteries and the veins and studied the pulse at different ages and under different bodily conditions. That the arteries contained the pneuma he believed in common with all other researchers of his time. For the first time he cleared up the question of the difference between nerves and tendons. Finally, he carefully worked out the

---

[1] One of the early Fathers, Tertullian, quotes, among other heinous acts committed by the heathen, that Herophilus tortured to death six hundred persons — a story on a par with much that is related nowadays in anti-vivisectionist literature.

anatomy of the genital organs. His physiological ideas were governed by the usual conceptions of antiquity — four different life-elements localized in corresponding main organs; they are therefore of no special interest. For the rest, Herophilus was a great admirer of Hippocrates, whose views on diseases and remedies he accepted without reserve.

Contemporary and in competition with Herophilus was ERASISTRATUS of Cheos, a small island in the Ægean. His dates and personal history are as unknown to us as those of Herophilus, and his writings were lost even in late antiquity. According to a late and unconfirmed report he was the nephew of Aristotle; it is certain that his teacher, Metrodorus, was the latter's contemporary and friend. Erasistratus began his career as court physician to the Seleucides of Syria, but he was called thence to Alexandria, where he founded a school of medicine. His anatomical works dealt chiefly with the circulatory system. The heart he studied with care and gave its valves the names they still bear. Further, he established the connexion between arteries and veins and explained the bleeding from the arteries in wounds by the assumption that their pneuma disappears, and in its stead the blood from the venous system penetrates into the arteries and then flows from the wound. Again, he investigated the lymphatic ducts and the secretion of chyle in live animals. He made important discoveries which increased the knowledge of the nervous system; he distinguished between the motor and sensory nerves and was the first to describe in detail the convolutions of the brain. As a physician Erasistratus was more practical than Herophilus; he utterly scorned the Hippocratic traditions, prescribed simple remedies, avoided venesection, and strongly advocated a hygienic mode of life.

Hostility between the medical schools of Alexandria

THIS opposition between the two Alexandrian anatomists had fateful consequences for science. They themselves impugned one another by polemics and intrigue, while there existed a still greater hostility between the respective schools. The "Herophilites" drove their master's conservatism and respect for Hippocrates to extreme limits, while the "Erasistratites" held up to scorn and counteracted the virtues of the medical tradition. This was naturally bound to prejudice science, and it was all the more disastrous as the cultural conditions in Alexandria became in time seriously impaired. The early enlightened Ptolemaic kings were succeeded by a line of degenerate scoundrels who neglected the interests of learning as they neglected all their other duties. The Museum declined, its grants were reduced, and the learned often fell victims to the tyrants' whims. Thus finally Alexandria became a provincial town within the great Roman Empire. The Museum certainly survived, but without the encouragement which the native rulers had given to it; it was eventually destroyed in a riot—the Alexandrine mob was known as the most unruly in the whole of the Roman Empire — and in the end the extremely fanatical Christian Church in Egypt wiped out the last vestiges of pagan scholarship.

Roman natural science

In Rome, which in time assumed Alexandria's position as the supreme capital of the world, there arose no equivalent to the Museum. It was not until a later epoch that the Roman people, with their decidedly practical mind, attained to the higher culture, and only in the juridical sphere did they make any independent contribution to the development of intellectual work; otherwise they appropriated Greek culture, special branches of which they converted to various — mostly practical — purposes. One applied science of this kind which the Romans created was the science of agriculture. In contrast to the Greeks they were agriculturists body and soul and early felt the need of having their experiences in this sphere collated and recorded. The old censor Cato actually wrote a treatise on agriculture, and after him there are mentioned a large number of writers on agricultural subjects. The foremost of these was undoubtedly Columella, whose writings contain sufficient of interest to biology to warrant his being mentioned.

LUCIUS JUNIUS MODERATUS COLUMELLA was born at the beginning of the Christian era in Spain, but he seems to have lived in Rome and there composed his treatise on agriculture in twelve books. Of biological interest is his account of domestic animals, their management and necessities of life, their races and areas of distribution. All the useful animals of his time, even the bee, are dealt with in his work, most sections of which are, as a matter of fact, of purely economic interest.

PLINY, too, shows a marked interest in the practical application of science. He was the most eminent of Rome's natural philosophers and, next to Aristotle, the most influential of the biologists of classical antiquity. Throughout later antiquity and the Middle Ages and far on into more recent times his *Natural History* has played an important part in the development of science; indeed, it may be said that even in our own day his influence has not entirely waned. In contrast to Aristotle, however, he has been harshly criticized in the biological literature of the present day — extravagantly so, because more has been demanded of him than he ever intended and more than he was able to offer. He has been characterized as a soulless compiler, because, more honest than Aristotle, he always quotes his sources; his superstition has been ridiculed because he tells of marvellous animals the existence of which none of his contemporaries doubted. Above all, the constantly repeated comparisons between him and Aristotle are entirely unjustified; the aims and methods of the one were not those of the other. A study of Pliny's life and work will confirm this.

GAIUS PLINIUS SECUNDUS was born A.D. 23 at Comum, now the Como of northern Italy. He belonged to a family of public officials, and his own career was in fact an official career after the typically Roman model. He received a thorough education from good private teachers in Rome and afterwards served alternately in the army and the civil administration. He spent a long time in Germany, where he held a military command, and eventually became commander-in-chief of a division of the fleet. On the first occasion known to history when Vesuvius was in eruption (A.D. 79), he lay with his vessels near Naples. In order to study the unusual phenomenon closer he ordered a boat to row him to the foot of the volcano, and there he perished. Both in his treatises and in his biographies he stands out as a man of the highest probity of the good old Roman type, brave, honest, and loyal. Constantly engaged in a round of official functions, he devoted every unoccupied moment, both at Rome and in the distant provinces, to study and authorship. His capacity for work is described as inexhaustible, and the writings he left bear witness to his remarkable erudition. He worked at the most widely different subjects: military science and military history, rhetoric, and linguistics. The only treatise of his which has come down to posterity is his great work of thirty-seven books, the *Natural History*, on which his fame really rests. It represents a veritable encyclopædia covering the entire knowledge of nature at that time, including its application to medicine, technology, and economy. It begins with an account of the universe and its laws and goes on to give an increasingly specialized description of various natural objects. Books VIII to XI deal with animals, and a number of scattered notes on zoological subjects are also to be found in other sections of the great work.

In his general conception Pliny was a Stoic. The Stoic philosophy was founded in Athens at the same time as the Epicurean, but its founders and earliest leaders all came from Semitic countries, and several historians have made an attempt to trace an oriental influence in its ascetic contempt for material life-values and its strong feeling for personal responsibility. However, Stoicism found its way to Rome, whose noblest men were attracted by its austere sense of duty and, as was the Roman habit, converted its system to practical uses. Stoicism laid greater stress than Epicureanism on a practical way of living; it was not so much concerned with the general conception of nature and its laws. Even Pliny's general ideas of nature constitute a not very interesting record of the dicta of earlier authors; here, for instance, we find the Aristotelean theory of a spherical universe, with the four elements as its essential components, with — on the Pythagorean, or rather Heracleitean, pattern — fire as the primary cause, the origin of the soul; a divinity governs the world, but it is folly to seek to discover its entity, though a greater folly still is polytheism. Oracular utterances and prodigies, on the other hand, are recounted by the score and without any expression of doubt of their value.

Pliny's description of animals

THE zoological section of his natural history is as encyclopædically treated as the rest. The various animals are enumerated without any sequence, but on the whole the largest and most remarkable are mentioned first; they are described with reference to their habits, their utility, and the mischief they do, the date of their first being exhibited and employed in Rome, and in general their relation to man. On the other hand, no attempt whatever is made to give a true description of their external and internal structure. The earlier biologists of antiquity, including Aristotle, were, as we know, not very conspicuous for any important criticisms on points of detail, especially where exotic animals are concerned, and Pliny with great goodwill collects all the marvels he can find in earlier writings and narrates them without reserve. Consequently his account teems with the most fantastic fables. As an example of the assertions he makes may be mentioned his description of the elephant, which, characteristically enough, is named first of all animals. "Amongst land-animals the elephant is the largest and the one whose intelligence comes nearest that of man, for he understands the language of his country, obeys commands, has a memory for training, takes delight in love and honour, and also possesses a rare thing even amongst men — honesty, self-control, and a sense of justice; he also worships stars and venerates the sun and moon. In the mountains of Mauretania it is said that herds of elephants move at new moon down to a river by the name of Amilo, ceremoniously cleanse themselves there by spraying one another with water, and after having thus paid their respects to the heavenly light return to the forests bearing their weary calves with them. It is also said that when they are to be transported overseas, they refuse to go on board until the master of the ship has given them a promise under oath to convey them home again." Further, they are so modest that they never mate except in very secluded spots, while adultery never occurs amongst them. Towards weaker animals they show compassion, so that an elephant when passing through a flock of sheep will with his trunk lift out of the way those he meets, for fear of trampling on them.[2] Besides these and similar childish statements accounts are given of the habits of elephants and the way to tame them, which are quite correct, as well as a number of facts of interest from the point of view of cultural history as to their employment amongst different peoples, when they were first exhibited at Rome, etc. Pliny likewise relates many wonderful stories about other lesser-known animals, such as the elk and the aurochs of northern Europe. On the other hand, the information he gives regarding the ordinary domestic animals of his own country is on the whole reliable and the

---

[2] It was probably this description that directly or indirectly induced a Danish king in the Middle Ages to found an Order of the Elephant, with the purpose of thereby exhorting its members to imitate the admirable qualities of that noble animal.

particulars of cattle-management at that period are quite sound. Among land-animals he includes even the cold-blooded vertebrates; after that he deals with birds, fishes (including molluscs), and insects. Details of all kinds of animals belonging to these groups are in the main similar to those mentioned above. Insects in particular seem to have attracted his attention; he never wearies of expressing his admiration for their perfect organisms, in spite of their small bodily size, and he gives an account of what was known of their organic systems. The bee he describes in great detail and he relates its habits, in many respects correctly, although he did not succeed, any more than the other ancient authors, in gaining any idea of its method of reproduction.

Pliny's anatomical ideas

AFTER Pliny has thus given an account of the animals known to him, he discusses the various organs of the human and animal body on the same plan; each organ is considered in reference to its qualities and occurrence in the various animals. Here we clearly find Aristotle to be the pattern and main source of information; but while the latter's description of the organs is given with a view to tracing the connexion and origin of the forms, Pliny's account still has the character of a work of reference, in which all the memoranda that he was able to collect out of his vast erudition are cited with no kind of theoretical purpose and without any deeper significance than the word that clothes them; for instance, the horn of the ox, of the horned snake, and of the snail are treated as all one. A wealth of valuable notes from the rich scientific store of the anatomical knowledge of antiquity has thus been preserved for posterity by Pliny, whose sources of information have been lost to us, but besides this he conscientiously notes down a number of ancient prodigies, which of course inspired fear in his time and have consequently been handed down to history — of sacrificial animals which had no liver, or of the Messenian champion of liberty, Aristomenes, whose heart the Spartans found to be covered with hair — without in any way hinting at the possibility of fraud on the part of the sacrificial priests. A mass of information regarding the medicinal use of animals or animal parts closes the zoological section of his *Natural History*.

Pliny himself states that he had recourse to two thousand books by various authors for the compiling of his work, which maintains also from beginning to end its character of a confused motley of notes. This many-sided learning, which has very much impressed past generations, seems in our day, when only first-hand knowledge is really respected, rather pitiful. Nevertheless, as previously pointed out, Pliny has certainly been underestimated. For fifteen hundred years his work was the main source of man's knowledge of natural history, and when during the Renaissance a Gesner or an Aldrovandi revived the pursuit of zoological research, they at once began where

Pliny had left off and carried on the work after his method. In this way present-day zoology, as regards the study of fauna and classification, takes Pliny as its starting-point, just as in the matter of comparative anatomy and morphology it is based on Aristotle, and therefore the services of the one should in all fairness be recognized as much as those of the other, even if they refer to entirely different fields of study.

CHAPTER VIII

THE DECLINE OF SCIENCE IN LATE ANTIQUITY

The decline of ancient civilization: its causes

IT HAS ALREADY been pointed out that the natural science of antiquity reached its zenith in Aristotle, and a number of reasons have been given for the fact that only in points of detail, but never in regard to the summarizing of the results achieved, did it advance beyond his standpoint. While Rome, first as a republic and then as an empire, was conquering and administering the whole of the civilized world, there began an era which, more than any other, should have been devoted to promoting the work of intellectual culture. The universal peace that prevailed during the first two centuries of the Christian era has never had its counterpart either before or since, for the border feuds and insurrections which disturbed it were entirely local and transient. And as there was peace, there was also prosperity; even up to the present day the ruins of buildings bear witness to the common and private wealth of those days throughout the length and breadth of the Roman Empire. And yet it was this very epoch which witnessed the decline of ancient science — indeed the whole of the culture of antiquity. It was not long before the best minds in the intellectual world of the time realized this fact. Pliny, for instance, is never tired of repeating that humanity is corrupt and that his age was worse than the era that had passed. The reason he gives is the increasing corruption of morals — an assertion with which innumerable other ancient authors are in agreement and which has therefore been repeated in more recent times. The cause cannot lie there, however; moral corruption is always a symptom and not a cause of cultural decadence. The cause is far more likely to be found in the change in the common conception of life which was a consequence of subjection under the Empire. The ancient provincial patriotism had lost its power to survive and there was no possibility of any fresh form of social community developing; instead the individual personality appears as struggling for freedom from external oppression and grievances. This self-assertion against an oppressive existence both Epicureans and Stoics sought to put into practice, each in their own way; but, as we have seen, their teachings formed no good soil in which to cultivate empirical research. In the long run, however, the purely negative insensibility to suffering which constituted the philosophy of life of these schools of thought could not suffice; in their place appeared lines of thought start-

ing out from the idea of leading the personality into an existence different from the earthly, of creating with the aid of some kind of higher,
secret knowledge a happier world for the soul to live in. There thus arose a
half-mystical, half-experimental psychology, which was nurtured by philosophical schools possessing sectarian organizations, like that of the Pythagoreans in the old days. One of these schools, and the most fantastic of all,
actually called themselves neo-Pythagoreans; another, more scientifically
serious, was the neo-Platonic, which sought to bring the human spirit, along
the mystical path of introspection, into contact with the world of ideas,
which Plato declared to be the only true world. Through this development
the very idea of philosophy became radically altered; the philosopher was no
longer a lover of wisdom, as the name implies, but a lover of piety. But as
such he retained no interest in natural phenomena; his spirit in fact lived in
supernatural regions of space, and if he devoted any time to the objects of
nature, it was merely in order to discover the secret divine powers which,
hidden from the eyes of the ignorant, dwelt in plants and animals.

For the belief in God awakened to new life during later antiquity; not
the old sacrificial faith[1] so indissolubly bound up with the inner life of the
petty states, but faith in a supreme power able to save the individual from
sorrow and suffering. Numerous religious brotherhoods were founded which
sought by mystical means to procure for their members peace and happiness
in this life, or at any rate in the life to come. Among these faiths appeared
Christianity, which was finally triumphant, thanks to the message of universal love and the sure promise of salvation which it offered to mankind, and
not least as a result of the strong community-organization which its first
followers set up, with unlimited charity within their ranks and stubborn
power of resistance against persecution from outside. But an epoch in which
the best of humanity sought their happiness in life beyond the bounds of
actual existence must inevitably be a period of decay, both materially and
within those spheres of the spiritual life which have to do with reality: exact
science as well as creative art.

As early as the second century of our era, when material prosperity was
still at its highest, there appear signs of this spiritual disintegration; during
this century lived the last of the great classical authors — the Latin poet
Juvenal, and the Greek Lucian, by the side of a mass of representatives of the
new era: miracle-workers, soothsayers, and necromancers, whom they
strenuously but vainly opposed. At that time, too, lived the last great biologist of the age of classical culture, the physician Galen, who in his writings

---

[1] The cult of sacrifice was also revived, it is true, in late antiquity, but it was not so much
the old national cult as one accompanied by mystical, impassioned ceremonies, originating from
the East. To the noblest minds of the time, however, it had very little, or at any rate a purely
conventional, value.

has strangely combined the many-sided biological learning of antiquity with the mystical trend of thought of the new era.

The last great biologist of antiquity

GALEN was born in 131 at Pergamum in Asia Minor, of Greek parents. After moving to Rome he latinized his name and called himself CLAUDIUS GALENUS, but continued to write in Greek, this being a characteristic example of the mixed culture prevailing at this time. His father, Nicon, was an architect; through a dream he learnt that his son was destined to be a physician, and Galen thus entered upon his medical career under what was thought to be divine instigation. Even before this occurred, he had been initiated by good teachers into the philosophy of his time: in his native town he studied under Platonists, Epicureans, and Stoics, but he was particularly versed in the writings of Aristotle and Theophrastus. Medicine he studied in his native country, then in Corinth, and finally in Alexandria, everywhere acquiring, besides medicine, philosophical knowledge from the best teachers of his time. Having thus completed his education, he returned home in the year 158 and was employed in his native city as physician in the temple of Æsculapius, as well as, characteristically enough, at the city's gladiatorial school. After six years, however, he moved to Rome and there began to give lectures on his own scientific subjects, as a result of which he won the friendship of men of repute and the envy of other physicians. In a still greater degree did the immense practice he acquired awaken feelings of bitterness, and as he himself never attempted to conciliate his envious fellow physicians, but on the contrary strongly resented the decline in the efficiency of medical practitioners, such a storm of hostility broke over his head that for a time he had to fly the field and return to his own country. He had, however, such an established reputation that after the lapse of only a few years he was summoned to become body-physician to the Emperor Marcus Aurelius, and thus, under the protection both of him and of his son Commodus, he was able to carry on his work in Rome unmolested. His last years passed quietly; it is not known when and where he died, but probably about the year 210 at Pergamum.

Galen's writings

As a writer Galen was very productive; he himself states that he wrote 256 treatises, of which 131 were of a medical character. Of the latter, 83 are still extant. His other works embraced philosophy, mathematics, grammar, and law, but most of them are now lost. He was thus a many-sided man of culture, with interests far above the specialized fields of activity of contemporary physicians and even of the classical Alexandrine doctors, and well capable of critically examining the various medical schools, which by working in opposition to one another with their dogmatically formulated programs brought medical science into disrepute. He also laid great store by

universality of knowledge, and in one of his writings which has been preserved to us he exhorts his professional brethren to devote themselves to the
study of philosophy as an essential foundation for acquiring a proper conception of man's nature in sickness and in health. For this purpose he refers
above all to Hippocrates, whom he extols with extravagant words in all his
works, declaring that his dicta should be interpreted as if they were the
utterances of a god. But both Plato and Aristotle also represent sources
whence he gained a true idea of nature and life, and on their ideal conception
of existence he has based his theory of biological phenomena. He has adopted
their fundamental principle of a divine intelligence as the origin and ruler
of all things, whose existence is proved by the finality of nature, and also
the theory of the soul as a purpose justifying the existence of the body. But
while Aristotle showed the finality of nature by comparing different forms
of life and pointing out the consistency displayed in their existence and
evolution, Galen deals only with the organs of the human body and seeks to
prove how in the smallest details they are constructed and applied exactly as
they should be. And in this perfect organization in the human body he sees
proofs of the power and wisdom of the Creator, whom he never wearies of
praising in words testifying to a deep personal sense of religion, and in a
tone which differs widely from the temperate scientific feeling with which
Aristotle shows the necessity for a supreme intelligence in existence. And
alternating with these pious expressions there are in his writings uncontrol-
lably violent outbursts against the representatives of the theory of the dominance of necessity in nature and of the atoms as the primary components of
matter, particularly against Epicurus and his disciple the physician Asclepiades.[2] In one other respect also Galen proves himself to belong to a new
era; not only are the old Greek philosophers quoted by him, but he also refers to the Mosaic story of creation, with which, it is true, he is at variance —
in fact, he believes matter to be eternal and denies the possibility of creation
out of nothing — but which nevertheless certainly influenced his conception
of nature. That, indeed, constitutes, one might say, one single hymn of
praise to the wisdom of the Creator. In every detail of the human system
does the divine Providence show its foresight; in the hand not only the number and length of the fingers, but even every tendon and muscle is a proof
thereof; likewise with the minutest details of the rest of the body. He scornfully rejects the assertion of the Epicureans that the organs develop with use
and weaken with disuse, saying that in that case energetic people would in
time acquire four legs and four arms, and the lazy only one of each. Again, if
rightly viewed, even organs which might appear to be useless are suited to

---

[2] Asclepiades was a famous Greek physician who lived in Rome during the first century
before the Christian era. His writings, now lost, were highly esteemed in antiquity, but they do
not seem to have included any biological investigations.

serve a certain purpose; if it is asked, for instance, why man has not long ears like the donkey, which would give better hearing, the answer is that man's ears are such as they are in order to enable him to wear a hat. And the wise Creator has not only taken utility into consideration when making man; He has even taken thought for human beauty, as may be seen in the distribution of hair on the face; the beard on the chin is a suitable adornment for a man, while the growth of a beard on the nose would give to the countenance a wild and barbaric appearance. But if the hair of the head is thus the work of the Creator, the hairs on the arms and legs are the work of chance; they are likened to self-sown weeds — the Greeks who went about with bare arms and legs considered these hairs disfiguring and therefore removed them. To such absurdities did Aristotle's theory of the finality of nature gradually lead, its application no longer being guided by his own sober and clear logic. However, the piety of Galen expresses itself in nobler and deeper thoughts when he leaves anatomical details and proceeds to ethical problems. There is a truly biblical tone about words such as these: "In my opinion true piety consists not in sacrificing hundreds of beasts or offering quantities of spices and incense, but in oneself knowing and learning about the wisdom, power, and love of the Creator." Equally noble are the words with which he exhorts his colleagues not to strive for profit, but to offer themselves to the service of suffering humanity. In this Galen shows the same noble and humane spirit as his lord and master, Marcus Aurelius, expressed in his *Meditations*, and there is every indication that, like the latter, he lived as he taught.

Galen's anatomical investigations

BUT if Galen in his general conception of life thus stood on the border-line between antiquity and the Middle Ages, he was as regards knowledge of anatomical detail the foremost philosopher of the classical period, and as such remained the undisputed authority in his own branch of learning up to the Renaissance, and strictly speaking even up to the time when Harvey discovered the circulation of the blood and thereby destroyed one of the foundation-stones of his theoretical system. Both as an anatomist and as a physician Galen had indeed the inestimable advantage of being able to build upon the work of brilliant predecessors, but he also realized the importance of his inheritance and considerably enhanced it by his own observations. These he carried out exclusively on animals, both dead and alive, especially apes, which he considered particularly suitable as material for investigation of human anatomy. There is never any question of his dissecting human bodies; the times had changed considerably since the days of Herophilus; old superstitions had again been revived and governed men's minds. He characteristically begins his enunciation of the anatomy of the human body with the hand, the most useful of all organs — that whereby the soul effects its will, for the whole body exists for the sake of the soul. The human hand is described in detail and with great thoroughness, but, as one can clearly see, as the result of investigating the hands of apes. Then he describes the rest of the extremities and afterwards the intestinal canal, the respiratory organs, the brain, the spine, the blood-vessels, and the genital organs. Galen stands highest as a brain and nerve anatomist, and in this sphere his anatomical and experimental investigations gave results which left all his predecessors far behind. He considerably increased the knowledge of the motor and sensory function of the nerves, which the Alexandrine anatomists had already observed, and he differentiated between the sensory, or, in his terminology, the "soft" nerves, and the motor, or "hard." The soft nerves go from the brain to the sense organs, the hard from the spinal marrow; as the nerves of the spinal marrow also show definitely sensible qualities, though Galen did not succeed in discovering the difference between the anterior and the posterior medullary nerves, he evades the difficulty by assuming a "mixed" consistency and function in certain medullary nerves. By experiments in severing different sections of spinal marrow in living animals he showed the connexion between these and corresponding parts of the body. The brain he likewise described in detail; of its nerves he traces seven couples, the ramifications of which are closely worked out. On the other hand, his idea of the function of the brain is confused, owing to speculations upon the "soul pneuma," which, produced in the cerebral ventricles, circulates through the entire nervous system and forms its most essential component, the basis of its functions. It may be mentioned in this connexion that he shared the ancient idea of the localization of the various qualities of the soul in various organs, which naturally gives rise to long expositions on the wisdom of the Creator and the finality of the creation. The account of the digestive apparatus and its function is in Galen, as in the anatomists of antiquity in general, one of the weak points. The human digestive canal is described after combining the results of dissections performed on various animals, both vegetarian and carnivorous, which fact does not help to make his idea of it clear. Digestion, which in Aristotle was the result of the cooking of the food, is ascribed by Galen to a special "transformation power" in the stomach; its products are transferred through the blood-vessels to the liver, where they are converted into blood; the useless parts of the food are absorbed by the spleen and converted by it into "black bile," which is excreted through the bowel. The kidneys serve to remove excessive water from the blood. This is afterwards conveyed through the veins of the liver, partly to the right chamber of the heart and partly out into the body.

Heart and blood-vessels

GALEN described the system of the blood-vessels in detail, and his opinion on this subject — with its errors as well as its merits — had a more lasting effect than that on any other subject. As one of his services to science it may be mentioned that he finally succeeded in overcoming the old preconception that the arteries and the left heart-chamber contain air. In his opinion they contain blood, with an admixture of "pneuma" — that half-airlike, half-firelike life-principle which gave rise to so much discussion in ancient times and on which the existence of living creatures depends. In one place Galen expresses the hope that the time may come when someone will discover the component in the air which forms pneuma, the substance which is the common precondition of life and combustion — a curious idea, which in fact the discovery of oxygen was eventually to bring to realization. He gives a detailed description of the heart, both its structure and its functions; on the other hand, like his predecessors, he lets both veins and arteries convey the blood from the heart to the rest of the body, in which it is consumed. He is not aware of any blood flowing from the body to the heart, while his idea of the movement of the blood is still further confused by his belief that the liver is to a certain extent the centre of the venous system, since the blood flows from it not only to the heart, but also to the rest of the body. The left ventricle receives through the *vena pulmonalis* "pneuma" from the lungs; from the right ventricle the excremental products proceed to the lungs, these products being "soot" from the combustion process in the heart, which is got rid of by exhalation. The wall between the right and the left ventricles of the heart is porous, permitting the blood to pass through. The walls of the blood-vessels are carefully described, and, generally speaking, Galen's detailed study of the construction of the individual organs is one of his strong points. He is aware of the connexions between the arteries and the veins, but, as is seen from the above, he has not realized the idea of circulation, and this fact, combined with the vagueness with which he explains his ideas on these organs, proved an obstacle to the development of biology for the next fifteen hundred years.

Galen carefully studied the respiratory process and on the whole described it correctly. With regard to the sense organs, in spite of his thorough investigations into the subject he made very little advance on his predecessors, and the same may be said of his description of the genital system and the embryonic process, in which he remains, on the whole, where Aristotle stood.

Splendid as Galen's scientific work was, it does not appear to have been highly appreciated by his contemporaries. He himself complains that but few understand him, but consoles himself with the thought that the Creator, in spite of man's ingratitude, never wearies of doing good. Here posterity has to an unusually generous extent made up for what his contemporaries failed to give him. The fact that Galen did not impress his contemporaries may have been due to the peculiar transitional attitude he adopted; to the survivors of the purely ancient school his romantic piety must have been repulsive, while the more mystically minded, who even then represented the majority, were on the whole not at all interested in exact natural science. The miracle-workers' laying on of hands and invocations were undoubtedly more relied upon than Galen's curative method, based on anatomical studies. On the whole, during this period interest in the study of nature waned more and more, at least in the sense in which the philosophers of old times understood it; the most one could do for educational purposes was to collect stories about natural objects. One such collection is the treatise still preserved in our day *On the Habits of Animals*, which was written by CLAUDIUS ÆLIANUS a generation after Galen. This writer, who was an orator by profession — that is to say, a public lecturer — lived in Rome in the first half of the third century; he is believed to have died in the year 260. His work is a collection of anecdotes about animals, gathered from various sources — thus following the method of Pliny. But while in Pliny the interest in nature is the principal motive, Ælianus is actuated by a feeling of pure edification. Pliny, it is true, can also edify his readers with the examples he cites of the virtues of elephants, but they are related in order to testify to the animals' great qualities of soul. In Ælianus even the lowest creatures are uplifted by a purely personal reverence for the Creator, so that the ecclesiastical writers of the Middle Ages had only to substitute the names of Christian saints for those of the gods quoted and they thus found ready to hand a collection of the most edifying sermons. Thus Ælianus tells of a cock that had one of its legs broken; the bird hopped on its other leg before a statue of a god, and stretching out the broken foot, crowed so pathetically that the god showed his mercy by miraculously healing the injury, whereupon the cock, gratefully flapping its wings, went on its way. Here we find the purely mediæval conception, and this more than a century before the final victory of Christianity; an example, *inter alia*, of the incorrectness of the frequent assertion that the Christian Church after its victory eradicated the culture of antiquity.

Neo-Platonists

WITH regard to the natural sciences an attempt has been made in the foregoing to throw light on the process of internal dissolution which gradually led biology from Aristotle's magnificent system of thought down to Ælianus' collection of legends. The interest in natural phenomena which had for so long been a living factor in the ancient world of culture had now entirely disappeared. What was left of the spirit of research turned to idealistic philosophy, Plato's creation, which was further developed by thinkers who adopted his name and made his theory of ideas their starting-point, proceeding thence in a curious direction, at the same time speculative and full of religious mysticism. In their relations with the outer world these neo-Platonists were as opposed to the men of antiquity as their Christian contemporaries. The founder of the school, Plotinus, was ashamed of possessing a body, while its last great thinker, Proclus (411–485), lived as a hermit; he dwelt in a cave, avoided wine, meat, and women, and saw visions of supernatural things. For to be uplifted into the supersensible by means of ecstatic rapture was to the neo-Platonists not only the whole object of life, but also the very foundation of science. With all its fantastic speculation this school nevertheless developed human thought in one important sphere; it discussed the idea of infinity as none of its predecessors had been able to do. To the ancient atomists infinity was really only an unlimited extension in time and space, akin to the custom of children and wild men, who when they are weary of counting, call the remainder "much" or "many." To the neo-Platonists, again, the infinite was equivalent to the inexpressible and the unknowable, that which exceeds all limitations and measures. And though their endeavour to attain to this infinity by way of ecstasy was naturally of no scientific value, there was nevertheless an indisputable truth to be gained as a result of their endeavours, seeing that the impotence of the power of knowledge in face of the infinite was established once and for all. The natural-research work of our time is based on the realization of the limitation of knowledge in face of the infinity of existence — a limitation which only unscientific dilettantism thinks it possible to override.

Destruction of the old culture

THERE were, then, among the thinkers of this period ideas which pointed beyond the limitations by which the ancient conceptions of existence had been surrounded. It is impossible to estimate how these aims might have developed in happier external circumstances. For as a result of the fall of the Roman Empire the external, purely material preconditions for the continuance of scientific research and for the progress of culture in general no longer existed. As early as the latter half of the imperial epoch the prosperity of those nations which formed the Roman Empire steadily declined as a result of misgovernment, civil war, and the inroads of neighbouring peoples. In the fifth century the world empire collapsed entirely owing to the invasion of the Germans, and a state of dire distress, economic, political, and moral, ensued. The new kingdoms which were founded by barbarous nations had great difficulty in firmly establishing themselves, and their rulers' utter lack of culture rendered impossible any kind of ordered system of government and, consequently, any high standard of prosperity. However, the inhabitants of western Europe gradually co-operated in reviving culture on a national basis. During the last hundred years of the Roman Empire, Gaul had been the most civilized country in the Empire, with numerous institutions founded for the study of classical learning. During the period of invasion many people fled from western Gaul to what was certainly a barbarous, but all the same a peaceful country, Ireland, and thus was founded a centre of culture which during the sixth and seventh centuries was the foremost upholder of the classical tradition and one of the starting-points for the future progress of civilization. In the eastern part of old Roman Empire the Byzantine power was still dominant with a despotic form of government and the Greek Orthodox Church as a binding force. There, too, efforts were made to develop national culture, which was expressed in literature, in the national tongue, combined with interest in Greek science. This was especially so in Syria, where the national movement was often associated with religious sectarianism; but in Persia too, under the Sassanid dynasty, Greek science was studied, particularly Aristotle. In these countries, however, there shortly arose a new cultural power, which took over and further developed the learning of the classical period — namely, the Arabian people, with their new religion, founded by Mohammed.

CHAPTER IX

The Arabian conquest of the East

WHEN MOHAMMED DIED, in 632, the religion he founded had already spread throughout Arabia, and his successors, the first caliphs, managed in the course of a few decades to bring under their dominion the old civilized countries of Babylon, Persia, Syria, and Egypt, to which were later added North Africa and Spain. War against the unfaithful was indeed the prophet's first commandment, and according to his injunction the heathen had a choice between death and conversion; such of the unfaithful, again, as possessed religious writings — Christians, Jews, and Persians — had their lives spared, but were subject to impositions and personal humiliation. The bedouins of the desert, who thus at one blow became the rulers of the most ancient civilized countries in the world, were themselves nothing but barbarians, it is true, but they were intelligent and susceptible to cultural influence, all the more so as in the course of the wandering life they led, they had already come into contact with their civilized neighbours. Their new religion was favourable for rapid cultural progress in that it was a legal doctrine with few and easily comprehensible rules, without, to be sure, the lofty ethical claims of Christianity, but also without the theological subtleties of the different ecclesiastical formulæ. And as, besides, the Arabs troubled themselves but little about social and political questions — they permitted the institutions of conquered nations to survive and contented themselves with appointing governors who collected taxes from them — they had ample time to devote themselves to purely intellectual interests. Indeed, they grasped the elements of the culture of the period with a rapidity which has been compared to that of the Japanese in our own day, and were able in many respects to build higher upon the foundations they found prepared for them. These foundations were Greek science, as the subject peoples produced it in Syrian and Persian translations; it was not until later that the Arabs learnt to read Greek writings in the original. They developed this material and thus created a science representing at the same time a direct continuation of the Greek and a reconstruction of it to suit the conditions which the peculiar Arabian view of the world required. According to Mohammed's theory, the Koran is the source of all learning and contains all the knowledge that man requires; but this claim, which would have rendered all research impossible, was evaded during more liberal eras, while obscurantist rulers continued to threaten the learned with its literal application. This was, however, the cause of a certain restraint invariably characterizing Arabian research, at least in form; the scientists preferred to give their works — even the most independent — the appearance of commentaries on the writings of some famous scientist of antiquity. In philosophy and natural science it was naturally Aristotle, in medicine Galen, who was made to represent the authority on whom the work was based, and at the same time the screen behind which the Arabian scientists saved themselves in the event of the authorities' finding the results of their research work inadmissible. During the most brilliant period of Arabian research it was certainly possible for original and great thoughts to be disguised beneath these commentaries on ancient writings, but the danger of slavish imitation lay in the method itself, and for more than five hundred years the science of the East was drowned in an utterly soulless amplification of ancient authorities.

Experimental method introduced into science

THE Arabian contribution to the development of the exact sciences has been most important in the spheres of mathematics and astronomy, in which they received impulses not only from Greek, but also from Hindu quarters — the so-called "Arabic numerals," which are now universally used, were borrowed by the Arabs from India — and, further, geography, a study which the Arabs applied to the investigating of several unknown regions, medicine, particularly pharmacology and, in connexion therewith, botany, and, finally, chemistry, which they were the first to raise to the rank of a science. Chemistry, indeed, is experimental science above all others, and with it experimenting as a scientific method was introduced and developed by the Arabs. This contribution alone is such as to ensure to Arabic science a place of honour in the history of research. Experimenting, in which the research-worker himself interferes with the course of events in nature and arranges that course with a view to having a specific question answered — this, the most certain method whereby the obedience of natural phenomena to law can be proved, was unknown to ancient research. Even Archimedes himself was no experimental physicist, eminent though he was as a practical engineer, while Galen's vivisections, as well as those of his Alexandrian predecessors, had rather the character of observations of live animals than of actual experiments. As a matter of fact, however, the experimental method is very ancient and has its origin in a number of experiences of various kinds which survived in different classes of people before science adopted their methodic system and employed it for obtaining results in exact research work. Thus every type of peoples has practised magical experiments based on the preparation of charms, which are concocted out of the most extraordinary ingredients and are used as love-potions, elixirs of life, enchantments, and pure poison. Such magic brews were prepared among the nations of antiquity by witches and wizards and are still concocted amongst inferior peoples and even amongst more primitive strata of higher types to this very day. That the enlightened scientists of antiquity refused to associate themselves with such magical preparations was but natural; it required that tendency towards the vulgar and the fantastic which the decline of ancient culture evoked in the scientific world, before the methods of popular sorcery could begin to be of interest to thinking and inquiring minds as well. Nor indeed does the earliest experimental science deny this origin; it appears in the form of alchemy, with its pronounced mystical aims, and above all in the conversion of base metals into precious metals, the discovery of elixirs of life and immortality, the reproduction of homuncules, etc. — aims to which it adhered throughout the Middle Ages, even after its means and methods had become characterized, at least in certain features, by a fair measure of exactness and an extensive knowledge of the inorganic objects of nature in particular. It was therefore at a later stage than any other branch of exact science that the experimental branch succeeded in freeing itself from its connexion with the supernatural world of thought, from which all science gradually broke away. It is only the research work of more modern times that has been able to enjoy to the full the advantages which experimental science offers.

Arabian natural philosophers

WITH biology the famous Arabian alchemists, one GEBER and others, had nothing to do; they occupied themselves only with inorganic nature. On the other hand, the East possessed a number of purely speculative researchers who dealt with the phenomena in living nature from a theoretical point of view and who exercised a lasting influence on the conception of them in succeeding ages. All these philosophers took Aristotle as the starting-point for their researches and, as already mentioned, they likewise gave to their own often quite daring speculations the form of commentaries on his works. Indeed, their position was always fraught with danger; they were looked upon with suspicion by the orthodox Mohammedans, who believed that all studies that did not concern the Holy Koran were prohibited. Against these constant persecutions they had no other support than the patronage of some science-loving prince, which had to be won and sustained by flattery and was at best an unreliable guarantee of life and maintenance. These philosophers held no posts as teachers — in the Mohammedan East there were colleges only for students of the Koran — but their scientific researches were always a private occupation; by profession they were frequently physicians, sometimes lawyers, officials, or courtiers.

Among these oriental thinkers there are primarily two who exerted some influence on the progress of science even in the West, their writings being at an early date translated into Latin and diligently studied in Europe, becoming at the same time the basis for continued philosophical research. Abu Sina, or, as he is called in Europe, with a latinized distortion of his name, Avicenna, was born at Bokhara in 980, of Persian stock. At that period Persia was divided into a number of major and minor states, ruled over by princes, who in mutual rivalry sought to win honour by exploits of war and peace. There prevailed a high standard of intellectual culture, and the conditions of the country have often been compared with those of Italy during the Renaissance. Avicenna indeed bears a great resemblance to personalities living at the time of the Renaissance; strictly speaking, he was a physician, but he was also mathematician, astronomer, philosopher, and poet. Chequered too were his fortunes; at one time he was an all-powerful minister at the court of some vassal prince, at another he was an exile fleeing from his enemies and in danger of his life. He died in 1037, his health shattered by his manifold exertions and his reckless love of pleasure. The most important of his numerous writings is his great *Canon of Medicine*, which, next to Galen's, remained the chief authority in the sphere of medical science. Its sections dealing generally with natural philosophy, anatomy, and physiology are of interest from the point of view of biological history. There is still extant a major work of his on general philosophy. As a thinker Avicenna takes Aristotle as his starting-point, but he is also to a certain extent influenced by neo-Platonism. His conception of nature is governed by the "purpose" theory of Aristotle and Galen. Entirely based on Galen, too, is his idea of the human anatomy. The Arabs were in fact even more afraid of dissecting human bodies than were the people of antiquity; it was forbidden in the Koran, and with however little prejudice the learned interpreted the sacred book, they dared not in this respect violate both it and public opinion. Avicenna, however, was more independent as a physiologist; here he could take advantage of the progress his contemporaries had made in the fields of physics and chemistry. But actually it was more for his excellence of form — brilliant style and well-arranged grouping of his subject — than for any original ideas that Avicenna won fame in the East and eventually, perhaps to a still higher degree, in the West.

Far more original as a thinker is the second of the great men of science in the East, Averroes, or Ibn-Rushd, as he was properly called in Arabic. He was born at Cordova in Spain in 1126, the son of an eminent judge. In his native city, which for several centuries had been the centre of Arabic culture in Spain, he studied philosophy, medicine, and jurisprudence, was for some years afterwards *cadi* of Seville, and was finally governor of a province. The fanatical religious reaction, however, which gradually spread among the Mohammedans in Spain towards the close of the twelfth century, once succeeded in bringing about his downfall, and, accused of heretical opinions and pursuits, he was imprisoned, stripped of his honours, and banished to a village near Cordova inhabited by Jews. Fortunately the ruling prince who committed this act of injustice died a year or two afterwards and his son and successor immediately repaired it; Averroes was recalled to court and resumed his honours, but died shortly after, in 1198.

As a natural philosopher Averroes followed Aristotle, and his principal work takes the form of commentaries on Aristotle's writings. Averroes's standpoint is, however, far more than that of his predecessors and even than that of any other mediæval philosopher, independent of his model. He bases his philosophy on the latter's ideas, but he develops them further on his own account. In particular he studied the relation between potentiality and reality in nature. Aristotle considered that marble is a potentiality, which becomes reality when a statue is made out of it, and he consistently applied this method to life in nature — the seed, the embryo, is a potentiality; the plant, the animal, reality. Averroes argued in opposition to this view that nothing in nature is potential that does not exist in reality, in however undeveloped and therefore disguised a form it may be; the plant already exists in the seed, in however undeveloped a state, just like the animal in the embryo. The simile of the marble and the statue Averroes considers inapplicable where nature is concerned; at best the simile would be admissible if the statue were to be found already shaped in the veins of the unsculptured block. By this method of speculation Averroes has carried science a long step nearer the present-day conception of natural evolution; Aristotle's purely abstract idea of potentiality is here replaced by something which approaches far nearer to our idea of energy. Averroes was the last great Arabic philosopher and the greatest natural philosopher of the Middle Ages; if anyone is worthy to be called the Aristotle of the Middle Ages, it is he. He resembles his prototype not only in the fact of his having lived in a decadent era and been subjected to religious persecution, but also in the fact that no one for centuries succeeded in developing his ideas further. Shortly after his death Arabic science succumbed to religious intolerance, while even the Christian schoolmen, who closely studied and highly honoured the Arabic philosopher,[1] saw in him only the interpreter of Aristotle and were not capable of realizing the great advance he made towards a more real conception of nature. He has not, however, been without influence; in the Middle Ages the opponents of ecclesiastical philosophy gathered round his name, and the ideas he evoked can thus be traced through the ages until they find confirmation in the natural science of our own day.

Arabic literature has produced, besides the natural philosophers mentioned, several authors who have dealt with zoology in a more restricted

---

[1] During his imaginary wanderings in the underworld Dante sees Averroes in the court of the heathen by the side of Aristotle and other philosophers of antiquity.

sense: faunistics and zoogeography. Such authors are mentioned as early as the ninth and tenth centuries, but their writings have not been preserved. On the other hand, there is still extant an account of the animals of Egypt, written by ABDALLATIF (1162–1231), which is manifestly based not only upon ancient authors, but also upon personal investigation. *Inter alia*, he gives a detailed description of the hippopotamus and the crocodile and also an account of the method customary in Egypt of hatching hen's eggs by artificial heat. A fairly large work entitled *Animal Life* by MUHAMMED EL DAMIRI, written at the end of the fourteenth century, has come down to us. He has described a great number of animal species — one statement declares it to be nearly nine hundred — partly based on his own observations, but partly also on pure imagination. Arabic literature possesses one author comparable with Pliny in the person of SAKARJA BEN MUHAMMED, called EL KASVINI after his own district of Kasvin in northern Persia. He lived in the thirteenth century and thus had at his disposal, besides Aristotle and Hippocrates, whom he freely quotes, a number of Arabian predecessors, of whose works he made extensive use. His important collective work, *The Wonders of Nature*, is based on Aristotle's natural philosophy of evolution from the lower to the higher; the capacity to feel and move differentiates the plants from the animals. His theory of fossilized animals is curious; he believes that they have been petrified by steam arising out of the ground on which they stood. For the rest, he describes a number of tropical animals which were unknown to ancient authors, for instance the orang-utan, which he pictures as having the human characteristics which the inhabitants of his native place ascribe to it, and, further, the flying dog, the dugong, and others.

On the whole, it is through their having promoted the knowledge of and the cultural influences between the East, even its most distant parts, and the West, that the Arabs have become best known among the peoples of the West; rather than by the really more profound cultural service they performed in having preserved and developed the remains of ancient culture at a period when the West was incapacitated from preserving the inheritance which nevertheless most directly devolved upon its peoples. Through the intermediary of the Arabian philosophers the few learned scholars of the West in the early Middle Ages acquired a knowledge of the products of classical culture; Aristotle, for instance, was long read at the mediæval universities in Latin versions of Arabic translations from the original writings, and the Arabic commentators, Avicenna, Averroes, and others, were the first to act as guides to an understanding of the treatises on nature and to help Europeans to penetrate that world of phenomena whose existence they had entirely forgotten. Thanks to Arabian science, the so-called dark centuries of the Middle Ages were at any rate culturally fruitful, and when oriental science, after flourishing for a brief period, died out, the people of the West had already laid the foundations of an entirely new cultural development.

CHAPTER X

The Eastern Roman Empire

IN A PREVIOUS CHAPTER mention has been made of how the culture of antiquity, itself already become decadent, received its death-blow through the transmigration of peoples which broke up the Roman Empire. The first political evidence of this dissolution was the splitting up of the mighty Empire in the year 395. The cultural world of the time was thereby divided into an eastern and a western half, which suffered essentially different fortunes. In the eastern section the old imperial constitution was still to survive for over a thousand years, maintained in power through the people's being so long accustomed to a despotic form of government, and upheld by an intimate connexion with the strangely established Greek Oriental Church — in actual fact the bond that held together the mixed populations which gave allegiance to the sceptre of the Emperor of the East. Greek was the prevailing language here and the medium for a peculiar form of culture, the Byzantine, which displayed extraordinary qualities of resistance to the pressure of hostile forces: the Mohammedans in the East, wild hordes of migratory peoples from the north and "Latins," as the western Europeans were here called, in the West. This constant struggle for cultural supremacy produced, as it invariably does, a tendency to strict conservatism, and the value of the Byzantine culture therefore lies not so much in independent creative work as in all that it did for the preservation of the ancient literature which, even for philological reasons, it already had some interest in preserving. The capital of the Empire certainly possessed valuable libraries, and educational establishments with highly complicated methods of instruction, but the studies pursued there consisted mostly in theological subtleties, the amplification of ancient authors, and the compilation of histories. The scholars of Constantinople cared little for natural science. On the other hand, the Byzantine physicians were famed for their great ability; they honourably upheld the best traditions of the medical science of antiquity. Their training was entirely practical, however — they received no academical instruction in the science of medicine — and they were in fact essentially practitioners; the theoretical branches of medicine, anatomy and physiology, they have done very little to promote. The principal medical work of the Byzantine era, written by PAULUS OF ÆGINA in the seventh century, deals only with practical

74

ULISSE ALDROVANDI

ANTONY VAN LEEUWENHOEK

medicine; its surgical section is celebrated for its excellence and has had great influence on the medical science of both Arabia and the Occident. — The Byzantine Empire and its culture eventually succumbed to the Turks, but before that it had had time to exercise considerable influence upon western European civilization, especially by spreading a wider knowledge of classical Greek literature and thereby paving the way for the great cultural regeneration of the Renaissance.

Western culture

THE western Roman Empire, unlike its eastern neighbour, fell a prey to hordes of migratory peoples and was dissolved by them into a number of minor states with constantly changing frontiers and unsettled internal conditions. The only one of the kingdoms founded under these circumstances which attained a successful development was that of the Franks, which at one time, under Charlemagne, embraced a large part of the western Roman territory and more as well. After his death his empire fell to pieces and out of its ruins gradually arose the national states of western Europe which still exist today. During the centuries of migration both material prosperity and intellectual culture in the western Roman countries were destroyed. The last remains of classical culture found a refuge in Ireland; there in the sixth and seventh centuries were read and copied not only Latin, but Greek authors, and thence culture spread to England, at that time conquered by the Anglo-Saxons. In Charlemagne's time these two countries possessed the highest intellectual development and it was from there that the Emperor summoned learned men, with whose aid he raised the standard of culture in his own country and created what was called a "renaissance" in the field of classical studies. After his death, however, western Europe was ravaged by a fresh barbaric invasion by the Danes, which destroyed culture exactly where it had hitherto been most highly developed — in Ireland, England, and France. The most decadent period of the Middle Ages really set in during the ninth and tenth centuries, just when the Arabic culture was most flourishing.

The one power that kept men together in that unhappy period was the Catholic Church; it gave consolation and support in time of trial and was able to induce minds broken down by misfortune to strive after ideals. As a unifying cultural force it came to take the place of what the Empire had once been, and so Rome became once more the capital of the world. But while the Church thus gave to culture fresh vitality, it at the same time set up narrow limitations for its development; it demanded absolute subjection, to the extent that not only did religious sentiment have to choose the paths the Church prescribed, but even the human intelligence had to adhere to its dogmas and doctrines as proved truths. These had been drawn up by the ecclesiastical Fathers of the first centuries of the Christian era, whose writings the priests and monks of the early Middle Ages read without interpreting them or adding anything fresh to them. Not until the latter half of the eleventh century did the first independent mediæval theologian appear, in person of ANSELM OF CANTERBURY. But not long afterwards a more liberal line of thought began to find expression, which, based on the trifling remains of classical literature still to be found at that time in the libraries of monasteries and churches, sought to establish rational principles of thought. During the twelfth century these ideas, expounded by the Frenchman ABÉLARD and his pupils, won widespread acceptance, in spite of strenuous opposition on the part of the Church, and received further stimulus from the influence of Arabian science, brought over partly by scholars who had studied in Spain and partly through the crusaders' contact with the East itself. In this way the countries of the West gained their knowledge of the great men of classical antiquity — Plato and Aristotle, Hippocrates and Galen, as well as of their Arabian commentators and successors.

The universities

WITH a view to the study of these sources of knowledge there was founded in the twelfth century a form of educational establishment which was to become of fundamental importance for the scientific development of the future — namely, the university. Antiquity had nothing equivalent to this kind of associations of teachers and pupils, for they rested on an ecclesiastical foundation. Charlemagne had already founded and attached to the metropolitan churches cathedral schools, in which some young priest gave instruction in theology, music, and other branches of learning necessary for men of the Church. As the number of pupils at these schools gradually increased, it became necessary to employ in them a larger and larger staff of teachers, *magistri*, who, in order to protect themselves against the dangers and insecurity that prevailed everywhere in those days, formed themselves into corporations. An association of this kind, *universitas magistrorum*, under its governor (*rector*) and with its large number of pupils grouped according to nations, represented at that epoch a considerable power, which, in the course of violent struggles with the civic and ecclesiastical authorities, endeavoured to acquire a wide measure of self-government and as a rule actually succeeded in doing so. When the number of both pupils and educational subjects increased still further, recourse was had to specialization in several faculties, a method of distribution which in its main features still survives. Instruction was given by means of pulpit lectures — a method based on the Church sermon and similarly adopted with a view to instructing large numbers of pupils at one and the same time. Further, the appearance of the university system involved a democratization of science, of which classical antiquity had no counterpart. Whereas the finest masters of antiquity could probably count their pupils only in tens or hundreds, the great universities of the Middle Ages, such as those of Paris, Oxford, Leipzig, etc., had thousands and even tens of thousands of students attending at one time. True, both mass education and self-government had their dangers; reactionary intellectual movements, equally with earnest strivings after knowledge, might, and in fact did at times, gain the mastery at the mediæval universities, just as, indeed, later liberal and reactionary aims alternately dominated university research and instruction.

Scholastic doctrine

THE science taught in mediæval schools and universities — scholastics, as it was called — was governed, as already mentioned, by ecclesiastical doctrine. The intellectual movements which were set on foot independently thereof and which were consequently persecuted as heretical were not founded, as they generally are nowadays, on natural science, but took their stand on purely speculative ground. The question which had been debated ever since the days of the Church Fathers of the relation of reason to faith, or, in other words, the right of the individual to criticize Church doctrine, was answered in the first place by the universities along fairly liberal lines, in spite of protests from the Church, but in the thirteenth century a religious reaction set in, evoked by the struggle against heresy and represented by the orders of mendicant friars founded for the express purpose of combating the heretical movement; finally these orders succeeded in usurping the control of university education, at least of theological instruction, which was thus compelled to adapt itself to ecclesiastico-political aims. This occurred, strangely enough, just at the time when a closer knowledge of Aristotle began to be disseminated in the universities, based on the Greek original and not merely on the Arabic translations of his writings. But the High-Church theologians who now held sway in the universities soon began to realize what a splendid ally they had in the Aristotelean philosophy, which they had originally mistrusted as mere heathen delusion. Aristotle's conception of the earth as the centre of the universe and yet as the home of all imperfection in contrast to the perfect heaven might very well be adapted to the Church's doctrine of sin and salvation. His strictly formalistic cosmic system and mode of thought, with its dominating intelligence and its denial of any material causality, was, like his conservative and authoritative view of human life, well suited to form a scientific basis for the hierarchical aims of the papal power. And if his writings did not agree in every detail with the revealed Word, the inconsistencies made apparent thereby could be explained away by reference to the author's paganism and ignorance of the way of salvation. Thus was created in the thirteenth century, mainly on the initiative of the greatest thinker of the Catholic Church, the canonized THOMAS AQUINAS, the curious and, in its way, fully elaborated system of thought which that Church has ever since then, held to be the only true one. According to this system, existence is divided

into three "kingdoms," those of nature, grace, and blessedness. In the first
dwell all men; the two latter are attainable only by members of the Church.
Knowledge of nature, therefore, even the heathen may acquire, and no
heathen has possessed deeper insight in this respect than Aristotle; he has
explored the kingdom of nature with unexcelled wisdom. Consequently the
Christian researcher may safely rely upon his explanation of nature and need
not engross himself in the subject — all the less so as the kingdoms of grace
and blessedness are open to him, the former in this life, the latter in the
eternal hereafter. In these circumstances the thinkers of the Middle Ages
devoted but little attention to natural-scientific research; they contented
themselves with the writings of Aristotle, which were closely commented
upon, even down to the smallest detail, without any effort's being made to
develop their subject-matter by actual investigation. There is a well-known
story of how the learned ecclesiastics disputed as to how many teeth the
horse should have according to Aristotle, instead of looking into the mouth
of a live horse to see for themselves. So much the more to the point were the
Aristotelean problems regarding the relation of ideas to reality; here the
dispute waxed hot between the realists, who believed that ideas existed
before things, and the nominalists, who declared that ideas exist only *in*
things. The view of the former was eventually given official sanction, but
their opponents refused to give in and so played their part in undermining
the reputation of the High-Church philosophy towards the end of the Middle
Ages.

There are no biological writings proper dating from the earlier Middle
Ages. The descriptive work on animals, the *PHYSIOLOGUS*, which is mentioned
in all zoological histories, can indeed hardly be included in this category; it
consists of a collection of edifying stories relating to the animal world,
intended to serve as examples for quotation in sermons and gathered together
from all quarters. Probably it dates from later antiquity, which produced
many such collections, as, for instance, that made by Ælianus mentioned
above. The *Physiologus*, which was an anonymous treatise revised and issued
in various editions, had a surprisingly wide circulation; it was translated
into Ethiopian, Icelandic, and most languages existing between these bor-
derlands of Christian culture. It abounds in fantastic stories; a number of
them have survived even to the present day.

Even in the Middle Ages, however, there existed people who had a
broader view of nature and a deeper interest in the life that stirs therein than
had the ecclesiastical legend-writers. Interesting evidence of this is to be
found in a treatise dating from about 1150 entitled *Physica*, written by the
nun HILDEGARD, of Bingen on the Rhine. The book contains notes on animals,
plants, and stones and on the benefit that man can derive from them. It is
entirely popular in style and without any pretension to learning, and for

that very reason it is of interest as a sample of the ideas about natural objects which people entertained in those days.

We find an author of another type in the person of the renowned Emperor FREDERICK II OF HOHENSTAUFEN. As is well known, he was one of the most remarkable rulers of the Middle Ages, Italian in his upbringing, half oriental in his habits and mode of thinking. In his south Italian kingdom he gathered round him learned men from the East and West. He had Aristotle's writings translated from the Greek into Latin and founded a school of medicine at Salerno, where for the first time since Alexandrine days human bodies were dissected. He himself wrote a book, still extant, on falconry, a sport to which princes and nobles were passionately addicted. Frederick's treatise is far more than a mere dissertation on hunting; in a lengthy introduction he gives an account of the anatomy of birds, in which he not only displays a knowledge of Aristotle's anatomical writings, but is also able to point out inaccuracies in his statements; further, he describes the habits of various birds, the movements of migratory birds, etc. Unfortunately Frederick lived during the period of ecclesiastical reaction in the thirteenth century, and after his death his Church opponents eradicated most of the cultural progress he had achieved; the dissection of human bodies was again prohibited and physicians had henceforth, as before, to rely on the classical authorities. The translation of Aristotle which he caused the learned MICHAEL SCOTUS to carry out was perhaps the most enduring evidence of his cultural aims; it was on this work, in fact, that the scientists of the later Middle Ages in general based their learned studies.

Of these scientists of the later Middle Ages none has won greater fame or survived longer in the popular mind than ALBERT VON BOLLSTÄDT, known both to his contemporaries and to posterity under the name of ALBERTUS MAGNUS (born about 1200, died 1280). He was of noble family, but from his earliest youth devoted himself to learned studies and afterwards became a member of the Dominican order, one of the then newly-founded orders of mendicant friars. His reputation for learning spread rapidly throughout the West; he was at one time a professor in Paris, being afterwards appointed to a school in Cologne founded by the Dominicans, and finally becoming Bishop of Regensburg. This last appointment, however, he did not hold for long; he returned to the quiet monastic life and devoted himself entirely to science. He believed his mission in life was to edit the writings of Aristotle — known by him only in the above-mentioned Latin translation — and to harmonize their results with the teaching of the Church. The majority of his many writings deal with theology and philosophy, though natural science appears to have occupied him most during the latter part of his life. As a natural philosopher he is principally a chemist. He was the first to produce arsenic in a free form and he made important discoveries in regard to particular combinations of metals; he also introduced into chemical terminology the word "affinity" as denoting chemical relationship. As a biologist he follows Aristotle, even where the latter's errors have been corrected by other ancient philosophers; he holds that the arteries contain air, that the brain is humid and cold, etc. The observations which he himself claims to have made are often purely fantastical, but they also sometimes bear witness to his powers of observation, which his chemical researches prove him to have possessed in a high degree. His greatest service undoubtedly lies in his having directed the world's attention to Aristotle's conception of nature and thereby also indirectly evoking an interest in nature itself — an interest which the succeeding centuries were able to cherish and widen.

Contemporary with Albertus and, like him, a Dominican friar was THOMAS, called CANTIMPRATENSIS, after the monastery at Cantimpré in France, where he worked. His home was at Liége, but he studied at Cologne and finally became a canon in the afore-mentioned monastery. His principal work, *De naturis rerum*, forms, like that of his master, a compilation of the nature theories of Aristotle and other classical authors, with a wealth of notes on animals, both real and imaginary. More than Albertus, Thomas has a penchant for weaving into his accounts of animals stories with a moral point to them, and also, on the whole, he enters more into detail and is less systematic than his master.

A third contemporary of these two, and a brother monk, was VINCENTIUS BELLOVACENSIS, who was likewise named after his monastery, at Beauvais in France. He wrote a work on nature entitled *Speculum naturæ — Nature's Mirror*. This work is compiled from various sources: Aristotle in Latin translation, Pliny, and Avicenna, as well as the Bible and the Church Fathers. Though more haphazard and less lucidly arranged than those of his colleagues mentioned above, it nevertheless had its influence on the age and the succeeding centuries.

It is not worth while recounting further examples of this kind of mediæval descriptions of nature — natural research it can scarcely be called. Those already cited sufficiently show their character, that of a compilation of the literary material of past ages in the service of that stock conservative theology which dominated science during these centuries. But even at this period there arose personalities whose ideas presage the intellectual liberation, the foundations of which were laid in the course of these centuries and which was destined later to overcome all obstacles during the Renaissance. One such man was ROGER BACON (born 1214, died 1294). By birth an Englishman, he studied at Oxford and Paris and entered the Franciscan order, in which he soon assumed a position of eminence. His liberal views, however, gained for him bitter enemies, and once he was arrested and had to spend years in prison, being deprived of every possibility of working until he was again released. No small cause of the mistrust he inspired was his interest in physical and chemical experiments, which resulted in his being suspected of witchcraft and necromancy. Although he appears to have been a clever experimentalist with a wide general knowledge, there is nevertheless no record of any epoch-making scientific discovery that can be attributed to him. His greatness lies in his general scientific ideas. He set himself up in determined opposition to the subtle mode of thinking of the schoolmen and urged that science should rather be based upon experience gained through observing natural phenonema — that is to say, upon a method harmonizing with that which has been adopted in natural-scientific research in more recent times.

Appearance of new ideas

THIS intellectual emancipation from the hidebound teachings of authority was inspired, however, less by the theoretical contributions of Roger Bacon and any of his successors than by the increasing knowledge of nature itself resulting from the discovery and exploration of new countries. The crusades had already made some contribution towards this expansion, but still greater was the influence of the knowledge of far-distant lands acquired through the journeys into the interior of Asia undertaken by MARCO POLO and several of his contemporaries, and further through the widely extended voyages of the Portuguese in the fifteenth century, and finally through the discovery of America. As a result of all these geographical discoveries biology also acquired a mass of fresh material, which it was impossible to deal with merely by studying Aristotle; it forced research rather to seek its own paths and research-workers to rely more upon themselves. Biology was thus compelled to abandon the purely literary method of compilation and classification, which had been the most characteristic feature of mediæval science, and instead had to rely for its progress upon working out its own observations and developing the results thereof. But before it could do this, biology had to free itself from the restrictions which the ecclesiastical authority of the Middle Ages had laid upon man's intellectual activities in general, and it thus came to take part in the great work of intellectual liberation whose various phases in history are generally summarized under the name of the Renaissance. The progress thus achieved in the knowledge of living nature will be dealt with in the next chapters.

THE HISTORY OF
BIOLOGY DURING THE RENAISSANCE

CHAPTER XI

THE END OF MEDIÆVAL SCIENCE

Revival of the study of ancient authors

THE UNIVERSAL SCIENCE of the Middle Ages, the philosophy of the schoolmen, was, as has already been pointed out, a system of thought complete of its kind, based on the infallible truth of the Catholic Church doctrine, with a strictly formalistic conception of nature founded on Aristotle. It was undoubtedly of service in its own time, especially in that it developed the formal sides of thought, but it lacked the possibilities of free expansion and it was thus inevitable that it should finally lose itself in barren subtleties. The intellectual movement which history calls the Renaissance was therefore hailed as a liberation of those in Europe who were true seekers after knowledge. This movement started in Italy, where the connexion with classical antiquity had never been entirely broken and where the system of the mediæval schoolmen had never really thrived; in the Italian colleges during the Middle Ages Latin, rhetoric, and medicine were studied rather than philosophy. The mediæval Italian felt himself to be the rightful heir of the old Roman people, and it was therefore natural that the cultural revival in that country should take the form of a close study of ancient literature; first of all it was the Roman writers of antiquity and later principally the Greek authors unknown to the Middle Ages who here attracted the interest which in other countries was devoted to the High-Church scholasticism and who offered in exchange an entirely new and freer idea of existence than mediæval philosophy had been able to offer—an opportunity of developing a more rich and many-sided human life than that which the Church of the Middle Ages permitted. It was also in this sphere — that of the general conception of life — that the great cultural revival in Italy exercised its greatest influence, an influence of unique depth in spheres of culture, art and literature, politics and economy. In the field of pure science this revolution was, at least in the beginning, less complete; the absolute value of truth, which the schoolmen ascribed to the formulæ of the Church, the scientists of the Renaissance, the Humanists, made over to the writers of antiquity. Aristotle was regarded by them with, if possible, still greater respect than by the mediæval professors, the only difference being that now one had access to the original writings of the master and could interpret them without the restriction which the Church had formerly laid upon them. There was no possibility, then, of any new conception of nature and its phenomena developing in this direction. But fortunately there were other points of departure for this development.

The fantastic speculations of neo-Platonism about infinity, and the alchemistic experimental science of the Arabs, formed the bases for a number of attempts at an explanation of nature unfettered by Church dogmas and scholastic systems, while, on the other hand, the great geographical discoveries, as well as the newly-found classical authors, offered ideas for special investigations in the sphere of biology which led to results far beyond those of either Aristotle or Galen. The Renaissance period, therefore, was for the science of biology a period of restless seeking and collecting, yielding results which the succeeding age utilized for the purpose of making a complete revaluation of the whole conception of nature common to the people of antiquity and the Middle Ages. It would seem most convenient, then, first of all to give a brief summary of the new philosophical speculations to which the Renaissance gave rise, and then to examine in detail the results which were achieved during that period by the science of biology.

CHAPTER XII

Opponents of scholasticism during the Middle Ages

EVEN DURING THE PERIOD when mediæval scholasticism was at the zenith of its power, there were not wanting movements hostile to it, the representatives of which, partly by way of logic, like the so-called nominalists, partly by exhortations to empirical observations, like Roger Bacon, previously mentioned, sought to undermine its thought-structure. These movements, besides, very often had points of contact with the mysticism which throughout the Middle Ages sought in the sphere of a holy life to induce a spirit of personal sincerity in contrast to the strictly formal piety taught by the Church. When, later on, scholasticism was discredited, owing to the reverence of humanism for antiquity, the field was left open for a philosophy in which all the above-mentioned elements — theoretical speculation, empirical observations, mysticism, both Christian and late classical — were included as fundamental components in a fresh conception of existence, out of which our own modern ideas of nature and life gradually developed.

The first important representative of this fresh view of nature was NICOLAUS CUSANUS. He took his name from the village of Kues, or Cusa, near Trier, where he was born in 1401. He received his education amongst the "brothers of the common life," a religious community having a pronounced mystical tendency — an education which had a decisive influence on his entire mode of thought. He had a brilliant career in the service of the Church, into which he shortly afterwards entered. He became a bishop, and later on, as a cardinal, he was one of the most trusted men of the papal supremacy at the time. As such, he acted constantly in the interests of humanity and enlightenment, ardently opposing the sale of indulgences, trials of witches, and other Church superstitions. He died in Italy in 1464.

New cosmic ideas

IN the course of his manifold practical activities, however, Cusanus found time for research work which places him in the first rank among the world's pioneer spirits. The problems he deals with in his numerous writings are, it is true, for the most part theological, but in connexion with them he touches upon the problem of man's place in existence, and it is here that he makes his most important contribution. Curiously interwoven with and derived from

84

his mystical speculations appear his new and audacious ideas on the structure of the universe and man's place therein. Basing his ideas on the mystical conception of infinity of the neo-Platonists, he asserts that it is impossible for the universe to have a spherical form, as Aristotle declares, for then it would always be possible to conceive of something existing outside the sphere, in which case it would not be the whole universe. Rather, the latter is infinite; it exceeds all form and all limitations. Nor, in that case, can the globe be the centre of the cosmos, for the cosmos has no centre, but man on the earth imagines he is in the centre of the cosmos, and he would believe the same were he to find himself on the sun or any other star. Cusanus thus maintained the relativity of mental observation. He derives this knowledge of his from what he calls "*docta ignorantia* (wise ignorance)," by which he means the knowledge that all contrasts as well as all change in existence finally become absorbed in an absolute maximum, infinite and unfathomable as God Himself. It was this "*docta ignorantia*" that Aristotle lacked, and therefore he believed in a finite world and absolute mental observations. For the rest, Cusanus employs his mode of thought quite as much in theological sophistry, as, for instance, touching the true nature of the Trinity; but while these subtleties are now long forgotten, through his ideas of nature he takes his place as one of the pioneer thinkers of the beginning of the new era, half mediæval mystic, half modern natural philosopher. His bold ideas seem otherwise to have attracted but little attention outside learned circles; it was not realized how revolutionary they were, all the more so as he did not concern himself with the details of our solar system; consequently he did not attack the theory of the earth as the centre of the sun's orbit. His high position in the Church undoubtedly saved him from such persecution as afterwards befell many of those who deduced the inevitable consequences of his theories.

If, then, Cusanus's ideas operated in silence, the views which about a hundred years later were expressed by COPERNICUS attracted all the more attention. Born in 1473 at Thorn in Poland, NICOLAUS COPERNICUS received his education at the Italian university of Bologna and finally became dean in his own native city, where he died in 1543. Even in his youth he was a keen student of mathematics and astronomy and already at that early age began his life's work, to think out a new cosmic system which, more easily than the Aristotelean-Ptolemaic, could be reconciled with the observations made in his own time upon the movements of the heavenly bodies. Their irregularities could in fact never be satisfactorily explained on the basis of the old solar system. Copernicus discovered a better means of accounting for the irregularities by letting the sun, in contrast to the direct evidence of the senses, represent the centre of the cosmic system, and the earth assume the place among the wandering planets which the sun held in the old system.

Otherwise Copernicus retained that system practically unaltered; he thus made the sun the immovable centre of the universe, the planets moving in circles round it and the whole surrounded by the sphere of fixed stars, such as the ancients imagined it. In reality, then, his theory was less subversive than Cusanus's and, as a matter of fact, not without precedents in antiquity; but still it attracted far more attention, because it was entirely at variance with what everyone was accustomed to see happening daily before his very eyes. Copernicus spent decades working out his theory, and not until the year before his death did he dare to publish a book on it. It aroused fierce opposition, particularly on religious grounds; the reformers as well as the Jesuits condemned its teachings, while its scientific influence was at first but small, all the more so as the proofs he offered of the truth of his new theory were really rather weak. Shortly after his death, however, a thinker was born who was able to reconcile Copernicus's ideas with those of Cusanus and thereby founded a theory of the universe which in its essentials still holds good today.

GIORDANO BRUNO was born at Nola in south Italy in 1548. As a young man he entered a monastery, but he was far from contented with the life there, was soon suspected of heresy, and saved himself by flight. After this he never found a permanent retreat; excommunicated and persecuted within the Catholic world, he nevertheless found no consolation in the Protestant countries which he visited. Upon returning to Italy he became a victim of the Inquisition and after many years' imprisonment was condemned as a heretic and burnt at the stake in 1600.

In numerous lectures, disputations, and published works he preached in the countries he visited the new doctrine which cost him his life. In this he takes as a starting-point Cusanus's speculations on infinity, Lucretius' atomic theory, and Copernicus's solar system. On the ideas which he found in these various conceptions he built still further with an originality which ranks him amongst the greatest thinkers of all time, in spite of the fantasy and mysticism with which he, like the other philosophers of the Renaissance, burdens his speculations. In agreement with Cusanus, but still more emphatically, Bruno maintains the subjectivity of mental observation; when a man moves, the horizon goes with him, from which we must conclude that there exists no absolute universal centre. On the contrary, both reason and faith demand an infinite world, infinite as God Himself. And, like mental impressions, place, movement, and time are relative and dependent upon the position in space from which they are observed. — Nor can the assertion maintained by Aristotle as to absolutely heavy and absolutely light bodies be true; this being so, there is no meaning whatever in the old belief that planets and fixed stars are lodged in spheres round the earth; on the contrary, they move in their orbits in space freely and by internal force. And like his cosmic system Aristotle's theory of matter as potentiality in contrast to form emanating from the divine intelligence is, in Bruno's opinion, also incorrect; matter is rather the essential in everything, the "divine essence" out of which all is evolved. In Bruno's view, Lucretius' atomic theory and the neo-Platonic ideas of the unity of matter are combined into a vision of the world at the same time mysteriously vague and grandly fantastic, as a single whole, one with God and one with itself, a combination of all the contrasts which human thought has speculated upon. It would take too long to discuss these ideas in detail, all the more so as Bruno, strictly speaking, does not touch upon any purely biological problems. His importance from the point of view of world history lies in the fact that he for the first time worked out, or perhaps rather guessed, the cosmic theory which has since come to be held in modern natural research. His influence has been great and has been widely felt through all the ages.

While, then, in the cosmological sphere Bruno was the pioneer of the new natural science, in a corresponding manner FRANCIS BACON (1561–1626) paved the way in the sphere of pure laws of thought. His life's activities and end were in all respects different from Bruno's. One thing, however, they had in common: restlessness, that diversified seeking after knowledge which was so characteristic of the Renaissance. Born in England in a refined home, Bacon received a thorough education, but lost his father at an early age and was not very successful in his official career, in spite of his brilliant gifts and his ruthless ambition. It was not until later on in life that he received higher appointments and eventually became Lord Chancellor in the reign of James I, whom he knew how to flatter. But he was shortly afterwards dismissed and condemned to pay a fine for bribery and corruption when in office, and his last five years he spent in retirement.

Bacon's reform of science

EVEN in his early years Bacon had planned to reform all human knowledge completely. This reform was to have been carried out in a work of mighty proportions entitled *Instauratio magna*. Bacon during his restless life found no time to carry out even approximately the great task he had set himself; the "great reform" remained but a fragment, of which the first two, and the best-constructed, sections are called *The Advancement of Learning* and *The New Method*. The latter section is that on which Bacon's fame principally rests; its title is chosen as a direct challenge to Aristotle, for whose theory of method, *Organum*, Bacon wished to substitute his own new method. Bacon's *Novum Organum* takes the form of a collection of aphorisms, intended to illustrate from various points of view the inaccuracy of the traditional scholastic mode of thought and the correctness of the new theory of thought, which it was necessary to set in its place. The defects of the Aristotelean philosophy are criticized in a number of strongly worded and merciless sentences; under its guidance human thought has been led into delusions, which are classified under four different categories denoted by the names "idols of the tribe," "of the cave," "of the market-place," and "of the theatre." By idols of the tribe are meant those fallacies which are incident to human nature, man's tendency to interpret the phenomena of nature according to human preconceptions. The idols of the cave are man's individual tendency to judge according to his own person or ego; it is as if a man sat in a cave and from there saw things in a one-sided light. The idols of the market-place are fallacies which arise out of human community-life, especially errors arising from the influence exercised over the mind by mere words, the confusing influence of traditional nomenclature upon the idea of things. Finally, the idols of the theatre are those which are induced by the power of tradition and result from received systems of philosophy and the tendency of their theory to captivate the senses. The criticism which in further developing these principles he directs against the philosophy of his time is in many cases extraordinarily sharp and holds good for every age. Thus he utters insistent warnings against the common tendency to regard natural phenomena as simple mechanical constructions, like those which man himself puts together and takes to pieces. Nature is, on the contrary, extremely complicated and one must be careful how one ascribes to its course of events the same order and regularity which man loves to ordain for himself. From this error arise fallacies such as the idea that the orbits of the heavenly bodies must necessarily be circular, just because the circle is the most regular figure. In contrast to the artificial and false idea of nature which the old philosophy creates by means of such modes of thought, Bacon sets up the true knowledge of nature, which is acquired by observation and experiment. Man overcomes nature by obeying her laws and learns to understand her by putting proper questions to her. Thus one arrives at the true scientific method, that which by careful observation of the peculiarities of existence and by a classification of them acquires knowledge of the general laws of nature. Bacon attaches the very highest hopes to the value of the knowledge of nature which he thus intended to create; throughout his long life he never ceased to contemplate with passionate enthusiasm the thought of the extraordinary life-values which awaited the human spirit in an enhanced knowledge of the true essence of nature. Such knowledge could be attained by means of a schematic procedure laid down once and for all, applicable equally to high and low in the realm of thought. To this art of deduction, however, based on a consideration of the temporal sequence of phenomena, their presence and absence, and their numerical relations, Bacon gave a value which it did not possess, and besides applied it in a manner which led to sheer absurdities. His knowledge of nature, moreover, was limited and by no means unprejudiced — he was, for instance, in opposition to Copernicus —

and the experiments he arranged were primitive. Further, he was no mathematician and he therefore was unable to employ the deductive reasoning
inherent in that science. Nevertheless Bacon has done an everlasting service
to the development of natural science, primarily through his activities as a
critic. It has already been mentioned how during the Renaissance ancient
culture in general and in the realm of science, primarily Aristotle, was treated
with unbounded respect. This uncritical and slavish attitude, which threatened to ruin all chances of further progress, Bacon combats with all the
severity of which he is capable; he throws overboard all respect for antiquity,
whose culture he considers to have led only to intellectual decay and vain
disputes. He maintains that the peoples of antiquity were really children in
comparison with his own age, which possessed far more of that experience
which to him was the one foundation of knowledge. And in this insistence
upon experience as the sole source of knowledge lies his other great service
to the development of science and in particular to that of natural research.
He realized more clearly than any of his contemporaries the necessity of
extending the knowledge of nature by accumulating the results of observations of its objects and of experiments carried out with its powerful forces,
and though he himself could give expression to his ideas only in clumsy
efforts, nevertheless these ideas, through their intrinsic theoretical truth,
exercised great influence in his day and have done so down to the most recent
times. Even in our own day one of the pioneers of research into the problem
of heredity, Johannsen, has openly acknowledged his debt to Bacon's
Organum as the source from which he obtained a clearer idea as to the objects
and means of natural research. And it is certainly no mere accident that the
country which gave Bacon birth should have led the way in the great pioneer
work that has been done in promoting the development of biology.

What Bacon thus theoretically conceived and insisted upon was brought
to practical realization independently of him by GALILEO, the creator of modern
physics and astronomy, and hence also the founder of the whole of modern
natural research and its conception of natural phenomena, so fundamentally
removed from Aristoteleanism.

GALILEO GALILEI was born in 1564 at Pisa, where his father held a good
post. At an early age he displayed mathematical and mechanical gifts, studied
at Pisa, first of all medicine and later mathematics, and when still quite
young was made professor of that science, first at Pisa and then at Padua.
In the latter city he worked as a teacher for eighteen years with brilliant
success; finally the university could find no hall large enough to seat all his
audience. And the results of his scientific work were still more brilliant;
especially after he had constructed a telescope and with it had begun to study
the heavenly bodies, his astronomical discoveries followed one another in
rapid succession: the globular form of the moon, the satellites of the planet

Jupiter, the sun-spots, the phases of Venus and Mercury. But it was impossible to reconcile all these new facts with the ancient Aristotelean-Ptolemaic cosmic theory and so Galileo early associated himself with the conception of the universe as enunciated by Copernicus and Bruno. His great fame procured for him the personally brilliant appointment of Astronomer Royal to the Medicean Grand Duke at Florence, with a high salary and no official duties. But in leaving the service of the powerful Venetian Republic he came under the influence of the power of the Roman Church, a circumstance all the more dangerous to him as his new discoveries excited the bitter hostility of the very parties which had condemned Bruno; moreover, he was himself a violent controversialist, who never spared his enemies. His end is a matter of common knowledge — how he was arraigned before the Inquisition on account of a "dialogue" on the solar system and under threat of death was compelled to make a public recantation of his "Copernican error," after which he lived in strict seclusion until his death, in 1642.

Galileo's theory

GALILEO's fundamental importance as a natural philosopher is not based merely upon his discoveries, epoch-making as they are; he has contributed in a still higher degree towards scientific progress through the principles which he laid down and which have become the basis of modern natural philosophy. As we know, Aristotle based his cosmic theory upon the contrast between form and matter, where form is assumed to be a realization of matter's powers of development; the higher the degree of its realization, the more perfect the form. Therefore the heavenly bodies, with their regular motions, are more form-perfect than the earth, with its many irregularities, while beyond the heavenly spheres is the world of pure form, God, the origin of all forms, the cause of all that happens in the universe. Galileo at once came into conflict with this system through his astronomical discoveries; according to Aristotle, the firmament, as existing nearest to the immutable divine intelligence, was itself immutably regular in its motions. Galileo discovered a great many irregularities; the sun-spots, Jupiter's moons, and all else that the newly-invented telescope brought into the light of day proved the firmament was not such a place of perfection and regularity as had been supposed. On the other hand, the phenomena of motion in bodies here upon earth showed an obedience to law of which the ancients had no notion. Galileo experimented with the free fall of bodies, with pendulous motions, and with the motions of bodies along an inclined plane, and discovered in all these phenomena ratios between weights, lengths of time, and rapidity of motion so mathematically regular that he could express them in the form of theorems as capable of demonstration as the old geometrical propositions formulated by Euclid. But it was just through this combination of natural-scientific experiment and mathematical calculation that, as he himself says, he created a new science. Instead of Aristotle's guiding reason, which was in reality nothing but an expression for the speculating philosopher's own inferences, deducted for the most part from purely human-cultural hypotheses Galileo brings the phenomena of motion on the earth under one common law, which operates out of mathematical necessity and whose manifestations can under given conditions be calculated in advance, just as at an earlier epoch it had been possible to calculate the regular path of the "divine" heavenly bodies. Galileo was, it is true, unable to find one common law governing the motions of terrestrial objects and the heavenly bodies — that was for Newton to find in his law of gravitation — but Galileo laid down the principle governing the natural-scientific treatment of terrestrial phenomena, a principle which he expressed in the words: "To measure what can be measured and to make measurable what cannot be measured." He seeks a mechanical reason for everything that happens — a force that sets things in motion. To refer to God as the cause of natural phenomena serves no purpose, in his opinion, for one can attribute anything whatever to the will of God, since no necessity underlies it. According to Galileo, natural science should compare material things merely with one another, not with supernatural things, and at the same time it has to be remembered that nature is itself a miracle, although its phenomena have a natural explanation. In actual fact, gravity is merely a word for something which we do not know; we cannot tell what it is that atrracts stones to the earth. Galileo sees clearly that it is useless to try to find out what the forces of nature are; the scientist can only discover how they operate.

Galileo's victory over Aristoteleanism

SUCH a complete revolution of the aims and methods of natural science as that carried out by Galileo could not of course penetrate men's minds all at once. He himself fell a victim not only to the Church's intolerance, but also to the superstitious respect in which the Renaissance held the culture of antiquity and its chief scientific authority, Aristotle. Actually another century was to pass before Aristoteleanism in every field of human knowledge was successfully eradicated from the ideas underlying the science of the present day. In order to rid natural science of Aristotelean fallacies it was, in fact, necessary to destroy Aristotle's entire thought-system, and this was first done during the seventeenth century by the great systematic thinkers of that period, Descartes, Spinoza, and Leibniz, who will be discussed later on. We shall now proceed to a survey of what the Renaissance period achieved in the way of pure biological research, not only in the purely descriptive sphere, but in the more speculative field as well.

CHAPTER XIII

DESCRIPTIVE BIOLOGICAL RESEARCH DURING THE RENAISSANCE

1. Zoography

THE EARLIEST ACTIVITIES of Renaissance research in the biological field were, in accordance with the general tendency of that period, purely philological. New editions of Aristotle, Hippocrates, Galen, and other natural philosophers of antiquity were published, their language commented upon, and attempts made to explain their contents. However, the actual historical course of events compelled the learned world to carry out independent work in regard to natural objects as well. Even the fauna and flora of central Europe were very imperfectly known to the ancient philosophers, and the information regarding many of them needed supplementing with innumerable facts, which entailed much independent research work. And this became all the more necessary when the great geographical discoveries acquainted mankind with the perfectly new and exceptionally rich nature of the tropics. All these circumstances combined to produce an abundant literature of a purely descriptive kind, both zoological and botanical, which, thanks to the art of book-printing, received such widespread publication as the biological works of antiquity could never hope for. Further, the methods of reproducing pictures, discovered in connexion with book-printing — woodcuts and copperplate engraving — now for the first time made it possible to utilize the illustration in the service of scientific literature — a means of extending human knowledge the importance of which can be appreciated only if we consider what it means in our own day and what would be the consequence if modern science were to be deprived of it. A review of some of the more eminent representatives of this branch of biological science during the Renaissance will give us some idea of the objects they aimed at and the respects in which they advanced this science. For this purpose we shall for the moment discuss only the results of zoological research during this period; the botanical results may perhaps more suitably be left to a subsequent chapter dealing with the history of biological classification.

EDWARD WOTTON (1492–1555) essentially represents the point of view of mediæval science. The son of a college porter in the University of Oxford, he studied medicine in his native city and became a physician with a wide and distinguished practice. His interest in nature he recorded in a lengthy work entitled *De differentiis animalium*, on which he worked for several decades. In this book he shows himself a faithful follower of Aristotle, whom he imitates both in his method of classification and in the field of anatomy. His division of the animal kingdom is entirely Aristotelean: sanguineous animals and non-sanguineous, viviparous and oviparous quadrupeds, and so on. Nevertheless he criticizes his classical predecessors to the extent that he does not accept without reservation the masses of fabulous animals which they invent, but on the other hand he has nothing to say about the many new animal forms which the explorers in his own century brought home with them and which otherwise excited general interest among his contemporaries, both educated and uneducated. Yet he contributes much information regarding the medicines which may be extracted from the various animal forms. As a profound exponent of Aristotle and representative of his ideas, Wotton came to exercise no small influence on his age, particularly upon the man who eventually became the finest zoological representative of the Renaissance, GESNER.

KONRAD GESNER was born at Zurich in 1516. His father was a Protestant artisan, who fell in 1531 at the famous battle of Kappel, in which the civic guard of Zurich, under the reformer Zwingli, were defeated by the Catholics. Young Konrad, who had previously been sent to a good school, was now unprotected, but his great reputation for zeal and energy brought him friends, who sent him to study at their expense in Basel, Paris, and Montpellier. At these places he studied such different subjects as classical and oriental languages, natural science, and medicine, and in general acquired in an unparalleled degree that many-sidedness in learning which during the Renaissance was particularly appreciated and admired. After having been for some time professor of Greek in Lausanne, he was appointed first town-physician at Zurich, which was at that time a moderately salaried post. There he died of a plague that ravaged the town in 1665 — that is, when he was still under fifty. Of a quiet and unambitious nature, he had a constant struggle against financial difficulties, which compelled him to wear out his strength in ill-paid hack-work. His energy was marvellous. He published and made commentaries on classical authors; he compiled dictionaries, wrote a lexicon of classical literature, which must have been a very fine work for his period, and was the author of works on popular medicine. Besides all this he found time for extensive journeys both for scientific purposes and for pleasure — he was one of the very first to be interested in mountain-climbing, and he had a keen feeling for Alpine beauty — and finally he had the time and leisure to carry out one of the greatest biological works the world has seen.

Gesner's *Historia animalium* comprises four immense folio volumes of about 3,500 pages in all. The animals are arranged according to the principles of Aristotle; the first part includes viviparous and oviparous quadrupeds, the second part birds, the third fishes, the fourth, which was published after the author's death, reptiles and insects. In each part he then describes one animal after the other on the lines of Pliny, but with far greater expert knowledge, based on his own experience and criticism of his source of information. Animals are arranged alphabetically "in order to facilitate the use of the work," though allied forms are grouped under one heading; all oxen under *Bos*, all apes under *Simia*, etc. Each animal form is discussed under eight sections, marked with letters of the alphabet and comprising (a) the name of the animal in different languages; (b) its habitat and origin and a description of its external and internal parts; (c) "the natural function of the body"; (d) the qualities of the soul; (e) the animal's use to man in general; (f) its utility as an article of food; (g) its utility for medical purposes; and (h) poetical and philosophical speculations about the animal, anecdotes and resemblances to be found in different authors. Thus the reader is able to find what he wants, whichever part of the work he turns to. This clearly shows its encyclopædic character, and actually it is far more reminiscent of Pliny than of Aristotle. As in Pliny, so in Gesner one seeks in vain for any idea as to the connexion in living nature, in vain for any comparison worked out between the different forms of life, regarding their organs or their functions. Gesner, however, surpasses Pliny in knowledge — in this respect, of course, he has the whole of the intermediate literature at his disposal, and indeed he has it at his finger-ends. True, he, too, brings in a great many stories of marvellous animals, but he certainly has not that absolutely unquestioning belief in the miraculous which the old Roman had. And, above all, he was able to record the results of his own research work, for he studied not only books, but also life. He was a keen collector of observations on animals, not only his own, but also those of other scientists with whom he corresponded.

Illustrations introduced into zoology

His most original contribution to science was his introduction of the illustration as an aid to the study of biology. He desired that, as far as possible, every description of an animal should be followed by an illustration so as to give the reader a clearer idea of the animal, and he spared neither trouble nor expense in procuring as accurate woodcuts as possible. His collaborators in this work were eminent artists, and he himself declared that the picture of the rhinoceros was done by no less a person than Albrecht Dürer. With all its weak points, Gesner's *Historia animalium* is at any rate the foremost purely zoological work of the Renaissance period, and its influence on the science of the succeeding age was considerable.

Somewhat younger than Gesner and partly his pupil was another of the foremost zoologists of the Renaissance, ULISSE ALDROVANDI. He was born at Bologna in 1522 of a respectable burgher family and was intended to be a merchant. Office work, however, attracted him but little, and so he went in for studying, first jurisprudence in his native town and then philosophy and medicine at Padua and Rome. When he was thirty years old, he took the degree of doctor of medicine and shortly afterwards, in 1560, he was made professor at Bologna, where he worked for forty years, resigning at the age of nearly eighty. He died in 1605. As a professor he lectured chiefly on pharmacology, and to aid him in his teaching he planted a botanical garden. This caused him to fall foul of the apothecaries of Bologna, who alleged that in cultivating medicinal plants he usurped their privileges. The controversy grew so fierce that it finally had to be settled by the Pope. Aldrovandi was, on the whole, a man who lived for his science; he spent his fortune on collecting natural objects and had recourse to the leading artists of his time to draw pictures of them. The Government of Bologna doubled his salary in recognition of his great services to science, and in return he bequeathed to the city his collections and library.

Aldrovandi's natural history

IN his energy and capacity for work Aldrovandi resembled Gesner, and as he lived longer and worked under more favourable conditions, he managed to achieve far more. His collected works on natural history consist of fourteen large folio volumes, besides which there are preserved in the University of Bologna quantities of unpublished manuscripts in his own handwriting. He himself published during his lifetime only four volumes, on birds; after his death his friends and pupils published the remainder: those on other animal groups, on plants, and on stones. These latter volumes, however, seem to have been in part radically revised by the editors, wherefore Aldrovandi should be judged only on what he himself published. His model was chiefly Gesner, whose work he diligently studied and it is from this point of view that his own work must be judged. His relation to Gesner is by no means in every respect that of an improver; he is far less critical, and similarly he has on the whole less stylistic ability; in his descriptions he piles up masses of like and unlike, so that one of his most eminent successors, Buffon, was moved to express the opinion that only one-tenth of the whole of Aldrovandi's works would be left if one extracted all that was useless and untrue from his writings. On the other hand, his illustrations, as well as his typographical equipment, are better than Gesner's, while, at least in some respects, he is in advance of the latter in regard to classification. Birds are classified according to certain groups: first, birds of prey; then wild and tame fowl (gallinaceous birds) — characterized as "*pulveratrices*"; i.e., those that bathe in sand — further, pigeons and sparrows, which bathe in both water and sand; then song-birds, baccivorous and insectivorous; and lastly waterfowl. Moreover, he has paid attention to anatomy, particularly osteology; and, finally, he cites a far greater number of exotic and hitherto unknown forms than Gesner. He too, then, has in his own degree contributed to the advance of biology, and though he by no means merits the vaunting eulogy which a contemporary artist wrote under his portrait — that, though not in his appearance, yet in his genius he resembled Aristotle — nevertheless his work has exercised a powerful influence, and it was not until Buffon's great zoological work in the eighteenth century that Aldrovandi's was definitely out-distanced.

Apart from these representatives of the knowledge of the animal world as then known, certain research-workers are worthy of mention who devoted themselves to the study of particular animal groups with which they dealt monographically. In the best of these monographs there is really far more evidence of independence in research and originality of ideas than in the great collective works; in the former is best seen that power of independent observation and investigation of natural objects which was a feature of the science of the Renaissance.

GUILLAUME RONDELET was born in 1507 at Montpellier in the south of France, where he also worked later as a professor. He studied first in his own district and then as body-physician to a distinguished personage on his travels in Italy, where among other people he made the acquaintance of the young Aldrovandi, who received much sound teaching from him. As a professor he established in his native city an anatomical theatre, but he had not been working there long as a teacher when he died, in the year 1556. His fame as a biologist rests on his work *De piscibus marinis*. In this book he describes and illustrates the aquatic animals he knows, for he regards as fishes not only seals and whales, but also crustaceans, molluscs, echinoderms, worms, and other marine invertebrates. He makes a very careful study of whales, fishes and cephalopods. According to his own statement, he dissected a large number of these creatures and he also gives a number of correct particulars which are sometimes at variance with the great authority Aristotle. He further compares, as far as was possible, the same organs in different fishes, giving exact accounts of different maxillary and dental forms, different *branchiæ*, etc. However, his comparative work practically gets stranded, owing to the impossibility of finding resemblances between the vertebrate and invertebrate forms discussed. His attempts at classification are likewise very primitive. He differentiates between selachians and osseans, which again are divided into "flat" and "high" fish; moreover, whales are dealt with in a group by themselves. He has as little notion of species in the modern sense as had Gesner and Aldrovandi, and therefore, like them, he had to begin the description of every form by recounting as many names for it as possible. On the other hand, he avoids for the most part the useless petty details of scholarship with which the two last-mentioned authors of collective works overburdened their accounts, and this at once gives to his work an impression of greater accuracy. And though he certainly does illustrate a number of marvellous creatures reported to have been seen, such as, for instance, a fish having "the appearance of a bishop," he does so with a reservation as to the irrational nature of stories relating to such phenomena.

Besides Rondelet, a younger countryman of his is worthy of mention, PIERRE BELON, in several respects a man possessing great ideas about the future. He was born in 1517 near Le Mans, in central France, of poor parents. His genius attracted the attention of the bishop there, who defrayed the cost of his medical studies in Paris. After that Belon went to Germany for a course of study. Upon returning to France he received, through the kindness of certain distinguished patrons, funds to enable him to undertake a still longer journey, through Greece, Turkey, Syria, and Egypt. Everywhere he collected material with great energy and made notes, not only on natural-scientific, but also on archæological and ethnographical subjects. On his return home he settled in Paris, where he was granted a pension by King Henry II. In 1564 he was murdered by highwaymen. His period of scientific authorship was thus brief, lasting not more than about ten years, but during that time he brought out ideas of great significance for the future. He was held in high esteem even by his contemporaries; he counted amongst his friends the famous poet Ronsard, who wrote verses in his honour.

Like Rondelet, Belon devoted himself to the study of marine animals and published two monographs: *L'Histoire naturelle des estranges poissons marins* and *La Nature et diversités des poissons*. The term "fishes" he makes even more comprehensive than Rondelet: not only whales and seals, crustaceans, molluscs, and actiniæ, but the hippopotamus, the beaver, and the otter are also described amongst the fishes. And even if all these animals could be classified by a faithful Catholic as fishes, just because the Church included them among the animals that may be eaten during a fast, it is hard to understand why the chameleon and the uromastix lizard are catalogued in the book — these beasts of the desert which have nothing whatever to do with water. Though the external grouping of the subject, then, leaves much to be desired, Belon has certainly endeavoured to introduce into the class of true fishes some kind of systematic division, based not merely on external, but also on internal anatomical characteristics. Cartilaginous and bony skeletons, Ovipara and Vivipara, constitute the bases of classification, which still hold good today, and on the whole his system of classification bears a somewhat more modern stamp than Rondelet's. Even attempts at an investigation of various forms on the lines of comparative anatomy occur in Belon's work. Whether and, if so, to what extent he was influenced by his immediate predecessor it is difficult to decide. Their works were published practically at the same time. As regards wealth of material, at any rate, Belon's work is superior. Moreover, thanks to his travels, he was able to include many oriental animal forms which were previously unknown to the Western world.

Far superior to the works on the fishes is Belon's second main treatise: *Histoire des oyseaux*. In this work he describes and illustrates all the birds he knows, arranged in groups according to their structure and habits — birds of prey, waterfowl, shore-birds, ground-pecking, wood-pecking, omnivorous, and small birds, divided into Insectivora and Granivora. The individual forms are characterized by a few names in Latin, Greek, and French; unlike Gesner, Belon scorns to extend his knowledge of languages. If this attempt at classification bears witness to Belon's keen powers of observation, there is still further proof of them in the attention he pays to the morphology and anatomy of the individual forms. The structure of beaks and claws is closely studied and compared in different forms, while anatomical relations are treated in the same way. Most noteworthy, however, is the detailed comparison in both text and illustration, in the first book of the work, between the skeleton of a man and that of a bird, the latter drawn in an attitude corresponding to that which the former assumes in his natural standing position. Although this comparison by no means agrees in every detail — for instance, the human clavicle and the bird's coracoid bone are made homologous — at any rate we have here a first attempt at a comparative anatomical investigation. The idea thus started by Belon was, it is true, for a long time neglected; it was not until two centuries later that it was taken up anew by Buffon, to be eventually developed by Cuvier into one of the most important fields of biological research. The fact that these two were both countrymen of Belon is indeed some evidence that his activities in this sphere did not pass entirely unnoticed.

2. Anatomy

THAT the age of comparative anatomy had not yet arrived was of course due to the fact that research was still fully occupied with purely descriptive anatomy. In this, as in other spheres, the Renaissance inherited from the great anatomists of antiquity, of whom Galen constituted the chief authority, in his reputation comparable with Aristotle, and like him regarded with infinite respect by the physicians of the Renaissance, who were philologically rather than biologically educated. However, it was not the physicians alone who required anatomical knowledge; even art, the result of the admiration of the Renaissance for antiquity, now began to demand a closer study of the structure of the human body. Amongst the pioneers in this field the first name that should be mentioned is that of the great universal genius LEONARDO DA VINCI (1452–1519).

Leonardo was a Florentine, and in his native city, which was the very centre of Renaissance culture, he was brought up to be an artist and at the same time a mechanician — the professions of painting and mechanics in those days were often combined. He afterwards led a restless life, carried out work in many places in Italy, and ended his days at the French Court. His world-wide fame he of course won as a pioneer in the art of painting. In this his greatest contribution was his introduction of a close study of human anatomy; he drew for the benefit of his pupils a vast number of anatomical figures, which are still preserved, and published a work on the proportions of the human body. He did not study only man, however; all sorts of natural objects and natural phenomena interested him. In a mass of roughly drafted notes, which were never combined into a connected whole and were not printed until our own day, he has recorded his observations and reflections on practically every sphere of human knowledge. He not only studied human anatomy; he also compared similar organs in different living creatures; he investigated optical sensations; he observed the structure of different geological strata, and maintained, in opposition to Aristotle, but in agreement with Xenophanes, that fossils were animal remains. In every field he shows himself an opponent, not only of scholastic traditions, but also of the slavish admiration for antiquity that characterized the Renaissance. Not the classical authors, but experience should be the source of human knowledge. Unfortunately his speculations were merely fragmentary and for that reason were unprinted. It was not until later that they were more closely studied, and Leonardo's influence upon science has been only indirect, as a result of the impression made by his personality and his art. The study of the human anatomy which he initiated was continued by other Renaissance artists and in that way reacted upon culture in general; these painters and sculptors were certainly not without their influence upon the impetus given to medical anatomy in the sixteenth century.

In the field of medical science the influence of the Renaissance was felt in the same way as in other branches of human knowledge; a return was made from the mediæval authorities to antiquity. The doctors applied themselves to the study of the classical languages; they severely condemned the barbarous Latin of the mediæval professors and formed their style on the best Roman and Greek models. And in conformity with this enlightened spirit the poor editions of the medical writers of antiquity which were based on Arabic translations were banned and were replaced by new editions of Hippocrates, Celsus, and Galen, which, published with careful textual criticism and sound commentaries, were, thanks to the progress of book-printing, widely dispersed throughout the universities. One of the most brilliant and at the same time most typical students of medicine during the Renaissance was Jacob Sylvius, of Paris. Born in 1478, he devoted himself from early youth to the study of classical languages, not only Roman and Greek, but also Hebrew. He was a fine stylist and was the author of several works on

French grammar. He was nearly fifty years old when he took up medicine, applying himself to the study and exposition of classical medical literature. In lectures, which were brilliant in their formal delivery, he expounded to the students of Paris the theories of Galen, which to his mind were infallible — "divinely inspired" — and could not be improved upon. These lectures were really exercises in classical oratory; there was no question of any empirical research. Practical instruction remained in all respects at the point to which the Middle Ages had advanced it. And in this respect the Middle Ages did actually advance beyond antiquity.

Mediæval dissection

As early as the middle of the thirteenth century dissections had to be carried out on human bodies at the Italian universities; the Emperor Frederick II, who had no prejudices, made it compulsory for students of medicine and surgery to attend these operations, while later on, the prohibition of the popes was powerless to prevent the development of these practices. At the universities of Salerno, Bologna, and Padua they were officially ordained and had, if possible, to be carried out regularly. As a result of this it should have been possible to leave the work of Galen at its worth, for, as we know, he had never dissected human bodies, and his anatomical dicta were very unreliable and highly misleading. This, however, was not to be; the Middle Ages were far too bound by respect for authority, and in particular the classical authorities. Moreover, the study of anatomy was rendered difficult on account of the antagonism prevailing between the physicians and the surgeons. Members of the faculty of medicine pursued their studies only on a literary and speculative basis and looked down upon the surgeons as merely a body of artisans. In dissections it was always a surgeon who wielded the knife, while the professor, staff in hand, pointed out and demonstrated what was brought to light. The results of this collaboration were also primitive. The surgeon's instruments and hold were the simplest imaginable; with a knife — the use of a saw and chisel, probe and canula was unknown then — the abdomen and chest cavity were opened and the internal organs laid bare for examination. After this the idea was that muscles, nerves, and blood-vessels should be exposed for study, but this was usually too difficult a task for the operators, nor did it amuse the students, who very soon marched off unless the proceedings ended with one of the professorial discussions that were so popular at that time. The professors of the faculty of philosophy, who were usually invited to attend the proceedings, taking Aristotle as their authority, would attack Galen, who would be courageously defended by all the medical professors present. The differences of opinion between these great authorities of the ancient world could be dragged out into endless discussions and give rise to the most absurd sophistical arguments. This was the course that anatomical studies were still taking in the sixteenth

century, so that nothing was to be expected from them in the way of biological development. Then a man came upon the scene who at once led anatomical research into a completely new direction, created an entirely original
method of procedure, and thus started a new era in the history of science.

ANDREAS VESALIUS was born in 1514 or 1515 at Brussels, of a family
which had taken its name from the district of Wesel in the Rhine Province and
which for several generations had been devoted to the medical profession. He
himself chose the same profession and prepared himself for it by a thorough
school-education. Though his studies were exclusively humanistic, for there
was no other kind of education given in the schools of those times, the young
Vesalius was able in his own way to satisfy his craving for biological knowledge; he studied ancient anatomical works which he found in the family
library and himself dissected animals of various kinds which he managed to
procure. At the age of eighteen he went to Paris in order to study medicine
seriously. There, however, Sylvius, whom we have mentioned above, was the
ruling spirit, with his classical-philological method of education. Vesalius
had again to rely upon his own resources. And his force of will enabled him
to make a way for himself. He began to collect bones from the places of
execution and went on with his dissection of animals. Soon he acquired such
a reputation that he was called upon by physicians and students to perform
public dissecting operations on human bodies in place of the surgeon, and he
fulfilled his task so well that not only the internal organs of the corpse, but
also the muscles, nerves, blood-vessels, and bones were completely demonstrated. After three years he left Paris, worked at his home for a brief period—
in the course of which he succeeded, *inter alia*, in putting together a complete
skeleton out of bones from the gallows — and then went to Italy. In Venice,
where there existed at that time a very keen interest in medicine, he increased
both his learning and his reputation, with the result that, immediately after
he had graduated, he was appointed professor at Padua, at the age of twenty-
two, after only four years of study. He could hardly have wished for a more
satisfactory field of activities. An enlightened government, an interested
audience, and a thoroughly educated public all equally favoured the attain-
ment of his ambitions. And, indeed, Vesalius surpassed all expectations. His
interest in his science was indefatigable and his enthusiasm for imparting
his knowledge inexhaustible. His demonstrations on dissection used to
bring together as many as five hundred listeners, and this in spite of what,
according to the ideas of his time, were unheard-of claims that he put upon
his audience, which he kept busy from morning to night for a space of three
weeks. Dissection lectures, which were always held in the winter so that
the material should not putrefy, began with a demonstration of the skeleton,
the bones of which were carefully gone over; then the muscles, blood-vessels,
and nerves of one corpse were prepared, and finally the internal organs of the

abdomen and chest of another body, as well as the brain. Vesalius himself used to perform the essential work in dissecting, assisted by students; the surgeons, who elsewhere had such an important part to play, had nothing whatever to do here. A mass of new surgical instruments came into use; the models had been partly invented by Vesalius himself and partly borrowed from among the tools owned by a number of artisans whom he visited in order to initiate himself into their technical ideas.

Vesalius's great anatomical work

In his demonstrations Vesalius was at first a faithful follower of Galen, for whom since his youth he had entertained the greatest respect. It soon became more and more obvious, however, that Galen's observations were incomplete and that his presentation of them was vague and self-contradictory. The more zealously Vesalius anatomized, the more did he realize how necessary it was to reproduce in print all his anatomical observations and, without reference to any authorities, to describe the structure of the human body as it really is. The result was his two literary masterpieces: *De humani corporis fabrica*, a large folio volume of over seven hundred pages, and a compendium of the same, *Epitome*, of thirty-one pages, both containing numerous illustrations by eminent artists after Vesalius's original preparations. These books were published in 1543 at Basel, where Vesalius spent a whole year's leave of absence superintending the printing. Through these two works Vesalius created the modern science of anatomy. They made an enormous impression on his contemporaries. Galen's followers were furious, particularly Vesalius's old master Sylvius. Many were the polemical treatises written in opposition to the man of dangerous newfangled ideas, and the rage of his opponents can still be perceived from the controversial methods they adopted. They were not content with merely declaring that Vesalius's work was absolutely inferior; the most abominable and absurd accusations were heaped upon his personal character — he was godless, he was sordid; like the ancient Alexandrian anatomists he had dissected men alive (sentimentality as regards animals was not so deep in those days as to make it worth while quoting the vivisection he actually performed on animals). Even after his death his memory was treated with contumely, especially in France, where the followers of Galen were to hold unrestricted sway for another hundred years. Vesalius could now no longer hope to enjoy the tranquil conditions under which he had worked in the past. In the year after the publication of his writings we find him relinquishing his professorship and accepting the post of court physician to the Emperor Charles V. What induced him to take this step is not known; it is assumed that after the completion of his anatomical masterpiece he wished to devote himself to practical medicine, but that he might just as well have done in Padua. It is more likely that he hoped that in his appointment at the court of the most powerful monarch in the world he would find protection from the persecution of his enemies. Besides, several of his ancestors had been court physicians. He accompanied his delicate master on all his many journeys through various European countries, in the course of which he had but little time for continued research. In 1555, however, he published a new and improved edition of his great work, in which he vigorously refutes his calumniators. Upon Charles's abdication he joined his son Philip II, whose notorious obscurantism offered but the smallest chance for his personal retainers to develop liberal ideas. In fact, after eight years we find Vesalius leaving the court; in 1564 he visited Venice, in the hope apparently of again taking up his old professorship, which then happened to be vacant. While waiting to be appointed he made a journey to the East, visited Jerusalem as a pilgrim, and never again returned to the West. The reason for his journey is not known for certain, nor indeed how he ended his life. Thus disappeared into the unknown one of the greatest scientists of modern times.

Vesalius's great anatomical work is arranged in the same order as that which he followed in his anatomizing, mentioned above: first he discusses bone-construction, then muscles, blood-vessels, and nerves, then the abdominal organs and those of the thorax, and the brain, and finally he devotes one chapter to an account of his vivisectional method. In his general conceptions Vesalius entirely adopts the standpoint of antiquity. His division of the component parts of the body into simple and complex is borrowed from Aristotle, as also are most of his physiological terms; the food is "cooked" in the abdominal cavity, the object of respiration is to cool the blood, the embryo arises out of the father's semen and the mother's menstrual blood. From Galen, whom he still highly respected, he takes his general conception of continuity in existence and of the causes that govern it. The Creator has, to His own honour and to the benefit of man, made the human body as perfect as possible; every part of it has been created just as it is in order to fulfil its specific purpose. In many important details also he adheres to Galen's ideas, particularly in regard to the circulatory system; he gives, it is true, an exhaustive description of the structure of the heart, but as regards its and the liver's relation to the vascular system he still retains the old traditional view. The greatness of Vesalius lies in his method and technique. In this he created the conditions necessary for the development of modern anatomy. Most of the technique which is practised in every anatomical theatre today originates from him; the instruments used at the present time are practically the same as those which he designed, and the majority of them he introduced into dissectional practice; the course of instruction as laid down in his works is still followed; the skeletons used for demonstration purposes are mounted after his method; and the plates used to facilitate instruction are for the most part merely improved editions of his own. But his great service to science is not confined to this. In almost every sphere of human anatomy he made important discoveries in matters of detail and, still more, corrected old fallacies. To enumerate all that he did in this respect would be impossible here; but he who cares to do so can compare, for instance, a picture of a skeleton made by any one of his predecessors with one of his. If we add to this his masterly and at the same time exact and highly imaginative descriptive method, with its splendid simplicity of arrangement in the midst of a wealth of detail, every impartial judge must admit that he was, in spite of his lack of original ideas, one of the greatest biologists that have ever lived.

Vesalius's influence on the development of anatomy was primarily of benefit to Italy. In France the disciples of Galen still upheld the authority of their master; in Germany the interest in the study of nature was being gradually suppressed by interminable religious strife. Italy, on the other hand, thanks to the impetus given by Vesalius's practice of anatomy in Padua, became throughout the succeeding century the centre of anatomical study. Vesalius's pupils followed in their master's footsteps and carried on his work by widening the field of detailed research. His prosector and successor in Padua, REALDO COLUMBUS (date of birth unknown, died 1559), made a special study of the organs of hearing and the blood-vessels in the lungs. He published the results of his experiments in a work entitled *De re anatomica*, in which he shows himself a well-informed anatomist, but a not very sympathetic personality, self-opinionated and overbearing, not least towards his old master. He was soon called away, however, to carry on other activities elsewhere and was succeeded in Padua by a man of far higher qualities, GABRIELE FALLOPIO. Born in 1523, Fallopio spent his youth in poverty, was for a time in the service of the Church, but afterwards had an opportunity of studying anatomy in Padua, probably during the very last years of Vesalius's professorship. His career was as rapid as the latter's; at the age of twenty-four he was a professor in Ferrara, whence he was summoned to Padua, where the Government maintained him in every way. He carried on the Vesalian traditions with honour, attracting to his lectures a large audience and at the same time working at an extensive medical practice. Unfortunately his life was short; he died in his fortieth year. During his lifetime he published only one, rather small, but useful, work entitled *Observationes anatomicæ*. In its introduction he speaks most highly of his master, Vesalius, and with the greatest modesty of his own observations. These are, however, in certain respects of fundamental importance. In particular, he increased the knowledge of the sexual organs — in this field the Fallopian tube bears his name — while his contribution to the knowledge of the structure of bone and of the organ of hearing was of considerable value. But he also made important discoveries in most other fields of human anatomy. Besides this his activities extended to other spheres of medical science; the results he achieved here were not published until after his death.

To Fallopio's professorship, which, as mentioned above, Vesalius had hoped to resume, was appointed after the latter's death another scientist who was also a pioneer in his branch — GIROLAMO FABRIZIO, usually called, after the place of his birth, FABRICIUS AB AQUAPENDENTE, to distinguish him from a contemporary German anatomist Fabricius. Born in 1537, he studied under Fallopio, was his prosector, and succeeded him as professor in 1565. In contrast to his famous predecessors he lived to a good old age: he died in 1619, having been for ten years emeritus professor. Besides anatomy he lectured on surgery; he raised that despised "handicraft" to the rank of a science and was himself an eminent practitioner, his profession bringing him immense wealth, which he generously utilized for the benefit of science. Anatomical research was in his time liberally patronized by the Venetian Government, which built a fine anatomical theatre and paid generous salaries to its staff.

Fabrizio was a very productive scientist, though more qualitatively than quantitatively. His predecessors had devoted themselves exclusively to human anatomy, and such contributions to comparative anatomical research as had been made in other quarters — Pierre Belon's, for instance — had passed practically unnoticed. Fabrizio adopted the method of comparative research, which really no one since Aristotle had applied with anything like original results, and he developed it further in one of the most important spheres of biology — namely, embryology. His treatises on the evolution of the egg and the embryo present in clear and concise form, with good illustrations, the process of embryonic development in a large number of vertebrates: birds and reptiles, mammals and sharks. He describes the anatomy of the embryo and the shape and appearance of the placenta and embryonic tissues, pointing out the similarities and differences between the various animal forms, with a wealth of hitherto unknown facts, which it would take too long to follow in detail. Fabrizio employs the same comparative method in a number of other spheres of biology. Thus, he describes the movements of animals from a comparative point of view; again he studies the noises of animals. This leads him to make an interesting attempt at animal psychology — certainly the first of its kind. Amongst his purely anatomical works may be noted his investigations into the structure of the ear, the eye, and the larynx. Of more definite value to posterity, however, was a three-page article on the venous valves, which he discovered experimentally — through binding the limbs of live human subjects for the purpose of bleeding — and which he afterwards closely studied with reference to their structure and distribution. In spite of this discovery, which was so obviously at variance with the Galenian theory of circulation, he could not abandon the latter; he explained away his discovery, and it was left to one of his pupils, the Englishman Harvey, using this fact as a starting-point, to formulate a true conception of the circulation of the blood.

Fabrizio was the last of the great anatomists of Padua — a line of great men in the service of biological research, such as scarcely any other university has been able to produce in an unbroken sequence. But besides these Italy possessed in the sixteenth century a great number of eminent specialists in the field of anatomy. Space, however, does not permit of our dealing with more than one or two of them as examples of the enthusiasm with which anatomical research was carried on in the country in which Vesalius stimulated such interest in that branch.

BARTOLOMMEO EUSTACCHI was a student of research possessing wide interests and deep knowledge, which, however, owing to the unhappy fate that befell his works, came to have but little influence on the progress of science. The date of his birth and the early circumstances of his life are unknown to us; in the middle of the sixteenth century we find him in practice as a physician in Rome and then as professor at a papal medical academy. He died in 1574. He had recorded his widely extensive anatomical investigations in a richly illustrated work, which at his death was ready for the press. It was withdrawn, however, and never published until 1714, when most of it was naturally out of date. During his life, Eustacchi found time to publish a number of smaller treatises, *Opuscula anatomica*, among which were several important investigations, as, for instance, that of the auditory organ, in which the Eustachian tube still bears his name, and of the blood-circulation and dental development in the embryo.

Another eminent anatomist was COSTANZO VAROLIO, of Bologna (1543–75), who in the course of a short life managed to carry out important investigations into the nervous system, in which the *pons Varolii* in the brain is named after him.

Far more remarkable than these two, however, is CESALPINO, a scientist who made weighty contributions in several different fields of research; in biology, as a speculative natural philosopher, and as a botanist. His life's work, however, is best described in another connexion, among the pioneers in the discovery of the circulation of the blood.

The position of MARC' AURELIO SEVERINO among the Italian anatomists is a curious one. Born in 1580 in south Italy, he came at an early age to Naples, where he studied the humanistic sciences and philosophy under the famous CAMPANELLA, known as a keen opponent of Aristotle and as a victim of political and scientific persecution. Soon, however, Severino began to devote himself to the study of medicine and was appointed professor of anatomy and surgery at Naples. He had, besides, a wide medical practice. At one time he was subjected to persecution by the Inquisition and had to flee from Naples,

but he was soon recalled and throughout his life enjoyed a great reputation.
He died in 1656 of a serious plague, which he had endeavoured to stamp out.
He wrote a handbook of human anatomy, a monograph on the viper, and,
finally, the work which made his name famous: *Zootomia Democritea*.

Zootomia Democritea

SEVERINO introduces his work with a defence of the comparative study of
the anatomy of different animals, the advantages of which he demonstrates
in a dichotomously arranged table, and then further dilates upon them with
a mass of quotations from other authors and arguments of his own. He finds
its best to begin the study of anatomy with animals, as they often have a
simpler and more easily accessible organization than man, with whom,
moreover, animal dissections offer interesting possibilities of comparison.
He submits a comprehensive plan of organization for the entire animal
kingdom, and even extends his interest to the invertebrates. He also discusses
the anatomy of plants. His special zootomical investigations, which com-
prise the fourth section of his work, actually consist of a miscellany of notes
on the anatomy of a number of different animal forms; he never records the
results of a radical anatomical study of any particular animal. The chapter
entitled "*Tetrapodographia*" recounts scattered observations on the anatomy
of domestic animals in particular, but also of the fox, the hare, the mole,
the tortoise, and the hedgehog. The "*Ornithographia*" contains similar information on birds and a special comparative study of their feet; details
(mostly external) are given of insects and spiders, and this zootomical handbook closes with a chapter on fishes, of which the ink-fish are dealt with in
greater detail. The last section of the work consists of an account of the
technique of the subject; the usual dissecting instruments are described, and
even the use of the magnifying-glass is recommended.

The title of the book, *Zootomia Democritea*, testifies to the tendency of
the work from beginning to end — antipathy to Aristotle, a feeling which
had been inculcated into Severino in Campanella's school. In the first
chapter he sets up the observation of nature in opposition to the theories of
Aristotle — the same system of natural observation on which Democritus
laid so much stress. Severino does not succeed, however, in creating any fresh
conception of natural phenomena in the place of the Aristotelean, and so,
like Campanella, he has to a great extent to fall back upon the mediæval
schoolmen, whose deductive method of argument and proof he employs in
his zootomical studies.

The death-blow to Aristotle's biological theories was destined to come
from quite a different quarter; curiously enough, from a man who had the
greatest respect for his teaching, but who at the same time established certain facts which rendered it impossible for him to follow it.

CHAPTER XIV

THE DISCOVERY OF THE CIRCULATION OF THE BLOOD

1. Harvey's Predecessors

Galen's system of blood-movement

BIOLOGICAL RESEARCH under the Renaissance, as the above narrative shows, considerably widened the knowledge of animate nature. The progress achieved was particularly great in the anatomical sphere; Vesalius and his school contributed not only to human, but also to animal anatomy a wealth of new facts which put the knowledge of classical antiquity completely in the shade. But as regards their general conception of nature these research-workers remained entirely on the ground that had been broken by Aristotle and Galen. Now, however, these newly-won facts could not be reconciled to the old system; the same thing had happened to Copernicus and Galileo in regard to astronomy. A definite break away from the ancient ideas of life was inevitable. In one field in particular was the influence of the ancient system fated — regarding the idea of the movement of the blood in the body and its importance to life. Hippocrates, Aristotle, and Galen had all held the same views on the heart and the vessels of the body in so far as they took the most important qualities of the blood to be the "vital spirits" which it was thought to contain; and in face of the speculations on these "spirits" the study of the movements of the blood in the veins was sadly neglected. Galen, who among the biologists of antiquity had the richest experimental material at his disposal, had worked up into a systematic whole all the knowledge of the vascular system which classical antiquity had accumulated. He had, as will be remembered, succeeded in destroying the old illusion that the arteries and the left ventricle of the heart contained air; he found in them a kind of blood which he believed to have acquired its light-red colour from the pneuma, the half-mysterious life-spirit, which it contained. The pneuma was conveyed to the blood in the arteries from the air, which was introduced by inhalation into the lungs and thence to the left ventricle of the heart. The non-pneuma-conveying blood — the venous blood — had its centre in the liver, where it was formed out of food from the digestive canal. From the liver the blood was conveyed through the veins partly out into the body, in which it was converted, by a process that was not very clearly explained, into "flesh," and partly to the right heart-

GEORGE LOUIS LECLERC DE BUFFON

WILLIAM HARVEY

chamber, from which "soot" was given off through the pulmonary arteries; the wall between the right and the left ventricles was full of fine pores, through which the blood oozed from the right to the left side, to be "cleansed" by the action of the pneuma. Galen had but vague ideas as to the movement of the blood in the vessels; in the veins, at any rate, the blood moved, according to his notion, alternately in both directions. Such was Galen's theory of the blood-vessels and their contents, and in this form it was still accepted by the great anatomists of the sixteenth century. All its vagueness and many contradictions would undoubtedly have been realized long before had not the blood-vessel system of old been considered the very centre of life itself; the mysterious pneuma was only one side of this blood's specific life-content; the different kinds of soul that man was believed to possess — the "vegetative," with the liver as its organ, and the "animal" in the heart — were also intimately connected with the blood and through it affected the entire body.[1] Speculation about these components in the organism certainly did not make the conception of the vascular system any clearer; moreover, it entailed the risk that any critical discussion of these organs might be interpreted as an attempt to call into question the immortal soul of man, which would inevitably have involved the scientific student in trouble with the theologians and the Inquisition. Typical in this respect is Vesalius's attitude regarding the pores in the dividing wall between the right and the left heart-chambers; he could not find any trace of them, but cautiously adds that all the same the blood might perhaps be able to ooze through the wall itself. His pupils adopted the same cautious attitude on this point, particularly Fabrizio, the discoverer of the venous valves.

To attack the traditional theory of the blood-vessels was thus a task that required courage. The man who was the first to grapple with an attempt to reform one detail of the old theory was in fact well qualified in that respect, a man whose whole life had been spent in a struggle against time-honoured ideas and who was at last to die for his principles. This was the well-known religious enthusiast and martyr, MICHAEL SERVETUS.

MIGUEL SERVET Y REVES, which was his real name, was born at Villanueva in north Spain, of noble parents. The date of his birth is not known for certain (1509 or 1511); how he spent his youth is also unknown. It was apparently at an early age, however, that he experienced that restlessness of spirit which throughout his life made it impossible for him to settle to anything permanent or find any definite mission in life. He became one of those passionate, revolutionary, and at the same time deeply mystical enthusiasts who were particularly in evidence during the Renaissance. Having visited

---

[1] The theory that the soul, not only of man, but also of animals, is in the blood occurs in the Old Testament: "Only be sure that thou eat not the blood: for the blood is the life; and thou mayest not eat the life with the flesh" (Deuteronomy xii. 23).

several places in Germany and Italy, he settled down in Strassburg and there published his first treatise, *De trinitatis erroribus*, in which he recorded the results of his mystical religious speculations. He disapproved of infant baptism and expressed a view of the Trinity which was regarded as Arian. The book evoked a storm of bitter criticism from both Catholic and Protestant theologians; Servet had to flee from Strassburg and subsequently reappeared under a different name. In Lyons he found refuge with a physician, who inspired him with a taste for medicine, and in order to continue his studies he moved to Paris and there practised anatomy with Vesalius. At the same time, characteristically enough, he lectured to the students on astrology. His theories of the influence of the heavenly bodies upon the health again brought him into trouble with the theologians and he had to flee from Paris. In the city of Vienne, on the Rhone, he found employment as a physician and spent there a few peaceful and happy years. During that period he recorded the results of his continued theological speculations in a book entitled *Christianismi restitutio*. He attempted by correspondence to win over to his views the reformer Calvin, but was rebuffed. When, in spite of this, he dared to publish his book anonymously in Vienne and concluded it with a venomous attack on Calvin, the latter became furious and had the author brought before the Inquisition in Vienne. Servet was cast into prison, but managed to escape, and this time sought refuge in Geneva, probably in order to co-operate with the anti-Calvinistic party which was just then planning an attack on the despotic reformer. Calvin, however, was on his guard; Servet was arrested and Calvin seized the opportunity offered by the trial of this sectarian, so hated by the whole Christian community, to strengthen his position. Having obtained the consent of several Protestant Church councils, the court at Geneva condemned Servet to be burnt at the stake, and the verdict was carried out on the 27th October 1553, to the eternal shame of Protestantism. Shortly before, the Catholic Inquisition in Vienne had caused Servet's portrait to be burnt in the absence of Servet himself. Through his death, however, Servet won such renown as neither his personality nor his writings in themselves warranted; the Catholics in particular have in latter times honoured his memory, in order to annoy the Calvinists. Statues have been erected to him in both Paris and Madrid.

Servet's investigation of the pulmonary system

SERVET's principal work,[2] *On the Restoration of Christianity*, is, as its title implies, purely theological and discusses from a mystical spiritualistic point

---

[2] Of the original edition of Servet's *Christianismi restitutio* there are, as far as we know, only three copies in existence; one in Vienna, one mutilated copy in Paris, and one defective copy in Edinburgh. An attempt to republish the work in England in the seventeen-twenties fell through owing to the opposition of the ecclesiastical authorities. In 1790 a new edition was at last published in Nuremberg; even this edition is somewhat rare.

of view the relation of God to the world and man. Every conceivable problem of life is drawn into discussion in this connexion — jurisprudence and statesmanship as well as astronomy, physics, and medicine. In the discussion on the Holy Spirit he points out that this cannot be properly comprehended without knowledge of the spirit of man, and the spirit of man in turn, if it is to be rightly understood, requires a knowledge of the human body. In this way Servet arrives at a discussion of the structure and function of the human body, and in particular the part played by the blood, which is so vital in its spiritual aspect. And here he pronounces the dictum that has given him, the religious idealist, a place in the history of biology; this was his exposition of the course of the pulmonary circulation. In order to gain an idea of the relation of the spiritual to the physical life we must, says Servet, realize the three vital elements in the body, which are: the blood, with its seat in the liver and the veins; "*spiritus vitalis*," in the heart and the arteries; and "*spiritus animalis*," which is a ray of light and is situated in the brain and the nerves. In all these dwells the power of God's spirit. The vital spirit is communicated by the heart to the liver, for in the heart dwells first of all the spirit communicated by God, as we see from the embryonic life, in which the heart is the first point that lives. On the other hand, the liver provides through the blood material to the spirit, which is formed by the union of the finest components of the blood with the inhaled air. This union takes place in the lungs, to which the blood is conveyed from the right heart-chamber, to be conveyed thence, purged of soot through exhalation and mingled with inhaled air, back to the left heart-chamber. That the blood does not, as is commonly imagined, pass through the heart wall is proved not only by the latter's solid consistency, but also by the powerful structure of the pulmonary veins, which cannot be explained simply by their function of feeding the lungs. All this is really obvious, concludes Servet, from the observations recorded by Galen, if only one understands how to interpret them aright.

The strange, strongly spiritualistic physiology which Servet expounds in his description of the importance of the blood, above referred to, is in itself nothing peculiar to him; on the contrary, it recurs often in the authors of the Renaissance and even of the seventeenth century; in Swedenborg, too, we find a similar method of speculation. Indeed, Servet reminds us of the latter in having arrived at his theories by way of speculation rather than through his own observations. True, he had dissected, as mentioned above, and that too under Vesalius himself, though he makes no reference to his experiences in that line, but tries to give a correct interpretation of Galen. What in these circumstances is surprising is that he gives such a clear idea of the pulmonary circulation — all the more so as his view of the blood-vessel system is otherwise purely Galenian, with the liver as the principal organ for the blood and the veins emanating therefrom. Nor indeed was it his object to gain any knowledge of the structure of the human body for any biological or medical purpose; just as his method was speculative, so his purpose was exclusively mystical-theological and he thus was content with the old tradition as it stood, except in that one point in which it was not consistent with his metaphysical construction of thought. However, he is undoubtedly the first to expound a theory of the pulmonary circulation agreeing with that confirmed by the research of later times.

It is hardly to be supposed that a treatise which was prohibited by contemporary and later governments, and to the best of their powers suppressed, would succeed in exercising any great influence on the progress of science; it was, in fact, to be more than a century and a half before anyone drew attention to Servet's contribution to the discussion of the circulatory system. Nevertheless, it would appear that Servet's ideas did after all have some influence on his contemporaries, since during the latter half of the sixteenth century one comes across in many authors statements, or at any rate hints, as to the blood-circulation between the right and left ventricles through the lungs. One or two of these writers, who had some influence on the final solution of the problem of the movement of the blood, should be mentioned here.

REALDO COLUMBUS, Vesalius's pupil and immediate successor in the chair of anatomy in Padua, to whom we have referred above, may claim to be named among the forerunners in this field, as he is the only author cited by Harvey, the great pioneer of research work on the blood. Columbus in his work on anatomy devotes a chapter to the vascular system. Here he presents the traditional theory of the liver as the centre of the venous system and the true blood-forming organ, from which the blood is conveyed to the different parts of the body. The arterial system originates in the heart. Its right and left ventricles are separated by an intermediate wall, which, contrary to the common assumption, is impenetrable; from the right side the blood is conveyed to the lungs, where it is mixed with air and, thus diluted, is conducted back to the right side of the heart. This, he adds, no one has hitherto observed or described, but it is none the less true and can be verified on experimental subjects, whether alive or dead. Columbus's work was published in 1559 — that is, six years after Servet's. There have been lively discussions whether both arrived at the same conclusion independently, and, if not, of which borrowed from the other. The question can of course never be definitely decided, but it is probable that Servet, who indisputably has the prior claim on the point, in some way influenced Columbus; the latter presumably read the dangerous heretical treatise, which he dared not quote even if he had desired to do so. There is no doubt, however, that Columbus, in a far greater degree than Servet, confirmed his statement by observation and experiment.

There was another whose opinions on the question of the circulation of the blood attracted far greater attention than the above-mentioned contribution to the subject. This was the Italian botanist and physician CE-SALPINO, who is still to this day extolled by his countrymen as the true discoverer of the circulation of the blood. ANDREA CESALPINO was born at Arezzo in Tuscany in 1519. He studied philosophy and medicine at Pisa, the latter under Columbus, who was called to that city from Padua. At the age of thirty he became a doctor of medicine and shortly afterwards professor of pharmacology at Pisa. In this capacity he devoted special attention to the study of botany and is reputed one of the pioneers of that science. His contributions in this field will be dealt with in another connexion. In his old age he was summoned to Rome, where he was appointed body-physician to the Pope, and where he died in 1603.

Cesalpino was a man of manifold interest; besides botany and pharmacology he studied anatomy, mineralogy, and metallurgy, but he was above all a natural philosopher in the true Aristotelean spirit. His theoretical speculations he published in a work with the characteristic title of *Peripatetic Problems*. In this book he endeavours to find a general explanation of nature along Aristotelean lines; in the purely philosophical aspect of his conclusions he goes beyond his master by deriving both form and matter from a single supreme principle, but as a physicist he takes his stand on the old ground, with celestial spheres and circular planetary orbits, heaviness and lightness as a quality of bodies — everything in fact which Galileo was intent on demolishing.[3] Even his biology, the subject of the fifth book, entirely follows the lines of Aristotle. It opens with the purely mediæval scholastic thesis that if the life in a being is one and indivisible, the body must also be one and its centre one, whence life emanates to the rest of animate things. Plants and lower animals, which are able to live even when cut up into bits, require no such centre point, but in sanguineous animals the heart without doubt constitutes this centre point — the heart, which is the first to begin to live and the last to die, and which is situated in the centre of the body. Thereupon Cesalpino endeavours, in a polemic against Galen interlarded with quotations from Aristotle, to prove that the veins originate in the heart and not in the liver, and that the nerves likewise originate in the heart and not in the brain, the latter point being proved, *inter alia*, by the fact that happiness and grief are felt first in the heart, while the function of the brain is to cool the blood, like the receptacle in a distilling apparatus. By thus swearing to the truth of his master's word, both good and evil, Cesalpino at any rate makes this point in regard to the circulation of

---

[3] Curiously enough, even Cesalpino, in spite of his loyal Aristoteleanism, fell into the hands of the Inquisition, but he saved himself by his dialectical cleverness, and perhaps also owing to his being in the papal service.

the blood, that the heart is actually the centre of the vascular system. And in regard to the relation of the lungs to the heart, he maintains with his teacher Columbus that the blood passes through the lungs from the right to the left side of the heart — a process which he for the first time calls *circulation*. But his servility to the authority of Aristotle prevents him from taking advantage of either his precursors' or his own progress in this field of research. He dares not abandon the theory of the pores in the heart wall, but, on the contrary, admits that some of the blood goes that way; he observes that when a vein is tied, it fills below and not above the ligature, but he does not venture to draw the conclusion that the blood-stream in the veins always leads to the heart — this he believes takes place during sleep, but not in a waking condition — and so the existence of the "vital spirit" in the blood naturally takes the first place in his investigations. His ponderous and involved presentation of his case — vague, too, in comparison with Servet's brief and explicit style — has enabled his admirers to interpret his statements as it suits their purposes, but just as none of his contemporaries saw in him one who had revolutionized knowledge of the vascular system — a fact which he himself, Catholic and papal favourite as he was, would hardly have dared to admit — so there must in truth be a partisan and chauvinistic spirit in those of posterity who would ascribe to him the honour of an idea which he himself neither clearly expressed nor ever definitely claimed.

2. Harvey

BESIDES those of whom we have given account above, there were during the Renaissance, as has been said, quite a large number of anatomical writers who made a study of the construction and function of the vascular system, in vain attempts to bring order out of the chaos to which the inaccurate conception of the ancient biologists had reduced the problem. The necessity of a solution was generally acknowledged; several had been on the right road, but had stopped prematurely. Then WILLIAM HARVEY took the decisive step and solved the hard problem in one stride.

WILLIAM HARVEY was born at Folkestone, on the south coast of England, in the year 1578, of respected and well-to-do parents, who gave their children a sound education. Having taken a philosophical degree at Cambridge, Harvey made a number of journeys and eventually came to Padua, where at that time Fabrizio had begun to attract pupils from far and near. Harvey joined them, and after four years of study took the degree of doctor of medicine. Returning to England, he settled down in London and started a medical practice. He practised in hospitals, was elected a member of the

London College of Physicians, and there gained such a reputation that he was commissioned to give lectures to his colleagues. Eventually he was appointed court physician to King James I and later to King Charles I. Afterwards he spent many years in peaceful and uninterrupted research work and in the duties of his medical practice in London, but then the great Civil War broke out and Harvey accompanied his king in his flight from London, while his house was plundered and his collections destroyed. He was then made a professor at Oxford, which was the headquarters of the King; when this city was also captured by the Parliamentary army after Charles's final defeat, Harvey, then sixty-eight years old, had to retire into private life. Fortunately he possessed private means and was also supported by his brother, a wealthy London merchant, so that his old age was free from care, and at the same time he retained the deep respect of his countrymen and colleagues. A stroke brought his life to a sudden and peaceful close in the year 1657. He left his fortune by will to the College, whose leading personality he had been during his lifetime, and ever since his death the College has continued to celebrate his memory, an annual festival being held in London in his honour. A fine monument has been erected over his grave.

Harvey's work on the circulation

THE work in which Harvey expounded his new idea of the circulation of the blood was published in 1628 at Frankfurt am Main in the form of a quarto volume containing seventy-two pages. Harvey, however, had spent his whole time, ever since, as a youth, he received his first lesson in anatomy in Fabrizio's school, in working out the ideas which were recorded in this modest volume. There are still extant the lecture notes dating from 1616, in which are expressed some of the thoughts which twelve years later assumed their final form, and it has thus been possible to check the careful research, the mature consideration, on which the work is based and which shows itself in the masterly style, at the same time concise and explicit, in which not a word seems superfluous. After giving an account of the old traditional theories on the subject, in which he sharply brings out their defects, Harvey presents his own observations on the movement of the heart. According to the old theory the walls of the heart were not muscular and the dilatation of the heart was its most important function; by this means the blood was conveyed from the veins into the heart. By careful experiments, of which he gives an account, Harvey found that the heart is muscular and that, on the contrary, its regular contraction is its most important movement, which drives the blood forward — that is, out into the blood-vessels — just as it is likewise during this movement that the heart beats against the thorax. In this movement not only the ventricles of the heart take part, but also its vestibule, the significance of which Harvey rightly emphasizes for the first time. He then gives an account of the course of the blood from the right to the left side of the heart through the lungs, and in this he acknowledges the services of Columbus in his explanation of this phenomenon. With regard to the part played by the lungs and the air in this circulation he has not much to add to the hypotheses of his predecessors. After having thus described the small circulation Harvey proceeds to a presentation of the blood's movement in the body itself and it is here that he brings out his most daring originality. According to the old theory, food was converted in the liver into blood, which was driven through the veins partly to the heart, in order to receive the "*spiritus vitalis*," and partly into the body. To this theory Harvey opposes a mathematical calculation; if the human heart contains two ounces of blood and gives sixty-five beats to the minute, then it drives in less than one minute ten pounds of blood out into the body. Such a quantity of blood cannot incessantly arise from the food consumed, but it must be assumed that the same quantity of blood incessantly circulates in the body; it is driven out through the arteries and returns through the veins. Harvey then collects a quantity of evidence in proof of this conclusion from the relation of the arteries and the veins in the body. He investigates the arterial pulse both in normal individuals and in those having calcinated veins; he opens a live serpent and ties up first the *vena cava* and then the aorta; while the vein is emptied between the heart and the ligature and swells up on the other side, the contrary is true of the aorta. He studies the venous valves in a man's arm, which were discovered by Fabrizio, and shows how they swell below a ligature; he severs a vein and an artery parallel to it and shows that the blood flows from the different ends of the wound. On these and several other grounds, deduced from the study of every possible animal form, he draws the conclusion that the arteries convey the blood from the heart out into the body; there it is transmitted into the ramifications of the veins and flows from these into the principal vein and thence back to the heart. The arterial blood, he considers, provides nourishment for the body, while that of the veins is impure. How the transition between the arterial and venous system takes place he could not explain; the capillary system he was unable to distinguish, not having access to a microscope, and he therefore assumed that some kind of ramified hollows formed the connecting link between the two. Another weak point in his theory was that he could never find a satisfactory explanation of how the components of the food are converted into blood, but he had to be content with the old hypothesis that the liver was the medium in this process. He lived to see the discovery by others of the lymphatic and thoracic ducts, but then he was no longer capable of realizing how well these experiences complemented his own discoveries; he desired to know nothing about them and on this point adhered to the old theory.

If we compare Harvey's account of the circulation of the blood with

the old vascular theory, we find fundamental differences in the two conceptions, both anatomically and physiologically. According to the old theory
the heart was not a muscular organ; it dilated purely passively and allowed
the blood to enter in order to be provided with "vital spirit," this being the
primary life-function of the heart, if it were not also, as Aristotle and
his followers until Cesalpino held, the centre of intelligence. Again, the
blood moved of itself owing to the specifically living qualities which the
"*spiritus*" lent it. Harvey, on the other hand, proves that the movement of
the blood is due to the purely mechanical function of the heart: the heart's
muscular contraction propels the blood out into the blood-vessels, through
the arteries out into the body, thence back to the heart through the veins,
and so farther through the lungs. In this contrast lies, one may say, the great
difference between the ancient and the modern biological conception. Even
Harvey's way of producing his proofs is purely modern; while Servet still
refers back to philosophical speculations and the interpretation of classical
authors, Harvey propounds a purely mathematical calculus on the volume
of the heart and vascular system and continues to prove his thesis by means
of observations and experiments on a number of both higher and lower animal forms. He thus fulfils in the sphere of biology the requirement which
his contemporary Bacon laid down as a principle of science: to explain nature by experience based upon observations and experiment. And even Galileo's fundamental principle governing natural research — to measure what
can be measured and to make measurable what cannot be measured — is
applied to living nature by Harvey for the first time. Galileo also thought
that science can only explain how the forces of nature operate; what their
innate essential quality is will never be known under any circumstance. In
his explanation of the circulation of the blood Harvey does indeed fulfil the
first half of this principle; on the other hand, by adhering to the ancient
belief in the vital spirits in the blood he remains in his theoretical conceptions entirely on ancient ground.

On the generation of animals

THIS conservatism of Harvey's displays itself conspicuously in a work which
he published in his old age: *Exercitationes de generatione animalium* (1651).
Like his work on the circulation of the blood, this book is the fruit of many
years' labour, but in contrast to the former it is somewhat lengthy and far
less perfect in form. In this Harvey gives a comparative account of the embryonic development in higher and lower animal forms. He is able here to
quote as his precursor his old teacher Fabrizio, and he does so with the
utmost piety. But above all he proves himself to be a follower of Aristotle,
whose conception of the true essence of life he has made entirely his own.
Along Aristotelean lines he endeavours to find a formal unity in the manifold aspects of phenomena, as displayed in the evolution of the embryo, and

he believes that he has discovered such unity in the egg, out of which all living creatures are evolved. His dictum: "All animals, even those that produce their young alive, including man himself, are evolved out of the egg" is well known. He was naturally not able to observe the eggs of mammals — such a study requires a microscope, which he did not possess — but he presupposes their existence on theoretical grounds, a conclusion which was confirmed long afterwards. Nevertheless, he is loath to abandon the idea of primal generation, though he limits this principle to the lowest animals. Out of the egg the higher animals are evolved by epigenesis, in that the organs are successively formed out of the indifferent matter in the egg, which thus constitutes, in harmony with Aristotle's theory, the potentiality out of which the individual is realized. The lower animals, on the other hand, are evolved by metamorphosis, a direct reconstruction of complete rudiments, as is proved especially by the evolution of the pupa of insects; Harvey, in fact, shares Aristotle's belief that the pupa is the insect's egg. On the subject of reproduction his ideas are entirely mediæval; the influence of the sperm on the development of the embryo he believes to be due to the vital force it contains, and this he compares with the secret force exerted by the heavenly bodies upon all life on the earth. That this last work of Harvey's should also contain a mass of remarkable detailed observations is not surprising; he describes with unprecedented care the ovary of the hen and its development, the nourishing of the chicken in the egg, and its growth from the very earliest stages; and of even greater interest are the comparisons he makes between the embryonic stages in different animals — mammals, birds, and lower types.

Harvey is without doubt one of the most remarkable figures in the history of human culture. His work is the most revolutionary that the development of biology has to show, for it undermines the foundations of the ancient conception of life and its manifestations, and nevertheless he himself retains this very conception as long as he lives. He thus brings to a close the great epoch in the history of biology which is governed by the ancient conception of nature and he initiates the modern development in the sphere of biology, just as Galileo does in that of physics. How an entirely new science of life has developed on the foundations laid by Harvey will be shown in the next section of this work.

PART TWO

BIOLOGY IN THE SEVENTEENTH
AND EIGHTEENTH CENTURIES

CHAPTER I

THE ORIGIN OF THE MODERN IDEA OF NATURE IN THE
SEVENTEENTH AND EIGHTEENTH CENTURIES

CLASSICAL ANTIQUITY gave rise to two explanations of natural phenomena, each splendid in its own way: that of Democritus and that of Aristotle. As will be remembered, Democritus attempted to explain all phenomena in existence, both physical and psychical, by the assumption that things were composed of a mass of particles, varying in size, shape, and movement, whose mutual interrelation caused all that is and all that happens, all, in fact, that is observable or conceivable. The weakness of this theory lay in the fact that it gave no explanation of the obedience to law which experience has proved beyond any doubt to exist in all that happens in nature. It was therefore supplanted by Aristotle's cosmic explanation, which maintained just this universal obedience to law, but based it upon the assumption of a divine intelligence which governs and gives form to what is in itself formless matter, controlling the latter in various degrees — less in inanimate nature, more in the animate, and most in the celestial spheres which hold sway over the imperfect earth. In animate nature this force appears as soul, vital spirit, which creates higher forms of existence the more it overcomes matter. This cosmic theory, which, owing to its logically consistent formulation, is unique in its greatness, has been characterized as dynamic and vitalistic in contrast to materialistic atomism. It has with greater reason been called æsthetic, since Aristotle really looked upon natural phenomena from the point of view of an artist who gives form to matter; it has even been called teleological, because according to it everything in existence has a purpose which is determined by the governing intelligence. In this latter characteristic we really find that quality in the Aristotelean thought-system which has proved most fateful both for that system and for man's conception of life in general. The divine intelligence which Aristotle invented in order to make possible the assumption of law-bound existence on purely speculative grounds became a welcome ally to the pious aims of late antiquity and still more so to the mediæval Church. One found indications of similarity between it and the "divine power" of the old myths of creation, and thus received an idea of the course of the world, apparently scientific, but actually based upon legends from the childhood of man.

Aristoteleanism in alliance with the Church

IT was just on account of its semi-scientific nature that it was extremely difficult to controvert this idea with reasons and proofs, seeing that it was at the same time cherished by the authority of the Church and protected by the latter's powerful resources, both spiritual and temporal. In the first part of this work it has been shown how the champions of natural science during the Renaissance took up the cudgels against Aristoteleanism, which was upheld not only by the authority of the Church, but also by the boundless respect that that age entertained for antiquity; how Cusanus and Bruno asserted the infinity of cosmic space in contrast to Aristotle's spherical universe; how Francis Bacon ruthlessly exposed the abstract structural system of the ancient philosophy and urged that research should be based on observations of nature itself; how Galileo, by means of observational material and mathematically conclusive proof, demolished the theory of the immutable regularity of the celestial regions and at the same time proved the purely mechanical obedience to law of the phenomena of motion here on earth; how Harvey through his discovery of the circulation of the blood proved a purely mechanical action in the life-process which the old theory considered to be the centre of animate life. But however many defects could be proved against the old system in detail, it nevertheless still remained unaffected, owing to its consistently carried-out construction; it required an entirely new system of thought in place of the old before the latter could definitely break down. Throughout the seventeenth century keen-minded thinkers set about creating such a system, and the strength that underlay Aristotle's cosmic idea, its unassailable consistency and perfect lucidity, had never been demonstrated so clearly as now, when, already condemned to fall, it made a stand against the assaults which finally shattered it. A survey of this struggle between Aristoteleanism and the new systems of thought is so much the more necessary as an introduction to the history of modern biology as it was actually during this struggle that not only the natural science of our own time, but the whole idea of life as conceived by present-day humanity in general came into being. Nevertheless the modern conception of nature by no means rests solely upon the purely mechanical foundations laid by Galileo. It is self-evident that in it there are very considerable elements of vanquished Aristoteleanism. But besides him there appeared also in opposition to the latter theory philosophers who from neo-Platonism and other similar systems of ideas adopted a purely mystical view of nature. This too has possessed its attractive sides for the human mind, especially the advantage of making possible a uniform conception of both the material and the ideal aspects of existence; during the Renaissance in particular it won many adherents — Bruno is the most brilliant example — and it has consequently left a strong impression upon modern natural science. In the following chapter an attempt will be made to illustrate how these elements in the conception of nature in our own time arose.

CHAPTER II

THE MECHANICAL NATURE-SYSTEMS

The period of the great systems

THE SEVENTEENTH CENTURY has been called the period of great systems of thought, during which all the knowledge that the Renaissance brought to light was summarized and classified. Order and system were, in fact, what this epoch strove to create in every sphere of life; in government the power was concentrated in the hands of despotic princes who by a rigorous exercise of power overcame all opposition on the part of their subjects and created ordered forms of administration instead of the universal unrest of the Middle Ages and the Renaissance; in the religious sphere the different denominations combined into stable churches which prohibited any divergence from the strictly formulated dogmas which they set up. Such a period was bound to be devoted to strictly delimited systems even in the scientific field, and indeed many such systems of different trends of thought, but all definitely formulated, more dogmatic than critical, based upon speculation rather than upon observation, saw the light of day during this epoch. Some of these which highly influenced the development of biological science deserve further mention.

The pioneer of modern philosophy

RENÉ DESCARTES (in his Latin writings he calls himself Cartesius) is commonly regarded as the pioneer amongst the systematic philosophers of the seventeenth century. Born in 1596 of wealthy parents, he was able to devote his whole life to research. His home was in Brittany, and he was brought up by Jesuits; he spent some years in Paris and was for a time an engineer officer in the service of foreign powers. In order that he might devote himself to his science undisturbed by the Catholic Church, he eventually settled down in Holland, where his most important work was done. He made a journey to Sweden, and died in Stockholm in 1650.

Descartes, like Pythagoras and Plato, was a mathematician, and like them, too, addicted to abstract speculations. His ambition was to place science on firm ground, valid for all possible phenomena, and excluding all accidental circumstances. Among these latter he counted, above all, mental impressions, and in order to exclude them he resolved to doubt everything in existence. But the very fact of doubt proved that he thought, and thinking gave him proof that he existed; "*Je pense, donc je suis*" was his oft-quoted starting-point. On this foundation he then builds up his entire conception of existence on the principle that the composite should be explained from its simple components. First he constructs out of the thought of man an idea of God, since man's finite and imperfect personality presupposes an infinite and perfect origin; once we believe this, we must also be right in relying upon our mental perceptions, for God cannot have given them to us without cause. Through the senses we are convinced that matter exists. The simplest and therefore the most essential qualities of matter are extension, divisibility, and mobility. On the other hand, form, which Aristotle, it will be remembered, made his main principle, is of momentary and therefore of secondary significance. Descartes also rejects the atomic theory, for it is inconsistent with the principle of divisibility; nor does space exist, for everything that exists must have extension. On these principles — extension, divisibility, and mobility — Descartes bases the whole of his theory of matter, both inanimate and animate, and he entirely rejects the theory of final causes, for it would be presumptuous to ascribe any limited purposes to unfathomable and infinite divinity. The only rational explanation of the universe is to regard the whole as a machine. Through vortical movements within the parts of matter the latter have accumulated and become heavenly bodies; and movement is all that takes place in nature.

Life-phenomena purely mechanical

ON this same principle he seeks to explain the phenomena of life — that is, the corporeal. These, in his view, occur purely mechanically, without the intervention of any of the spiritual forces which the Aristoteleans assumed, whether animal or vegetative. Confirmation of this idea of the living body as a mechanism Descartes found in Harvey's discovery of the circulation of the blood, which he enthusiastically upheld and to the acceptance of which he powerfully contributed. This fact in itself would be sufficient to ensure him a place in the history of biology. And in drawing conclusions from Harvey's observations which the latter, faithful Aristotelean as he was, could himself never have perceived, he formed a theory of the human body as a mechanism which may be regarded as the foundation of modern physiology. In particular, he sought to explain mechanically the function of the nervous system. He believed that from the brain the so-called animal spirits are conveyed through the nerves to the muscles, which are thereby set in motion through the impulse given them from the brain. It is not at all necessary that these impulses should be conscious; they may take place in the complete absence of thought "just as in a machine." Thus mental impressions can immediately call forth movements through the nerve currents' being "thrown back" — that is, reflected. He has thus recognized, described, and from his own point of view explained the phenomenon of reflex motion. In regard to animals he believes that all their manifestations of life are the result of such reflexes; it is not possible, he thinks, to ascribe to them the possession of a soul. That man has a soul, on the other hand, Descartes considered to be proved by the fact that man has consciousness, and this soul he regards as a substance, the existence of which, however, is entirely independent of the body. Only at one point, he considers, is there any co-operation between soul and body — namely, in the *glandula pinealis;* there the currents from the nervous system react upon the soul and impart to it a share in mental impressions; there, on the other hand, the soul substance makes impressions upon the nervous system, which give rise to conscious actions.

Thus Descartes has created a purely mechanical cosmic theory in which everything that happens takes place out of mathematical necessity; in which neither the accidental movement of the atoms nor the direct intervention of God is needed to keep the course of events on the move. It is manifest that the great discoveries made during the Renaissance have conditioned his theory. But he himself would not acknowledge any precursors — his acknowledgment of Harvey constitutes the one exception — he was, in fact, a very cautious man, and Galileo's fate had made a deep impression on him. He anxiously avoided offending the Church, for which he always showed a deep respect; his manner of escaping from controversy was more adroit than courageous, as when he gives an assurance that his theory of the creation is merely a game of thought; it might be conceivable that the universe arose as his theory declares, but one knows all the same that the Church maintains the true theory of creation. This, together with his constant assertion of the immortality of the soul of man, was, however, the reason why so many eminent ecclesiastical personages dared to embrace his theory, and thus the mechanical explanation of the cosmos wormed its way, one might say, into the consciousness, thrusting out Aristoteleanism. And naturally the new explanation of the cosmos — Cartesianism, as it was called — had great advantages over the old — above all, in that it rendered possible the application of the newly-achieved results of research in the fields of physics, astronomy, and biology. But even the new theory had its weak points. In particular, the relation of the consciousness to material phenomena, or, in other words, the soul's relation to the body, was a problem which worried Descartes and which, as we have seen above, he finally solved, though not very successfully. For Aristotle this problem did not exist; he had in fact made the soul equivalent to form and thus evaded the point. For the rest, it seems that the individual life did not concern him very much, any more than it did the other philosophers of early antiquity. Late antiquity, and in a still greater degree Christianity, had, on the other hand, devoted earnest attention to this problem, and now that material phenomena were given a purely mechanical explanation, it became extremely acute and was for a long time to be the main point of discussion in the philosophical agenda. This question has naturally been of only indirect importance to biology, which deals mostly with the material phenomena of life, but all the same it has had its influence on many purely biological problems and therefore cannot be entirely neglected here.

It was just in this sphere of science that Cartesianism experienced the strongest opposition on the part of such scientists as did not accept Aristotle's views. In France it was PIERRE GASSENDI (1592–1655) who stands out most conspicuously as the opponent of Descartes. He was born of poor parents, but rose to high dignities in the Catholic Church. As a philosopher he sought to revive the ancient atomic theory and wrote a defence of Epicurus, who in the Middle Ages was the object of universal execration. Against Descartes he argued that his conclusion that thought was a proof of existence was incapable of realization. For the rest, Gassendi was a great admirer of Galileo for his discoveries in the realm of physics, which he partly improved upon; being a priest, however, he was forced to deny the Copernican cosmic system. He conceived warmth to be the soul in existence. The relation between matter and human consciousness he tried to explain in the same way as Lucretius, but he admitted that there were insoluble difficulties in the way; besides, as a priest he had of course to maintain the existence of an immortal soul.

Another thinker who held a markedly mechanical view of existence was the Englishman THOMAS HOBBES. He had studied in Oxford and travelled a great deal in Europe, and afterwards spent the greater part of his long life as a private scholar. He died in 1679 at the age of ninety-one. He regarded all that happens as motion; mental impressions were in his view motions in the nervous system, which arose as a reaction to motions in the external world. Hobbes speculated most, however, upon problems of ethics and statesmanship; he had no interest in biology.

The same may be said of another philosopher, who nevertheless, curiously enough, came to play a not unimportant part in the history of biology — BARUCH SPINOZA. Born of Jewish parents at Amsterdam in 1632, he was brought up to become a rabbi, but as he failed to follow the teachings of the synagogue, he was excommunicated and afterwards lived in the closest retirement, making a livelihood by polishing eye-glasses, until the time of his death, in 1677. Only a few liberal-minded people dared during his lifetime to acknowledge acquaintance with the outcast, and although in some respects he was highly admired, it was not until later that his writings won any general acceptance. He himself, thanks undoubtedly to his mild and unassuming temperament and his retired life, escaped falling a victim to the religious fanaticism of the age, for even in Holland, which was a comparatively liberal-minded country, tolerance towards heterodox persons was a rare thing in those days.

Spinoza's system of thought is one of the most magnificent and consummate in human history, one of the most ingenious attempts to reconcile the opposition between consciousness and matter which Cartesianism brought out. In it he is governed by a feeling inspired by the Jewish faith of his childhood — a religious awe of the infinite, eternal, immutable, "that which is in itself and is comprehended out of itself." This he names substance: that into which all things that exist enter as parts. This immutable substance has an infinite number of forms in which it appears, of which we human beings can distinguish only two: the material extension, and the spiritual consciousness. These cannot in any way be explained out of one another, but, on the other hand, both revert to the substance out of which they arose, and man can therefore conclude from the laws that govern the one that they also govern the other; the laws of human reason have absolute force in nature as well. From the immutability of the substance it follows that the development that seems to take place is only apparent; everything, after a brief individual existence, reverts to the substance, like a wave that sinks back into the sea, giving place to new individuals of equally ephemeral existence. To acquire knowledge of the substance is the highest aim of man; it cannot, however, be attained by way of thought, but only by direct introspection. Spinoza thus ends in mysticism — that, too, probably induced by his Hebraic-oriental origin. It is strange that, in spite of this and of his utter denial of any kind of development, his system has been deeply admired by the very students of nature of more recent times who have made development the principal aim of their researches. Goethe was strongly influenced by it, and in more recent times Haeckel and his monist disciples have given it enthusiastic support, in reality perhaps more on account of the religious persecution suffered by Spinoza than on account of the subject-matter of his extremely involved writings.

In most respects his somewhat younger contemporary GOTTFRIED WILHELM LEIBNIZ forms a sharp contrast to Spinoza. Leibniz was born at Leipzig in 1646, the son of a professor. He was a veritable infant prodigy; as a boy he had read practically the whole of the classics and at the age of seventeen he delivered his doctor's dissertation. Mathematics and jurisprudence were subjects of particular interest to him; he became one of the pioneers of the former science, while the latter provided him with an income as a government official and diplomatic representative at the courts of several German minor princes. He died at Hanover in 1716. Throughout his life his energy was remarkable and his interests incredibly many-sided. In the course of his travels in most countries with any standard of culture, he had made the acquaintance of the most eminent men of his time, and in questions of culture his advice was sought from all sides. Peter the Great of Russia, as well as the most learned men in western Europe, corresponded with him. And as his temperament was as pacific as his interests were universal, he endeavoured everywhere to reconcile and to unite. At one time he speculated upon a universal science in which all human knowledge was to be represented by short symbols; on other occasions he worked for the union of the different Christian Churches. The same efforts to reconcile opposed views likewise govern his natural philosophy. Thus, he seeks to show that the Church's doctrine of the omnipotence of God, and natural science's mechanical explanation of the universe are by no means mutually exclusive, but are capable of being harmonized.

Leibniz's monad theory

As a natural philosopher Leibniz took the atomic theory as his starting-point. As he found it impossible, however, to derive consciousness and the manifestations of the soul in general from the movements of atoms, he sought a way out of the difficulty by assuming a universe composed of units of an ideal, not of a material, character. The idea for this theory he found through using the microscope, which had then just been invented; by this means it is possible to see that every drop of water swarms with life and that life exists everywhere, even where the eye cannot see it; it was thus but a short step to the conclusion that the smallest particles of matter are life-principles — not dead atoms, but living "monads," as Leibniz called them. These monads he conceives as being of infinite variety, some of a higher type, others of a lower. The human soul is one such monad, which has consciousness; the life of animals consists of lower monads, unconscious, but percipient; the monads of plants live, but are not percipient; the monads of inanimate nature are in an indifferent state, as in a dreamless sleep, the human body being composed of these latter monads. The activity of the monads is not motion, as the atomic theory supposed, for motion is something relative, but their ultimate quality can only be conceived as force — *conatus*, as Leibniz calls it. By this means they each, in a higher or lower degree, obtain some notion of existence. On the other hand, they do not react upon one another, their interrelation being governed by a harmonious cosmic order, originally created by God, who is the supreme monad. Thus, the human body functions by force of the harmony of existence parallel and in tune with the soul, like two clocks which go exactly alike. The kingdoms of nature and of grace act similarly towards one another. — All this extremely abstract speculation might at first sight appear foreign to all that is meant by natural science. Leibniz has, however, actually exercised great influence on natural research, partly by awakening interest in life, both in the great multiplicity of its manifestations and in its most minute forms, and, above all, by insisting upon the idea of force as the basis of natural phenomena instead of movement, which even Descartes believed. And his endeavour to reconcile the kingdoms of nature and of grace, which may appear foreign to the ideas of natural research of our own day, would have seemed by no means unattractive at a time when the foremost natural philosophers were at the same time pious Christians and subservient each to his own Church. This even Galileo had been, as also those two of his successors who have contributed more than any others towards giving natural research its modern character — Boyle and Newton. These two are all the more worthy of mention here as through their activities they have, each in his own way, powerfully, if only indirectly, affected the development of biology.

ROBERT BOYLE (1627-91) is generally looked upon as the first modern chemist, in so far as he definitely broke away from the mystical speculations of alchemy and made the object of chemistry the breaking up of complex substances into their simplest elements. Thus he freed experimental effort from the fantastic and semi-magical aims and means of the Middle Ages and created a natural-scientific method based on rational calculations. On the other hand, he had little bent for purely speculative problems and accepted the general cosmic views of the Church without reserve.

Far more renowned and of far greater influence on the intellectual progress of man was ISAAC NEWTON, one of the greatest pioneers of natural science that the world has seen. Born in 1642, of a peasant family, he studied in Cambridge, was for many years professor there, became in his old age Director of the Royal Mint in London, and died at the age of eighty-four, honoured and respected as few scientists have been. He was known everywhere for his liberal-minded political views, deep religious sense, and modest, lovable nature.

The gravitation theory

NEWTON's important discoveries in the sphere of mathematics and optics are universally known. Most famous and most vital from the point of view of cultural development is, however, his theory of gravitation. Galileo had, it will be remembered, established the fact that the movement of bodies on our earth takes place on fixed, mathematically calculable principles. Newton now proves that the same laws governing the movement of bodies at the earth's surface also govern the movement of the heavenly bodies in their relation to one another. All the world knows the story of how in his youth, at the sight of a falling apple, he began to ponder the question whether it was possible to calculate the movement of the moon round the earth according to the same law of attraction as that governing the fall of the apple. He spent twenty years working out his idea and finally laid down the well-known principle that bodies attract one another with a force directly proportional to the mass and in inverse ratio to the square of the distance. The extraordinary importance of this discovery was by no means immediately clear to everyone; it was at variance with the speculations of the Cartesian philosophers as well as with the doctrines of the theologians, and it was not until after the lapse of several decades that humanity realized that here was a new foundation on which to base the conception of the universe. This was, of course, due partly to the fact that Newton himself remained in certain respects at a somewhat antiquated point of view. Like Galileo, he was quite aware, as he himself says, that it has not been possible to discover "the cause of the qualities of gravitation from phenomena, and I form no hypotheses," and he maintained that it was justifiable to conclude from the existence of properties in those bodies which had been investigated the existence of the same properties in all bodies. At the same time he was firmly convinced that finality in nature presupposes a personal God as the Creator of the universe, and even as the Maintainer of the whole, since the irregularities in the course of the heavenly bodies must some time be adjusted through the personal intervention of the Creator. The latter assumption, which induced Leibniz to liken Newton's cosmic system to a clock which now and then had to be regulated in order to go properly, testifies more clearly than any other factor to that mixture of childlike innocence and intellectual keenness in Newton which gives to the whole of his personality the character of old and new in conjunction, such as is so often to be found in philosophers at the turning-point in the scientific history we are here discussing. It was left to the eighteenth century entirely to shake off the traditional ideas of the structure of the universe and in their place to create that theory of existence which has been maintained ever since. The man who more than any other exerted a decisive influence in this respect is usually, and rightly so, not counted a scientist at all, yet he has in a greater degree than most affected the progress of science; that man was Voltaire.

FRANÇOIS MARIE AROUET DE VOLTAIRE is one of the best-known and most discussed figures in cultural history — uncritically vaunted to the skies by his admirers, violently calumniated by his enemies. In the course of his long life (1694–1778) he exercised, more than perhaps any other has ever done, a purely cultural influence in every sphere of life. His literary, political, and religious activities are universally known. As a man of letters of middle-class origin he had acquired a name in Paris for his cleverness and love of opposition, when he was suddenly ordered by the Government to leave the country and took refuge in England. After a three years' sojourn in that country (1726–9) he returned full of ideas which he had assimilated there, and devoted the rest of his life to making them known to the world. Among these new conceptions were Newton's discoveries in physics and astronomy. By combining these with a number of ideas gathered from Leibniz's speculations he produced a theory of the universe which was not only purely mechanical — that the Cartesian theory had already been — but which was also based upon mathematically calculable facts and was therefore bound to work with indisputable authority. He worked with indefatigable enthusiasm to get this theory known, using his brilliant literary powers and popular gifts to make it attractive; indeed, it was mostly thanks to him that Newton's discovery became known to the world of culture in Europe within a space of a few decades. Thus it came about that Voltaire to a certain degree stands in the same relation to Newton as Haeckel does to Darwin. Voltaire also reminds us of Haeckel in that he made his natural-scientific theory the basis of a comprehensive view of the world, to which he unceasingly refers in his struggle against ecclesiastical authority, whose doctrines of the creation and of miracles he despised and ridiculed from that standpoint. From his time originates the custom of citing "natural laws" as proofs controverting the Church's traditional cosmic theory. Otherwise Voltaire's notions of the universe constitute in themselves no really radical break with the old tradition; he believed both in a personal God and in the causal finality of nature, which to a certain extent contributed towards making the transition from the ancient to the modern cosmic theory less of a shock to the great majority. Furthermore, his doctrines had the rare consequence of bringing about a radical revaluation of the whole of man's ideas of life.

With the coming of this so-called "period of enlightenment" introduced by Voltaire we may regard the conception of nature created by antiquity and handed down through the Middle Ages and after as definitely shattered. It would, however, be an exaggeration to assume that Voltairianism reigned supreme in his period. Besides the adherents of the time-honoured cult of antiquity, of whom there were always a great number, there were to be found during the enlightened period, in ever-increasing numbers, supporters of the mystical-speculative tendencies which have been mentioned above. Throughout the whole era here under discussion a not unimportant part was played by natural-scientific mysticism, influencing even otherwise quite critical people in the scientific world and everywhere attracting adherents, who devoted themselves entirely to its aims and purposes. Its roots, as has already been mentioned, lay far back in time, while its ramifications can be traced even in the field of modern natural research, exact though it apparently is. Its development thus deserves a chapter to itself, the beginning of which must take us back to the days of the Renaissance.

CHAPTER III

Magic during the Renaissance

IN THE FIRST SECTION of this work we have pointed out how important was the role played by mystical speculation in science during the Renaissance, even in the theories of its principal representatives, such as a Cusanus or a Bruno. As a matter of fact, through the break-down of scholasticism in science, the field was left open for all those wild fantasies that seem to be common to all times and generations, although they are at times thrust out of sight and dare not show themselves for fear of learned authority and the derision of critics. Seldom indeed is it that mystical speculation and magical experiments have gone so far and had such scientific pretensions as during the Renaissance.

All this Renaissance magic was based on a great many preconceptions: primitive superstition, cabbalistic interpretation (originating from the East) of the books of the Bible and a number of apocryphal appendices, Arabian experimental science and its Western development. Finally, the neo-Platonic philosophy, striving to gain by means of direct introspection a mystical, uniform conception of the whole of existence — spirit and matter, animate and inanimate things — provided a common speculative framework in which to fit all these various elements. Ideas taking this as their point of departure and objective, foreign though they really are to both the aims and the methods of natural science, have nevertheless had a deep influence on its development; they have induced a striving after a uniform view of life at periods when science threatened to disintegrate into aimless detailed research, and they have produced a love of nature during epochs when humanity had otherwise turned to abstract philosophic speculation.

The Renaissance produced a number of personalities of this mystical, half-experimenting, half-brooding type: the Italian PICO DE MIRANDOLA, the Germans HEINRICH VON NETTESHEIM — called Cornelius Agrippa — and TRITHEMIUS, and many others. They are, however, of but little interest to biological history, although their speculations may have in certain cases indirectly influenced the development of biology during the succeeding era. One of their contemporaries, however, far more radically than they, furthered the progress of biology; a man, moreover, who, owing to the life he lived, is of more than common human interest; namely, Paracelsus.

ALBRECHT VON HALLER

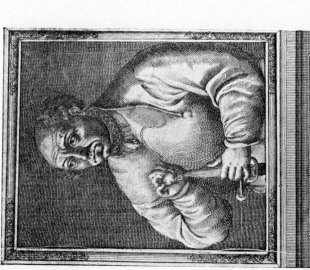

PARACELSUS

THEOPHRASTUS HOHENHEIM was born in 1490, or somewhat later, at the famous monastery for pilgrims at Maria-Einsiedeln in the Canton of Schwyz in Switzerland. The name PARACELSUS, by which he is best known, he himself adopted later.[1] His father, who is said to have been the illegitimate son of a Knight of St. John of the noble family of Bombast von Hohenheim, was a physician at the above monastery; his mother, who was probably a peasant woman, was before her marriage a sick-nurse there. Young Theophrastus, who was called after the great Athenian botanist and disciple of Aristotle, grew up in poverty and remained throughout his life, in spite of his ancestors' nobility, in all respects a child of the people. Nevertheless he received a sound education, partly from his father and partly from two priests, friends of the latter; and as a youth he was a student at Basel. However, he soon wearied of scholasticism, studied alchemy for a time under Trithemius, referred to above — an abbot who had established a laboratory in his monastery — and afterwards became an apprentice at a mine in the Tyrol, where he studied metallurgy and was initiated into the professional secrets of the miners. Even this work, however, did not appeal to him, and he soon joined the hosts of learned adventurers who ever since the Middle Ages had wandered about Europe under the name of *scholares vagantes* or mendicant students. Young Hohenheim took his profession more seriously than most of his colleagues; having wandered through Germany, Spain, and France, he joined the army with which Christian II conquered Sweden in 1520, as a field surgeon. He thus came to Stockholm, proceeded thence to Moscow, from there again — by which route is not known — to Constantinople, and finally returned home. In the course of these journeys he naturally had an opportunity of visiting several universities, but their official learning was of far less interest to him than the experiences he was able to gain of people who were at that time believed to be familiar with the occult sciences: barber-surgeons, witches, gipsies, and executioners. He made such good use of the medical knowledge thus gained that in 1526 he was appointed first town-physician at Basel, with the right to revise the city's pharmacopœia and to hold lectures at the University. As a practitioner he was brilliantly successful with daring cures and simple, cheap medicines, while at the same time he harassed his colleagues by his extremely overbearing manner. He gave his course of lectures in German instead of in Latin and started by ceremoniously burning the classical treatises on medicine, which naturally increased the hatred of the medical profession. When, moreover, the apothecaries, embittered by his sharp criticism, intrigued against him, he had to

---

[1] The name Paracelsus he used as a nom de plume in some of his writings; it probably means "higher than Celsus" (the Roman physician). Another name which he adopted is Aureolus Bombastus, the former meaning "golden," and the latter having been borne by his grandfather. He also sometimes calls himself "Eremita," after the place of his birth.

leave Basel after some years. Thus once more he began his wanderer's life, tramping from town to town in Germany for about ten years, everywhere winning popularity for his wonderful cures, and hatred for his lack of consideration and imperiousness towards his colleagues and patients. There seemed no possibility for him to settle down in peace; rather he had to flee for his life time and again, until finally Archbishop Ernst invited him to settle at Salzburg, which he did about the year 1540. Now at last it seemed that better days were in store for him, but it was not to be for long; in 1541 he suffered a violent death — his calumniators declared it was through an accident when under the influence of drink, but his friends said it was a result of a hostile attack. Before his death he had bequeathed his few possessions to the poor.

Paracelsus's personality

BOTH in life and after death Paracelsus has been very variously judged: by some he has been represented as a bare-faced scoundrel, an impudent trader upon the good faith and superstition of humanity; by others — of early times as well as in our own day — he has been highly extolled as one of the boldest spirits that ever lived and one of the greatest promoters of science. In actual fact, one can find, in his writings and in his life, support for both of these judgments; on the one hand, uncritical superstition, fantastic paradoxes, boundless self-conceit, and shamelessly scurrilous language in controversy; on the other hand, penetrating criticism of his predecessors' theories, and audacious ideas of his own, aiming far into the future. To brand him as a conscious cheat would in any event be utterly unjust; rather he is of a type that has been very common during the *Sturm und Drang* periods of human culture. Throughout his writings we come across the same naïve self-satisfaction, the same pugnacious temperament, as one finds in several romantic writers of the beginning of the nineteenth century; like them he was undoubtedly induced by an honest intention to "explain the whole of nature and to reform the whole world." People with such grandiose aims easily acquire something of the charlatan and humbug in their natures, which, however, does not exclude the possibility of really splendid traits in their character. And such Paracelsus certainly possessed; his kindness towards the poor, his earnest desire to help suffering humanity, his often very poorly requited loyalty towards his friends, are sufficient proof of that. In his writings, moreover, he often praises the medical profession as a high and noble calling, claiming of its members not only knowledge, but also goodness and morality. Side by side with this there appears in him a feeling of self-respect, which sometimes finds worthy expression, such as in his motto: "*Nemo sit alterius qui suus esse potest*,"[2] but more often manifests itself as high-sounding

---

[2] "Let no one be another's who can be his own."

self-praise, as when he says: "My back hair knows more than you and all your hack writers; my shoe-laces are more learned than your Galen and Avicenna." And his confidence in success is unbounded. "I shall be king and the kingdom shall be mine ———. I wish I could protect my bald head from flies as easily as I do my kingdom, and were Milan as secure against its enemies as my kingdom is against you, no Swiss nor *Landsknecht* would find his way there." The coarse expressions with which he spices his controversial writings are such that they could not possibly be quoted. Otherwise his language is vigorous and original; he wrote in German, like his great contemporary Luther, for whom in his writings he expresses the greatest admiration.

Alchemistic conception of the human body and its functions

PARACELSUS's scientific theories are far more difficult to characterize than his personality. As the essential part of his activities he always regarded his medical practice, and his general theories of life invariably have direct reference to diseases and how to fight against them. And, being originally an alchemist, at bottom he despised anatomy as he did all detailed research in general. He sought rather to get the human body and its functions regarded as a part of the world in its entirety and therefore also as dependent upon the cosmic process, as it goes on both on the earth's surface and in the firmament that surrounds it. This was, indeed, just what Aristotle strove after, but while the latter sought to solve the problem by means of a theory which had primary reference to the forms of being, to Paracelsus, who from his early youth had lived in the thought-world of mediæval alchemy, existence represented one single mighty chemical process. The alchemistic thought-structure was, of course, based upon experiments of an essentially magical character and upon speculations of neo-Platonic-oriental origin, especially upon the Hebrew cabbala, with its belief in the secret power of words and graphical signs and their mystical connexion with the things they denoted. All these elements of learned speculation Paracelsus interwove with all that he had learnt of folk-magic during his years of wandering, into a natural-philosophical system of unique character. Its guiding thought probably emanates directly or indirectly from the cabbala — that is to say, the intrinsic connexion that Paracelsus believes is to be found between everything that exists: the celestial bodies, the things on the earth, and human beings. This occult connexion, which Paracelsus considers it to be the function of science to investigate, leads him, the more involved he becomes, into a mysticism which a modern reader will find extremely difficult to grasp even in its main features. And the systematic divisions of the subject, with which Paracelsus is excessively generous, certainly do not make the matter any clearer. In one of his principal works on science in general, which he himself published under the title of *Paramirum*, he divides the causes of sickness, which he always takes as his starting-point, into five classes: *Ens astrale, veneni, naturale, spirituale, deale.* What these different *"ens"* really are one never gets to know; they are apparently mystical powers which produce diseases and which have different origins. *Ens astrale* proceeds from the stars, which have life and can poison the atmosphere, precisely as a person in an unventilated room pollutes the air in it with his breath. *Ens veneni* is a cause of sickness originating in the digestion; each living being has, in fact, his given food, which is made by that being partly into sustenance for the body and partly into a poison, which is expelled through the excretive organs; besides the true excreta, which are the specific poison of the body, quicksilver is excreted through perspiration, sulphur through the nose, and arsenic through the ears — the yellow colour of the cerumen in the ear probably reminded Paracelsus of certain arsenic associations. Thus the ox consumes grass in its own way and man the flesh of the ox in his. Every being has in its body an "alchemist," who directs the work; if he gets out of order, the body becomes sick. In another treatise he is named Archeus, and is apparently to be regarded as a spiritual being, though no exact description of him is given. Paracelsus describes in greater detail the third cause of sickness, *ens naturale*, and here he expounds his real theory of life and the universe. The human body is a microcosm, possessing elements corresponding to all the phenomena of the exterior world, particularly to the heavenly bodies; thus the liver corresponds to Jupiter, the gall-bladder to Mars; the heart is the sun, the brain the moon, the spleen is Saturn, the lungs Mercury, and the kidneys Venus. All these organs perform planetary movements in the body, and if they come into an unfavourable position, disease arises. On the other hand, they are all independent of food and therefore also of the poisons derived from it. Moreover, there are included in the body the four elements, as well as the basic substances of the four temperaments, which are like the gustatory impressions, sour, sweet, salt, and bitter. All these likewise circulate and give rise to disease. In regard to *ens spirituale* Paracelsus emphasizes the difference between soul and spirit: the soul is a work of God, but the spirit is created by the human will and by means of it man can influence his fellow men. Thus disease can be occasioned by men's hatred; if an enemy makes an image of wax and maltreats a part of it, his action induces suffering in the corresponding part of the person he desires to persecute — a method to which witches are particularly partial. *Ens deale*, finally, is the divine will itself, which gives sickness and health as may seem good to it; against the divine will medicines are of no avail, but only piety and prayers.

Paracelsus's influence

IN other writings Paracelsus expounds his theory of the connexion in nature in different directions; one describes the doctrine of the "signatures" in plants and their connexion with diseases, as, for instance, that Hypericum, owing to its perforated leaves, cures wounds from stabs; the peony, owing to its cerebral-shaped pistil, cures paralysis of the brain, etc. To go deeper into these fantastic ideas is hardly worth while; what has already been stated may seem to many readers more than enough of such nonsense. Nevertheless, Paracelsus's influence has been both wide and deep. In medicine he was in many respects a pioneer; he did his best to treat wounds hygienically and otherwise to leave them in peace; owing to his belief in specific causes of diseases, he sought for a single cure for each disease, and several of his methods proceeding on these lines still hold good, especially the use of mercury in the treatment of syphilis; moreover, this aim of his formed a contrast to Galen's and his followers' disastrous attempts to create universal medicines to be used for all diseases and composed of everything imaginable. Even modern biology is based, at least in one respect, on a principle laid down by Paracelsus; his conception of life-phenomena as fundamentally chemical processes has without doubt paved the way for modern physiology, which certainly could not develop before chemical science had been freed from the primitive mysticism in which it was veiled in the time of Paracelsus and for a hundred years after him, but which in any case represented a more stimulating starting-point for the modern idea of substance conversion in the body than the Aristotelean "cooking" theory, which even Vesalius accepted. Further, the conception of life as a mystical force uniting the whole of existence, which Paracelsus, it is true, did not actually found, but developed and stamped with his own original personality, has never, in spite of its many fallacies and absurdities, succeeded in being entirely suppressed. Time and again it has been thrust aside by some more exact research based upon actual facts and referred to the vast public of the dilettanti and mountebanks, but it never really died out, so that at certain times — for instance, during the romantic period at the beginning of the last century — it revived with renewed vigour. And during such periods Paracelsus's reputation has been freshly enhanced. At all events, history cannot but acknowledge the fertile genius, the splendid character — in spite of its many exaggerations — and the force of will with which Paracelsus throughout a life of adversity and distress fought for what he considered to be the supreme aim of science.

Paracelsus was not very fortunate in his disciples. Cultured people could not endure his presence for long, and the riff-raff he gathered about him during his wanderings was not suitable material for a scientific school. His chief influence was exerted by his writings, which were read eagerly and produced a mass of imitations, wherein the defects rather than the merits of the true Paracelsian writings were conspicuous, and which, published under the master's name, contributed more than anything else to lower his reputation. The principal successor to Paracelsus appeared about a generation after his death, a philosopher who extracted from his writings and still further developed his peculiar conception of life and its functions.

JEAN BAPTISTE VAN HELMONT was born at Brussels in the year 1577, of a noble and wealthy family. Left fatherless at an early age, he became a precocious child, and by his seventeenth year he had already completed his university studies in philosophy. This, however, failed to satisfy him; he entered a Jesuit college and there studied theology, especially that of a mystical character, and this puzzling over life's problems entirely obsessed his mind. He eagerly studied the neo-Platonists and Paracelsus, for whom throughout his life he expressed a keen, but by no means uncritical admiration. At the age of twenty-two he took the degree of doctor of medicine, spent the next few years in travelling through different countries, and then contracted a wealthy marriage and settled down on an estate in his home district, dividing his time between scientific research and splendid acts of benevolence. He carried on his medical practice simply for charity and without any fee; every offer of permanent employment, even the most brilliant, he firmly declined. He died in 1644.

As already mentioned, van Helmont regarded Paracelsus as his master and undoubtedly had a certain affinity with him. True, he possessed none of his precursor's intrepid geniality, but he was far more cultured, both scientifically and socially. In his personal character he was mild and lovable, but he seems to have been of a nervous disposition, which gave his mystical speculations a peculiar quality of exaltation. He had spiritual visions, which he produced by auto-suggestion through gazing at some strong source of light; he employed in his researches for obtaining scientific results a direct introspection which he achieved by exalted concentration of thought, both during the day's work and in the night's rest, when in a state between sleeping and waking he received inspiration, by which he set great store. This inspiration, however, often led him widely astray, as when he believed that he had succeeded in producing rats in a vessel in which some rags and bran had been kept, or when he imagined he had converted quicksilver into gold, a discovery which so delighted him that he had his son, who was born just at that time, christened Mercurius.[3] But fortunately he had also better inspirations. One of these was his determined opposition to the classical authorities Aristotle and Galen. Their theories he believed it was his mission in life to challenge, both because they led to practically worthless results and because they were pagan. This latter reason is characteristic. While the man of the Renaissance period, Paracelsus, scorned the classics because they were antiquated authorities and stood in the way of his personal ideas, the emotional disciple of the Jesuits, van Helmont, felt himself moved first of all to substitute a Christian for the heathen science. This induced him, however, to make

---

[3] This Franz Mercurius van Helmont devoted himself even more exclusively than his father to purely mystical speculations. The fact of his having published the latter's complete works constituted his greatest service to science.

what in many respects was a sound criticism, particularly of Aristoteleanism, the weak points of which he brought out very clearly. In his writings he closely examines Aristotle's theory of the relation of form to reality. Similarly, he rejects the theory of the four elements, for, said he, fire is not an element at all. Again, when Aristotle believes the fertilizing property of the sperm to be due to its heat, van Helmont satirically asks how it is that the cold-blooded fishes are more fertile than any warm-blooded animal. However, it proves here, as often, to be easier to destroy than to build up; van Helmont's own theories cannot compete with Aristoteleanism at least in clearness and consistency. This, indeed, is due partly to the peculiar mystical principles on which he bases his views, and partly to his lack of stylistic ability; his writings are extremely obscure and difficult to read. His conception of nature, like that of Paracelsus, is chemical, with a strong dose of mysticism, which belonged to chemistry at that time.

Van Helmont's fermentation theory

THE "fermentation process" plays an important part in his theories on natural phenomena; he had thoroughly studied this phenomenon and had shown that fermenting must produces a kind of air which is identical with that given off by burning charcoal and that which sometimes renders the air of caves irrespirable. For this element he invented the name of *gas*, a word which has since been accepted by science. He distinguished several kinds of gas, of which, however, only the above-mentioned "gas sylvestre," or what we call carbon dioxide, has been fully described. According to him, digestion and every kind of conversion of substance in general are due to ferments. The many different processes of fermentation which he thought he had discovered in the human body were for the most part mere creations of his fancy, but he had certain ideas which were to prove to be correct, as when he points out the part played by acid in the cavity of the stomach in the digestive process, and shows that the undue acidity of the digestive juices is neutralized by the gall. He was not content with merely establishing facts of this kind, however; like Paracelsus, he sought to get on the tracks of life itself and, like him too, saw its innermost essence personified in an *archeus*, which is situated in the region of the stomach and controls a number of subsidiary *archei* in the other parts of the body. This central *archeus*, however, regulates only the material conversion in the body and exists in various forms in all beings; man has, besides his immortal soul, "*intellectus*," which makes the soul participate in blessedness and would wholly control the body had not the Fall intervened. After the Fall man received a lower soul, "*ratio*," which makes it bound to earth and liable to its impulses, and finally to death. Beings in the universe have the power of reacting upon one another by a kind of force operating at a distance which he calls "*blas*"; in particular the *blas* proceeding from the heavenly bodies has remarkable properties, but it would take too long to enter into fuller details. Van Helmont regards water as the material fundamental element underlying the whole of existence; out of water arises everything on earth, both inanimate and animate substance. As a proof of this theory he cites an experiment: he filled a bowl with 200 pounds of dry soil and in it planted a willow, weighing 5 pounds, and watered it with rain-water. After five years the willow-tree weighed 164 pounds while the soil, when again dried, weighed about 200 pounds less 3 ounces. Thus, he argued, the entire willow-tree was formed of rain-water. This experiment, which indeed was perfectly correct in both plan and execution, although the conclusion he drew from it was wrong, testifies better than anything else that van Helmont was nevertheless a true pioneer in the field of natural science; to have thought out the first biological experiment based on quantitative calculations is a service to science which well compensates for many mistakes in the theoretical sphere. Even as a medical practitioner van Helmont showed the same curious mixture of fancifulness and foresight: on the one hand he worked at a mysterious universal medicine which he called "alkahest," while on the other he strongly protested against the abuses common at that time in connexion with violent blood-letting and strangely concocted, excessively strong medicines. Moreover, like Paracelsus, he undoubtedly exercised a strong influence, for both good and evil; his fertile, far-seeing ideas proved of value far into the future, and traces of his mystical fancies can even be found in scientists of subsequent generations, a number of whom will be mentioned in the next chapters.

CHAPTER IV

BIOLOGICAL RESEARCH IN THE SEVENTEENTH CENTURY

1. Harvey's Successors

IN THE FOREGOING have been described the two entirely contrasted natural systems which appeared independently in opposition to Aristoteleanism: the mechanical conception of nature, and the mystical view of life. As has been shown in the first part of this work, the foundations of the mechanical view of natural phenomena were laid by Harvey, who proved that the circulation of the blood, which had up to that time been regarded as an expression for certain vital spirits, goes on as a purely mechanical process. Although himself a convinced disciple of Aristotle, Harvey thereby laid the foundations of that modern scientific theory of the phenomena of life which follows the same methods in dealing with them as those applied to the investigation of phenomena in inorganic nature. This discovery of Harvey's created an immense sensation; during the immediately succeeding decades after its publication (in 1628) it was the one great question of the day and occasioned a vast quantity of literature both for and against it. Its overwhelming truth, however, soon silenced all opposition; the conservative adherents to the old system gradually died off and the young research-workers were easily won over to the new view and devoted themselves to gathering fresh proofs of its validity. How successful it was is best evidenced by the extraordinary stimulus given to the study of anatomy during the middle and latter half of the seventeenth century. This period, perhaps more than any other, can be regarded as one of brilliant anatomical achievement, to which the preceding era, beginning with Vesalius's revolutionary inventions in the field of technique and methods of observation, impresses one mostly as being a period of introduction. A comparison between these two epochs also produces a remarkable contrast of a national character; while during the Renaissance Italy was the sole centre of anatomical research, its range had now spread northwards: now for the first time England, Holland, and Scandinavia begin to make definite contributions to the development of biology. And simultaneously with this shifting of the centre of biological research we find another change appearing in its conditions, first in Italy and later the farther north we go. During the sixteenth century the natural philosophers were still mostly university teachers; such had been both Vesalius and Galileo. In the seventeenth century, on the other hand, and still more so in the eighteenth century, the universities cease to be the centres of scientific progress and become instead the seats of unproductive conservatism, mechanically repeating the formulæ inherited from the Middle Ages; the real pioneer scientists are now private scholars. Descartes, Spinoza, Leibniz, as well as Harvey and van Helmont, all worked, as we have seen, independently of the universities, as we shall also find did several of the leading scientists among their successors, in both the seventeenth and the eighteenth centuries. A new type of bond of association between men of learning came to be established in connexion therewith — namely, the scientific societies. Such "academies" were founded during the seventeenth century all over Europe, earliest in Italy, afterwards in all countries north of the Alps. Princes and distinguished people allowed themselves to be nominated as patrons or to be elected honorary members, thereby acquiring an interest in the study of nature. To promote this study they established laboratories and made collections of natural objects — so-called "curiosity cabinets" — mostly, it is true, as the name implies, for their own amusement, but still in many cases for the benefit of science, owing to the possibilities they offered for research and the grants of money made by scientists in connexion therwith. All this naturally increased, as it were, the social reputation of science and in this respect offered a decided contrast to the Renaissance; whereas then the students of nature had to live in inferior positions, during the seventeenth and eighteenth centuries many of them held important posts in the community. The period now to be described was thus in all respects a brilliant one for natural science — a period which has no counterpart until we come to the latter half of the nineteenth century.

Discovery of the lymphatic system

As Harvey's immediate successors it is fair to regard those scientists who, like him, studied the vascular system in man and the higher animals and thus continued along the path he initiated. Mention has already been made (Part 1, p. 113) of how, even during Harvey's lifetime, a hitherto unknown type of vessel was discovered — namely, the lymphatic duct system — and how Harvey adopted an attitude of complete ignorance on the subject and adhered to the ancient tradition. As a matter of fact, a year before the publication of Harvey's treatise on the circulation of the blood there appeared a work on "the lacteal veins — a new discovery." — The author, who had died the previous year, was an Italian physician named Gasparo Aselli (1581–1626). He had begun his profession as an army surgeon and afterwards became for a time professor of anatomy at Pavia, but finally practised in Milan. There he once carried out, together with some of his colleagues, a vivisection operation on a dog which had just before had a substantial meal, and thereupon found the peritoneum and intestines covered with a mass of white threads; he casually cut one of them off and saw that a fluid oozed out; they were thus not nerves, but vessels of a kind hitherto unknown. Influenced, however, by the traditional Galenian idea of the liver as a blood-former, Aselli assumed that these vessels, "chyle vessels," as they are now called, extended from the intestines to the liver. Nevertheless, the discovery aroused general interest. Aselli's book was reprinted several times, and a number of contemporary anatomists interested themselves in this new discovery. Among them there is one who deserves special mention — JOHANN VESLING (1598–1649), who, though born in Germany, became a professor at Pavia and in a handbook of anatomy, printed in 1647, gave a detailed account of the chyle vessels (lacteals). Twenty years after Aselli's work had been published, however, a young student by the name of JEAN PECQUET (1622–74) made a discovery which considerably enhanced the knowledge of the newly-discovered vascular system. While performing a dissection he found the canal which forms the common trunk of the lacteals and the lymphatics — the so-called *ductus thoracicus*. This contradicted Aselli's and his immediate successors' belief that the lacteals lead from the intestinal canal to the liver. Pecquet was born in Normandy, studied at Montpellier, and eventually became body-physician to the minister Fouquet, who was all-powerful in the early youth of Louis XIV. When Fouquet was sentenced to imprisonment for fraud, his physician had to go with him, and after that he disappears from history. He is said to have had a blind confidence in the power of brandy to cure all manner of diseases — an illusion which rapidly brought him to ruin. It was instead in Scandinavia, hitherto unknown in science, that the problem of the lymphatic duct system was finally solved, and it was a strange coincidence that two scientists from different countries should have quite simultaneously and independently of each other attained the same result.

THOMAS BARTHOLIN was born in Copenhagen in 1616. His father, Caspar Bartholin, was professor of anatomy and a distinguished scientist of the old school. Having matriculated in his early youth and learnt all that his own country could teach him, young Thomas at the age of twenty started out on his travels, which lasted for nine years. First he studied for three years at Leyden and there became acquainted with Harvey's discoveries, after which he worked for two years in the anatomical theatre at Pavia, moved on to Naples, where he was a pupil of the old Severino, then gave a dissertation for his doctor's degree at Basel, and did not return to his native country before a professorship was assured to him. As professor of anatomy he did splendid work, which within a short time made the unknown Danish university famous throughout Europe; foreign pupils flocked to him, among whom may be specially mentioned the German MICHAEL LYSER, who as prosector used to perform Bartholin's most important dissections, and afterwards became professor at Leipzig. A number of valuable works were published by Bartholin in the sixteen-fifties, but his productive powers very soon waned. After Lyser's resignation from the post of prosector (1652) little occurred there in the way of fresh anatomical results; as early as 1660 Bartholin was completely relieved of all instructional and examination work, and then he had only his membership of the board of the university, of which he seems to have taken advantage mostly for the purpose of providing his relations with good appointments. With this end in view he is believed to have passed over deserving pupils, including Steno, who is mentioned later on. Other not very attractive characteristics of his are also mentioned, such as that, physician though he was, he once fled from the city during a plague for fear of infection and retired into the country; again, the way in which he procured extra salary from the Government caused him to be censured on several occasions. His writings abound in expressions of the most extravagant self-praise, though he displays in them a really genuine respect for science — undoubtedly the most attractive trait in his character. He died in 1680 and was succeeded by his son Caspar, who carried on his work, reaping a number of external honours, but in no wise possessing his father's merits.

Bartholin discovers the lymphatic system

THOMAS BARTHOLIN's most important achievement is generally considered to have been his clearing up of the mystery of the lymphatic duct system. He was not aware of Pecquet's discovery when he began to study the chyle vessels (lacteals), and therefore believed, with Aselli, that they led to the liver. By observation and experiment, however, he soon found that this was not so and at the same time discovered that these vessels were connected with a vascular system distributed throughout the entire body, and containing a fluid clear as water. These discoveries he described in a treatise which came out in 1653, in which he declares that the liver cannot perform the blood-forming function which classical anatomy had ascribed to it; nor, since the chyle vessels do not connect with the liver, but on the contrary a number of lymphatic ducts go in the opposite direction, can the food be converted into blood in the liver, as the anatomists up to the time of Harvey and Aselli, and the two latter as well, had concluded. He closes his essay by erecting a monument to the liver, the body's now dethroned ruler, in the form of a parody of the high-falutin memorial speeches which it was the custom of the time to make in honour of the distinguished dead.

Bartholin set great store by his discovery of the lymphatic system and wrote several fresh articles on the subject, without, however, making any very important additions to his first account. The rest of his scientific literary work, which is rather extensive, cannot be compared from the point of view of interest with that dealing with the discovery which he rightly regarded as his greatest contribution to the development of anatomy. Consequently he must have been extremely chagrined to find at the very start a competitor for the honour of having made the great discovery — a youth into the bargain, without any previous successes to his credit, either scholarly or literary.

OLOF RUDBECK was born at Västerås in 1630, the youngest but one of eleven children. His father was the imperious Bishop JOHANNES RUDBECKIUS, a man of deep learning, great powers of leadership, and corresponding ambition. In his diocese he had founded a college, which, as far as education was concerned, could compete with the University. It possessed both a library and a botanical garden. There young Olof received a thorough education, and there too he undoubtedly acquired that love for natural science which induced him, immediately after his entry into the University, to take up the study of medicine, which was at that time in not very high favour. Having matriculated at the age of seventeen, he felt himself at once moved to begin to work on his own account, which the poorly equipped faculty of medicine rendered it absolutely necessary to do. Like Vesalius, he devoted himself to the dissecting of animals and in doing so was initiated almost at once into the then newly-discovered and interesting chyle system. The observations which he had made within a very short time in this field created such a sensation that Queen Christina herself desired to become acquainted with them. In 1652 young Rudbeck was given an opportunity of demonstrating his experiments before the Queen and as a reward was given a grant to enable him to travel abroad. Before starting he published his observations in the form of a dissertation, in the year 1553, and then went to Leyden, where he studied for three years. When he returned home, he was appointed professor of anatomy and now devoted himself with great energy to reforming the system of medical education, which had till then consisted mostly of lectures on the writings of the ancient authorities. Rudbeck built after his own design a splendid anatomical theatre, which still exists, and there carried out, as often as material was available, dissections of human bodies. Exercises of this kind had never been seen before in Upsala and they consequently aroused bitter opposition, but Rudbeck did not allow himself to be deterred; on the contrary, he openly showed his contempt for his opponents' prejudices. To ridicule them he once caused the remains of a criminal which he had dissected to be buried with great pomp and drew up for the ceremony a program in which he delightfully parodied the academical rhetoric of the period. However, this educational work made too great demands on even his extraordinary energy to allow him time for scientific research; a book on general animal anatomy which he intended to publish was — undoubtedly to the immense loss of science — never written. His childhood's interest in botany bore fruit, for he devoted much of his time to the production of a large work composed of botanical engravings and entitled *Campi elysii*, but this was never completed ,either, owing to the fact that he allowed himself to be attracted to a new sphere of activity; he became engrossed in that extraordinary, colossal, linguistic-archæological-patriotic work *Atland*, in which he seeks to prove that Sweden is the oldest civilized country in the world. He died in 1702, having shortly before seen a great part of his scientific production go up in flames in the conflagration which destroyed Upsala in the same year.

Discovery of the "vasa serosa"

THE anatomical work of his youth is, however, all that justifies Olof Rudbeck's being regarded as the earliest in the long line of eminent naturalists that Sweden has produced. In his first dissertation in 1652 he gives an account of the circulation of the blood in the true spirit of Harvey and presents a number of theses, one of which denies the existence of any spirit in the body other than the animal, while another denies the property of the liver as a blood-forming organ — proving that he was far in advance of Harvey's standpoint. In a dissertation printed in the following year he describes the lymphatic duct system, which he had independently discovered when attempting to ascertain the structure of the lacteals; he describes the course of these "*vasa serosa*," as he calls them, not only within the trunk, but also in the extremities; he gives an account of their distensions, the lymphatic glands; he observes the nature of the lymphatic fluid — that it is salt to the taste and coagulates in cooking; he tries to ascertain the movement of the fluid in the vessels by observing their valves; and, finally, he endeavours to work out a theory of the significance which the entire system has for the body.

Between the two rivals for the honour of having discovered the lymphatic duct system in its entirety, Bartholin and Rudbeck, there ensued a struggle for priority, carried on with the aid of pamphlets, and breaking out into mutual recriminations of a not very attractive character. National chauvinism took a part in the game, and the question was debated long after the death of the parties to the dispute, until finally an impartial examination of the documents put an end to the controversy. After having investigated the matter R. Tigerstedt came to the conclusion that Rudbeck was the first to make the discovery, but that Bartholin had the prior claim as regards the date of publication. That these two eminent scientists both came to the same result independently of each other would now appear to be beyond all doubt.

The discovery of the lymphatic duct system constituted a great and important work supplementing Harvey's discovery of the circulation of the blood. It was possible now for the first time for research to grapple with the problem of how the food substances in the digestive canal are utilized by the body, a subject on which the biologists of antiquity had as confused ideas as they had on the movement of the blood in the veins. Both Rudbeck and Bartholin, indeed, were prepared to draw the obvious conclusion from their discovery: since the lacteals did not, as even Aselli believed, lead to the liver, but just the opposite, it was impossible for that organ to play the all-controlling part in the digestive process which the ancients had ascribed to it. Both Rudbeck's thesis and Bartholin's funeral eulogy on the liver were confirmations of this. It must be admitted, however, that in this both scientists to a certain extent overshot their mark; research work in recent times has revealed that a large portion of the substances from the intestinal tube — carbo-hydrates, albuminous substances, and others — are received by the ramifications of the cystic vein in the digestive canal and conveyed to the liver, where they are converted. It would of course be absurd, however, to ascribe to the opponents of Rudbeck and Bartholin greater prescience in this respect; conservatism which can give no other reason for adherence to the past than respect for tradition has no historical justification in face of the pioneer who bases his ideas on newly-discovered facts, even though he often overestimates their fundamental value.

As has been mentioned, anatomical research during the middle and close of the seventeenth century was practised with special keenness in England; Harvey's achievement acted as a stimulus particularly on his own countrymen. Space does not allow of a discussion of all the eminent discoverers in this field of science which England produced during the era in question; one or two of the most representative will be given here as examples.

FRANCIS GLISSON (1597–1677) was the son of a landowner; he studied at Cambridge, first philosophy and afterwards medicine. He took his doctor's degree in 1634 and as early as two years later became a professor. The Civil Wars, however, soon compelled him to abandon his educational activities; he moved to London, where he became a practitioner of repute and one of the first members of the Royal Society, the scientific association founded in 1660, which has ever since been the most distinguished centre where the scientists of England have gathered. Besides some purely medical works, which were excellent for the period in which they were written, Glisson published two pioneer works on anatomy, the one dealing with the anatomy of the liver, the other with the stomach and intestines. The first gives a monographical account of the liver which, for the then prevailing conditions, was an exemplary work and laid the foundation on which the modern knowledge of the anatomy of this organ rests. In memory of this, the subperitoneal tissue of the liver is to this day called Glisson's capsule. But the author does not merely give a detailed description of the liver; he also expounds a general biological theory in connexion therewith, in which he entirely adopts Aristotle's standpoint. In the component parts of the body he distinguishes matter and form and describes them entirely in the Aristotelean spirit: matter, that out of which something is produced; form, the change in matter which serves a given purpose; that which produces form is either nature or art. He emphatically protests against the "physical" conception of form and matter which the "philosophers" advance. His conclusions are formed on the scholastic model, and the physiological problems that he sets up he solves by a purely abstract method. Nevertheless, there are among his ideas one or two which remind one of modern lines of thought, such as his belief that the evacuation of the gall-bladder is caused by nervous irritability. And when it comes to the lacteals, he shows a knowledge and a comprehension of Pecquet's and Bartholin's discoveries. He thus exhibits in his views, as often happens in transitional periods, a mixture of old and new.

Glisson's younger contemporary and friend THOMAS WHARTON was a specialist pure and simple. Born in 1614, likewise the son of a landowner, he took the degree of doctor of medicine at Oxford and then practised in London, finally becoming head of a hospital there and a highly reputed member of the College of Physicians. He died in 1673. His fame as an anatomist rests on his *Adenographia*, a work in which is given for the first time a comparative account of the glands of the body. In this book Wharton seeks in the first place to explain the actual term "gland" and assumes secretion to be the essential criterion of it; he lays special stress on the difference between the "viscera," or intestines, and the glands. The tongue is not a gland, but a muscle; nor is the brain a gland, but a special, "precious" substance. As belonging to the true glands he characterizes the digestive, lymphatic, and sexual glands. Wharton discovered the exit of the submaxillary gland — it now bears his name — and he also for the first time gave a detailed description of the pancreas. The kidneys, the testes, and the thyroids are also carefully described. With regard to the *glandula pinealis*, Wharton denies its quality of a soul-organ, as maintained by Descartes; he considers it to be an excretal gland, to which the nerves from the brain drain off waste products, which are then removed by the blood-vessels — a curious conclusion in regard to internal secretion, formed more than two centuries before this process was established in our own time. The hypophysis, according to Wharton, possesses similar functions, though with a different kind of ejecting apparatus. Wharton does little in the way of theoretical speculation; on the other hand, investigation and discussion of disease conditions in the glands play an important part in his work.

A more weighty personality we find in THOMAS WILLIS. The son of a farmer, he was born in 1621, studied in Oxford, and during that time fought in the ranks of the Royal Army against the Parliamentary troops. Having taken his medical degree, he worked as a medical practitioner until, in 1660, upon the victory of the King's party, he obtained a professorship as a reward for his loyal conduct. This appointment, however, he did not retain for long, but he moved to London and again set up in practice; there he made a great reputation, became a member of the Royal Society, and published several treatises. He has been described as an honest man of firm character, towards the end of his life not very popular at court, owing to his outspoken criticism of its corrupt morals. He died in 1675.

Willis's principal work is his investigation of the anatomy of the brain and the nervous system. His exposition of the structure of the brain and of the nerves proceeding directly therefrom is the first that may be said to fulfil modern requirements; to his description of the outward configuration of these parts posterity has had very little to add. He performed a still further service to science in having paid special attention not only to the human brain, but also to that of other vertebrates: in the introduction to his work he expressly points out that comparative anatomy alone can provide a fully satisfactory explanation of the structure and functions of the organs. The account of his investigations into the brains of different animal forms is illustrated with very fine engravings, a number of which were drawn by his friend, the famous architect, Sir Christopher Wren, the designer of St. Paul's Cathedral in London. With regard to the functions of the nervous system, Willis, in contrast to the Aristoteleans Harvey and Glisson, associated himself entirely with the theory advanced by Descartes. The manifestations of life are induced by currents in the nervous system, which penetrate the brain and according to their nature are distributed into the different parts thereof; the world of ideas and the memory he places in the cortex of the great brain. Willis thus forestalls Swedenborg's radical investigations into the localizations in the brain, which are certainly considerably deeper and more detailed than his precursor's. Willis also carefully studied the functional spheres of the different nerves; thus, by binding up the vagus nerve in a live dog, he established their influence on the lungs and heart.

In a work published at a later date Willis deals with the soul of animals. This work has a far wider range than its title implies, for it contains a mass of information of various kinds. Thus, he gives an account of a number of investigations he made into the anatomy of invertebrates, which would have been of interest to his age had not at the same time Malpighi and Swammerdam given far better accounts. The main purpose of the work, however, is a comprehensive study of the vegetative and sensitive soul, which in his view — and in that of Descartes — is common to the animals and man. Man has, indeed, his rational soul as well, which is immaterial and can therefore survive after death; the soul under discussion here is, on the other hand, that material vital spirit which finds expression in currents in the nervous system and which produces all the manifestations of life in animals and those that are purely animal in man. How it happens that animals can in some cases perform actions which indicate a conscious intelligence is a problem which causes the author considerable difficulty; true reason cannot exist in animals, for then they would soon come to resemble man and would, moreover, be immortal; it must therefore be the material vital spirit that, either instinctively or mechanically, directs their actions. In endeavouring to apply this theory Willis becomes involved in a maze of speculations that it would be hopeless, and indeed would take too long, to follow in all the subtleties to which they give rise; the numerous learned authorities whom he quotes merely give additional confirmation of the state of helpless confusion into which psycho-physiological speculation had drifted ever since it left behind it the safe harbour of Aristoteleanism. When a scientist of Willis's rank can seriously discuss the question whether the vital spirit can be compared with *spiritus vini* or hartshorn oil, it is easy to realize to what hopeless lengths the natural-scientific speculation of that age could go. Nevertheless, there were even at that time students of nature who applied with far more substantial results the newly-discovered exact method of research in the sphere of biology. Examples will be given in the next section of this chapter; first, however, we may give one more example of radical anatomical research during this epoch.

RAYMOND VIEUSSENS was born in 1641, of a military family; he devoted himself to medical studies at Montpellier and after having graduated became a doctor at a hospital there. He paid specially keen attention to the study of the structure of the nervous system and after many years of preparatory work published his great *Neurologis universalis* in 1685. This brought him immediate fame. He was summoned to the court in Paris and was for a time physician there, but afterwards returned to his old post, which he retained until his death, in 1715. His description of the nervous system is remarkable for its unprecedented accuracy and completeness in its anatomical details; for us it is of special interest as representing the foundation upon which Swedenborg based his ingenious speculations upon the connexions of the nervous system, which Vieussens studied and illustrated with great exactitude. Unfortunately he, too, became involved, with but little success, in physiological speculations upon the "*spiritus*" of the nervous system, which he believes to be secreted in the brain from the blood circulating through it to the latter, as well as upon a "*spiritus nitro-aerius,*" contained in the blood itself; a kind of acid component thereof, which is drawn from the air and from the constituents of food. As a result of these ideas he became involved in a tedious controversy. He also brought out a couple of anatomical works on the heart and the vascular system, but they are of less value than his neurology.

2. Attempts at a Mechanical Explanation of Life-phenomena

GIOVANNI ALFONSO BORELLI was born at Naples in 1608. His father had been an officer in the service of Spain. At an early age young Alfonso showed a marked genius for mathematics; in order to complete his education in this subject he went to the University of Pisa. There Galileo had once been professor and he now lived, an honoured and influential man, as court astronomer in the neighbouring city of Florence. It was not surprising, therefore, that Borelli was won over to his physical and astronomical theories, and he entered with ardour into the new field of research which they opened up for him. After a period of deep study he was elected professor at the University of Messina by the Government of his country, and there he gave instruction for some years. Conditions at that university, however, were limited and offered him no scope, wherefore Borelli returned in 1656 to Pisa, whence he was summoned to Florence in the following year. In the latter city the disciples of Galileo had founded a free academy, called "Accademia del cimento"; here Borelli found employment and worked for ten years. He went in seriously for medical studies with a view to applying Galileo's physical principles to medicine. Unfortunately for him, however, he was induced by the promise of a higher salary to return to Messina. As a matter of fact, shortly afterwards, in 1674, the inhabitants of that city formed a conspiracy against Spanish rule in Sicily. The rebellion having been quelled, Borelli, who had leagued himself with his countrymen, had to save himself by flight. Ruined, the already ageing philosopher arrived at Rome and there at first obtained employment in the service of Queen Christina of Sweden, who in her exile, as once she had done in her native land, loved to surround herself with distinguished scientists. For some years he remained her private physician and actually dedicated the best of his writings to her. But a fresh misfortune befell him; through the dishonesty of a subordinate he was again ruined, and the Queen, whose own affairs were, as is well known, in a bad way, was unable to assist him. He had to take refuge in a monastery, and there he died, in 1679.

The movements of animals

BORELLI's restless life may possibly have been due to his temperament, which is said to have been reserved and morose. However, he enjoyed a universal reputation as one of the foremost scientists of his time, and his literary production was of an exceptionally many-sided character. He carried on his teacher Galileo's work in physics and astronomy. But his books on these subjects, however valuable they may have been, were entirely overshadowed by his great biological work, *De motu animalium*, which was published in the same year in which he died, dedicated, as mentioned above, to Queen Christina, who, according to the preface, defrayed the cost of printing. Through this work he ranks by the side of Harvey as one of the leading pioneers of modern biology.

"As is generally done in other physical-mathematical sciences, we shall endeavour, with phenomena as our foundation, to expound this science of the movements of animals; and seeing that muscles are the principal organs of animal motion, we must first examine their structure, parts, and visible action." In these words Borelli states his views on the function of biology and thereby declares his starting-point to be Galileo's conception of nature, and the work reveals the fact that here, as with Galileo, we are face to face with "a new science." Even the very arrangement is original: by means of short sentences, which, on the model of Euclid, are called propositions, with accompanying proofs and corollaries, the inquiry is led from the simplest element of the motory system, the individual muscle, gradually to more and more complicated organs and organic systems, until finally the whole of the being's power of movement is described in the form of a summary. First of all, of course, he discusses the movements in man, to whom the lion's share of the work is devoted, after which he studies the movements of mammals, the flight of birds, the swimming of fishes, and even the characteristic movements of insects and others of the lower animals. The first propositions are introduced by an analysis of the actual muscular substance, Borelli maintaining that these elements of bodily movement are identical with the flesh — which the Aristoteleans denied. Then he describes in detail the different mechanical functions of the muscles, this being clearly explained in schematic form, with figures appended. On this basis he then proceeds to a study of the different forms of movement: first the individual extremities, then the movement of the whole body under different kinds of action, lifting, walking, running, jumping; and even, what was for a southerner an unusual form of motion, that of skating is observed and analysed. After the movements of mammals, as above mentioned, have been examined, he describes birds' power of flight in comparison with the foregoing movements, and finally he analyses the action of swimming, in which he pays attention not only to the motory system of fishes, but also to the possibilities of man and other land-creatures in this respect. In connexion herewith Borelli puts forward suggestions for a diver's dress and a submarine vessel; whether he was ever in a position to make practical tests of these inventions history does not relate.

Muscular physiology

In another part of his work Borelli seeks to explain the causes of muscular action. For this purpose he first of all tries a number of purely mechanical alternatives, which he rejects, among them being the possibility of the muscle's being shortened merely by contracting its mass, produced by the concentration of its smallest particles. Such takes place, according to Borelli, when a piece of red-hot wire becomes shorter on cooling, but this cannot be assumed to take place in muscular contraction, which, indeed, occurs as the result of impulse from the nervous system — that is to say, through fluid flowing to it from without. Here obviously comes in Descartes's theory of the currents in the nerves; on the basis of this theory Borelli concludes that the swelling of the muscles upon contraction is caused by a process of fermentation, which arises when the blood in the muscles is mixed with the nervous fluid flowing into it. For the fact that Borelli was unable, with the means available at that time, to explain the extremely involved co-operation of physical and chemical processes which constitute muscular contraction, he cannot, however, be blamed; on the contrary, his assumption that a complicated chemical action and not a simple mechanical one is here involved must be admitted to be an inspired presentiment of what our modern science has at last definitely established. After having then given an account of a thorough investigation of the muscular mechanism of the heart and the respiratory system, Borelli concludes his work with some speculations on the subjects of digestion and fertilization, which are partly based upon the opinions of his precursors and are otherwise not very successful, so that they may be passed over here. The same is true of his purely medical speculations, such as his theory that fevers do not originate in the blood, but in the nervous fluid, and other assumptions in connexion therewith.

Borelli creates experimental biology

BORELLI was above all a mechanician and his greatness lies in his having created experimental biology operating with purely mechanical forces. In the introduction to his work, it is true, he gives the assurance that all the mechanical phenomena in the living body, which he proceeds to describe, are produced by the vital spirit — had he not admitted this, he would certainly never have obtained the papal authority to publish his work, which now adorns the first page of his book — but having once made this theoretical reservation, he carefully avoids in the discussion that follows the inclusion of any other points than the purely mechanical. And it is just for this reason that, in spite of occasional weaknesses, his work stands out as the first to apply all through the fundamental principle on which modern biology is based.

Borelli was highly appreciated by his own and the succeeding age; thus, the great Dutch physician Boerhaave advises every doctor to read the work *De motu animalium*. Although he was certainly the foremost, he was not the only scientist of his kind to tackle biological problems from a purely mechanical point of view. Of those of his contemporaries who distinguished themselves in this respect two in particular are worthy of a detailed account.

CLAUDE PERRAULT was born in Paris in 1613, the son of a lawyer. He studied at the university there; first of all, mathematics and the classical languages, and then chemistry. Having taken a doctor's degree he practised for a time, but became more and more attracted to architecture. It is in this sphere that he became best known: as the designer of the Louvre colonnade he is mentioned in every guide-book to Paris. However, the interest in anatomy that he acquired as the result of his medical studies he maintained throughout his life; he dissected animals of every available kind and compared the results he achieved. Finally he fell a victim to his own zeal: he died (in 1688) of blood-poisoning contracted when dissecting a camel that had died in the Royal Zoological Garden.

The work in which Perrault recorded his biological speculations bears the characteristic title of *Essais de la physique*. The third of the four volumes that the work comprises is called *Méchanique des animaux*, and in it he has developed his ideas on the functions of animal life. The work was published in 1680 — that is to say, at the same time as Borelli's, and, of course, quite independently of the latter's. As already mentioned, Borelli was a disciple of Galileo; Perrault, on the other hand, shows himself in his writings to be manifestly influenced by Gassendi, although accessible biographies make no mention of either any personal or any literary contact between them. Gassendi based his conception of nature on the ancient atomic theory, such as it has been preserved in literature, mainly through Lucretius; moreover, he was an admirer of Galileo and an opponent of Descartes. In Perrault we find views on all these questions in full accord with Gassendi. In his first chapter he states matter to be composed of individual particles, at the same time hard and elastic; the air in particular is composed partly of finer, spherical and partly of coarser, cubiform particles. On the other hand, when it comes to a question of gravity, Perrault shows himself familiar with Galileo's discoveries in that field, and finally he sharply criticizes Descartes, particularly his theory that animals lack consciousness. In opposition to this assertion Perrault maintains the independent and peculiar intelligence of animals, citing numerous examples.

Perrault's philosophical method

WITH regard to the knowledge of animals, as also with regard to physics, Perrault lays down two scientific methods: the "historical," which is purely descriptive, and the "philosophical," which seeks to ascertain the causes of what takes place in nature. Following this philosophical method, Perrault deals with special phenomena in the animal kingdom, wherein he endeavours always to find out the mechanical connexion of causes, declaring that this is all that it is possible for man to discover. He entirely disagrees with both the older, idealistic philosophy, which scorns to have anything to do with natural phenomena, and the younger philosophy — that is, that of Descartes — which denies all manifestations of soul in animals. Perrault then tries to ascertain the mechanical course in quite a number of vital functions; particularly sense-impressions, the digestion, and the external movements of the body. On all these subjects he expresses original opinions, constantly pointing out comparisons with mechanical organizations in animate nature, and in doing so he displays throughout his capacity as a practical technician. Thus, he compares the muscles which act counter to one another in an arm with the shrouds on a boat-mast which counterbalance one another; in another connexion the valves of the heart are compared with the mechanism of a sluice-gate. His study of the mechanism of the auditory organ is full of striking observations; in regard to the visual organ, too, he makes some interesting comparisons — for instance, between the lenses in different animals — although naturally the results of modern optics, founded by Huygens and Newton, are unknown to him. In every case he tries in the first place to ascertain the nature of the movements performed by the different parts of the body; consequently he pays special attention to the structure and function of the muscles; in contrast to Borelli, however, he entertains the false idea that it is not "the flesh," but the interposed lengthwise and crosswise fibres — that is to say, the connective tissue — which expands and contracts. He gives a detailed description of this process; he believes that the muscle in its natural position is contracted; when it relaxes, it does so because the nerves convey to the muscular fibres a "*substance spiritueuse*," which expands them just as metals are expanded by heat. The motory impulses therefor come from the brain; of its parts he considers the *medulla oblongata* to be the most essential and the great brain the least important; he had, indeed, observed that it is possible to remove the great brain from a live dog without the animal's dying, but if the *medulla oblongata* is injured the dog dies at once. This fact the physiology of our own day has, of course, confirmed, although the conclusion which Perrault draws from it as to the lesser importance of the great brain is wrong. To "peristaltics," by which he means the movements within the body, he devotes a special chapter, which likewise contains many keen-sighted observations, the process of nutrition as a whole being of particular interest to him; naturally, however, he has a number of false ideas as to its chemistry, such as his belief that the air contains directly alimentary constituents, proved by the fact that some broods of serpents, which were kept in a jar without food, developed "on air." Owing to the limited possibilities of investigation in those days, we cannot blame him for being unable to discover that animals have an embryonic reserve nutriment to live on.

In spite of occasional fallacies, Perrault may thus be regarded, side by side with Borelli, as one of the pioneers of modern biology. Besides these, the Danish philosopher Steno, well known for the strange fate that befell him, is worthy of mention as having been active in the same direction.

Nils Steensen, known under the latinized form of his name, NICOLAUS STENO, was born in Copenhagen in 1638, of a wealthy family of goldsmiths.

He early showed an inclination for medical studies, which he was able to carry out first at home under Bartholin, and then in Amsterdam and Leyden. He soon made important anatomical discoveries, which gained him a European reputation. After this he spent some years in his native city seeking employment at the University, but as he was constantly passed over in favour of the relations of Bartholin, he grew weary of waiting, and having received his paternal inheritance, he journeyed to Paris. There he studied cerebral anatomy for a time, publishing a treatise on that subject, and then went to Italy, where he worked at Pavia and Florence. The Grand Duke of Tuscany was very gracious to him and gave him money to continue his studies and his publications. Under these new conditions, however, Steno underwent a severe spiritual crisis, resulting in his being converted to Catholicism, a step which brought him brilliant worldly advantages, but soon completely upset his intellectual balance. After spending some time in his native country, where the authorities had now learnt — too late — to estimate him at his true value and offered him a position that would bring in a good income, though at the same time they naturally did not look with favour upon his religious conversion, he returned to Italy, took holy orders, and was soon appointed bishop and chief organizer of Catholic propaganda in north Germany. In the latter capacity he displayed fanatic zeal; he addressed a letter to Spinoza, amongst others, with whom he had been acquainted in his youth, urging him to become converted, which the latter declined to answer. At the same time he gave himself up to violent asceticism, which rapidly undermined his health. He was only forty-eight when he died (in the year 1686), and he was buried with great pomp at Florence, where a fine monument in the Church of San Lorenzo perpetuates his memory.

As an anatomist Steno devoted himself principally to the study of two organic systems: the glandular and the muscular systems. With regard to the glands, Glisson and Wharton had, of course, been the pioneers, but Steno at any rate made fresh and important contributions to the knowledge of these organs; he discovers the exit of the parotid gland, which has been given the name of "*ductus stenonianus*"; he thoroughly explained the anatomy of the other glands of the mouth, and, lastly, found the exit of the tear gland. As an anatomist of the muscular system his chief aim was, as he himself says, to apply to anatomy the laws of mathematics and thus to create a geometrical system for the muscles. His main work on muscular investigation, in which he carefully analyses, along the lines just indicated, a number of muscles, starting from their simple component parts, was published twelve years before Borelli's important work. In this book Steno's theory is expounded, though Borelli expressly states that its rules apply only in certain special cases. In actual fact Steno's work deals with a far more limited field of investigation than Borelli's; in method it is more speculative and less experimental than the latter's, and therefore, though published earlier, it has not the same universal application as the work *De motu animalium*.

Steno as a palæontologist

BESIDES the above work Steno found time during the short period which he devoted to natural science to make a number of important contributions. Thus, he discovered the ovaries in the shark, which, as is well known, produces its young alive — a discovery the importance of which he himself fully realized; up to that time people had thought that ovaries existed only in oviparous animals. Steno's greatest service to biology, however, is his creation of modern palæontology. Already during his first visit to Florence he had had an opportunity of studying a kind of stone images found in great numbers there, which the inhabitants called "*glossopetri*" or stone-tongues, and by means of comparative study he proved that they must have been the teeth of sharks. During his later sojourn in Tuscany he carried out a systematic study of that district's geological strata, and thought himself justified in concluding from their position and appearance that they had been stratified out of water, a fact which he believed to be still further confirmed by the quantity of animals and plants found in them. Supported by these facts, he outlines a theory of the origin of the earth's strata which is a presage of present-day geological science. He never got further than this rough outline, however; the results of his geological investigations would not, as the times were then, accord with the Church doctrine that he had so zealously embraced, and, indeed, this was one of the reasons why he completely abandoned a science which he had initiated and studied with such splendid results. A tragic fate indeed, although by no means without its counterpart in scientific history. In this his last natural-scientific work, however, Steno deals also with other problems besides the purely geological and palæontological. His geological stratification theory is really only one link in a general theory regarding transubstantiation in nature, according to which all things that exist have originally been and are still being precipitated out of fluids. He thus comes to the question of the crystallization process in the mineral kingdom, which he investigated with great thoroughness and good results, and in connexion therewith he discusses the transubstantiation and organic formation in animals, which he likewise conceives to be a stratificational process similar to that which takes place in inanimate nature, a precipitation from fluids, of which he distinguishes various kinds in the animal and plant organism. Thus he reconciles the changes in substance in animate and inanimate nature under the same point of view and without giving any idea of the essential difference between the growth of a crystal and that of a living organism. This, then, clearly denotes the limitation in the mechanical conception of nature in the seventeenth century — a limitation which undoubtedly contributed to the fact that the promising ideas which such investigators as Borelli, Perrault, and Steno had produced were not followed up. The insignificant progress made in the sphere of organic chemistry at that period in fact rendered impossible the expansion of experimental biology beyond the purely mechanical sphere in which the scientists here mentioned achieved such splendid results. Another reason why biological research took a new direction, however, proved of still more decisive importance — the discovery of the microscope, and its constant improvement, resulting in the opening up of hitherto unguessed possibilities for biology. We shall now proceed to discuss this method and its representatives.

3. Microscopics and Microtechnology

THE fact that ground lenses magnify the vision seems to have been already established in classical antiquity. Eye-glasses and simple magnifying-glasses came into use in the sixteenth century; the inventors of complex lenticular systems are commonly said to have been two Dutch spectacle-makers, Janssen by name, father and son. These earliest microscopes must have been extremely primitive: a tube with a plate for the object, without any adjusting apparatus, and the lens or lenses at the other end of the tube; the tube was held to the eye and directed when in use towards the light like a telescope. The magnification was, to start with, not more than ten times, but it nevertheless excited general wonder, especially when tiny live creeping things were put under the microscope and could show their movements. It was considered particularly fascinating to watch fleas, from which the earliest type of microscope received the name of "*vitrum pulicare*," or flea-glass. During the seventeenth century, however, the construction of the microscope, chiefly the system of lenses, was considerably improved, with the result that good individual instruments made by clever masters in the art, such as the Dutchman Leeuwenhoek, mentioned below, could magnify up to 270 times. But throughout the eighteenth century and for a good part of the nineteenth, microscope construction underwent but few changes, except for isolated improvements, such as the introduction of a stand and mirror, and it was not until the thirties that the long line of new inventions that have gradually made the microscope what it is today had their beginning. Microscopy has had, therefore, two periods of brilliant achievement in the course of its history: the seventeenth century, and the latter half of the nineteenth century up to the present day.

Even Harvey, according to his own statement, used a "*perspicillum*" when studying the circulation of the blood in insects. The first scientific treatise that is based exclusively on microscopical investigations was the

Italian FRANCISCO STELLUTI's study of the structure of the bee, which was published in Rome in 1625. But foremost among those who systematically based their research on magnifying apparatus must be mentioned the Italian Malpighi.

MARCELLO MALPIGHI was born in 1628 at Cavalcuore, a place near Bologna, where his father owned an estate. Here Marcello spent his childhood. At an early age he became a student at Bologna and devoted himself to the study of Aristotelean philosophy. He had to break off his studies, however, upon the death of both his parents, in 1649, and had to leave the University for a year or two in order to settle his father's affairs and look after his younger brothers and sisters. With this latter end in view he returned to the University, after a short time graduated as a doctor (in 1653), and then devoted himself to medical practice and university teaching. His brilliant gifts were soon apparent, but certain intrigues delayed his advancement, and when at last — in 1656 — the Senate of Bologna instituted a professorship for his special benefit, he preferred to accept an appointment to a chair of medicine at Pisa. There he got to know Borelli, and their acquaintance developed into a lifelong, firm friendship, Borelli in the beginning possessed a tutorial influence over his colleague, who was twenty years younger, and taught him to realize the defects in the Aristoteleanism which he had till then embraced. Malpighi, however, considering that the climate of Pisa was bad for his health, returned once more to Bologna, but shortly afterwards he was called, upon Borelli's recommendation, to be professor at Messina, at a good salary. After four years, however, he relinquished this appointment, owing to intrigues and troublesome interference on the part of the authorities. So for the third time — in 1666 — he returned to Bologna, where a professorship awaited him, which he held with honour for twenty-five years. In 1691, being then in his sixty-fourth year and in failing health, he went to Rome and became private physician to the Pope, and he died there of apoplexy three years later — in 1694.

In contrast to most of the biologists of the earlier period, but like so many of those of the present day, Malpighi published his observations not in large consecutive works, but in the form of short reports, sometimes comprising only a few pages, usually sent in to the Royal Society of London, of which he was a member and which undertook the printing of them. Practically every one of these small papers contained some important discovery in different branches of biology. The connecting link in this literary work is not any common idea running through it all, but is represented by microscopical technology, which Malpighi with hitherto unrivalled genius applied to every imaginable object in living nature that came within his range. Thus Malpighi was the founder of microscopical anatomy in both the animal and the vegetable kingdoms. One reason for the disconnected way in which he recorded his experiences may perhaps be found in his lack of capacity as a writer; indeed, from the point of view of style his papers were not of a high standard, his exposition being often unclear and at times almost impossible to understand.

Malpighi's investigations

OF Malpighi's writings the first in point of time and undoubtedly the most important as to contents is the short account, in the form of two letters addressed to Borelli, of his investigation of the structure of the lungs. In the first of these essays he declares that the substance of the lungs had till that time been regarded as "fleshlike," which was incorrect; the lung consists rather of a network of extremely thin-walled cells, which are connected with the finest ramifications of the windpipe. This, he states, can best be observed by flushing out with water the blood from a fresh lung, then inflating the lung through the windpipe, and afterwards drying it. In connexion with this discovery he advances some speculations with regard to the function of the lungs, which he assumes to be to keep the blood flowing and to prevent it from coagulating, which happens when it has run out of the veins. He also discusses the high temperature which fever produces in the blood and considers it to be due to a process of fermentation. In the second letter he gives an account of the finer structure of the lung of the frog, and in connexion therewith he describes his discovery of the capillary circulation as a connecting link between arteries and veins, which he also observed in the frog. In order to demonstrate this vascular system he recommends that a frog's lung be inflated, then dried, and in that state examined under a magnifying-glass. He himself emphasizes the importance of the fact that the transition between the venous and the arterial blood had been discovered, and posterity has confirmed the truth of his discovery. The achievement that comes next in importance is his investigation of a series of organs which he placed in the category of the glands. These investigations he carried out partly with fresh material, partly with such as had been hardened by cooking, besides which, by means of injections into the blood-vessels and the preparation of the tissues, he endeavoured to trace the minutest elements of the organ. In the liver, with which he started his investigation, he thus followed the blood-vessels up to their finest ramifications, which he connected with a mass of small protuberances which may be brought up on the cooked liver. By establishing the existence of these he considered the liver's glandular character proved, which modern science has shown to be correct, although the small protuberances are actually pure outgrowths without any equivalent in the true structure of the liver. Malpighi also includes in his investigations of glands his observations of the cortex of the cerebrum. He observed in this organ the pyramid-cells, which he believes to be glandular elements that secrete the "*fluidum*" whereby the muscles are

moved to contract. The nerves, which are hollow, form the passages outwards for this fluid, the nature of which Malpighi does not further describe,
although, like Willis, he seems to regard it mostly as some kind of fugitive
liquid. For the rest, he has made contributions to the knowledge of the
blood-vessels' ramifications in the brain; on the other hand, his speculations
in regard to the function of the cerebral cortex are not very enlightening,
just as, on the whole, Malpighi was more of a practical observer than a
theoretician. It was left to Swedenborg, a couple of generations after Malpighi, to work out an explanation — partly based on the latter's observations — of the localizations in the cortex of the brain, which even our own
age might well think remarkable. — Finally, Malpighi studied the kidney
and the spleen, using the same methods as those applied to his observations
of the above-mentioned organs, and in this field, too, he achieved valuable
results; in the kidney he established the course of the blood-vessels and of
the tubules and has in general given a good description of the inner structure
of the organ in man and in several other mammal forms; the glomeruli of
the kidney still bear his name, and likewise the name of the Malpighian
follicular bodies in the spleen testify to his powers of observation. Extremely
useful has been Malpighi's monograph on the tongue, the muscles and nerves
of which he explained and the papillæ of which he described and characterized as gustatory organs. And finally he published an account of the development of the hen's egg, which forms a creditable supplement to the
investigations previously carried out by Fabrizio and Harvey. In the sphere
of invertebrate biology Malpighi has also performed a service by investigating the structure and history of the development of the silk-worm; he
discovered in this subject the excretal organs characteristic of the Tracheata,
which are now called the Malpighian tubes, and in other respects, too, he
laid the foundations of our knowledge of the anatomy of insect larval forms
and likewise made valuable observations regarding the butterfly's evolution
out of the pupa and its anatomical structure.

Malpighi's works on vegetable anatomy

THERE still remains to give an account of Malpighi's activities as a pioneer
in a quite new field — vegetable anatomy. Biology, as a universal science
of life and its manifestations, has for obvious reasons been based principally
on the study of those creatures which have stood in the closest relation to
man — that is to say, first and foremost man himself, and secondly the higher
and lower animals; for the purposes of this science plants have as a rule come
last. There are two fields of biological study, however, in which plants have
from the beginning been more useful for standardizing purposes than animals — namely, classification and the cell and tissue principle. The fact that
plants have proved a more convenient starting-point in this latter sphere
is, of course, due to their having, on account of their cellulose formations,

such extremely easily discernible elements. Thanks to the resultant structural conditions, which in their main features are distinguishable even to the naked eye, plants have constituted the starting-point for the study of the elementary nature of living matter as a whole. And the honour of having introduced this study into science is due to Malpighi, even though he may have had to share it to a certain extent with another, the English physician Grew.

The results of Malpighi's investigations into the subject of vegetable anatomy were, after ten years of preparation, submitted to the Royal Society of London and were there published. They consist of a comparative study of the anatomy of different plants, both ligneous plants and herbs. First the structure of the bark is described, then that of the wood and pith, and finally the buds, leaves, flowers, and fruits. The different parts of these plants are composed of small "utriculi" or cells, which can be distinguished by means of a magnifying-glass and which in their turn form a larger connective group. The cuticle and bast of the bark, the vesicular system of the wood and its fibres are analysed, special interest being devoted to the spiral vessels, whose inner spiral thickening induces a comparison with the tracheal system of insects, in regard not only to structure, but also to function. Upon this chance similarity Malpighi now bases a universal theory of respiration applicable to all living creatures — which, for all its conjectural ideas, represents a shrewd guess as to the uniformity of life-phenomena in all organisms. He believes that the more perfect the living beings are, the smaller their respiratory organs are: man and the higher animals do with a pair of lungs of comparatively small size, whereas fishes have numerous closely ramified gills, and the tracheæ of insects spread throughout the entire body, while again the spiral vessels in plants develop in such quantities that they fill up even the most insignificant ramifications of the individual plant. Plants, he supposes, take up air out of the soil through the roots; the leaves possess no openings that could serve this purpose. With regard to the significance of respiration for living beings, he believes that it consists in promoting the mobility and "fermentation" of the alimental juices. On the whole, fermentation plays much the same part in Malpighi's physiological speculations as "cooking" does in Aristotle's; at any rate, it constitutes an advance, in spite of the indefiniteness of the idea, which was inevitable considering the stage of development which chemistry had reached in those days. In connexion with his account of the elementary constituents of plants Malpighi advances a number of general physiological speculations, all intending to demonstrate similarities between vegetable and animal organisms and their functions. In doing so he only follows, it is true, a principle which was peculiar to a botanist of that time and which had its origin in Cesalpino's physiological speculations on plants, to which we shall revert later on. It is, however, but natural in the circumstances that these comparisons should lead to false conclusions, and, as a matter of fact, they did to a great extent prevent Malpighi from taking advantage of the promising material for study, which otherwise he might possibly have been able to do. Thus he compares the buds, out of which gradually sprout leaves and branches, with the ovary and the uterus; then he deals with the flowers and carefully compares their special parts in different plants; as he fails to clear up the question of their sexuality he advances the theory that the flowers serve to purify the juices of the plants before germination, just as menstruation precedes pregnancy. He studied the evolution of the vegetable substance in a number of different seeds, but seeks to identify therewith the uterus, the Fallopian tube, the umbilicus, and the amnion, which naturally leads him to extravagant conclusions. Malpighi devoted special attention to the study of gall-formations in a number of vegetable forms; he is fully convinced that they are produced by insects, but on the other hand he found that the tubercles on the roots of pulse plants are not produced by insects, though he failed to find any other explanation of their origin.[1] He also studied and speculated on a number of other malformations in plants; with regard to the tubers in many plants, he is of the firm opinion that they contain reserve nutriment. He is again tempted, however, by the theory of the nutriment of plants, with which he closes his work, to make dangerous comparisons with the conditions obtaining in the animal kingdom.

The other creator of plant anatomy

At the same time, however, as Malpighi submitted the first part of his vegetable anatomy to the Royal Society, that society had sent to the printers another work on the same subject compiled independently of Malpighi by an English doctor, NEHEMIAH GREW. Born in 1628, Grew was the son of a clergyman who during the great Civil War joined the opponents of the Crown and so, upon the return of Charles II, was deprived of his benefice. His son, who was then an undergraduate at Cambridge, went (presumably for the same reason) to continue his studies abroad. In 1671 he graduated at Leyden, with a dissertation on the fluids of the nervous system. He then settled down as a practitioner in a provincial town, but, thanks to the reputation he gained by his work in vegetable anatomy, he was able within a short time to move to London, where he applied himself to both medical practice and research, eventually becoming secretary to the Royal Society. He died in 1712.

As a scientist Grew concentrated almost exclusively upon vegetable anatomy, whereby his investigations at once acquire a different character from those of Malpighi, in that the latter's constantly repeated comparisons with human and animal anatomy are altogether lacking. Grew also studied the

---

[1] It is only in our own time that it has been established that they are produced by bacteria.

construction of the fruit substance and the germination of the seed in a number of plants, but in doing so he employs a terminology of his own, nor does he borrow anything from animal anatomy; the word "parenchyma," which he invented, has been retained in vegetable anatomy. He describes plants organ by organ; cells and vessels in the stem he discovered independently of Malpighi and on the whole describes the anatomical details more soberly and in greater detail, though with less fertility of ideas, than the latter. He advanced a theory that the pistil in plants corresponds to the female, and the stamen, with its pollen, to the male, and pointed out their hermaphroditism, but, on the other hand, he entered into speculations upon male and female "juices" in plants, which are of no interest nowadays except from the point of view of mere curiosity. He voluntarily abandoned in favour of Malpighi any claim to priority in regard to the discovery of the vascular system in plants; on the other hand, Malpighi undertook a Latin translation of Grew's writings. These two scientists improved vegetable anatomy so far it was to take more than a century before any important addition could be made to their work. Through them biology acquired its knowledge of organized matter as being something peculiar in its structure; the idea of tissue was established — for the time being, it is true, only in the sphere of botany — and in the vegetable kingdom, also, the simple elements of the tissues — the cells — were observed and described. It was, however, to be nearly two centuries before the fundamental value of these achievements was fully appreciated; true, both their contemporaries and the immediately succeeding age admired the exactness of their investigations, but it considered the results more from the point of view of curiosity. All the greater admiration is due to those scientists who at any rate guessed that here was to hand information of the highest importance for the future of science. The fact that their contemporaries failed to continue along the line they had laid down was undoubtedly due mostly to the microscope's having at the same time opened up a field in the sphere of animal anatomy of such considerable scope and of greater immediate interest. The anatomy of the lower animals in particular was an entirely unexploited field, possessing vast possibilities for development, of which, indeed, splendid advantage was taken just about that time.

ANTONY VAN LEEUWENHOEK was born in 1632 at Delft in Holland and was sent as a boy to Amsterdam to be trained for business. Having worked for a time in the cloth trade, he returned to his native town and got an appointment with the municipal authorities,[2] which must have taken up very little of his time, since he was able to devote the greater part of his days to indulging his interest in the study of nature. As a research-worker Leeuwenhoek

[2] His title was "Kamerbewarer der Kamer van Heeren Schepen van Delft."

was a self-taught man; he had never received any scientific training, and as he knew no Latin, in which language most natural philosophers of that time generally published their works, he was unable in his old age to come into contact with the scientific life around him; he had to depend entirely upon himself. It was the remarkable phenomena revealed by the magnifying-glass that fascinated him from the very beginning; he taught himself to grind lenses, and by diligence and having a naturally delicate touch, he developed this art further than any of his contemporaries. Sparing no pains to find out new methods and combinations, he gave to his magnifying apparatus all sorts of forms, some of them very strange; he tried glass, rock-crystal, and even diamonds for his lenses; but the greatest advance he made was in the manufacture of simple lenses with strong magnification; one such lens, which is still in existence, is said to magnify as much as 270 times. As often with self-taught men, he was extremely jealous of his inventions; he never sold a magnifying-glass nor even lent one to anyone; on the other hand, scientists who visited him were permitted to use a number of his instruments, though never the most powerful. It is said that among his property there were found more than four hundred microscopes and magnifying-glasses. A number of them he had bequeathed to the Royal Society of London, of which he was a member and which published most of his observations. Busily occupied to the last, Leeuwenhoek reached the age of over ninety; he died in his native town in 1723.

Leeuwenhoek's investigations

LEEUWENHOEK's collected writings have quite an extensive range, and their contents are extraordinarily varied. The only connecting link that unites them is the microscopical method; this Leeuwenhoek applied to literally everything that came within his range of vision: crystals and minerals, plants and animals. With respect to the last he developed no special microtomical technique, but he studied and illustrated the details of what he observed. This detailed study, however, he advanced further than anyone of his time, and if he possessed the most powerful magnifying lenses known to his age, he certainly had also the keenest eye. He took exact measurements of everything that he examined; unfortunately there was in his time no unit of measure which could have served his purpose, so that he was compelled to select such objects of comparison as he thought suitable — a hair, a grain of sand —and to state his measurements in fractions, often thousandth parts, thereof. He took careful notes of everything that he examined and sent them in the form of letters to the Royal Society, to which he had been introduced by his friend de Graaf, and of which he soon became a member. It often happened that one and the same letter contained a mass of different notes on various observations he had made. It was undoubtedly to his great advantage that he so seldom engaged in theoretical speculations, but only described and illustrated what he saw; if at any time he starts theorizing, he generally fails, but usually he appears conscious of his limitations and holds to the realities which he knew so well how to master.

Biology has Leeuwenhoek to thank for a long series of facts of fundamental importance. His studies on the circulation of the blood deserve first consideration. He has explained and completed the knowledge of the capillary system which Malpighi originated, while he clearly proved that the veins and arteries, each separately, are continued on immediately through the capillaries and thus through them merge directly into one another. Moreover, Leeuwenhoek for the first time clearly recognized the blood corpuscles and described them, first in the frog and then in man and a number of animal forms. Malpighi had thought that he could distinguish in the blood "fat globules," but he did not investigate the matter further, so that to Leeuwenhoek is due the honour of having really solved the problem. The same is the case with the spermatozoa, which, it is true, a Dutch student by the name of HAMM was the first to observe, but which Leeuwenhoek at any rate studied more closely in a number of animal forms. In this connexion he made thorough investigation into the fertilization of various animals, especially fishes and frogs. In the frog he noted the spermatozoon's association with the egg and believes — like Aristotle, as a matter of fact — that it is from the male that the actual life comes; the female only provides, through the egg, nourishment and powers of development. This he tries to prove by pairing different-coloured rabbits: if a white female is paired with a grey male, all the young will be grey like the father. Had he continued the experiment through several generations, he would certainly have obtained other results. — He has further observed a number of histological details of various kinds: the stripes of the striated muscles, the structure of dental bone, the construction of the optic lens in man and the higher animals. No less remarkable are his discoveries in the lower animal world. He thus discovered the Infusoria and the Rotatoria in water; he explained the reproduction of ants, found their true eggs, and showed that what had hitherto been called ants' eggs are really the pupæ of the insect. He very definitely opposes the hitherto prevalent view that minute creatures of all kinds arise through putrefaction or fermentation in inanimate matter. Instead he declares that even the smallest animals possess reproductive powers and propagate solely by means of them. In proof of this he demonstrated particularly the evolution of fleas and aphids. If, finally, we add to this that he demonstrated the difference between the structure of the stem of monocotyledons and dicotyledons in the vegetable kingdom, this — by no means complete — sketch will have given some idea of a life of activity which, without making any important theoretical contributions, advanced the knowledge of nature in an unusually high degree.

On the whole, Holland during the latter half of the seventeenth century proved a centre of biological research. Of the many prominent scientists who lived and worked in that country during that period it is possible to mention only a few of the most important — those who in one way or another led research into fresh directions.

JAN SWAMMERDAM was born in Amsterdam in the year 1637. His father was an apothecary who by saving had accumulated a considerable fortune and was, moreover, interested in natural science. He possessed a natural collection, which he augmented and looked after with great care. He had intended his son to take orders, but as the study of nature seemed to be his sole interest, he was permitted to study to become a doctor. After preliminary studies in his native town he entered the University of Leyden in 1661. Even then he had already proved a clever technician in the anatomical sphere, and he rapidly acquired fame for his splendid dissecting and injection-work. He formed a lifelong friendship with Steno, who was about the same age and who happened to be visiting Leyden at the time; they worked together and travelled together to Paris in order to continue their studies there. Here Swammerdam found a new friend in the person of the King's librarian, Thévenot, a friend who throughout his life loyally assisted him in every possible way. Returning to his native country, Swammerdam graduated at Leyden in 1667 with a dissertation on respiration and then settled down at his father's place in Amsterdam. He had already earlier applied himself to the study of the anatomy of the lower animals, and this interest now engaged all his powers. During the short years that remained to him he achieved results which not only left all his predecessors far behind, but actually remained unexcelled for the space of more than a hundred years. In the mean while his fortunes took an extremely unhappy turn. He contracted a malarial fever, which, except for occasional intervals, never left him for the rest of his life. At the same time the strain entailed on him by his work impaired his health. Besides, he was of a passionate nature; his writings are full of bitter controversy, and his quarrels about questions of scientific priority brought him many enemies. On the other hand, he had several loyal friends, who stood by him to the end. Worst of all, however, he fell out with his own family; his father, who seems to have been an economical and surly old man, thought that it was about time that his son, whom he had supported for more than thirty years, applied himself to medical practice or some other profession that might provide him with an income. In preparation for this, young Swammerdam was sent into the country to recover his frail health, but he spent night and day engrossed in his investigations, so that his health went from bad to worse. After repeated quarrels his father finally deprived him of all financial support. Swammerdam found himself in dire need and sought in vain to sell his collections in order to buy his daily bread. Even his intellectual powers had now waned; in about the year 1673 he ceased to work at his science and became abosrbed in religious contemplation. His old friend Steno sought to take advantage of this state of affairs: at the price of the same religious conversion which he himself had just undergone, he offered Swammerdam splendid prospects in Florence. The latter refused, but instead sought to cure his distress of soul by visiting Antoinette Bourignon, who was very notorious at that time. She was an extremely gifted but hysterical woman who in virtue of personal revelation desired to reform Christianity on ascetic and mystical lines, and who, persecuted by both Catholic and Protestant priests, wandered from country to country surrounded by a small band of believers. Swammerdam joined this band, but was unable to find the peace he sought; after leading a roving life for a couple of years he returned in the deepest spiritual and bodily misery to his native country. There at last he obtained, through his father's death, which occurred at the same time, financial independence, but then quarrelled with a sister over the inheritance, which still further embittered his mind. In the year 1680 he found repose in death, when not yet forty-three years old. In 1880 a beautiful monument was raised over his grave and there was created to his memory a fund, which is used for the purpose of giving prizes for research work carried out in the spheres of learning in which he had studied.

Swammerdam's scientific activities thus lasted for only about six years, during which period he published a few works of great value — particularly, besides the dissertation above mentioned and an essay on the genital organs of woman, a work on the anatomy of insects, in which he recorded his earlier researches on that subject. His still unpublished manuscripts he bequeathed to Thévenot; after the latter's death they passed through many hands until they were finally purchased by the famous Boerhaave of Leyden and were published by him together with the already printed work on insects under the title of *Bijbel der Natuure*, in 1737. Although it thus came out more than half a century after it had been written, this work was by no means out of date; in fact, it was to be some time before its detailed anatomical descriptions were improved upon — a proof of Swammerdam's incomparable genius as an anatomist of invertebrate life. The title of the work was probably given to it by Boerhaave, but fully reflects the state of mind in which the author found himself towards the close of his life. Nevertheless, religious observations do not form any disturbing element in it; on the contrary, his presentation is purely natural-scientific with the exception of a few contributions of the religious moralizing character, particularly one that is a reflection upon the short life of the day-fly. The undoubtedly valuable collections on which Swammerdam based these studies were after his death sold by auction and dispersed.

What still strikes the reader of Swammerdam's works is his mastery over even the most complicated details in the minute creatures he investigated. This he could not possibly have acquired without a high standard of knowledge of the technique of dissection, and, indeed, it was this knowledge which excited the admiration of his contemporaries; visitors from far and near were amazed at his fine instruments and the skill with which he handled them — glass tubes drawn out to points as fine as hairs, by means of which organs were spread out and canals injected; scalpules so fine that they had to be ground under a magnifying-glass, and so on. Extraordinary lightness of touch and unique powers of observation enabled him to utilize the methods which he worked out, to which, finally, we must add his love of research, for which he literally gave his life.

Anatomy of insects

SWAMMERDAM's great work in part contains a collection of anatomical monographs on insects and other invertebrate animals; particularly well known is his exposition of the anatomy of the bee, which even Cuvier considered to be unsurpassed, and further the head-louse, the day-fly, the rhinoceros-beetle, the *Helix pomatia*, and many more. These monographs, however, are all based on one theory of the evolution of insects and in connexion therewith that of all living creatures. Supported in his investigations by the development of a number of different insects' larvæ to pupa and imago and adopting a sharp controversial attitude towards Harvey, Swammerdam declares that the insect does not undergo any transformation, but that merely growth takes place of parts which already existed before. Again and again this statement is emphasized, that no generation, but only an excrescence of parts takes place, wherefore accident plays no part in the evolution of the insect, but what takes place is predetermined. This evolutionary principle is then applied to the development of the frog from the egg through the various larval stages, and finally, though quite summarily, to the evolution of man, which is likewise made dependent on predetermined necessity. Lastly, the evolution of the bud of plants to leaf and flower is compared in detail with the metamorphosis of insects. In order to facilitate his analysis insects are divided according to their metamorphosis into four groups: (1) those that come from the egg with all their feet complete — spiders, lice; (2) the animal that has all its feet when it is hatched, but whose wings develop later on — as, for instance, day-flies; (3) those from whose egg comes a larva, either with or without feet, which becomes a pupa after chrysalizing — as, for instance, ants, bees; (4) those in which a larva, like the foregoing, becomes a pupa without chrysalizing — certain flies. In this method of grouping Swammerdam laid the foundations of modern insect-classification, which, as is well known, still rests to a great extent upon the evolutionary history of insects. That he grouped spiders and even snails and worms under his first category is not to be wondered at; all invertebrates were at that time lumped together, and Swammerdam's point of departure was from first to last not morphological, but evolutionary. But at any rate he performed a service — as also did Leeuwenhoek — in awakening interest in these organisms, which had hitherto been regarded as existences not only of a lower type, but also utterly incomparable with the higher; they still arose, according to Harvey, by spontaneous generation, and this alone was a proof that no conclusions could be drawn from them touching the life of the higher animals.

Swammerdam's preformation theory

SWAMMERDAM showed that on the contrary it was just the life-conditions of the lower animals which, if viewed in a proper light, gave fresh stimulus to the knowledge of life in its entirety. Particularly does he insist upon this being realized in embryonic development. The theory which he advanced on this subject — growth of previously created parts instead of new formation — came to exercise immense influence during the immediately succeeding period: under the name of the theory of preformation or evolution it entirely supplanted Harvey's theory of epigenesis. True, in its application it was in its turn driven to sheer absurdities, particularly by certain scientists who will be named later on, but when it first appeared, it was certainly called for and marked an advance in biological science. In fact, it resulted in the assertion for the first time of the obedience of ontogenetical evolution to law; it definitely invalidated the old ideas of the spontaneous genesis of lower animals; it established the fact that according to nature the offspring must resemble the parent, whereas in earlier times, practically speaking, anything could arise out of anything — the legendary tales of women who under the influence of witchcraft were delivered of kittens and puppies instead of children had at any rate been discussed by certain scientists — and, finally, it satisfied, as far as embryology was concerned, the contemporary demand for a mechanical explanation of nature. But it is true that a century later the epigenetical theory was again to appear in a form that justified its acceptance — a change of which an account will be given further on.

There is a name that is worthy of mention by the side of Swammerdam — that of his contemporary and friend FREDERIC RUYSCH. He was born at The Hague in 1638 of a highly respected family, his father being secretary in the States General. Having when still young contracted an advantageous marriage, he was in a position to apply himself at his own option to medical research; he became a doctor at Leyden and a professor in Amsterdam, in an appointment which he held for sixty-three years. Moreover, he had an extensive medical practice. He died in 1731 in his ninety-third year. His long life, active to the last, is reminiscent of Leeuwenhoek's, as is also the fact that his greatest service to the world consists in the employment of technical methods, which, though he had not invented them, were nevertheless improved by him. From his friend Swammerdam he had learnt the art of using coloured wax for injections, and he acquired a masterly skill in this method, such as few attained after him. He was able to fill out the finest capillary vessels without either bursting or deforming them, and, besides, he preserved the preparations thus carried out in a wonderfully natural manner. And he was as jealous of his method as Leeuwenhoek was of his microscopes, though really with far less excuse than the latter; the microscopes survived the man who made them, while the method of injection went down with its inventor to the grave. Even with regard to the value which his discoveries had for science, the learned professor is no match for the untaught functionary, but Ruysch, in the application of his method, certainly did succeed in providing science with a mass of new facts, particularly in the sphere of human anatomy. He discovered the bronchial arteries and the arachnoids of the brain, besides which he studied and extended the knowledge of the iris and retina of the eye; and, further, he compared the male and female skeleton and investigated the changes made by age in the structure of bone. He made a splendid collection of anatomical preparations, of which he published a richly illustrated description. The objects were arranged in groups — human organs, shells, minerals, and other things all together — in a manner which in our time would be considered not only highly unscientific, but also utterly lacking in taste.[3] His contemporaries, however, were ecstatic over it; foreigners visited the museum, and poets lauded it in verse. Tsar Peter of Russia, who, as is well known, entertained almost childish admiration for all products of technical skill, finally purchased the entire collection for thirty thousand guilders, but naturally neither he nor any of his subjects could make any use of it. A second collection, which was made later, was purchased by the opponent of the Tsar, the Polish king Stanislaus Leszczynski. There is now nothing left of these collections: it is only through Ruysch's books that we in modern times can gain any idea of what he did. And it cannot be denied that there was in him but little in the way of ideas, yet at the same time extraordinary technical ability and quite a lot of humbug.

There was another Dutch physician of the same age as he, REINIER DE GRAAF, who possessed far sounder qualities as a scientist. Born in 1641 of Catholic parents, he studied at Leyden and at Angers in France, where he took his doctor's degree. When still quite young he had won a great reputation, but owing to his faith he was prevented from obtaining a professorship at Leyden, which was a strictly Protestant university, and he therefore settled down as a practitioner in Delft. His unusually promising career was cut short in 1673, when a serious illness deprived him of a happy domestic life and the possibility of carrying on his intensive research-work. He had

---

[3] Thus there was amongst the groups the skeleton of a child holding a piece of injected peritoneum like a handkerchief before its eyes.

already had time, however, to make important contributions to biology in both the anatomical and the physiological sphere. His doctor's dissertation deals with the pancreatic secretion. In it he shows how, by introducing a canula into the duct of the pancreatic gland in a live dog, it is possible to obtain some of the secretion for the purpose of closer examination -- a method which has since then been generally adopted in physiology.

The work on which de Graaf's fame principally rests, however, is his study of the sexual organs, both male and female, but chiefly the latter. The ovaries, of course, had already been described before, of both the higher and the lower vertebrates; that they produced eggs in birds was known, but a great many contradictory theories had been advanced on the subject of what kind of function they possessed in man and the other mammals. The Aristoteleans naturally supported their master's doctrine that the sexual product of the woman is the menstrual blood and that otherwise the male semen is the essential origin of the embryo, which the woman then nourishes and produces. De Graaf, on the other hand, after making a comparative study of the ovaries of mammals and birds, came to the conclusion that the cell-like protuberances already observed by Vesalius and Fallopio in the ovary of mammals corresponded to the egg of the bird ovary, and that the process of fertilization is similar in every animal type; just as a bird's fertilized egg in the ducts of the ovary acquires albumen and shell, the egg of the mammal becomes fertilized through the Fallopian tube, finds its way to the uterus, and there develops further. The very word "ovary" was suggested by him; hitherto the female sexual gland as well as the male had been called testis, a word which he still employs alternatively with the new one. He definitely rejects the assertion of the Aristoteleans that the embryo originates from the man alone; in disproof of that assertion he cites many cases in which demonstrably purely external characteristics have been inherited by the embryo from the mother, both in human beings and in animals; even cases of extrauterine gestation are cited by him as proof that the embryo is derived from the ovaries and not from outside. Likewise in regard to several other details in the structure of the sexual organs he records valuable fresh observations.

These investigations of de Graaf's proved of fundamental importance, although he was wrong in his assumption that the follicles in the ovary, which now bear his name, correspond to the eggs in the ovary of a bird — the true eggs of mammals were not discovered until a century and a half after his death. Nevertheless, the explanation he gave of the actual phenomenon of fertilization was of decisive significance for the future development of the knowledge of this phenomenon. It was impossible, however, either for his own or for the immediately succeeding age to reconcile his claim as to the significance of the egg in embryonic development with the important part that the spermatozoa should be assumed to play in the same process. And so we find developing, especially in the eighteenth century, the controversy between ovists and animalculists, as the champions of the importance of the egg and the spermatozoa respectively in fertilization were called — a controversy which was to set science by the ears for decades. In some of the following chapters more light will be thrown on this controversy as well as on the dispute between the respective champions of epigenesis and preformation which was raging at the same time.

The close of the great period in the history of anatomy

WITH this, our account of the brilliant epoch in the history of biology represented by the seventeenth century comes to a close. It is perfectly natural that towards the end of that century, and during the decades immediately following, interest should wane in just those spheres in which progress had been greatest; the forced march had to be followed by a period of mustering of forces and reflection, during which the results achieved had to be weighed from the theoretical point of view and classified. It is therefore worth while considering what were the solutions which the next age sought to give to the theoretical questions that had arisen in connexion with the great practical advance made in the field of anatomy and experimental biology. In the following chapter some samples will be given of theories of this kind.

CHAPTER V

BIOLOGICAL SPECULATIONS AND CONTROVERSIAL QUESTIONS AT
THE BEGINNING OF THE EIGHTEENTH CENTURY

A S HAS BEEN POINTED OUT in the foregoing, the power of the authorities of antiquity was broken during the seventeenth century as the result of a series of brilliant scientific discoveries; in its stead natural scientists based their researches upon the knowledge of the mechanical subjection to law which prevails in nature. However, the need was felt for a definite and uniform conception of nature such as Aristoteleanism undeniably possessed and which was lacking in the new systems of thought which took its place. In actual fact these systems, whether they emanated from Descartes, Spinoza, or Leibniz, were quite as dogmatic as Aristoteleanism; they were pure thought-structures, which, although based on the new natural science, were yet by no means capable of satisfactorily solving the problems to which that science gave rise. As far as biology is concerned, while it is true that physicists like Borelli or Perrault had been able with the aid of the newly-discovered mechanical laws to find solutions to a number of pure problems of motion, yet as soon as more complicated processes in the organism, such as the digestion, the circulation, or sense-impressions, had to be considered, the mechanical principle was found wanting; nor had the other branches of physics and even chemistry as yet reached such a state of development that they could be employed as a means of explaining such phenomena as those just mentioned. In these circumstances many a scientist was content merely to study the new facts which had been brought to light as a result of improved experimental technique, but there were others who devoted their lives to seeking firm ground on which to base a uniform explanation of life-phenomena. In modern times it is not easy to appreciate the difficulties with which these biological thinkers had to contend in their efforts to reconcile the individual results of past research work under one common point of view. Uniformity in the conception of nature in our day, of course, rests essentially upon the law of the indestructibility of energy, to which may be added, in the field of biology, the doctrine of the cell as a unit of life. But the theoretical natural science of the seventeenth century tended, instead of to these ideas, to the assumption of the existence of an unknown force as the origin of life and a basis for its continuance. This force could then be conceived of as something either purely mechanical or more idealistic; in the former case one was bewildered, since mechanics cannot provide the answer to more than a small fraction of the questions which the new discovery brought to light; while in the latter case there was the risk of reverting to mysticism in one form or another. These natural-scientific speculations from the beginning of the eighteenth century, which we shall now discuss, originated, curiously enough, less from the anatomists and biologists than from the medical practitioners, who sought to base their medical treatment on a general theory of the functions of the body. Of these latter scientists some few have exercised a radical influence even on the general development of biology and therefore deserve to be mentioned in this connexion.

THOMAS SYDENHAM lived, it is true, entirely in the seventeenth century — he was born in 1624 and died in 1689 — but his influence did not really make itself felt until after his death, and, indeed, it has increased still more since then. He belonged to a good country-family and studied for a time at Oxford, but upon the outbreak of the Civil War he joined the Parliamentary party and became an officer in its army. Afterwards, however, he continued his medical studies, took a low medical degree, and settled down in London as a pracititoner; he did not obtain his doctor's degree until he was over fifty years old. Personally Sydenham enjoyed a great reputation; he counted among his friends such people as the chemist Boyle and the philosopher Locke. On the other hand, opinions differed as to his capacity as a physician; his audacious ideas required time before they could penetrate the ordinary mind. Nowadays he is universally regarded as one of the pioneers of medical science.

Sydenham's medical doctrine

IN the seventeenth century London was a very unhealthful city; one plague followed another in rapid succession. It was these epidemics that inspired Sydenham to work out his medical theories; he studied the symptoms of the various diseases and endeavoured by that means to characterize the disease itself in the same way as the botanist describes a plant-species. "That botanist would have but little conscience who contented himself with the general description of a thistle and overlooked the special and peculiar characteristics in each species." He places a higher value on this exact study of nature than on any theories; this study should take into account all factors affecting the disease in its entirety. Even the season of the year when the disease is most widely dispersed should be carefully observed, as indeed all other conditions that may influence the plague as a whole. On the other hand, purely individual variations in particular cases are of minor interest. Like Hippocrates he considers that it is the nature of the patient that cures disease; it is therefore not so much worth while worrying about trying to diagnose the disorders in the fluids of the body on each occasion as to try to discover a treatment that may assist the working of nature. What he really means when he talks of "nature" is not at all clear — whether it is a combination of the individual's life-manifestations or some special life-force; similarly, one does not gain a very clear explanation of the ideas he borrowed from Hippocrates concerning the fluids of the body and the balance or disturbances therein. His general conception of nature is on the whole purely empirical — in this he was influenced by Bacon, whom, indeed, he quotes with admiration. In certain cases, it is true, he can form quite daring hypotheses, but as a rule he consistently applies his principle as to observation's being the only source of knowledge in disease.[1] This principle has indeed been adopted by posterity, but he also exercised a powerful influence on the medical and biological thinkers of the immediately succeeding age, although these latter could not restrain themselves within the limits which he laid down for research, but went further afield in the world of hypothesis.

Among these medical researchers who formed general theories of importance to the development of biology, two men are conspicuous at the beginning of the eighteenth century who, born in the same year and working in the same town, yet proved in all essentials strangely contrasted. These two were Hoffmann and Stahl.

FRIEDRICH HOFFMANN was born in Halle in 1660, the son of a wealthy physician. At the age of fifteen he had the misfortune to lose both his parents, who died of the plague, as well as his inheritance, as a result of a fire, and thus early had to fend for himself. He was given an opportunity, however, of studying medicine at the University of Jena, where a highly reputed representative of the chemical and medical research of the period, G. W. WEDEL, was his teacher. Having further studied at Erfurt, he took his degree at Jena, spent some time in England, where he made the acquaintance of Boyle, and then set up as a practitioner in a couple of small German states until, in 1693, he was called to an appointment at the newly-founded university in his native town of Halle. There he spent the rest of his life as a professor, with the exception of a couple of years which he spent at the court in Berlin. His work both as a teacher and as a physician was crowned with extraordinary success: among his numerous pupils were included even old doctors who sought to complete their training with him; as a practitioner he was resorted to by high and low and was overwhelmed with consultations and loaded with brilliant honours. Considerate even towards those of different opinion, of an affectionate, sympathetic nature, he was himself universally beloved. He died in 1742, in harness to the last.

---

[1] In his own circle his contempt for theories seems sometimes to have been expressed in a somewhat original way; thus, a colleague who asked for advice regarding the choice of medical literature was told to read *Don Quixote*.

Hoffmann's practical work and his theory

UNDOUBTEDLY Hoffmann's services to science lie principally in the sphere of practical medicine. He described several diseases hitherto unaccounted for; both in theory and in practice he insisted upon accurate diagnosis based upon natural-scientific principles, considerate treatment of the sick, and simple medicines. He himself made up and sold at a great profit quite a number of preparations, which still play their part in popular medicine. He was a very productive writer on medical subjects in every conceivable specialized sphere, but he also tried to combine in one general theory of the functions of the body the principles at which he had arrived in the course of his work, and this theory has not been without its importance for the general development of biology. It takes as its starting-point the so-called chemiatric theories prevalent in the seventeenth century, which ultimately originated in Paracelsus's fantastic speculations as to the human body's being composed of quicksilver, sulphur, and salt, and in conformity with its original sought to explain the functions of the body as essentially phenomena of chemical change, for which purpose recourse was had either to the mechanical theories of the movements of the body, described in the foregoing, or else to the mass of mystical speculations still available at that time, to fill up the gaps in the proposed system. Hoffmann takes his stand at the very start on chemico-mechanical ground. He himself was a clever chemist and besides possessed a complete mastery of the anatomical literature of his age, in which sphere both Borelli and Perrault had some influence on him; and, finally, he had not neglected the discoveries of either Newton or Leibniz. He began with the principle that matter and motion form the foundation of existence; the body is a machine, which is kept going by the circulation of the blood. Life is thus a purely mechanical process, from whose functions the activities of the soul can be excluded; when the body dies, it is not the soul that leaves the body, but the body that abandons the soul, so that the latter can no longer use the organs of the body as its tools. The movement of the blood is caused by the heart; the latter's action, again, is regulated by the movements in the nervous system, in the fibres of which there circulates a fluid, *"spiritus animalis,"* which is formed of extremely light ether-particles and is produced in the brain and by its movements induces and regulates the muscular functions, sense-impressions, and alimental processes. The power of the blood to maintain life is due to the fact that it contains a *"spiritus"* formed of the ether constituents of the air and the sulphurous element in the blood. Chemically, in fact, the components of the blood are partly sulphurous, partly ethereal, and partly earthy; the sulphur element is the cause of the warmth of the body, both the natural warmth and that increased by inflammation, which is induced by the sulphur particles easily becoming extremely mobile through the action of the ether. The function of the lungs is to mingle the component parts of the blood, besides which the inhaled air conveys to it fresh ether particles, which augment its power to keep the mechanism of the body working. It would take too long to record here the complicated accounts of the production and dispersion of the nervous fluid; it may just be mentioned that the finest and most vitally essential part of the fluid is said to emanate from the cortex of the great brain. The male semen is closely akin to the nervous fluid, and its function is thus to give life to the egg, so that it may start developing.

Hoffmann, having thus described the mechanism of the body, declares that man naturally possesses an immortal soul, given him by God; the will of this soul controls the movement of the body, and through it we understand, think, and act. Following many other old authors and supported by the Holy Scriptures, he divides man into three *"principia"*—namely, *corpus, spiritus,* and *anima*—that is, the body, the above-described nervous fluid, and the consciousness. But besides these man possesses a higher "substance," which the ancient philosophers called *mens* and which Scripture names the image of the spirit of God; this substance makes use of the consciousness's impression of things and forms them into ideas; false sense-impressions may be rectified by clear reason, but a mass of confused sense-impressions causes madness. Concussion of mind may also disturb the circulation of the blood and produce a condition of sickness in the body. But Hoffmann resolutely denies that the movement and function of the body originate in the soul; "although the human soul possesses a certain limited influence over the bodily parts, nevertheless medicine both in theory and in practice is pure mechanics, in that it is based upon purely mechanical principles — namely, motion and matter." The inconsistencies and the arbitrary constructions of thought in this attempt to form a mechanical conception of life-phenomena will be easily realized by the modern reader, but should not in any way detract from the respect due to this attempt — which at any rate is based on very substantial experiences, considering the age — to find a natural connexion in the life-process. The assumption of an immortal soul is explained by the fact that Hoffmann was a devout Christian with a markedly pietistic temperament; Halle was the source and centre of pietism, and Hoffmann was a warm friend of its founders, Spener and Franke. He also displays in many places a naïve childlike piety, as when he dedicates one of his books to "The Holy Trinity, the Supreme Physician." Thus we have here a proof that a mechanical conception of life and ancient theological dogmas were formerly capable of being reconciled, which would hardly be considered possible in our own day.

One of Hoffmann's first steps when elected professor at Halle was to bring about the appointment of an old fellow student from Jena, GEORG ERNST STAHL, to be assistant professor of medicine; Hoffmann retained for himself the position of teacher in practical medicine, while Stahl took over the theoretical side. Stahl was born in 1660 of a Protestant family at Ansbach in Bavaria, receiving a strictly religious upbringing, which left its mark on his entire life. He studied at Jena, where be became a doctor and for a time gave lectures. After having been for some years court physician at Weimar, he came, as mentioned above, to Halle and taught there for about twenty years. At first his relations with Hoffmann were in every way friendly, but gradually the good feeling between them changed, and, finding that Hoff-mann's personal superiority excluded all possibility of competition, Stahl resigned from his professorship and in 1716 accepted an appointment as physician to the court in Berlin. He died there in 1734. Hoffmann and Stahl possessed their pietistic devoutness in common, but otherwise they were highly contrasted: Hoffmann, of stately build, lovable, and popular; Stahl, in his appearance insignificant, in his manner austere and inaccessible, in-tolerant towards his opponents, and bitter in controversy. At any rate he was a sincere seeker after truth, who was honest enough — a quality other-wise not very common amongst scientists — when he changed his opinion, openly to admit the incorrectness of his former views, and he likewise possessed that rare habit of gratefully acknowledging his predecessors' contributions to the problems he dealt with.

Reformer of chemistry

As a scientific writer Stahl was, like Hoffmann, extraordinarily productive and he dealt with a considerable number of different medical problems. As a scientist he was undeniably superior to his rival; in fact, his name is one of the foremost in the history of the natural sciences — principally on ac-count of his work as a chemist. At the close of the seventeenth century there was still being commonly taught at the German universities the subject of alchemy — belief in the transformation of metals, in the philosophers' stone, and all the rest of the mediæval mysticism which in western Europe, thanks to Boyle and his successors, had already been disestablished. Even Stahl had begun as an alchemist, and in his earliest writings he discusses the usual alchemistic problems, but by his own efforts he undeceived himself and thereafter never hesitated to point to the treatises of his youth as a warning. That uniform conception of the changes in nature which the alchemists sought to produce by means of their mystical speculations Stahl now endeavoured to attain by a comparison of those processes that take place in combustion on the one hand and the calcination of metals on the other. Finally he obtained a common ground of explanation for these phe-nomena by postulating the existence of a fluid substance, phlogiston, in both combustible substances and metals; in combustion phlogiston disappeared from the burnt material, as it did also from the metal in calcination; the metal calces were thus like the metal, minus phlogiston. If the metal calces were heated with a substance containing phlogiston — as, for instance, coal — the metal was recovered by the reintroduction of phlogiston. —

This theory rendered possible a uniform conception of a number of processes of conversion in inorganic nature; it constituted a working hypothesis which had a great influence upon the science of chemistry in succeeding ages and made the eighteenth century a period of brilliant achievement in chemical history; names such as Priestley, Bergman, Scheele, bear witness to the progress made in chemistry when the phlogiston theory was dominant; and when eventually Lavoisier, by introducing the weighing method, proved that the theory was untenable and substituted the idea of oxidation for calcination, the new theory could be applied directly to the discovery that had been made when the old theory prevailed. Were it only for the advance he thus brought about in chemistry alone, Stahl would deserve a place in the history of biology, which has been so essentially dependent upon the progress of chemistry, and indeed will always be so.

Stahl's medical theory

WHAT constitutes Stahl's principal claim to be mentioned as a biologist, however, is the theory of life which he expounds in his great work *Theoria medica vera*, in which he seeks to formulate a general theory of the human body and its functions, both in its normal state and in sickness. He himself has declared, and it has been repeated after him, that his chemical theories exercised no influence upon his ideas on the subject. This is true in so far as he does not — like his predecessors amongst medical chemists, Paracelsus, van Helmont, and others — base his entire conception of the human body upon speculation as to its chemical composition, but, on the other hand, the essential part of his work gives ample proof that chemistry is the science on which he bases his ideas. Above all, he is no anatomist; he scorns the result of ordinary macroscopical anatomy and he can hardly find words to express his contempt for Leeuwenhoek's and de Graaf's microscopical investigation of the sexual products; he likewise strongly contemns the discovery of the capillary system, the existence of which he simply denies. On the other hand, he displays a very keen interest in the "mixing (*mixtio*)" of the body and its parts — that is, their chemical composition — and he believes that a true conception of the phenomena of life should be based on the knowledge of this "*mixtio*." Indeed, it is in this direction that he has performed his greatest services to biology.

The first chapter of Stahl's principal work, mentioned above, is entitled "An Examination of the Difference between Mechanism and Organism." This title might well hold good for the whole of Stahl's literary work on general science; the contrast mechanism–organism is to him the main point in both biology and medical science; he discusses it from every conceivable point of view, and in support of his views thereon cites a number of arguments, both good and bad. The main argument, which he repeats again and again in proof of his theory, is that organism is something fundamentally different from mechanism, that consequently the mechanical physiology which his contemporaries universally accepted must be utterly repudiated. In the living organism the soul is the essential part; the body exists for the sake of the soul and is controlled by the soul. As a proof of this assertion he quotes, to start with, a number of ancient Aristotelean arguments on the finality of the structure of the body. Further, he declares that the existence of the body is due to a thing which is in itself foreign to the essence of the body, but, on the other hand, is akin to the essence of the soul, owing to its immateriality — namely, motion. The soul's function consists in going from object to object and comparing them, and the maintenance of the body by means of mental activity and constant moving goes on, subject to the will of the soul, as the result of motions suited to the objects that the soul requires. The fact that Stahl thus calls motion "thing" and compares it with the soul in contrast to the body proves that at any rate he had learnt nothing from Galileo and Newton. If, then, we find in this and other similar arguments the utter hollowness of Stahl's philosophical speculations, he has on other occasions an exceptionally keen eye, trained through his chemical studies for the essential in the composition of organism. As something essential to all the constituents of the body he points out the extreme easiness and rapidity with which they are chemically decomposed. This property evidently made a great impression on him; he constantly reverts to it and searches for an explanation for it, but it is obvious that, with the fundamental ideas that he once embraced, it is always the soul which ultimately keeps the body together and prevents it from disintegrating. This easy decomposability is considered to be due to a very complicated chemical combination in its constituent parts — a fact that differentiates it from ordinary chemical associations. The chemical quality is different in different forms of life and peculiar to each individual. Finally, the constituent parts of the body possess, besides their chemical quality, a special "texture" and "structure": the former an arrangement of the smallest parts of the body, the latter a combination of the elements thus formed, these two factors being characteristic for every living being. "Living body is nothing else than that which has structure," he declares. It is hardly necessary to lay special stress on the fact that as a result of all this investigation into the chemical nature of organism Stahl advanced science a long way; both the complex composition and the resultant easy decomposability of the constituent parts of the living body are indeed facts of fundamental importance for modern biology, and of still greater importance is his postulate that structure is something peculiar to the living organism in contrast to dead natural objects. Here Stahl has without doubt had some presentiment as to the significance of tissue structure as a basis of life in all its forms; that he was unable to follow up the idea to a conclusion of immense value to science was certainly due to his lack of interest in anatomy. Nor, indeed, did his contemporary age realize the importance of this question; it was not until sixty years after Stahl's death that Bichat, basing his results on anatomical studies along many different lines of inquiry, established the vital part played by the tissues in maintaining the functions of the body, but, as we shall see later on, he had come from a school in France that adopted and developed Stahl's ideas.

Doctrine of the soul as cause of life-phenomena

THE theory of Stahl's which aroused most interest in his time — that is, which evoked most applause and most controversy — was his doctrine of the soul as the cause of all life-phenomena, as their one supreme condition and their final aim. This "animistic" conception of the structure and functions of the body, according to which every manifestation of life, whether it is a question of the absorption of food, the blood-circulation, the processes of secretion and excretion, or simple movements from one place to another, muscular activity and sensations, takes place exclusively for the sake of the soul, is induced by it, controlled by it, and pursues its normal course thanks to it — this theory, so utterly opposed to the contemporary mechanical conception of life, was in reality the one main factor for Stahl, the very foundation on which he built up his medical system. For Stahl aimed at creating a new medical science, and his speculations in common biology were intended merely to lay the foundations of that science. Naturally, the diseases of the body are also caused by the soul; if it relaxes its control of the body or any part thereof, there at once ensues general or local decomposition of the inconstant chemical associations of which the body is made up, and sickness or death results. And Stahl does not hesitate to follow up this theory to its ultimate conclusions: if the soul desires to do so, it can naturally keep the body whole, but, as it happens, the soul is wayward, inconstant, and inconsiderate, and the body has to suffer for it. The soul of animals possesses in this respect less freedom of action than the human soul, with the result that animals are less often sick. One would suppose that in these circumstances any kind of medical treatment would be superfluous, since it has to deal with the body, which is in any case powerless, but Stahl does not draw this conclusion; like the homœopaths of a later period, however, he prescribes remedies having a mild action, with which he believes it possible to help the soul in its functions to the improvement of the body; violent remedies, such as quinine and opium, he deprecates. There is one conclusion that he draws from his system which does him honour — namely, when he prescribes mild treatment in mental cases; otherwise the physicians of his age, even the most humane, generally employed violent and sometimes brutal methods in their attempt to drive out the mental disease from the unfortunates who were thus afflicted.

Those of Stahl's contemporaries who adopted his ideas were at any rate not compelled to associate themselves with the peculiar theories referred to above. His criticism of that age's mechanistic conception of life is indeed often of such penetrating keenness that it must have proved attractive to those who sought to probe the contemporary controversial problems in that sphere. Especially does he inveigh against the theories of these "vital spirits" on which his opponents' explanations of the phenomena of life rested and which they could not possibly do without. Compared with these theories his soul-theory was at least simple and easy to comprehend; it cleared up satisfactorily enough the question of the relation of the psychical phenomena to the material, a problem on which all previous attempts to explain mechanically the phenomena of life came to grief. Stahl also had a sharp eye for other weaknesses in the contemporary explanations of life and demonstrated their inanity, as, for instance, the pan-sperma theories that were so common at the time. Besides his above-mentioned keen analysis of the contrasts between living and inorganic natural objects, which is only briefly summarized here, these critical contributions relating to the controversial biological questions of his age constitute Stahl's great service to science. This is, it is true, counterbalanced by his vague, yet subtle, natural philosophy, which has also been but briefly recounted here, and the understanding of which is rendered all the more difficult by a very obscure and badly arranged method of presentation. He gained many followers among his contemporaries; several of his own pupils gave practical demonstrations of the dangers of regarding the soul as an instrumental component in the functions of the body and the treatment of disease by indulging in extravagant speculations along mystical and theosophical lines. The valuable parts of his theories were most strictly adhered to and most faithfully developed at the University of Montpellier, where an entire school of physicians embraced his ideas. Among his opponents may be especially mentioned, besides his old friend Hoffmann, Leibniz, who in a contentious pamphlet sharply inveighed against his contempt of anatomy, chemistry, and other exact methods of research, and, from the standpoint of his own monad theory, rejected Stahl's theories of the soul and motion as being separate from the material part of living beings and as factors operating independently thereof. The influence that Stahl had on the development of biology in later times may at first glance seem small; indirectly, however, he has certainly been of greater significance than many of those who are more frequently quoted. Among those who have openly acknowledged their indebtedness to him may be mentioned such a comparatively well-known scientist as the embryologist Caspar Friedrich Wolff.

The man who, of the medical and biological theorists of that time, undoubtedly enjoyed the highest reputation among his contemporaries was, however, HERMANN BOERHAAVE. He was born in 1668, the son of a country parson, near Leyden in Holland; and there he studied, first of all, theology; but after becoming acquainted with Spinoza's theories he soon put an end to all idea of entering the clergy. So he had to look about him for a new means of livelihood. After taking a degree in philosophy at Leyden he moved to the small university of Harderwijk and there very quickly passed a medical examination, after which he settled down in Leyden as a practitioner and teacher. At first he had a hard struggle, but he assiduously carried on his profession, and his reputation rose year by year until he was finally elected to the first chair of medicine at Leyden and became universally acknowledged as the foremost physician in Europe. In that position he acquired an influence such as few have ever possessed before or since; his advice was sought not only from every corner of this hemisphere but even from the most distant parts of the East. He made a vast income and died a multi-millionaire. These successes were made possible owing to his brilliant gifts and, in spite of lifelong physical ill-health, his unfailing energy. But above even these merits his contemporaries valued his noble character; he lived extremely simply, while he used his great wealth to render help to the poor and sick and to give generous support to science; thus, he rescued Swammerdam's writings from destruction and enabled Linnæus to carry out his work in Holland; he was friendly and modest in society, but when the necessity arose, he could stand upon his dignity against even the highest in the community. He died in 1738, having during the last years of his life had to give up his professorial duties owing to ill health.

Boerhaave's theory: limitation of natural-scientific research

BOERHAAVE's attitude towards the general biological problems of his time was undoubtedly dictated by the fact that he had studied Spinoza in his youth and was throughout his life a keen admirer of Sydenham. Reminiscent of the former is his clearly and vigorously expressed characterization of the relation between body and soul. In man everything that involves thought is to be ascribed entirely to the soul as its starting-point. Whatever, on the other hand, involves extension, impenetrability, form, or motion, must be referred entirely to the body and its motion. Again, he is reminiscent of Sydenham in his realization of the limitations of natural-scientific research. "The investigation of the ultimate metaphysical and the primary physical causes is neither necessary nor useful nor possible for a physician. Examples of these causes are: the elements, the first forms, the origin of procreation, movement, etc." This quotation, moreover, shows his decidedly practical nature, also resembling Sydenham's. He actually placed it as the foremost aim of his science to create capable practical physicians. To gain this end, however, he considered that a grounding in general science was indispensable, and he therefore made a close study of the general structure and functions of the body, his work being based on what was a rare thing in those days, a thorough knowledge of the entire range of the then known medical and biological literature. The common biological, or, as he termed it, physiological, section of his principal work, *Institutiones medicæ*, gives the impression, owing to the mass of second-hand information that he imparts when quoting his sources, of being to a certain extent a compilation — in fact, it was published with the expressed intention of providing a handbook for instructional purposes — but the ideas he presents in it are at any rate thought out on entirely original lines, and the whole work seems, in comparison with Hoffmann's or Stahl's theoretical speculations, strikingly modern. The abstract theories are, as a matter of fact, entirely thrust into the background in favour of a close analysis of all the known facts relating to the functions of the body. First he describes the digestion, starting with a detailed account of mastication; then the functions of the digestive canal and its glands; then the circulation of the blood, and respiration, the brain and nervous system, several glandular systems, the muscles, the skin, sensations, and reproduction. From a purely anatomical point of view the presentation does not on the whole differ from the results achieved in modern times; we find here that he has taken full advantage of every step of progress made by such people as Borelli, Malpighi, and Ruysch. In particular Ruysch's careful dissections and injections Boerhaave, who, indeed, was a personal friend of his, was able to take advantage of in a masterly way.

His mechanical conception of life

WHEN he comes to explain the functions of the different organs, he bases his ideas on a strictly mechanical conception: the action of the body is motion; "the power to exert movement is called function, which takes place in accordance with mechanical laws and only by them can be explained." Thus both the disintegration and assimilation of food in the body are purely mechanical — he denies that the gastric juices have any chemical reaction — the principal agent is the body's own heat and the constant movements of the digestive canal and its surrounding organs, but the nervous fluid also plays a predominant part in the functions of the body. With regard to the question of the "cooking" of food in the digestive canal, as assumed by ancient authors, Boerhaave takes up a somewhat sceptical attitude. On the other hand, he believes that acrid and unsuitable food-substances become excluded by contracting the openings of the chyle vessels into the bowel. Such food as has been taken into the chyle vessels is conveyed through the thorax to the venous system; there blood and chyle are mingled, and this mixture becomes complete through the blood's passing into the lungs, whose porous structure serves to render the mixture as thorough as possible. In a contentious article written against Borelli, who believed it to be the case, he denies that the air from the lungs passes into the blood; Boerhaave is unable to explain why it is that living creatures cannot breathe in an unventilated room. The brain also serves to purify the blood, which passes through it for that purpose. Moreover, the cerebral cortex collects from the blood its finest constituents, which give rise to the fluid that is conveyed from the brain through the tubular nerve-threads out into all the different parts of the body and induces movements in them. In particular Boerhaave inquired deeply into the problem of muscular contraction and its relation to impulses derived from the nervous system; he gives an account of an experiment to show that muscular action is dependent upon the nerve and he considers that this influence of the nerve is due to the flowing of fluid from the brain. With regard to the mechanical action of the muscles, Boerhaave highly commends Borelli's mechanical investigations; the affluxion of the nervous fluid he believes takes place in accordance with Mariotte's law.[2] Boerhaave gives a detailed description of the structure and function of the genital organs, which is based on the discoveries of Leeuwenhoek and de Graaf. He holds that the sperm is "refined" blood; its small, living "animalcula" contain rudiments of the organs of the future embryo; as eggs he regards the follicles in the ovary, in this following de Graaf; conception takes place as a result of the "living elements" of the sperm penetrating the pores of the egg.

As a whole Boerhaave's biological theory must be considered to come far nearer our modern ideas than either Hoffmann's or Stahl's — this both on account of what he knows and above all on account of what he considers it impossible to know. His insight into the limitations of natural science really testifies more than anything else to his greatness; as regards facts, we cannot expect of him more than it was possible for his age to attain. But it is just his deliberateness that it has been difficult both for his contemporaries and for posterity to understand; the desire to solve the ultimate riddle of life has again driven the philosopher beyond the limits of what science can attain with the means available. We shall leave Boerhaave, clear-sighted and conscious of his own limitations, and shall proceed to consider a scientist who sought to solve the riddle of life along speculative lines and who expended on this endeavour one of the richest and most fertile geniuses known to history — namely, Swedenborg.

The son of Bishop Jesper Swedberg, a famous hymn-writer and preacher of the Swedish Church, EMANUEL SWEDENBORG was born in 1688 and received a thorough school and university education at Upsala, where he grew up; having completed which, he spent several years in England and on the Continent, studying principally natural sciences, both theoretical and applied. Having returned home, he served as a military engineer during the last fighting years of Charles XII, then became assessor of the board of mines,

---

[2] Boerhaave undoubtedly refers to the hydrostatic experiment which goes by the name of Mariotte's bottle; how its phenomena are to be applied to the nervous and muscular functions is, however, not clearly stated.

and was elevated to the nobility,[3] displaying during the next decades indefatigable energy as an official, member of the House of Nobles, and scientific writer. Then during the years 1744–5 he underwent a severe spiritual crisis; after repeated phases of alternate depression and exaltation he beheld in a vision the Saviour Himself and learnt from Him that he was henceforth to devote himself entirely to spiritual matters. He at once resigned his post of assessor and devoted his whole life to spreading the new doctrine that he believed he had received direct from heaven through repeated spiritual revelations. Pestered by the priesthood in his native country, he lived his last years mostly abroad, and died in deep poverty in London in the year 1772, misunderstood by his own age, but honoured as a religious founder by a small group of believers.

Swedenborg's natural-scientific works are extraordinarily extensive; he published books on mathematics, physics and chemistry, geology and cosmology, anatomy and physiology, and, besides this, much of what he wrote remained unprinted and has not been published until our own time, as, for instance, his anatomical work *De Cerebro*, which contains his most important investigations regarding the brain. All these works are full of ideas and genius, the true value of which was not appreciated until our own day, but which, on the other hand, contain very little in the way of original observations. He himself considered that he possessed more talent for thinking about already existing facts and their interrelation than for making observations of his own; but for the very reason that he did not support his speculations upon facts which he himself had observed, he ran the risk of letting his thinking be influenced by that attraction for the mystical which he had always felt and which had been encouraged by the religious environment of his childhood. Among the students of nature who thus impressed him must especially be mentioned Olof Rudbeck, who in Swedenborg's youth was the predominant figure in the University of Upsala and from whom he learnt not only his love of nature, but also a tendency to many-sided activities and fantastic conclusions. As we have already seen, Rudbeck was an upholder of the seventeenth century's mechanical conception of natural phenomena, both inanimate and animate, and this conception was also adopted by Swedenborg. It was developed in the course of his foreign tour by studying both the philosophy of Descartes and the writings of contemporary physicists and anatomists.

Swedenborg's views of the life-problem

His views of life are at first much the same as those we found in Hoffmann; the body is a mechanism, to which is added a vegetative life-force, *animus*, which consists of a fine material substance, and finally a higher soul, *mens*.

---

[3] After being ennobled he called himself Swedenborg, having previously borne the family name of Swedberg.

But while Hoffmann leaves the latter to the metaphysicians, Swedenborg becomes involved in speculations upon it; he holds that it likewise consists of a fine material substance, which leaves the body at death and continues to live in space; during life it receives mental impressions from the *animus* and forms them into knowledge. But what interests him most deeply is the question why knowledge is limited; like van Helmont he concludes that this is due to the Fall, for before the Fall Adam was omniscient, and it now became Swedenborg's aim to acquire this omniscience. He mainly sought to gain it by studying the function of the brain and its relation to the life of the soul.

Swedenborg's investigations of the brain really constitute the principal part of his activities as a natural scientist. In this field he succeeded, by brilliant comparison of conclusions drawn from the results of clinical post-mortem examinations and from contemporary anatomical works — mainly Malpighi's and Vieussens's researches referred to above — in creating a theory to explain the function of the central nervous system, which is far superior to any that the anatomical specialists of his time were capable of forming. Thus he localized the functions of the soul entirely in the cortex of the great brain and was of the opinion that the corpuscula of the latter (the pyramid-cells) discovered by Malpighi are connected by means of threads with the various parts of the body and with one another, so that definite parts of the body and definite parts of the cerebral cortex are conjoined to one another and form the substructure for the functions of the soul; it is through this apparatus that the sensations are put into motion. This theory of the brain, the value of which has been appreciated only in modern times, was, however, made the basis for the most fantastic speculations on the soul, which Swedenborg now believes to consist of a "*fluidum spirituosum*," a substance of exceptional fineness and directly derived from the eternal light. It is impossible, owing to the existence of sin, for man during his earthly life to come into contact with this supreme soul-substance, "*anima*," which possesses entirely ideal qualities, but he must be content with such lower experiences as his *mens* and *animus* give him through the senses. Swedenborg himself sought by way of desperate spiritual struggles to acquire that ideal knowledge which man, in his view, had inaccessibly preserved within him, but when he thought that he had attained his object after the vision mentioned above, his victory led merely to an initiation into the secrets of the spiritual world, which has certainly conduced to the edification of the few members of the Church he founded and of the far more numerous followers of spiritualism, but which has proved absolutely useless to science and to humanity at large, and which besides was the cause of the really splendid contributions he made in the field of natural research being considerably underestimated for a long time afterwards. It has been left to our own time to do him justice in this respect and to give him the place due to him in the history of scientific research.

As will have been realized from the above, the theoretical speculations to which reference has been made here led, on the whole, to poor results. The general theories of life and its manifestations which were formed at the period under discussion received a decidedly dogmatic stamp and became as numerous as those who formulated them.

On those lines, therefore, it was impossible in the long run to achieve any satisfactory results. Simultaneously with these efforts, however, there appeared others which succeeded better in satisfying humanity's craving for knowledge, and which during the immediately succeeding period won a very large number of adherents — those works which comprised a systematic description and classification of living creatures on earth. To these, then, we shall now proceed.

CHAPTER VI

THE DEVELOPMENT OF SYSTEMATIC CLASSIFICATION
BEFORE LINNÆUS

Primitive systematic categories of animals and plants

As LONG AS MAN'S KNOWLEDGE OF NATURE is limited to what he can observe in his immediate vicinity, he has little difficulty in controlling the objects of his knowledge, but when his range of vision widens, there arises the irresistible need for combining the individual objects that have been observed under general expressions, which serve to fix the knowledge of them and to impart it to others, "since no language would suffice to denote everything individually, and since in a language which did so, no understanding, no common knowledge, nor retention of such an infinity of terms would be possible" (F. A. Lange). Those categories in which natural objects are thus grouped by the most primitive peoples, out of sheer practical necessity, are naturally based on such qualities in animals and plants as well as the inanimate things that are observed as are easily comprehended, striking to the eye, and of special importance to the observers, and such terms are also used and invented even today among civilized peoples by all those who are concerned with nature in a purely practical way. On the other hand, a grouping of natural objects based on scientific principles has taken a long time to develop. In this respect the ancient Greek natural philosophy was content with the primitive popular nomenclature. Practically the first to devote scientific study to these groupings were, as far as we know, Plato and Aristotle. From Plato originates grouping in species and genera — that is to say, laterally arranged and superordinated terms — and his school still further extended this grouping of terms: the dichotomical determination-tables which even today play such an important part in plant and animal systematization originate from his school. But the further this grouping of terms went on, the more abstract became the result; the higher one came in the series of terms arranged one above another, the further away has one come from the things which one started from. This is a fact which the biological systematicians have not always realized; the practical advantage of systematic categories has led to the zoologist's and the botanist's forgetting how artificial their system has really become.

System of Aristotle

In this direction Aristotle did not go beyond what Plato had initiated; in his biological works there are, as is well known, only two systematical terms:

MARCELLO MALPIGHI

GIOVANNI ALFONSO BORELLI

eidos or species, and *genos*, the family, in which are included all combinations of forms which come above the notion of species. Nor indeed has he given us any really worked-out system; the animal system which is counted for his has been compiled by others from his writings. His knowledge of forms was also so slight that there seems to have been no difficulty in following the simple grouping which he employed. As a matter of fact, during the centuries that followed there was no need for a more detailed classification; the animals and plants which became known in late antiquity and the Middle Ages were not so numerous that they could not be covered by the Aristotelean natural philosophy. It was not until the great geographical discoveries of the sixteenth and seventeenth centuries introduced the knowledge of a great number of new life-forms that it was an inevitable necessity to widen the biological classification if the material collected was not to accumulate into an absolutely intractable mass.

The classification of plants especially demanded revision and expansion. Actually it was long after zoology had done so that botany attained the rank of an independent science. In antiquity and the Middle Ages botanical knowledge was essentially supplementary to pharmacology. Aristotle's botanical writings are, except for a few fragments, entirely lost. His disciple THEOPHRASTUS' great work on plants was adopted by later writers as a model; in it he thoroughly discusses the difference between plants and animals, higher plants and higher animals being exclusively compared and the comparison developing into abstract and fruitless speculations. The old primitive division into herbs, bushes, and trees is the only one to be found here. Besides Theophrastus' work there was during classical antiquity a purely pharmacological account of plants which was very celebrated and which was ascribed to a philosopher named DIOSCORIDES, whose character and period are unknown (he probably lived at the beginning of the Christian era); it was on Theophrastus and him that Pliny based the account of plants which is included in his great *Natural History*. In the Middle Ages these writings, which were believed to contain all the plants in existence, were closely studied and commented upon; attempts to find the plants from central Europe in these works, which applied only to the Mediterranean countries, led to the most absurd speculations. Only some few Arabian authors ventured through all this long period to describe new plants. It was not until the Renaissance that a change took place in this respect. One pioneer in this field was OTTO BRUNFELS, born, probably in 1488, in south Germany. In his youth he was a monk; then he became a Lutheran and a schoolmaster at Mainz; he died at Berne in 1534. He published an important work entitled *Herbarum vivæ eicones*, which inspired Linnæus to call him the father of botany. In this work, which was illustrated with excellent woodcuts, Brunfels describes all the plants he knows. In his botanical descriptions he still partly takes his stand by the old point of view; he begins each description with a list of names in different languages, followed by an account of what ancient authors have said of the plant in question; finally he gives his own "judgment" on the plant and ends with a statement as to its "powers." Compared with Gesner's exposition of the individual forms of animals (Part I, p. 93), this is certainly clumsy, but as being the first of its kind the work at any rate deserves respect. There is no system in it whatever; the book begins with Plantago, plantain, "because it is common and because more than any other plant it bears witness to God's omnipotence."

Thus it was at all events the medicinal powers of plants which most interested Brunfels, and the same is true of his numerous successors in the sixteenth century. The most interesting of these is LEONARD FUCHS (1501–66), who after working at humanistic studies under Catholic guidance went over to Protestantism, devoted himself to medicine, and finally became professor at Tübingen. His important botanical work *Historia Stirpium*, profusely and beautifully illustrated, was published in 1542. Its chief interest lies in the fact that he gives a list of all the terms he uses: an enumeration followed by short descriptions of the names of the different parts of plants. Curiously enough, the word "flower" is entirely absent. His description of individual plants, as compared with Brunfels's, indicates an important advance; of every plant an account is given of the (1) form, (2) habitat, (3) season (when it should be collected), (4) "temperament," (5) powers. It is only under the last heading that the views of the ancient authorities are referred to. Occasionally also the author, after the fashion of Aristotle, differentiates to some extent between species and genus.

Cesalpino's plant-system

THE first to deal with botany as a truly independent science, however, was ANDREA CESALPINO (1519–1603). His life was described in the first section (Part I, p. 113), as also his general scientific point of view — strict Aristoteleanism. His great work on botany, *De Plantis*, is based on the same system. Not only the fundamental ideas, but even the actual formal treatment of the subject is entirely on the Aristotelean model: exhaustive comparative analysis of the forms, concisely worded theoretical definitions, and, based on these, abstract conclusions, without any idea of such practical utility as was the main point with the old herbalists of the type of Brunfels. He begins a definition of the difference between plants and animals in the true Aristotelean style: plants feed, grow, and produce offspring, but lack the sensibility and motion of animals and therefore also need smaller organs than animals. Then follows a comparison between vegetable and animal organs, which, owing to its abstract one-sidedness, leads to curious results: the alimental organs of plants are the roots; thus these correspond to the stomach and intestinal canal in animals. Stalk and stem produce the fruit; thus they belong to the reproductive system. The plant is composed of several layers: bark, liber, wood, pith; of these the pith is the innermost and thus corresponds to the intestines of animals and is physiologically the most important. He is at much pains to discover which part of the plant corresponds to the heart in animals — we have previously pointed out (Part I, p. 113) the great importance which Cesalpino attaches to the heart as the centre of the body and of life. Finally, the plant's centre of life is found to be the collar of the root — the place where the stem and the root system join. Thence extend the vessels of the plant, of which the lacteal vessels especially are observed and compared with the veins in animals. Propagation by means of cuttings shows, however, that the central point of the plant is not as absolute as that of the animal; with true Aristotelean terminology it is maintained that this central point *"actu"* (actually) is in the root-collar, but *"potentia"* (potentially) can be everywhere. Cesalpino, moreover, is particularly interested in the fruits of plants, in which he sees the equivalent of the animal embryo; the function of the leaves is to protect the fruits, and the flower-petals are modified foils — an idea which was later adopted by Goethe. But Cesalpino does not admit the existence of sex in plants: the fruit is formed from buds and these again are produced out of the pith and the liber; the pith, which is the most vital part of the plant, provides the actual ovule, and the liber gives rise to the flower-leaf. Different kinds of fruits are carefully analysed and the plants are classified in accordance therewith, though the traditional division into trees, shrubs, half-shrubs, and herbs is retained as the main division. These four categories are then divided in their turn, according to the nature of the fruit, into a number of subdivisions. Cesalpino, however, like Aristotle, makes no summary of his system, not even in the form of chapter headings; nor is there any special systematic nomenclature. The mulberry-tree, the hazel-bush, and other fruit-trees are thus described each by itself; nevertheless, there sometimes occur divisions into lower categories than those named: of the carrot, Daucus, for instance, three forms are mentioned, Creticus, Montanus, Campestris, a division which has the character of a determination of species, or rather of variety. Nevertheless, these and other categories occurring in Cesalpino are not sharply defined; he was undoubtedly more concerned with anatomical and physiological than with systematic problems.

Cesalpino's system, in spite of its deficiencies, is the first to have been really based on the comparative study of forms; in this connexion Linnæus, who made a summary of it, expresses the opinion that Cesalpino is the first to lay down a definite basis for plant classification. In later times, however, this basis has been regarded as artificial, since it rests merely upon the consideration of one single organ, and in contrast thereto have been adduced contributions to a natural system of classification made by certain of the old, and otherwise not particularly systematic, herbalists. Independently of Cesalpino, plant classification was actually developed in a new direction through CASPAR BAUHIN. He was born at Basel in 1550 and studied medicine and botany, as did also an elder brother, under the above-mentioned Fuchs at Tübingen. He afterwards worked for a number of years as a professor in Basel, until his death, in 1624. His chief botanical works, *Prodromus* and *Pinax theatri botanici*, constitute the first attempts at a critical compilation of all the then known scientific names and descriptions of plants.

Bauhin's system

BAUHIN is entirely independent of Cesalpino; he bases his principles on his master Fuchs and those like him, the semi-medical herbalists of the sixteenth century. But he differs from the latter in his keen eye for the natural affinity of plants; he groups together such plants as resemble one another generally in their external form and discusses them in order, starting with those he considers the most primitive: the Graminaceæ, then the Liliaceæ, the Zingiberaceæ, after which the dicotyledons, and finally shrubs and trees. These groups are, however, neither characterized nor given names. Only the individual plants are described, which are combined under one genus-name, after which they are characterized in respect of all the forms that belong to each one of those names. These diagnoses are brief and concise and are accompanied by short accounts of earlier authors' statements on each plant. On the other hand, the actual genus-names are not in any way characterized, any more than the larger groups mentioned above; there is therefore no justification for the assertion that is sometimes made that Bauhin clearly grasped the contrast between genus and species. With greater reason he has been called the originator of natural plant classification based on the common likeness between the plant forms, as opposed to the artificial systematization founded by Cesalpino, which is based on an individual organic system — a contrast that has proved of great significance in botany, whereas in zoology it has not been of such consequence. And above all as a critic of earlier botanical literature Bauhin carried out a work of lasting value.

JOACHIM JUNG, generally called Jungius, holds a peculiar position amongst the botanists of the seventeenth century. Born at Lübeck in 1587, he became, while still young, professor in mathematics at Giessen, but soon relinquished his appointment, and thereafter, for more than ten years, he lived a somewhat restless life, until in 1628 he became rector of a gymnasium in Hamburg. He displayed extraordinarily keen and many-sided activity both as a scientist and as a tutor, but eventually he came to work in rather difficult circumstances, partly owing to quarrels with the Hamburg priests, who accused him of heresy.[1] For these and other reasons most of what he

---

[1] In the course of his education in Greek, Jung had studied, besides the New Testament, profane classical authors; when challenged on this point, he defended himself by saying that

wrote remained unprinted and was partially dispersed after his death (in 1657). Some few treatises were published by his pupils, among them one entitled *Isagoge phytoscopica* (*Handbook of Botanical Study*). This work, comprising a volume of forty-six quarto pages, must be regarded as one of the pioneer works in botany. It gives a concentrated account of the theory of botany, under the obvious influence of Cesalpino's, but without the latter's profitless Aristotelean speculations; to begin with, the plant is characterized as such, after which an account is given of the various organs, each of which is briefly diagnosed in a manner that is striking, though abstract. "A leaf is that which stretches out from its place of attachment in height and length so that the surfaces of the third dimension are dissimilar to one another; it is the leaf's inner surface that is differentiated from the outside." — The whole exposition, with its concise, vigorous sentences and its analyses of different parts of the plant drawn up in tabular form, is more reminiscent of Linnæus's work than that of any other of the early botanists. Linnæus, in fact, mentions Jung as his precursor as far as the drawing up of rules for the description of flowers is concerned and actually took up the characteristic description of plant-organs at the point where Jung had finished and certainly brought it up to a far higher standard.

One who in his time was of considerable importance as a classifier of plants was AUGUSTUS QUIRINUS RIVINUS (1652–1723). Born in Leipzig of a family of scholars, which really bore the name of Bachmann, he studied medicine in his native town, ultimately becoming a professor there. He was a many-sided scholar, working in widely differing spheres; his chief fame, however, rests on his great botanical work *Ordo plantarum*, which he published in two large folio volumes, illustrated with fine copper engravings, entirely at his own expense. He was the first to insist that the old division into trees, bushes, and herbs should be done away with; in its place he would classify plants exclusively according to their corolla, and he thus created an artificial system, which, however, was not very practical. He likewise urged the adoption of a simplified nomenclature for the plants themselves, but in this, too, his criticism of the old system was more successful than his attempts at reform.

A far greater service to classification was rendered by JOSEPH PITTON DE TOURNEFORT, at just about the same time. He was born at Aix in the south of France in 1656 and was destined by his father for the priesthood — much against his will. When his father died, therefore, he gave up theology and

---

the latter wrote purer Greek than that of the New Testament, whereupon the priests in Hamburg and theologians at Wittenberg accused him of blasphemy, because he had reproached the Holy Spirit, which had inspired the words of the Bible, with a deficient knowledge of languages. Jung had to abridge his school education, but, thanks probably to his high reputation, escaped the sentence of excommunication with which he was threatened.

applied himself to botany, which had always interested him. In order to be able to earn his living he started by taking the degree of doctor of medicine. His botanical works soon gained him a wide reputation; he was appointed professor at the Jardin des Plantes in Paris and had an opportunity of making many long journeys for research purposes. He died in 1708 as the result of an accident.

In the introduction to the important botanical work in which he summarized the results of his research activities he expounds his principles of plant classification. He defines the plant as an organic body, which always possesses roots, practically always seeds, and nearly always stalk, leaves, and flowers. He bases his ideas of the structure of plants on Cesalpino and Malpighi. When later it comes to classifying and giving characters to plants, he maintains, under the manifest influence of Cesalpino, that only the flowers and fruits can come into question; he seeks far and wide for proofs as to why root, stalk, and leaf do not provide reliable characters. In particular, the plant genera should be based on similarities in the structure of the flowers and fruits, but as the same genus includes forms whose remaining parts are different, so the genera must in their turn be divided into sub-categories. Tournefort pays great attention to his description of the genera, and his diagnoses of them are often so striking that subsequent systematicians, up to our own time, have been able to accept them, though they are only based on the characteristics of flower and fruit; on the other hand, the "species" into which the genera are divided are mentioned with only a few words regarding the form of the stalk and the leaf, without any further description. His method of procedure is thus the exact opposite of Bauhin's. But over and above this, Tournefort works out for the first time a systematic classification of categories higher than the genera — that is to say, he divides the plants into a number of classes, which again are severally divided into sections; each of these is characterized in a few words, but is not given a name. The characters of these higher categories are derived from the peculiarities of the flower; several categories of flowers which still to some extent hold good today are determined by him: with and without corolla, with or without a gamopetalous corolla, and, again, cruciform, lingulate, and other flower-forms. The division into herbs, bushes, and trees abolished by Rivinus he himself, however, was never able entirely to reject; his system comprises seventeen classes of herbs and five classes of bushes and trees. With regard to anatomy and physiology Tournefort has not much to offer that is new; in the course of his journeys he had observed the artificial fertilization of date-palms practised in very remote periods and already described by Theophrastus. They are, as is well known, both male and female, and the cultivators facilitate fertilization by suspending male clusters over the females, but Tournefort is unable to derive any theoretical conclusions of importance from the fact. It was left to another scientist, Camerarius, to prove the sexuality of plants.

Sexuality of plants

IT was known of old that in certain plants the individuals are of two different kinds, both of which must concur before any reproduction by means of fertilization can take place. The classical example of this, known to all the natural philosophers of antiquity, is, as mentioned above, the date-palm, the fruitful specimens of which have been quite correctly called, by the peoples who cultivate them, females, while those that are required for fertilization have been called males. But other plants of the same kind have also been known since ancient times, though many plants that resembled one another, but were differentiated by varying size and development were taken for females and males. A well-known instance of this was that of the two ferns *Filix mas* and *Filix femina*, which are still retained as names of species in two different fern-genera. But these ideas mostly belong to popular belief; scientists, both those of the classical period and, on their authority, those of the sixteenth and seventeenth centuries, denied, or at any rate overlooked, the existence of sexuality in plants, owing mostly to the fact that the great majority of plants are hermaphrodites. When no difference can be found in the male and female specimen, what is the use of assuming sexual reproduction? Grew was the first to believe that plants reproduce themselves sexually, "like snails" (these are, of course, also hermaphrodites). His opinion in this case, however, was based mostly upon theoretical speculation, and, as a rule, such speculations are, of course, less convincing than direct observation. The scientist who proved the sexuality of plants as the result of convincing experiments was RUDOLPH JACOB CAMERARIUS (1665–1721). He belonged to an old scholarly family, known since the Renaissance period, which had originally been called Cammerer, and he worked throughout his life at Tübingen, where he was for many years professor of medicine. He generally recorded the results of his work in small articles, frequently written, according to the custom of the period, in the form of letters to other scholars. The essay which alone justifies the mention of his name in a history of biology is a "Letter on the Sex of Plants," dated 1694. In this article he gives an exhaustive account of all the ancient authorities' ideas of the reproduction of plants and of the parts of flowers; he himself arrives at the conclusion that the pollen is the male, and the ovary is the female, element and discusses in connexion therewith a number of theories on sexuality and fertilization in general, without, however, contributing anything of special value from a theoretical point of view. Of all the greater significance are the experiments by which he proves his theory of the sexual properties of plants. He cultivated for this purpose a fairly large number of both monœcious and diœcious plants and found that if the male flowers are picked off in time, there will be no fruit, while fruit will certainly develop if the pistils of the female flowers are provided with pollen. These proofs had undoubtedly a convincing effect, if not on all his contemporaries, at any rate on succeeding ages. Linnæus in particular has acknowledged the contribution he made to the development of plant physiology.

Animal system neglected

WHILE, then, during the first two centuries of the new era plant classification was splendidly reorganized, during the same period animal classification on the whole made no progress. The zoography of the Renaissance period has already been described (Part I, pp.92–8); it was, generally speaking, not very systematic; in the best event one adhered to Aristotle, and in the latter's *Historia animalium* zoology had, in fact, an old and sound foundation, which contemporary botany lacked — a careful comparison, based on unique powers of observation and sense of form, between the individual animal forms, the value of which is manifest from the fact that most of the groups into which animals are there divided still hold good in the present system of classification. Particularly in regard to vertebrate animals, which have for obvious reasons been of primary interest to humanity, Aristotle had, as has already been pointed out, a keen eye for the natural affinity between the different forms, which is based upon agreement in the general structure and functions of the body. Thus there was opened up to animal biology during this period an important and fruitful field for research in the anatomical and physiological sphere, and this, again, caused the comparison between the life-forms in the animal kingdom to receive a different character from that between the life-forms in the vegetable kingdom; in the former a comparison between internal organs, the complex structure of which it was possible to make out only after exhaustive investigations; in the latter, a study for the most part of problems of the purely external form. In zoology, too, however, it was absolutely necessary to develop form classification, mainly owing to the fact that the different categories into which the known animal world is divided required a more definite determination than that given it by Aristotle and his successors. And this undoubtedly demanded co-operation between zoology and botany in order to find a common ground of comparison and valuation for all the forms in which life on earth manifests itself. The very first to make an attempt to deal with vegetable and animal classification on similar principles was Ray; the scientist who finally worked out a uniform system for all living creatures was Linnæus.

JOHN RAY was born in 1627 or 1628 at Black Notley, a village near Braintree, in Essex. His father was a well-to-do blacksmith who could afford to send his eldest son to college. In 1644 young Ray went up to Cambridge and at first studied the classical languages and theology; but he was also interested in mathematics and natural science. He gave lectures to the under-

graduates on Greek and mathematics alternately, and was eventually ordained, after which he held many college offices. His university period was
not to last long, however; the reactionary Government of Charles II required
the English clergy to subscribe to an Act of Uniformity drawn up with a
view to suppressing liberty of conscience; and Ray was one of those who
preferred to give up office rather than to submit. It thus came about that,
like so many of England's best scientists, he had to spend the greater part of
his life following the profession of a private scholar. This Ray was enabled to
do thanks to his connexion with FRANCIS WILLUGHBY, a very wealthy young
man of noble family, who, eight years younger than Ray, had been a pupil
of his at Cambridge and was his constant companion throughout his life,
their friendship being based on a common interest in natural science. After
Ray's resignation the two friends went for a several years' tour through
Europe, during the course of which Ray applied himself especially to botany,
Willughby to zoology. Having returned home laden with collections, they
settled down in Willughby's country-house in order to work up the material
they had collected. In 1672, however, Willughby's death abruptly terminated
their collaboration; by his will he appointed Ray one of his executors and
left him sixty pounds a year for life, with the charge of educating his two
sons, for which purpose Ray remained for some years in his friend's family.
Having married, he finally settled down in his parents' cottage, which he
had inherited, and there for several decades he continued his researches,
universally respected in scientific circles in England and contented with his
lot in spite of his modest circumstances. He died in 1705, three daughters
surviving him.

Ray's Methodus plantarum

RAY's literary work was extensive and many-sided — sermons and religious
essays, handbooks on the classics, treatises on folk-lore, and, finally, the
works on natural science on which his fame entirely rests. The greatest of
these, in both volume and importance, is his *Historia plantarum generalis*,
a work of 2,860 closely printed folio pages, in which he summarized the entire
botanical knowledge of his time. At an earlier date he published a résumé of
the system in which he arranged the plants in his *Historia*, under the title
of *Methodus plantarum*. This great history, which contains a systematic description of all the then known plants, starts with a general survey of the
nature and conditions of plants. He quotes Aristotle's principle as to the
division of the organs into simple and complex, similar and dissimilar. As
regards the various parts of the plants he bases his system on Jung's definitions and terminology, which are regularly quoted, but are in each individual
case considerably extended and thoroughly investigated. He cites the plant
classification which Cesalpino originated, according to fruits and seeds, but
he points out that the form of leaves and other parts must also be taken into

account, so that plants which resemble one another are grouped together although the seeds are different. Above all, he reminds us that nature never makes any jumps; on the contrary, the extremes are connected by middle forms, just as the zoophytes come between the vegetable and animal kingdoms. In regard to the anatomy of plants, in all essentials Ray follows Malpighi; Grew's ideas on the sexuality of plants are also accepted, without, however, being further developed; Ray was ignorant of Camerarius's observations. On the other hand, he describes the germination of plants, making original observations of considerable value; the difference between plants with one and those with two cotyledons was established by him. Ray discussed the notion of species more thoroughly than any previous biologist. In his view, plants belong to the same species if they give rise through their seed to a new plant similar to themselves, in the same way as bulls and cows are the same species because in mating they produce creatures which resemble themselves. The number of species is invariable, for God rested on the seventh day from all his work — that is, from creating new species. On the other hand, the different-coloured flowers in plants should not be regarded as separate species, any more than the different-coloured calves born of cows; in the former this is proved by the fact that the colour variations are not reproduced through seed, but only through cuttings. The invariability of species is, however, not absolute; plant species can be varied through the "degeneration" of the seeds — thus it has certainly occurred that the seed of the cauliflower has produced leaf-cabbage and that from the seed of *primula veris major* has arisen *primula pratensis inodora*. Ray even includes in the discussion a number of ancient stories as to grain's having degenerated into weed: wheat to Lolium and maize to other kinds of weed. True, he doubts the truth of a number of these statements, but he nevertheless believes the thing to be possible. This belief of his in the variability of species has been cited as proof of an unprejudiced view, in contrast to the theory that arose later as to the absolute constancy of species. The examples quoted rather go to show clearly enough that Ray was unable to rid himself of a certain amount of primitive superstition.

As far as actual classification was concerned, Ray retained the division into herbs and trees, or, more correctly, herbaceous plants and ligneous plants, maintaining that the latter are differentiated from the former by the existence of winter buds — in actual fact, an incorrect assumption. In a later edition of his *Methodus*, however, influenced by Rivinus, he abandoned this division. Herbs are then divided into: *imperfectæ* (fungi, algæ, lichens, and corals) and *perfectæ* (plants bearing flowers, which are again divided into those having two and those having one cotyledon). The sub-groups under these categories are numerous, some natural and well characterized, others composed of all sorts of plants, massed together owing to some purely accidental character. Trees are also divided according to the number of cotyledons, and then again into sub-groups. The actual genera are mentioned by one name and are described with a short diagnosis; the species into which they are divided are characterized in a few words, followed by a more detailed description. Ray was thus the first to describe both genus and species at the same time.

Ray's zoological system

As a zoologist Ray left no comprehensive work corresponding to the botanical work above mentioned. During his period of collaboration with Willughby the latter took over the zoological side, and after his premature death Ray published in his name a couple of works on birds and fishes; how much in these works originated from the one or from the other of the two friends it is not easy to decide. In his own name, on the other hand, Ray promulgated two zoological works: a survey of the quadrupeds and reptiles, and a work on insects. The former of these, a small octavo volume, is his most important contribution to the knowledge of the animal kingdom. He begins with some general reflections on the characteristics of animals; he defines the animal as a body having life and powers of perception and of independent motion, and he then discusses Descartes's assertion that animals lack sensibility, the incorrectness of which is proved. As regards the reproduction of animals, he denies spontaneous generation and then deals with the theories of epigenesis and preformation, ovism and animalculism, without making any very important contributions in that connexion. The theory of fabulous creatures, which had always up to then been included on the authority of classical authors, is examined and entirely exploded. In regard to systematic classification, which comprises the greater part of the work, Ray follows Aristotle in essentials, and this for good reasons, since the latter's division of the quadrupeds is on the whole both natural and well founded. Ray, however, did not venture to follow up the consequences of the comparative anatomical method which Aristotle founded; like the latter he refers whales to the fishes, although he is quite well aware of their closer anatomical affinity with the mammals. On the other hand, in certain respects Ray goes deeper into the characteristics of the individual animal groups which he adopted; above all, he takes account of the structure of the circulatory organs and on this basis divides animals first of all into sanguiferous and bloodless — he quite realized that the last-mentioned group possesses blood of a kind, though colourless, but he prefers to retain the Aristotelean nomenclature. The sanguiferous animals are divided into those that breathe with lungs and those that breathe with gills; those provided with lungs are again divided into animals having two heart-ventricles and animals with only one. To the last belong oviparous quadrupeds and reptiles; the first is divided into oviparous (birds) and viviparous (partly land animals — mammals —

partly aquatic animals — whales). Land animals are also characterized by their hairy covering, as the result of which the manatee, which lives in the water, can be included among them. The bloodless animals are divided into small (the insects) and large (molluscs, crayfish, crustaceans). In examining this system we may pass over the "bloodless" animals, of which Ray himself only studied the insects; aquatic animals were, on the whole, of no interest to him, and as, moreover, Willughby had devoted himself to birds and fishes, there remained only the quadrupeds, which, as mentioned above, formed the subject of his principal zoological work. The hairy quadrupeds are divided into Ungulata or hoofed animals, and Unguiculata or clawed animals. Among the former are reckoned one-hoofed (equine), pair-hoofed (Ruminantia and swine), and multi-hoofed (rhinoceros and hippopotamus). Amongst the Unguiculata are included pair-clawed (the camel) and multi-clawed: (1) with claws grown together (the elephant); (2) with separate claws, of which there are: flat claws (apes) and narrow claws (carnivorous animals and Rodentia). Moreover, a number of mammals are classified as "anomalous" — namely, the hedgehog, the mole, the shrew-mouse, the armadillo, the sloth, and the bat. The oviparous quadrupeds are finally divided into frogs (including tortoises), lizards, and snakes. In this system each genus is then characterized with a diagnosis — for instance, the genus Ovis, the genus Martes — and the species of the genera are likewise given each a separate diagnosis. On the other hand, the genera of frogs, lizards, and snakes are not diagnosed, only a common characteristic being named, followed by diagnosis of the species.

To measure Ray's work as a systematician by modern standards would naturally be entirely unhistorical, but his system can by no means bear comparison even with that of Linnæus. And yet for his age it constitutes an extraordinary advance, primarily in that he clearly realized the difference between species and genus, secondly on account of his possessing what was undeniably — in comparison with his predecessors — an extremely keen eye for the similarities on which the assumption of affinity in its wider sense may be based; several of his larger groups, both in the vegetable and in the animal kingdom, are "natural" in the best sense of the word. In the sphere of botany, also, the difference discovered by him between mono- and dicotyledons is of essential importance. On the other hand, several of the subdivisions which he formed are highly artificial, as will be clearly seen from a glance at his division of mammals, according to claws and nails. And in any case he established no common systematic categories to cover all living creatures. The one who by doing so paved the way for a completely uniform conception of life-form on this earth was Linnæus, the founder of modern plant and animal classification.

CHAPTER VII

LINNÆUS AND HIS PUPILS

Linnæus's life and work

NILS INGEMARSSON was a peasant lad from Sunnerbo, in the province of Småland in Sweden, who was destined for the priesthood. When at school, not having previously had any family name, as was the case with the country people in general in Sweden, he adopted the name of Linnæus, after a mighty linden-tree growing near his home, which was regarded by the country folk as a sort of sacred tree. After a long period of study at Lund University — frequently interrupted, owing to his poverty — he was ordained priest in 1704 at the age of thirty, and two years later he was appointed curate at Råshult. At the same time he married Christina Brodersonia, daughter of the Vicar of Stenbrohult. Some years later he succeeded his father-in-law as vicar of that place. While following his vocation he also devoted himself with keen enthusiasm to horticulture and the study of herbs; in his large garden grew many a herb that was not to be found in his neighbours' gardens and with the peculiar properties of which he was well acquainted. The eldest of his large family was a son, CARL, born on the 23rd May 1707. Even in his earliest childhood Carl displayed the same keen interest in botany that his father had done; his greatest joy was to work in the small garden he had had laid out and there to cultivate as many remarkable plants as possible. At his school, at Växiö, however, he was, as he himself relates, far from happy; "crude schoolmasters in a crude manner gave the children a mind for sciences enough to make their hair stand on end." In humanistics, which at that time were the most important, he likewise made but little progress, but he was all the more successful in the physical-mathematical subjects. His teacher in physics, Rothman, quickly recognizing his great gift for natural science, gave him Boerhaave's and Tournefort's works to read and urged Carl's family to accept his plan to devote himself to medicine instead of studying for the priesthood. In 1727 he became an undergraduate at Lund, where he found a paternal friend in Stobæus, professor of medicine. On the advice of Rothman, however, he removed for the next academical year to Upsala, where the medical teaching was considered to be of a higher standard according to the requirements of the age, which, however, is not saying very much. Linnæus had for the most part to carry on his studies by himself. During his first term at Upsala he lived in dire want, but he soon succeeded in procuring patrons there: the dean, Celsius, who was likewise interested in botany, took him into his family and undertook to procure him further advancement. Even as a young student Linnæus had always shown that capacity which never left him throughout his life, of exciting the admiration and sympathy of those he met who possessed interests similar to his own — a quality based on the keenness with which he himself embraced the work he had made his own. When once he had acquired friends at the University he gained one success after another. Though not yet a graduate, he obtained permission to lecture on botany and he used to attract large audiences. He received a number of grants, and with the aid of public funds he made journeys of exploration to the Lapp district and Dalecarlia, in the course of which he collected material for research consisting not only of natural objects, but also of human customs and habits. During the latter expedition he made the acquaintance of his future wife, daughter of the wealthy town-physician of Falun, Moræus. In order to secure further advancement in the career he had chosen, Linnæus had to obtain the degree of doctor of medicine, but there was no such degree in Sweden at that time. He accordingly made a journey, with the financial support of his future father-in-law, to Holland, where at the small university of Harderwijk, which never attained to a very high standard of scholarship, he took his doctor's degree in a couple of weeks. By that time, however, the money he had brought with him had become exhausted and Linnæus had no other resource than to chance his luck elsewhere. He accordingly went, in company with a fellow-countryman, to Amsterdam and thence to Leyden. There he became acquainted with several scientists and people interested in science, chief of whom was Boerhaave, who treated him with paternal kindness. With the assistance of one or two patrons Linnæus was able to print his most epoch-making work, the *Systema naturæ*, which he had already begun in Sweden and which brought him immediate fame. He then spent three years visiting the principal centres of learning in Holland, publishing one work after another with marvellous rapidity, supported by patrons and often almost persuaded to settle in Holland for good. He longed to return home, however, and after paying visits to both England and France, he returned to Sweden with a European reputation, but without any very brilliant prospects for the future. He succeeded, though with some difficulty at first, in making a living as a physician in Stockholm, until in 1741 he won the position for which he had striven so long — the professorship of botany at Upsala. During his Stockholm period he had taken part in the founding of the Academy of Science and had been its first principal; at Upsala, from the day of his arrival, he became the foremost member of the University. His time and capacity for work sufficed for everything — for his teaching, which went on summer and winter, before ever-increasing audiences, both of Swedes and foreigners; for the reorganization of the botanical garden (which existed in Rudbeck's time, but which had now fallen into decay), making it one of the finest in Europe; for the production of extraordinarily fine scientific works and an extensive correspondence. As a founder of schools and an organizer of work he has had few equals in the history of biology. Every year he sent out pupils on research expeditions, whose collections and observations were afterwards worked up under the master's own guidance. He himself was acknowledged throughout the whole civilized world as an authority on natural-scientific questions, his advice being sought by governments as well as private individuals. His native country also learnt to appreciate him; he received several high honours; among other things he was ennobled and took the name of von Linné.

The climax of Linnæus's greatness falls within the period of the seventeen-fifties; then he published the last of his great works, and then, too, he received his highest honours. The quarter of a century of life that still remained to him was a period of decline. The hardships suffered in his youth and the cares of his maturer years had undermined his health. By the beginning of the fifties he had already become seriously ill, but he still managed to work during the succeeding decades — in part producing results of considerable importance — although his powers of movement began to fail. During the seventies, however, he was subject to repeated paralytic strokes, which dulled his intelligence and finally paralysed him entirely. In 1778 death brought release.

In 1763 Linnæus had taken a step which was certainly the most unfortunate he ever took in his life; he had obtained from the Government the right to recommend his successor and he appointed his only son, Carl von Linné the younger, who thus at the age of twenty-two became aspirant to the professorship, possessed no brilliant gifts and had never passed any tests of scholarship. Although his promotion was by no means so ridiculous in the eyes of his contemporaries as it would have been in modern times — it was quite usual for people to purchase a "survivance" to an official post similar to that which young Linné obtained on account of his father's services — nevertheless this step had the most unfortunate consequences. The feelings entertained by the large crowd of far more competent pupils were naturally very bitter and were enhanced the more the worthlessness of young Linné's character manifested itself, as it unfortunately did very soon, no doubt hastened on by his unmerited promotion. And, to make matters worse, it caused also a division in the Linné family. On his father's death the son laid claim to his collections, which his mother and sisters, supported by a will, refused to allow. The quarrel was finally settled by arrangement, and shortly afterwards Linné the younger died, at the early age of forty-two, after a life which brought little honour to the name he bore and which died out with him. The unlucky position into which he got himself as regards both his family and his colleagues was also no doubt responsible for the sale (ignominious indeed for his country) of the collections of Linnæus — his herbarium, library, and correspondence — to England, where they are still preserved by the Linnean Society, which was founded for that purpose.

His fame

LINNÆUS has in the course of years been very differently judged. Already in his youth he had been hailed by his contemporaries as the "*princeps botanicorum*," a title that he succeeded in holding, not only throughout his life, but long after his death. But the reverse came in connexion with the acceptance of the descent theory in the middle of last century, for the opponents of this doctrine quoted Linnæus as their chief authority, and that not only on scientific grounds but also from motives which lay far removed from all that natural science means: his primitive Christian piety was thrust into the breach by religious and social conservatism against the "unbelief" of the new biology. It was naturally inconceivable that in such circumstances Linnæus and his works should be judged with impartiality; in the eyes of many he became simply the arch-enemy of the new science, and the judgments passed on him at the time were often not only spiteful, but also utterly absurd. Towards the close of the century, however, a calmer atmosphere prevailed, as was clearly manifested when in 1907 the bicentenary of Linnæus's birth was celebrated by the entire civilized world as a red-letter day in the annals of human culture.

Linnæus is universally reckoned among the examples of early scientific maturity, and it is true that by the time he had reached about his twenty-fifth year, he had already fully worked out the principles on which his subsequent work rests. Less remarked has been the steady development which he underwent so long as he was generally capable of working; the Linnæus whom we meet in the first edition of the *Systema naturæ* and writings contemporary therewith is not in the least the same person as the one who composed the final editions of that work. This may to some extent explain why such contradictory judgments have been passed on him; the one has sought support for its opinion of him in the work of his youth, the other in that of his old age.

His general conceptions of nature

IF on the basis of Linnæus's writings we were to try to form an opinion as to his general conception of nature, we should soon discover that he never formulated any elaborate theory of the phenomena of life in their entirety, such as Hoffmann, Stahl, and Boerhaave did, each in his own way. In the works of his youth there appears only a naïvely popular conception of nature, which, as a matter of fact, he retained, practically speaking, throughout his life: nature is created by God to His honour and for the blessing of mankind, and everything that happens happens at His command and under His guidance. No other explanation of natural phenomena is looked for. How little Linnæus actually interested himself in the general scientific questions that occupied the minds of his contemporaries is at once shown by the fact that even in the twelfth edition of his *Systema naturæ* he still lets the universe consist of the ancient four elements, fire, air, water, and earth; he seems not to have been aware of the fact that many decades earlier Stahl had published his new theory of the process of combustion. On the other hand, in this and in other works of his later years there occur a number of ideas contributing to a mechanical explanation of life that are in striking contrast to his romantic piety. In the above-mentioned edition of *Systema naturæ* he defines (on p.15) animal life as a hydraulic machine which is kept going by an ethereal-electric fire maintained by breathing;[1] on the other hand there comes in here the universally known, sublimely poetical description of God's omnipotence: how he saw the Eternal wherever he went, and how his brain reeled when he saw traces of Him in everything, from the life of the minutest creatures here on earth to the movements of the heavenly bodies, "which are upheld in their empty nothingness by the first movement, the essence of all things, the mainspring and director of all causes, the Lord and Master of this world; should we call Him Fate, we should not be wrong, for everything hangs upon His finger; should we call Him Nature, we should not be wrong either, for all things have emanated from Him; should we call Him Providence, we should likewise be right, for everything happens according to His nod and His will." The strange, half-pantheistic conception of God that is here apparent occurs in Seneca, whose *Quæstiones naturales* Linnæus cites in this connexion and often elsewhere; besides which he quotes in the work in question the Bible, Aristotle, Cesalpino, and van Helmont in support of his theory of the universe and of life-phenomena. To Galileo's physics, Newton's astronomy, and Stahl's chemistry, on the other hand, he has paid no attention; at any rate, there are no quotations that would indicate his having done so.

His gifts as a systematician

At the time when polemics were levelled at him, Linnæus was accused of Aristoteleanism in a derogatory sense. This accusation may have a certain amount of justification, but it is likely in all ages to be laid at the door of everyone desirous of arranging things according to formal principles, and that was what Linnæus desired, just as it was exactly what biology in his time needed. Far from blaming him for it, therefore, posterity should, on the contrary, be grateful to him for having, instead of working out specu-

---

[1] Is it possible that the "fire-machine" constructed by Triewald, which Linnæus saw in his youth in the mine at Dannemora, may have been recalled to his mind and have given rise to this curious definition?

latively some doubtless unproductive system of thought, devoted himself entirely to ascertaining the relation of forms to one another — an investigation which was so suited to his peculiar gifts. That in doing so he accepted the old biblical conception of nature was as natural in his day as for a systematist of our own day to embrace the theory of descent without going closely into the question of its justification. He was thereby able, unhindered by any theoretical barriers, freely to develop and take advantage of that extraordinary capacity for observing natural objects and summarizing his observations which was peculiar to him, and thus to establish the mastery over research-material on which modern biology is based.

Linnæus was, as has already been mentioned, essentially autodidactic, in so far as the education he received from others was highly deficient and fragmentary. Nevertheless, the conditions under which he was trained for his life's work were particularly suited to his natural genius. The powers of observation which formed one of his most conspicuous characteristics had received from his very earliest years in his father's garden, under his guidance and under the influence of the love of the vegetable world imparted by him, such stimulating exercise as to afford every opportunity for the full development of his extraordinary sense of form. During his youth his teachers gave him a knowledge of such biological literature as then existed, without at the same time burdening him with any theories out of which he would afterwards have had to work himself up, while at an early age he gained that liberty of action which is the indispensable condition for anyone who, in whatever sphere his work may lie, wishes to create something new. The works of his youth, the small lists of plants which he drew up and which were not printed until our own day, already give clear evidence of where his chief interest lay; he enumerates the plants he collected in various places, with observations as to their occurrence, and by means of them he tests the various systems which he found amongst his predecessors, principally Tournefort, but also Ray and Rivinus, without, however, finding any real satisfaction in them. On the contrary, we find from his notes that he felt himself called upon to reform the science of botany, which he considered to have seriously degenerated. Thus he became aware, through a criticism in a journal, of Camerarius's discovery of sex in plants, which a French naturalist had accepted, and he was so excited by the news that he at once devoted himself to making a close study of the problem. He immediately realized that in the hitherto neglected stamens and pistils one had to do with the flower's most vital organs, and from that point of view alone their employment as a basis for systematic classification was justified. Thus arose his sexual system, the first step towards the realization of the ambition he had set himself to attain: a general system for natural objects. And at the same time he made himself quite clear as to the principles on which such a systematic classification would have to be worked out, the result being the second of the important works written in his youth, *Methodus plantarum*, wherein he presented most of the principles which have since then been the common property of plant and animal classification. He first of all laid down an explanation and a definition of the various parts of the plant, after the model of Jung and Ray, whom, however, he far surpasses in the matter of precision, both of observation and of expression. Further, he worked out in an incomparable manner the principles of nomenclature, synonymy, and characteristics of the various categories of the system, all of which have since then been the common property of all systematicians of any ability, but which in his time reacted with all the overwhelming force of a novel idea. With all these ideas partly written down, partly in his head, Linnæus came to Holland, and was able, under the unusually favourable conditions which he enjoyed there, to make them fully available for research. But before giving an account of those, the greatest works of his life, we must devote a few words to a man with whom he closely collaborated and who made a strong and lasting impression on him.

PETER ARTEDI was born in 1705 at Anundsjö, in northern Sweden, the son of a priest named Arctædius. He entered Upsala University in 1724 and, like Linnæus, had difficulty in obtaining his father's consent to his studying medicine in preference to theology. It was natural science, however, that chiefly attracted him, and in that field he too, like Linnæus, had for the most part to study on his own. At the time when Linnæus came to Upsala, Artedi was considered the most promising naturalist in the University and there soon arose a firm friendship between them, resulting in a co-operation which proved of great benefit to both. Artedi was especially interested in zoology, chiefly in icthyology, while Linnæus applied himself to botany, so that there was no necessity for them to encroach upon one another's fields of activity, but at the same time they could exchange ideas and observations. In their characters, too, they were fortunate in being able to complement one another; Linnæus was lively and enthusiastic, Artedi calm and critical. Financially they were both in an equally bad way and they had recourse to one another's assistance. In 1734 Artedi received a grant to enable him to travel abroad and he went to London, where he studied zoology, mainly icthyology. A year later he came to Amsterdam without resources and without the slightest prospect of getting home. Linnæus, who had already acquired some connexions in the city, introduced his friend to a wealthy apothecary who possessed a large collection of fishes. This museum Artedi was now commissioned to catalogue and was able at the same time to complete a large work on fishes on which he had long been engaged. His career, however, was short; one evening, upon returning from a visit to his benefactor, he fell into a canal and was drowned (autumn, 1735).

The icthyology of Artedi

His work was published by Linnæus with the assistance of a Dutch patron. It was probably in all essentials the work of Artedi, though Linnæus made some additions here and there. The work purports to be a complete monograph on fishes; the anatomical section, however, is of minor importance. The chief interest lies in the presentation of the theory of the system, which is incorporated in the part entitled "*Philosophia icthyologica.*" In this are discussed with sharp criticism the various systematical categories. He starts, after the model of Tournefort, with the genus, which is defined as a collection of species that, as regards the shape, position, number, and mutual relation of the parts, agree with one another and differ from other genera. The species, moreover, he bases, not like Ray and Linnæus on common origin, but on dissimilarity in the same genus in respect of some individual part of the body, a principle the weakness of which in comparison with Linnæus's becomes at once apparent. As higher categories he adduces classes and orders; the classes should be "natural" — that is, be based upon agreement in several essential parts and not upon unessential factors, such as occurrence, size, and the like. Fishes form one such "natural" class, owing to the shape of their body and their fins, whales nevertheless still being counted in the "natural" class. The orders into which the class is divided are on the whole the same as those still in use today — a proof of Artedi's systematical acumen; selachians, acanthopterygian and malacopterygian osseans are categories invented by him. Linnæus adopted his icthyological system unaltered in his *Systema naturæ.*

Linnæus's first great work: Systema naturæ

This great work of Linnæus, the natural system "in which nature's three kingdoms are presented divided into classes, orders, genera, and species," was published, as already mentioned, in Leyden in 1735. At the same time was printed the above referred to *Fundamenta botanica,* and three years later the important work *Classes plantarum.* These three really contain all that is essential in the reform of classification which Linnæus carried out. Like Ray, but in contrast to Tournefort, Linnæus as a systematician takes as his starting-point the idea of species. He adopts Ray's theory of the species as created from the very beginning and immutable, laying this down as a fundamental principle without limitations or exceptions. "We count as many species as have been created from the beginning; the individual creatures are reproduced from eggs, and each egg produces a progeny in all respects like the parents." Thus there is no room for spontaneous generation, no possibility for the seeds of one plant to give rise to a plant of a different kind. Rather it was expressly maintained that in the beginning there was created of each species one single pair, one of each sex, so that all individuals of the same species possess a common origin. Again, there exist as many genera as there are, among the natural vegetable species, flowers (or fructi-fications, as they are termed in these works) differing in number, shape, and position. Classes are defined as a collection of genera that agree in regard to fructification in certain main features. The order, again, is a subdivision of the class which embraces a number of more easily summarized genera. The application of these principles is Linnæus's universally known sexual system, in which the classes are essentially determined according to the number of stamens, and the orders according to the number of pistils. The practical utility of the system is sufficiently evidenced by the fact that it is used to this day in school education, although it was long ago abandoned in actual scientific work. That this system, based as it was on only one organic system, was one-sided, Linnæus was very well aware, and in several instances he departed from the fundamental principle merely in order to preserve the con-nexion in certain groups which he found to be natural, as, for instance, the classes Didynamia, Tetradynamia, and Gynandria, which, it is true, are characterized by the stamens, but not only by their number, and which com-prise forms that even the system of classification of our own day keeps to-gether. For Linnæus was fully aware that what should really be striven for in botany was a "natural system": a classification of genera into groups on account of a common similarity, not merely on account of the relations of certain organs. He spent his whole life working out this natural system; the results he achieved will be mentioned later on.

Linnæus's systematic classification of the animal kingdom cannot be said to have turned out as successfully as his plant system. He divides animals into six classes: (1) Quadrupedia, (2) Aves, (3) Amphibia, (4) Pisces, (5) In-secta, and (6) Vermes. Quadrupeds are characterized as follows: "the body hairy, four legs, females producing live young, which they suckle"; birds: "the body feathered, two legs, two wings, beak, females laying eggs" — that is to say, purely external characteristics. Linnæus's precursor Ray based his system, as we have seen, essentially upon anatomical characteristics: the structure of the respiratory organs and of the heart; moreover, he differen-tiated, although with faulty characterization, between vertebrates and in-vertebrates; the latter he divides into four groups, while Linnæus has only two — all details in which Linnæus was undeniably inferior to his pre-cursor. Fishes Linnæus has dealt with entirely in accordance with Artedi's system, which he in fact acknowledges. Of the lower animals the only ones that interest him are the insects. Moreover, Linnæus has not laid down any general principles for animal classification similar to his *Fundamenta botanica*. Artedi's "*Philosophia icthyologica*" might certainly be said to have filled the gap, but the latter's method, as we have already seen, differs not a little from Linnæus's, primarily in the fact that his system is based on the genus and not on the species — and in these circumstances it only remains to show that the Linnæan reform was from the beginning more adapted to vegetables than to animals.

The conception of species: their immutability

BUT, all the same, Linnæus's contribution to the development of biology has been of vital importance to science as a whole. In the first place, by fixing the term "species" as he did, he laid the foundation for the system of classification as it exists today. At the time when the dispute on the descent theory was raging at its hottest, Linnæus, it is true, was exposed to the severest censure just because he had declared the species to be immutable, as they were created from the beginning — the dispute, in fact, raged just as much over the belief in the creation as over the constancy of the species itself — but in spite of the fact that the immutability theory is now abandoned, the Linnæan species is used in practice by systematic science even today, because it has not been possible to find any better substitute for it; a species is regarded as the sum total of those individuals which resemble one another as if they had a common origin. The other systematical categories which Linnæus created also remain to this day, although some new ones have come into existence as well. And quite as remarkable is Linnæus's influence on what may be called the technical side of the classification system, which he himself actually founded, exactly as it is applied today: his rules regarding nomenclature, description, characterization, and synonymy have really proved so complete that in principle posterity has had but little to add to them. Moreover, the whole of this radical reform was carried out at one stroke by a hitherto unknown young man after only a few short years of utterly inadequate scientific training. This wonderful result was rendered possible only by the fact that Linnæus combined exceptionally well-trained powers of observation with an unparalleled natural genius for the formal side of science. This latter gift was, so to speak, in his very blood: he had a passion for classifying everything that came within his grasp; his medical writings consist of groups of diseases in tabular form; his predecessors in science he likewise classified under various headings, and once he even arranged, mostly as a joke, all his contemporary botanists according to military rank, with himself as their general. For the fact that this mania for classification never degenerated into mere dull pedantry he had to thank his extraordinary love of nature and his passion and gift for observing life in all its manifestations. It was this quality that prevented him from stagnating at the point to which he had so rapidly attained, instead of which he spent his whole life striving to extend and perfect the science that he had already so thoroughly recreated. These efforts, which, with the aid of his pupils, he continued as long as his powers lasted, consisted in improving the system he had already created, extending and perfecting the natural vegetable system that he had already made it his ambition to work out, and, finally, in making a number of observations of life in nature and the interrelation of its different phenomena.

The binary nomenclature

THE most important purely formal improvement of the system which Linnæus effected was the binary nomenclature, introduced in 1753 into the classification of the vegetable kingdom and somewhat later into that of the animal kingdom. Previously he had followed the example of Tournefort in characterizing every genus by a single term, while the character of the species was designated by a short diagnosis of some few words. Now he introduced instead one single character word for the species also, so that every plant or animal received its character and its fixed place in the system by means of only two words. This reform is certainly the most important of his contributions in the purely formal sphere. Thanks to this alone biology has been able to master the vast amount of form-material that has been collected up to the present day, which could certainly never have been handled if it had been necessary to employ diagnoses in order to denote the species. His other reforms in connexion with the system applied not so much to botany as to zoology; he left his botanical sexual system for the most part undisturbed and contented himself with incorporating into it the new species which were sent to him from all over the world. Of the improvements which he introduced into the animal system the most worthy of mention is the fact that he at last associated whales with the quadrupeds, which resulted in the latter's receiving the name they have since borne — Mammalia — that is, animals which feed their young. The orders into which he divided this class, mainly after the dental structure, were, however, still somewhat artificial and were long ago rearranged; on the other hand, his method of associating man with the apes in the order Primates has been retained. Birds, which were essentially classified according to their beaks, have, as is well known, been still further regrouped. His transfer in the last editions of *Systema naturæ* of the Cartilaginei to the amphibians, which was at variance with Artedi's system, was extremely unfortunate, based as it was on a misconception of those fishes' gills. On the whole, Linnæus disliked cold-blooded animals; as a motto for the amphibians he chose the words: "Terrible are Thy works, O Lord," and he assures us that there are not many who would wish to collect these animals. In regard to the invertebrate animals he let his system stand unaltered; only the species, in particular the insects, were reduplicated like the plants.

The "natural method of plants"

IN spite of the enormous amount of work entailed in describing these new animals and plants from all parts of the world, Linnæus found time to apply himself to theoretical problems of great importance. Chief among these should be mentioned his work on the natural vegetable system. As early as in the work *Classes plantarum*, published in Holland in 1738, he promulgated what he called "fragments of a natural method of arrangement": a list of sixty-five "orders," each embracing a number of vegetable genera, but without any characterization of the peculiarities that warranted their being grouped together. By way of introduction he describes the natural system as the highest, but hitherto unattained, aim of botany, which he exhorts all truly distinguished botanists to strive after. For the creation of such a system no particular parts of plants or flowers should, he maintains, be used as a standard, but only the common agreement existing between all parts of the plant. Several of the groups which he founded, such as palms, grasses, Liliaceæ, Umbellata, are still regarded as entirely natural. Throughout the whole of the rest of his life Linnæus never let the natural system out of sight, although he never thought that he would complete it. In his *Philosophia botanica* (1751) he again cites a number of natural groups, now provided with names, and in doing so points out that the vegetable groups everywhere border on one another, like the countries on a map of the world. In point of fact, his realization of the difficulty of trying in a comprehensible way to present the natural affinities of living creatures was a proof of his keen eye for the infinite multiplicity of nature; his caution might well be borne in mind by many a biologist of our own time who has rashly drawn up a genealogical tree for some animal group or other. In connexion with this feeling of Linnæus for the difficulty of determining natural affinity, it is worth mentioning that in his later writings he discusses with far greater caution than in his earlier years the question of the bordering of the species on one another. It was not only that he had seen masses of varieties overlapping one another, but he had also observed the altered forms produced by hybridizing — he himself was very successful in hybridizing in his own garden — and as a result of all this the delimitations of species, which he once felt to be so certain, began to be obliterated. The doctrine of the original creation he certainly could not abandon, but he began to consider the possibility of the genera's having been created and only one or a few species of each, and afterwards new species' being able to arise out of the old. In the final edition of *Systema naturæ* he has omitted the definite assertion that no new species arise. He who has so often been accused of dogmatism was really less dogmatic than many modern scientists who have proved themselves ready to accept blindly the prevailing theories of the day.

In the above-mentioned *Philosophia botanica* Linnæus has also expounded an organic theory in respect of the vegetable kingdom. A great many of his clearly formulated characters of the various parts of plants are still valid to-day. Many consider the anatomical section of this work to be weak, even as compared with the investigations of the earlier botanical anatomists Malpighi and Grew. This may be true, for Linnæus was, generally speaking, no anatomist; true, he did not fail to urge the study of anatomy as well, but his own gifts lay far rather in work upon living nature than at the dissecting-table. Of his purely morphological observations, on the other hand, many are of lasting value; he thus established the fact that all leaves, both plant leaves and flower-petals, go through a common process of develop-ment, a discovery very often attributed to Goethe. But when he tries his hand at comparative anatomy, he usually fails, as when he com-pares the parts of plants and animals: marrow and spinal marrow, skin and bark, etc.

Phenological and geographical biology

ON the other hand, Linnæus's contributions to the knowledge of the con-ditions under which plants and animals live in their natural state are excep-tionally many-sided. These natural observations of his, which occur scattered throughout his disputations and platform speeches, bear witness not only to his keenness of observation, but still more to his ability to combine and draw conclusions from what he observed. Thus in the course of a graduation speech "On the Rise of the Habitable Earth," which begins with a discussion of how all vegetable species were able to grow at once in paradise — they must have existed there, for otherwise Adam would not have been able, as stated in the Bible, to give them names — he expounded a theory of the propaga-tion of plants, based on universal research-material, which was so well ar-ranged that it still has its value at the present day. In other dissertations he has contributed to the knowledge of the "stations" of plants (nowadays termed "locations") and has described the influence of external conditions upon the size, florescence, and distribution. All that is now called pheno-logical, ecological, and geographical zoology and botany has consequently its origin in him. Finally, in the disputations "*Politia naturæ*" and "*Œconomia naturæ*" he gives a radical explanation of all that we moderns call harmony in nature: that all living creatures are adapted to certain conditions of life and that the various plants and animals through their activities keep nature in equipoise, so that "every vegetable species has been given its special insect for the purpose of keeping her under control and to prevent her from spread-ing too much and ousting her neighbours," while the Hymenoptera Para-sitica and small birds look after the insects, and birds of prey after the small birds. That he lets all this take place under God's constant guidance, to His honour and for the benefit of man, should not in our time detract from the value of the observations and the wealth of ideas expressed in these works.

The balance which Linnæus thus found in nature he sought also in the ethical sphere through his well-known speculations upon the "*Nemesis divina*," which, however childish they may be in their detail, are neverthe-less typical both of the man himself and of his time; both Leibniz and Voltaire[2] likewise ruminated over the righteousness governing the cosmic process. And in this, too, Linnæus perceived the divine guidance that he so earnestly sought in natural phenomena. He was an optimist all through — one of the few happy beings who could see harmony everywhere because they have had such a harmonious disposition themselves. He regarded his life's work with a mixture of naïve self-satisfaction and humble gratitude to the Almighty, under whose guidance he was always conscious of living. And he may well have been satisfied, for in the science he so faithfully served, few have exercised so great an influence as he.

Pupils of Linnæus

It has been pointed out above that Linnæus possessed an extraordinary power of gathering pupils round him and interesting them in facts and ideas in the science he represented. Naturally they were for the most part Swedes, but a number of foreigners also came to hear him. His Swedish pupils, after receiving their education, were generally sent to various foreign countries in order to make collections and to describe the places they visited. Linnæus had drawn up for them special instructions, which might serve equally well today as a guide for research-workers in a foreign country. These pupils were travellers and collectors; as a general rule, they made no independent discoveries. Several of them fell the victims of hardship and disease, some obtained distinguished appointments abroad, and others returned home. As examples of these collectors may be mentioned F. Hasselqvist, who travelled for three years in the East and died in Smyrna in 1752, and P. Löfling, whom Linnæus called his most beloved pupil and who, on the invitation of the Spanish Government, worked first on the Pyrenean peninsula and then in South America, where he died in 1757. Further, Per Kalm (1716–79), the first biologist in Finland to work independently, who in the course of a three years' sojourn in North America made valuable contributions to that country's natural history and national economy, and who afterwards acted as professor of economics at Åbo, strove incessantly by the application of natural science to practical life to further the material development of his country. Kalm's fellow-countryman Peter Forskål (1732–63) first studied

---

[2] Voltaire, it will be remembered, after the terrible earthquake that took place in Lisbon in 1755, wrote a poem in which he asks Providence of what those innocent men were guilty who perished in it. Linnæus takes up the same question; he consoles himself with the consideration of the many innocent people whom the Inquisition had burnt at the stake in that city and recalls that the earthquake took place on All Saints' Day — the same day on which the autos-da-fé used to be held. A genuine Old Testament-like idea; the city had sinned and was punished accordingly. Undeniably more attractive are some of the innumerable examples of Nemesis which Linnæus cites from private life, such as the story of the lady who struck her servant for having fallen downstairs and dropped a precious china bowl; the same evening the lady herself fell down the same stairs and broke her leg. Generally speaking, Linnæus possessed a strongly democratic turn of mind, a lively sympathy for the poor and oppressed.

natural science at Upsala, then philosophy at Göttingen; owing to an essay he wrote attacking the Wolffian philosophy, which was predominant at that time, he was unable to obtain an appointment in Sweden, with the result that he went into Danish service as natural scientist on an expedition to the East. There he died, leaving behind him singularly valuable collections. Another who took a post abroad was DANIEL SOLANDER, who had an appointment in the British Museum, London, and died there in 1782; and finally may be mentioned KARL PETER THUNBERG (1743–1828), who held Linnæus's professorship from 1784, after having travelled for nine years in eastern Asia, particularly in the then unknown country of Japan, and collected a rare amount of material in the way of plants and animals. More independent than any of these others, however, was the Dane, JOHAN CHRISTIAN FABRICIUS (1745–1808). The son of a physician, he became an undergraduate in Copenhagen in 1762 and afterwards spent two years at Upsala with Linnæus, with whom he formed a lifelong friendship. After returning home he published several valuable works on entomology, in which he applied Linnæus's method to the insects — the titles of his works correspond to Linnæus's, as will be seen from the bibliography at the end of this book — and greatly increased the knowledge of that class of animals. Abroad he was highly esteemed; at home, on the other hand, he received but little encouragement. After a long period of waiting he eventually became professor at Kiel, on very poor terms, wherefore he spent most of his later years abroad, chiefly in Paris, where he had many friends. Of Linnæus's personal pupils he was perhaps the one who, besides strictly applying his master's system, likewise understood best how to employ it for his own researches, which were of lasting value.

Development of systematic biology after Linnæus

FOR it was not long before the Linnæan natural science began in a general way to degenerate into a spiritless task for collectors and describers, who merely aimed at discovering and incorporating in the system as many fresh species as possible, at the very highest in the hope of being able to use them to some practical purpose for the benefit of mankind — an idea which very much attracted that "age of utility" and which, it is true, Linnæus himself also strongly emphasized. On the other hand, they neglected to cherish and develop those ideas for the future by which Linnæus himself set such store — the natural system and the study of the conditions of life in nature. The result was that the system of descriptive classification, which has so often been called Linnæan science, actually became an expression for a quite limited part of the master's high aims; it certainly became, and has remained so to this very day, a necessary basis for the future progress of biology, as it is also a pedagogically indispensable introduction to that science; but it has also been possible to practise it without any deep insight into the phenomena of life and has thereby not infrequently acquired the character of merely a collector's hobby.

In fact, this seems to have been the fate which threatened biology as a whole during the era at present under discussion; that it did not actually do so is due to a large extent to the fact that contemporaneously with Linnæus there was another scientist at work who was aiming at leading science into a direction entirely different from his, and who, at any rate partially, succeeded in his efforts. This man was Buffon.

CHAPTER VIII

BUFFON

His studies and career

GEORGES LOUIS LECLERC DE BUFFON was born in 1707 at Montbard, in Burgundy. His father was councillor of the Burgundian parlement at Dijon, the capital of the province, and thus belonged to that bureaucratic nobility which was so influential in France in earlier times and which gave to the country many of its finest men, who were often remarkable for their cultural interests and their wealth. Both existed indeed in the home in which young Buffon was brought up; he received a thorough education in his native city and had good prospects in the career which his family had long followed, when chance turned his footsteps into a different direction. He made the acquaintance of a young Englishman, Lord Kingston, who was travelling on the Continent accompanied by a tutor who had studied natural science, and travelled with him through France and Italy. During this tour Buffon's interest in nature, which was to be the dominating factor in his life, ripened. He accompanied his friend to England and spent a year in London studying, particularly mathematics, physics, and botany — sciences which were at the height of their development in the country that gave birth to Newton and Ray. Having returned to France, Buffon published a translation of Newton's *Fluxions*, as well as of the English botanist Hales's *Vegetable Staticks* — two works which presaged the direction that his own activities were to take. As he was wealthy, he was able to devote himself to regular scientific labour, first of all directing his attention to mathematics and physics. In 1739 he was elected an associate of the French Academy of Sciences and in the same year was appointed "keeper of the Jardin du Roi," a post of some distinction, which was still further enhanced by his activities that resulted in the Jardin du Roi, now the Jardin des Plantes, becoming the centre of biological research in France. In the period that followed, his great gifts proved of benefit both to himself and to the science he had chosen: he succeeded in arousing a general interest for natural science amongst the leading circles in France, so that even the King, Louis XV, who was so indifferent and such a stranger to all ideal interests, granted large sums for the improvement of his garden and to assist the scientific work carried out there, while many other eminent personages likewise patronized this science and its distinguished representative. Buffon himself was created a count, was made a member of the French Academy, and was in many other ways honoured by the great. It was also his fortune to play a brilliant part at a time when brilliant qualities were valued more than usual. He was of handsome person and stately presence, which was further enhanced by the exquisite care devoted to his dress and outward appearance; he was an excellent stylist and orator, and, although comparatively reticent in society, he knew how to entertain in a manner befitting his social position. Naturally he also had his enemies in the scientific world, who caused him much annoyance both in public and in private. Even the theological faculty in Paris was not satisfied with him, as his views did not seem to be entirely orthodox, and there was once a question of arraigning him. The distinguished man of the world, however, who naturally had not the least inclination to become a martyr, parried the accusation with a few elegant courtesies about the infallible authority of the Church, and so the matter was allowed to drop. At the same time, in private letters to trustworthy friends he expressed extremely sceptical opinions, which place him in utter contrast to Linnæus, with his childishly naïve piety. These two were destined to become antagonists in other spheres also. — Active to the last, Buffon attained an age of over eighty, dying in 1788. His only son, whom he desired, but in vain, to become his successor, died by the guillotine during the Revolution.

His Histoire naturelle

At quite an early age Buffon had believed it to be his mission in life to write a general natural history, an account of all the knowledge of nature that could be amassed, and in 1749 the first part of his *Histoire naturelle* was published — a work upon which he was engaged for the rest of his life. In the preparation of this work he associated himself with the eminent anatomist Louis Daubenton (1716–1800), who was for a long time conservator under him and afterwards became a professor at the Collège de France. He carried out the anatomical and morphological detail-work, while Buffon had the management of the whole and was responsible both for the ideas incorporated in it and for the method of presentation. The original edition comprised fifteen volumes, the first of which dealt with general natural science, and the remainder with man, the mammalians, and the birds. Buffon got no further in the sphere of biology. In some supplementary volumes, which came out later, a number of general natural-scientific questions were discussed, including mineralogy, a science on which Buffon was weakest, as he was no chemist. Later on, several editions of the great work were published — a proof of how popular it was, in spite of its expense. The main reason for this was undoubtedly Buffon's brilliant style. Not only did he produce vivid descriptions of the nature and habits of animals, the like of which had never been read before and seldom have been equalled since, but he also succeeded in dealing with the most difficult physical and cosmological problems in a

clear and easily comprehensive style. Moreover, besides these external qualities, his work possesses immense practical advantages: it contains the idea
of a new and magnificent conception of natural science, and particularly of
biology, and its influence on the future development of the latter science has
been exceedingly great.

Buffon began as a physicist; as we have already seen, he translated a
work of Newton's, and he also studied Leibniz; he had at once been struck
by the wonderful obedience to law that, according to the then new physical
and astronomical discoveries, governs the universe. In the light of these
discoveries the cosmos appears as a mighty piece of mechanism, which works
according to given laws, and in which both the past and the future can be
mathematically calculated. Is it not likely that in such circumstances the
phenomena here on earth, both in inanimate and animate nature, would also
be subject to a similar obedience to law? That is the question Buffon has put;
he has answered it in the affirmative and he has tried to give proofs of it.
His lasting service to science lies in the fact that he thus endeavoured to
incorporate biological phenomena in their entirety as a link in the great law-
bound world-process; thereby he made a great advance towards the goal that
our modern natural science has set itself, and progressed far beyond the mechanistic biologists of the seventeenth century, the Borellis, the Perraults, and
others who only sought to apply the laws of mechanics to the human body,
without any more universal objects in view. That Buffon, with the limited
material of facts available, could not succeed in creating a theory capable of
passing the test of modern knowledge is quite obvious, but that does not
prevent us from acknowledging the greatness inherent in his very ideas, and
the ingenuity with which he attempted to carry them out.

His general views

BUFFON introduces his natural history with an account of the general principles on which he considers such a history should be written. Here he at once
expresses his view of nature as one whole, all of whose forces gear into one
another and all of whose manifestations stand in mutual causal connexion.
But at the same time he utters a warning, in words reminiscent of Bacon,
against bringing the multiplicity of nature under too simple points of view;
with an obvious allusion to Linnæus he warns us against those who speak,
for instance, of a mineral growing, and who compare in detail the organs of
animals with those of plants; it is, he says, trying to compel nature to come
under our arbitrary laws, not ascribing to the Creator more ideas than we
ourselves possess. The vast wealth of nature must rather be realized and acknowledged from the beginning; the first causes of its phenomena will always
be hidden from us, and what remains to be done is to observe a number of
particular phenomena, compare them, and in them try to find a regular course
of events. It is thus impossible to create any universal system covering all

natural phenomena; its forms and manifestations imperceptibly merge into one another, wherefore vegetable systems in particular, such as those set up by Tournefort and Linnæus, are utterly unnatural. Buffon is very keenly opposed to Linnæus; he asks ironically what is the use of the sexual system when the plants have ceased to flower. In fact, the whole of the Linnæan system of species classification was intrinsically repugnant to Buffon; it seemed to him to be an arbitrary decimation of unified nature into little bits; Linnæus's efforts to create a natural system and his emphasis upon the incompleteness of the classification system, in which he did not vary very much from Buffon's own ideas, were simply neglected by him; apparently they appeared to him merely as slight inconsistencies in a falsely founded view. He criticizes Linnæus's animal classification with similar asperity and undeniably touches its weakest point when he rejects the two great classes, insects and worms; no one, he says, can imagine that crayfish are insects, and shells worms. Instead of six Linnæus should have set up twelve classes, or even still more, for the more groups there are the nearer we arrive at the truth. In fact, in nature there are only individuals; genera, orders, and classes exist only in our imagination. In this Buffon is undoubtedly right in theory, but he overlooks the practical advantage of the "imagined" categories, without which the various life-forms could not possibly be dealt with by science. Instead of the artificial classification-system which he thus rejects, Buffon presents an introduction to the study of nature that, to some extent, is reminiscent of the modern intuitive method of instruction; the description of nature should follow the course which a man ought to pursue if, after having forgotten all that he ever knew, he were put in a place surrounded by natural objects; he would first learn to differentiate between animals, plants, and stones, and then, as regards animals, he would observe the most essential features in their habitat and mode of life and would group the individual animals accordingly in his mind and finally would learn to compare the different animals with one another in greater detail, first distinguishing the tame animals from the wild, then among the wild those who lead the same mode of life and resemble one another in their structure. He glorifies the ancient biologists Aristotle and Pliny, just because they followed a similar natural plan of dealing with living creatures. In his opinion, however, modern research should in no way confine itself merely to observing and describing; the scientist should rather confirm his observations by means of experiment; he should know how to combine observations and to generalize facts, to make individual phenomena obedient to general laws, and, finally, to compare the most comprehensive phenomena of nature with one another. The ultimate aim is to bring all phenomena under the general laws of physics, those laws whose causes remain incomprehensible to man, while only their effects are perceptible. Here Buffon has undoubtedly learnt from

Newton and, either through him or directly also, from Galileo; at any rate, he here displays an insight into both the aims and the limitations of natural science that few scientists before his age possessed and that many have lacked even in our own time.

His history of the earth

BUFFON, having thus given an account of the general principles on which naturalists should work, proceeds to give a theory of the earth and its development into a habitation for living creatures. This problem was manifestly of very great interest to him; in fact he has dealt with it repeatedly — in an essay at the beginning of this great work, called "*Théorie de la terre*," and in another, more extensive essay, written considerably later, entitled "*Des époques de la nature*." In this sphere, it is true, he had precursors; Steno, whose geological works have been mentioned previously, Ray, who wrote a treatise on the changes in the earth, and, the greatest genius of them all, Swedenborg;[1] but Buffon must nevertheless be mentioned as the first who thought to investigate the earth's history with special reference to the development of living creatures.

Moreover, in his latter work he had the temerity to reject the biblical six-thousand-year age of the world and to attribute to the earth a far higher age — certainly small in comparison with the length of the epochs which modern geology assumes, but at least entailing a break with the till then incontrovertible theory of the creation — a break induced by the impossibility of fitting the geological and biological evolution on the earth within such a short space of time as six thousand years. Steno had already faced this dilemma, but, pious Catholic as he was, he preferred to abandon geology rather than Church doctrine; Buffon courageously took the step in spite of his previous *contretemps* with the French theologians. Even as early as in his "*Théorie de la terre*" he expounds his ideas as to the origin of the earth, expressly emphasizing, however, their purely hypothetical character. He assumes, in agreement with Leibniz, that the earth evolved from an incandescent state, but while the latter believed that the earth itself had from the beginning been a "sun," Buffon derives it from the sun's mass, assuming that once upon a time a comet collided with the sun, with the result that pieces broke off which gave rise to the earth and the other planets. This hypothesis has often been cited as a proof of Buffon's extravagant imagination; really, it is no more eccentric than many contemporary cosmogonies, in which comets in general quite often played a part, and in fact Buffon's is presented with much more reservation than the others. After the incandescent state there followed a period when the seas covered the earth, when the tide exercised great influence upon earth-formation. As a proof of this theory of

---

[1] Buffon is said to have known Swedenborg's cosmological theories; he never quotes them, however, though he does quote Steno and Ray and some other, less important authors.

the seas' wide distribution Buffon cites the discovery of fossilized marine animals, especially shells, up in the mountains, and even the stratified nature of the geological formation in general.[2] In the essay "*Époques de la nature*" Buffon divides the history of the earth into seven periods: (1) when the earth and the planets were formed, (2) when the great mountain-ranges were created, (3) when water covered the mainland, (4) when the water subsided and the volcanoes began their activity, (5) when elephants and other tropical animals inhabited the North, (6) when the continents were separated from one another, (7) when man appeared. It would take too long to give a more detailed account of his description of these periods. In this geological theory he makes Vulcanism in general play a more important part than in the earlier work, and the tide becomes of less significance. The greatest service he rendered, however, is that he clearly realized the change in the animal and vegetable kingdoms from epoch to epoch; he tried to work out a "natural history of creation," based on law-bound evolution; he speculates upon the origin of the various life-forms and the place of their appearance and combines these two circumstances with calculations as to climatic changes and other purely physical conditions — in all this a pioneer of the conception of nature which was not generally accepted until a century after his time.

His biological theory

BUFFON has expounded his biological theories in a volume bearing the title *Historie naturelle des animaux*. It begins with an investigation into the difference between animals, vegetables, and minerals and establishes the fact that there is no absolutely definite boundary between the animal and the vegetable kingdoms, but that transition forms may exist; common to both kingdoms is the individuals' power of giving rise by means of reproduction to new individuals like themselves. Another common property is the power of growth; this shows that a fundamental agreement prevails amongst all living creatures, in spite of differences in detail, whereas only matter as a fundamental substance is common to animate and inanimate things. Both animals and vegetables arise as species, the criterion of which is that they propagate; on the other hand, there is no question of a common creative origin. On the whole Buffon refuses to see in the origin of life the result of a

---

[2] As an example of the lengths to which it was still possible to go during the "enlightened" eighteenth century in explanation of natural phenomena, it may be mentioned that Voltaire, who, however, would pass as a disciple of Newton's, declared that Buffon's theory of the origin of fossils up in the mountains was irrational; the shells that had been found there had probably been left there by pilgrims who took them from the East. The two geniuses consequently came into serious disagreement; but later they were reconciled and Voltaire declared Buffon to be a second Archimedes. Buffon capped this compliment with the assurance that no one would ever be called Voltaire the Second. Voltaire's flattery shows, however, that Buffon was considered, and indeed wished to be considered, primarily a physicist.

particular act of creation; life, says he, is not a metaphysical characteristic of living creatures, but a physical quality of matter.

Since, then, the most vital quality of life is the power of reproduction, Buffon devotes very close study to it. For this purpose he does not start from the most highly organized creatures, but begins his investigation — this, too, a modern feature — with the most primitive form of reproduction — that by means of division in plants and primitive animals. Why is it that a severed branch of a twig of a tree grows up into a new tree, that a piece of polypus gives rise to a new polypus? Buffon answers this question by assuming that the plant and the animal are composed of a mass of particles formed like the individual in its entirety, and which therefore, when they become detached, can develop further and form a new individual of the same kind. This theory of independent particles, the idea for which he undoubtedly got from Leibniz's monad theory, Buffon further develops to form the basis of his conception of all the phenomena and functions of life; just as inanimate matter is composed of an incalculable mass of minute particles, so there exists in nature a vast number of organic particles that are animate and formed like animate beings. "Just as there may be required perhaps a million minute salt cubes to form one grain of sea-salt, so it would take millions of organic particles similar to the whole to form a bud containing the individual of a tree or a polypus." By making this assumption Buffon also seeks to get rid of the preformation theory, which was generally embraced by his contemporaries and which he keenly criticizes, maintaining among other things that it would presuppose an infinite number of daughter individuals contained in the original mother animal, which in itself is an entirely irrational supposition. But when it comes to setting up an acceptable theory of sexual reproduction in place of the preformation theory, Buffon comes to realize, as he himself openly acknowledges, that it is easier to destroy than to build up. He founded a general physiological hypothesis according to which animals through the food absorb a quantity of these ubiquitous organic particles and in the various organs of the body assimilate from these what the body requires; whatever is left is collected in the genital organs and gives rise to individuals like the parents. That the embryo is thus formed by a combination of a mass of minute independently living particles he believes to be proved by the spermatozoa existing in the seminal fluid; that the female sexual product actually consists of similar minute beings he also believes he had proved by a microscopical study of the ripe follicles in the mammalian ovary; in their fluid he believed that he had found mobile life-elements similar to those in the semen which he actually illustrates[3] and which in his

---

[3] What Buffon and his collaborators — he quotes several, including the English microscopist Needham — actually saw in the follicular fluid it is difficult to say; perhaps detached cells from the follicular epithelium; maybe also coagulation products,

view combine with the spermatozoa to form the new individual. In connexion with the question of evolution, reproduction, and growth, Buffon sets up a very abstract and difficult hypothesis concerning the mutual connexion of the different parts within the organism; he believes that each individual represents a "*moule intérieur*," by which he apparently means the constant form of every living creature, attained with the co-operation of the organs, which are fed and grow by assimilating these living particles that fill the whole of nature and are the one essential in the assimilation of food — of both animals and vegetables — in growth, and in reproduction. This theory of living particles thus forms the very corner-stone of Buffon's biological speculation and is both its strength and its weakness; with its aid he avoids the difficulty of explaining the origin of life without the assumption of a supernatural act of creation — he does not expressly deny such an act, it is true (that would have been too daring for his age), but it is quite obvious that he will have nothing to do with it — on the other hand, he had to make good with assumptions which very much resemble the ancient spontaneous-generation hypotheses, which had already been rejected by the biologists of the seventeenth century. At any rate, of greater value than the results of these speculations is his criticism of the actual method of natural research, which even in modern times makes profitable reading; his own theories he in no wise propounds as if they were proved truths, and his warnings against confusing hypotheses and facts many a modern biologist might well take seriously to heart.

Besides these purely biological questions Buffon also discusses psychological problems. His speculations on animal psychology are, however, of little importance; he certainly admits the existence of intelligence in animals, in contrast to Descartes, but he denies that they possess memory and reflection. On the other hand he has some striking observations to make on the domestic animals' intellectual dependence upon human training, as well as on their sense-impressions and the varying power of the latter.

On man

LIKE Linnæus, Buffon treats man as a natural-history subject and gives a detailed description of man's evolutional history, alimentary conditions, and habits of life, which has justly become famous, not merely for its important formal merits, but also as being the first attempt at anthropology in the modern sense. Though human anatomy had already been thoroughly dealt with in the sixteenth and seventeenth centuries, nevertheless a universal treatment of man in regard to his entire relation to nature was something quite new. From an anatomical and physiological point of view he certainly has not very much that is fresh to relate, but he conscientiously and critically summarizes the existing scientific material; he gives an account of the development of man from embryonic life through the various ages; he tries to analyse the growth of speech in the child, and the influence of mental emotions upon facial expression; he discusses the power of mental perceptions to reproduce reality and insists on science's dependence upon them. He investigates the circumstances of death at various ages; he gives statistics showing the mortality in certain French provinces and seeks with the aid of "probability calculations" to ascertain the longevity of various ages; he compiles data with regard to the peculiarities of wild tribes and abnormal phenomena in civilized peoples. Above all, he maintains that man has his bodily functions in common with the animals, but that, on the other hand, there is a fundamental psychical difference between them, which renders it impossible to compare human and animal intellectual qualities.

His zoography

Of the animals Buffon, as already mentioned, had time to complete only the quadrupeds and birds, which are dealt with in detailed monographs covering each species. Naturally, these are each of varying value; all of them however share in common a brilliant exposition and a universal treatment of the subject, in striking contrast both to the earlier zoographers' motley mass of notes and to Linnæus's brief, summary diagnoses. As a describer of nature Buffon is of fundamental importance, in certain respects still unexcelled. It would take too long to enter into the peculiarities of his zoography; it need only be pointed out that it is not merely formal services that have earned it its well-merited fame, but in many of his descriptions, particularly in those of birds, there are a number of keen and striking detailed observations as to mode of life, reproduction, and other biologically interesting factors.

Daubenton's comparative anatomy

To each monograph on mammalian animals Buffon's collaborator, Daubenton, has added an account of the animal's anatomy. By way of introduction he sets forth the principles on which to base this kind of general — nowadays we should say "comparative" — anatomy, in contrast to the descriptive, which had hitherto been practised. He considers that all animals should be investigated in respect of their most vital organs — bone-structure, heart, brain, respiratory, digestive, excretive, and sexual organs — and the results thus obtained compared. Following this principle, an account is given of every mammal's anatomy, particularly the bone-structure: the bone-structure of the horse is compared in detail with that of man, and the bones of other animal species are mutually compared. Such a comparative examination of the anatomy of various animals carried out on a uniform plan was at that period something new and proved of great significance for the future; the part played by comparative anatomy in modern biology is too well known to need any special emphasis here.

Thus Buffon carried out his plan of presenting nature, both inanimate and animate, as one whole, evolved and held together by purely mechanical laws. In this, however, he was not entirely consistent. In a curious treatise entitled *Homo duplex* he has described man as composed of two principles, fundamentally distinct from one another, one spiritual and the other material, of which the material develops first and predominates in the embryonic stage and during childhood, while the spiritual appears later and is developed by means of education and training, without which it would lead to stupidity and vain delusions. This dualistic conception of man might well appear to be fully in accordance with the principles that had till then been and were still at that time officially recognized — it might be suspected that Buffon here made a concession to the ecclesiastical authorities who had persecuted him — had not the style of the whole been so utterly different from all that is understood by conventional religion. Instead we here come across a trait in Buffon which we should not expect to find in that exceedingly brilliant and successful man — namely, a deep pessimism. To his mind, the contrast between spiritual and material appears most marked in those attacks of melancholy and listlessness when one lacks all power of decision, when one "does what one would not and would do what one does not": when one feels that the personality is divided into two, of which the one part, reason, indicts the other without being able to overcome its resistance; sometimes reason wins, and then one performs one's duties gladly; sometimes the flesh wins, and then one indulges in pleasure, but sooner or later these unhappy hours and days return when disharmony prevails. Especially vivid is the passage in which Buffon describes how love, which makes animals happy, simply makes men wretched; in words of wild despair he depicts the vainness and folly of this passion, which certainly brings with it bodily satisfaction, but is morally valueless and only calls forth jealousy and other degenerate feelings. This melancholy conception of life was, as a matter of fact, in no way peculiar to Buffon; it was, on the contrary, as we shall see later on, a widespread view during the epoch to which he belonged.

Buffon's influence

BUFFON has played a fundamental part in the history of biology, not on account of the discoveries he made, but on account of the new ideas he produced. Those ideas that he brought out, which he was able only imperfectly to realize in detail, have since then been taken up by others, who, having better opportunities for obtaining actual scientific material, have applied them in a wider sense: thus, Cuvier, the pioneer of comparative anatomy and palæontology, adopted many of Buffon's fundamental ideas; similarly Bichat, the originator of the tissue theory, in his sphere, as well as Lamarck, with his theories of the evolution of living organisms, in that field, has manifestly felt the influence of Buffon's speculations. Through these scientists many of the ideas produced by Buffon have now been incorporated in the general knowledge of natural science. If in spite of this he has often been depicted, especially beyond the borders of France, as a talented dilettante, a witty popular scientist and writer, this is mainly due to the relations in which he stood, both in his lifetime and long afterwards, to the representatives of the Linnæan system of classification; these latter, who for a long time felt that they were the sole upholders of a truly exact natural science, looked compassionately down upon Buffon's unsystematic descriptions and imaginative speculations. When, then, the dominion of Linnæanism fell, the comparative and speculative lines of research which succeeded it already possessed entirely different intellectual material to build upon, and Buffon's theories thereafter necessarily appeared vague and childish. His services, however, must in all fairness be duly acknowledged; in the purely theoretical sphere he was the foremost biologist of the eighteenth century, the one who possessed the greatest wealth of ideas, of real benefit to subsequent ages and exerting an influence stretching far into the future.

CHAPTER IX

INVERTEBRATE RESEARCH IN THE EIGHTEENTH CENTURY

Successors of the great seventeenth-century biologists

THE EIGHTEENTH CENTURY displays on the whole great activity in the sphere of the natural sciences. In physics and chemistry the successors of Newton and Stahl worked at extending the spheres to which their masters had opened the way. In the realm of biology Linnæus and his disciples held sway and were fully occupied in incorporating known species into the system and in discovering fresh ones. As already mentioned, Buffon's activities belonged rather to the future. But besides this there worked during the eighteenth century a number of naturalists whose achievements connected them more or less directly with the great biologists of the preceding century. Towards these pioneer anatomists, microscopists, and physiologists their successors during the eighteenth century have to some extent the character of Epigoni: they made no such epoch-making discoveries as Harvey's or Malpighi's, but, on the other hand, they took advantage in many and various ways of the discoveries that had already been made; the problems which had already been of direct importance were discussed from various points of view, while at the same time ideas were expressed in more than one field of research which gave a presage of future ends to be gained. Malpighi's and Swammerdam's investigations into the anatomy of the lower animals were thus resumed and carried a step further, as also Borelli's and his successors' physiological work; Leeuwenhoek's and de Graaf's discoveries in the field of reproduction were elaborated, and the discussion between the animalculists and the ovists went on, particularly in the first half of the eighteenth century, with undiminished liveliness; the epigenesis and preformation theories were also keenly debated, although at first with the balance decidedly in favour of the supporters of preformation. Later in the century, however, contributions were made in these very spheres which put a different complexion on those questions. And even for several other spheres of biology the latter half of the eighteenth century represents a period of decisive preparation for the development that took place during the nineteenth century. — In the following paragraphs we shall review, to begin with, some of the more important contributions to the biology of the lower animals, and afterwards the advances made in the spheres of anatomy, physiology, and evolution.

RENÉ ANTOINE FERCHAULT DE RÉAUMUR was born of noble and wealthy parents in the year 1683. He received his education at a Jesuit college, afterwards studying jurisprudence in Paris, but he soon abandoned that career and applied himself whole-heartedly to natural science. Having inherited a fortune, he was able to lead the life of a private scholar; membership in the French Academy of Science was the only distinction he obtained. He died in 1757.

Réaumur was active in many branches of natural science, both theoretical and applied. He invented improved methods of iron-refining and made important contributions to our knowledge of the expansion of gases and fluids and of specific heat. He is best known for his invention of the eighty-degree thermometer scale, which bears his name and which is still used in many countries. In biology, too, his activities have been many-sided and important. His greatest and most famous work is his *Mémoires pour servir à l'histoire des insectes*, a work in six large quarto volumes. This work is undoubtedly of fundamental importance in insect biology and is in fact one of the most monumental works written in this field of research. It offers a number of extremely valuable contributions to the knowledge of the anatomical structure of the insects, their evolutional history and conditions of life. His chief master is Swammerdam, whose system he in the main adopts, but he considerably widens the sphere of the latter's researches. True, he did not possess the master's incomparable ability in the work of preparing material, but instead he had at his disposal a greater wealth of material for his researches, while a long life made it possible for him to carry out lengthy and laborious series of observations and experiments on the living habits of insects. The community life of the social insects, in particular of the bees, the development of the parasitic Hymenoptera, and the activities of leaf-mining and gall-forming insects may be specially mentioned among the subjects dealt with by Réaumur in his great work — subjects to which he made important contributions. Besides these his book contains a mass of valuable detailed descriptions of larval and imaginal forms from practically all insect groups.

Réaumur's physiological researches

MOREOVER, outside the sphere of insects he has presented biology with the results of important discoveries. Thus, he has established the fact that the shell of molluscs is formed by means of a secretive process, and in connexion therewith he studied the formation of pearls in mussels. He studied also the movements of a number of primitive animal forms, he investigated the electric phenomena in the ray, and he further observed the regeneration of the extremities and other parts of the body of crayfish, in regard to this latter phenomenon producing a theory reminiscent of Buffon's hypothesis as to the body's being composed of organized particles. And, finally, he carried out several interesting experiments on the digestion, principally on the influence of the gastric juices; he obtained gastric juice from a chicken by letting the bird swallow a sponge attached to a piece of thread, with which the sponge was after a time recovered from the stomach drenched with gastric juice, which was afterwards used for the purpose of acting upon various kinds of food substances. Among his contemporaries he justly enjoyed a great reputation; Linnæus cites him often and with recognition, while he had many pupils, of whom de Geer in particular at once carried on his work.

CHARLES DE GEER was born in 1720 at Finspong, in the Swedish province of Östergötland. He was a descendant of the rich merchant and manufacturer Louis de Geer, who had emigrated from Holland, and consistently with his family's origin he received his education in that country. He studied at the University of Utrecht, where he devoted himself to both physics and biology. As a child he inherited Lövsta Foundry, in Uppland, and as soon as he came of age he took over its management. He introduced a number of improvements in the iron-manufacture and thereby acquired considerable wealth. Regarded as one of the richest and most brilliant noblemen in Sweden, he became in course of time Court Marshal and baron and received many other distinctions. He showed great consideration for his workers, founding schools for their benefit and improving their wage conditions. He was highly reputed in scientific circles in Europe and was a member of several learned societies. He died in 1778.

At an early age de Geer had been interested in entomology. In this field he continued the investigations begun by Réaumur, and published under the same title a sequel to the latter's great work, which it in every way equals in value. It comprised seven volumes, containing observations upon the systematic classification of insects, their habits of life and evolutionary history. Although contemporary with Linnæus, de Geer did not adopt his nomenclature, but retained the old method of characterizing the species by means of diagnoses. Otherwise he was a keen observer, who in more spheres than one made contributions of lasting value, not least in regard to the lower and hitherto neglected insect-forms.

Among the naturalists who during this period made valuable contributions to the knowledge of the lower animals should also be mentioned ABRAHAM TREMBLEY (1700–84). He was born at Geneva, studied first of all there, then in Holland and England, was for a time private tutor in certain distinguished families, and finally became a librarian in his native town. His reputation as a biologist is based upon his important monograph on the fresh-water polypi. In this work he gives a careful account of a number of "polypus-forms" — he includes both Hydra and Plumatella in the same genus. He closely studied their habits, particularly their movements and food, and was, properly speaking, the first who clearly realized their animal character. He observed their natural propagation, but above all he carried out systematic and extensive experiments in regard to their power of re-generation, thereby opening up for research a field that has been cultivated, especially in modern times, with splendid results.

AUGUST ROESEL VON ROSENHOF (1705–59) was born in Thuringia, but worked mostly at Nuremberg, first of all as a painter and afterwards as a naturalist. Under the striking title of *Monatliche Insectenbelustigungen* he published in the seventeen-fifties a series of observations on the life of the lower animals, illustrated with beautiful engravings done by himself. A number of sound detailed observations regarding the life-habits and develop-ment of insects are given in these writings, but he paid special attention to the evolutionary history of frogs, from their mating and egg-laying through all their larval stages, and has thus given to posterity valuable additions to the knowledge of these creatures, which, as is well known, are much used in modern experimental physiology.

PIERRE LYONET (1707–89) was also a highly reputed biologist among his contemporaries. Born at The Hague of French parents, he was given a very extensive education; he was a brilliant linguist and at one time followed the career of a diplomat. As a biologist he applied himself most actively to the sphere of insect-anatomy, in the spirit of Swammerdam; an admirable work of unsurpassed brilliance even in our own time is his great monograph on the larva of *Cossus ligniperda* or goat-moth caterpillar, the anatomy of which he studied and illustrated with extraordinary conscientiousness and keenness of observation

CHAPTER X

EXPERIMENTAL AND SPECULATIVE BIOLOGY IN THE EIGHTEENTH CENTURY

BESIDES THESE MONOGRAPHISTS, who were highly regarded in their own day and are still well worth reading even nowadays, there lived during the eighteenth century many scientists whose works embraced fields of research of wide extent in regard to both the material investigated and the problems dealt with. In particular, experimental biology and theoretical questions in connexion therewith were developed on a considerable scale during this period by scientists who have merited the attention both of their own age and of posterity. Foremost among these should be mentioned Haller, a great man in his own age and a scientist for all time, famous as a botanist, anatomist, physiologist, statesman, and poet.

ALBRECHT VON HALLER was born at Berne in 1707. His father was a wealthy and highly reputed lawyer, who gave his son a thorough education, at first in his own home with a private tutor, then at the University of Tübingen, and finally at Leyden under Boerhaave. Young Albrecht was an infant prodigy; at the age of ten he had a thorough knowledge of Greek and Hebrew, at fifteen he had written an epic poem and some tragedies, at nineteen he was a doctor of medicine. It was obvious that a young man thus equipped would in time become something quite out of the ordinary; unfortunately, as so often happens, none of the successes that he actually attained fully reached the height of his dreams. Having taken his degree, Haller studied for a time in Paris, afterwards settling down in Berne as a physician. He there became universally known as a botanist and poet and in 1736 was appointed a professor of medicine at the then newly-founded University of Göttingen, where he did splendid work; he laid out a botanical garden and built an anatomical theatre, founded a still existing and much thought-of scientific society, and besides found time for scientific authorship of an extraordinarily many-sided character. But he never won contentment; he was troubled with melancholy and a longing for his native country, and finally he resigned his professorship and returned home to Berne (in 1753). There he was elected to the municipal council and made a name as a distinguished official; his services were utilized as a diplomat and he performed his duties in that capacity with honour. Meanwhile he continued his scientific writing with undiminished zeal; his productivity was nothing

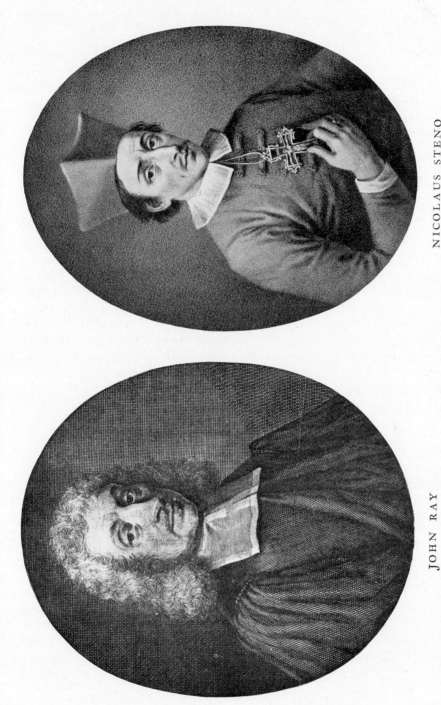

NICOLAUS STENO

JOHN RAY

short of amazing; a list of his writings gives a total of 650, among them many very extensive works. But during this period the melancholy that had pursued him since his youth increased. He felt dissatisfied with the results he obtained, as also with the new ideas that were becoming more and more common. In particular, the increasing free-thinking in the world troubled him and called forth a number of pamphlets in defence of Christianity from his hand. He himself, however, in spite of his firm belief in the Gospel, had no internal peace, but ruminated over the rightness of what he had done in his life; the vivisections which he had performed, and which always troubled his sensitive mind, now appeared to be specially repugnant to him. After some years of decline in health he died in 1777.

In his youth Haller devoted himself principally to botany and verse-writing. This, of course, is not the place to criticize his poetry; this much, however, may be mentioned, that he is considered to have discovered the poetical value of Alpine beauty; otherwise his poems are now read apparently only by students of literature. As a botanist Haller appears in conscious rivalry with Linnæus; he seeks to set up, in opposition to Linnæus's artificial system, a natural system based primarily on the character of the fruit. It was not successful; Linnæus's investigations into the possibilities of the natural system clearly proved that the time was not yet ripe for such a one, whereas the sexual system in every respect fulfilled the requirements of the period. Haller was embittered by defeat; and although his criticism of his successful rival may be partially justified, nevertheless his disappointment over his failure is clearly apparent.[1] In actual fact the two rivals were incommensurable; in contrast to Haller's magnificent but divided many-sidedness is Linnæus's consummate limitation — the former never touched supremacy at any point; the latter possessed only one sphere, but there he was master.

Haller's physiological researches

THE branch of biology in which Haller made his finest contribution is undoubtedly physiology; in this field he has not only developed the method, he also established new and important facts, made valuable additions of a purely theoretical nature, and finally compiled the results hitherto attained in a comprehensive manner, which should be a pattern for all time. His writings on this subject consist partly of a mass of articles written for journals, in which he recorded the results of his direct observations, partly of an immense physiological compendium, and, finally, of a smaller and extremely concise handbook on physiology, which was still in use for educational purposes until the nineteenth century. In the foreword to the last-mentioned

---

[1] Curiously enough, Haller published his bitterest attacks against Linnæus in the form of a series of disputations (*Dubia ex Linnæi fundamentis hausta*, Göttingen, 1751-3) which purported to have been written by his son, a young medical student.

work he defines physiology as "*animata anatomē*," a vitalized anatomy, and it is in fact the phenomena of life that he seeks to discover in his special investigations. The most remarkable of these is without doubt his treatise on the irritable and sensible parts of the body, which he published simultaneously in several languages. In this investigation he first establishes the fact that the organs of the body are partly irritable, partly non-irritable; why this is so, science cannot discover; it can only show that it is so. As irritable (*irritabilis*) he mentions such a part of the body as contracts upon being touched; as sensible (*sensibilis*), again, he defines a part of the body, contact with which induces an impression in the mind. Which organs belong to the one or the other category is a question which can be answered only by experiment. The performing of such experiments on live animals Haller finds highly revolting, but in the interests of truth it cannot in this case be avoided. Thus he has proved that, of the two layers of skin, the epidermis is non-sensitive, the cutis on the other hand has feeling, and adipose tissue is non-sensitive. Muscles are sensible, but this is due not to the actual muscular substance, but to the nerves which are in connexion therewith; the tendons, again, are non-sensible, because they are not connected with nerves. Bones and periosteum are insensible, the cerebral membrane, the peritoneum, and the veins likewise. The intestines are sensible, but not the liver, the spleen, or the kidneys. Irritability exists in muscles, but is induced through the nervous system; thus the diaphragm has been made to contract by irritating a severed nerve. Therefore the irritability cannot have anything to do with the mind, for that is indivisible. Haller then enumerates a number of irritable organs — veins, intestines, sexual organs. Finally he discusses the question of what it is that induces irritability. Muscle is composed of lime and earth; if it is asked which component part is irritable, the answer must be the lime. Lastly he deals with the question of vital organs, which serve the unconscious manifestations of life, and voluntary organs, which serve the will.

The investigation here referred to must without doubt be regarded as one of those that have led biology into new directions. Not only the scheme of the work, but also the method of presentation he employs and the conclusions he draws are each of fundamental importance; irritability and sensibility are facts, which hold true to this very day,[2] and the experimental method by means of which the phenomena have been established is still used today. There occur, indeed, in the examples just given one or two actual mistakes — thus the peritoneum is, as a matter of fact, sensible and the intestines insensible — while the actual theoretical treatment suffers from the

[2] Instead of irritability the characteristic property of muscle is nowadays termed "contractility." Haller's irritability theory was later applied without distinction to various organs in the body, thereby causing considerable confusion.

fact that Haller did not succeed in producing any term for tissue — muscles, intestines, and other viscera are quoted as organs of the same category. Moreover, the chemical basis of his muscular theory, the composition of lime and earth, is extremely primitive. But in spite of all this Haller, as a result of his work in this field, has won a brilliant name in the history of science.

His compendiums of physiology

THE compendiums which Haller produced still further extended the services he has rendered to the development of biology. In the two works on general physiology cited above he has summarized all the then known physiological facts in a concise and easily accessible form. He starts by taking the simplest component parts of the body, which are divided into solid and fluid. The simplest elements of the solid components are, according to him, the fibres, the composition of which has already been mentioned — lime and earth. By cell-tissue, a word which often occurs, is meant what modern histology terms adipose tissue. Haller considers the most vital part of the organism to be the blood-vessel system; it represents the element that connects together his whole physiological theory. In his description of each organ he always starts with its blood-vessels. The more blood-vessels an organ has, the more important it is. Of the thyreoidea he says that we do not know its function, but it must be an important one, since the organ in question is so rich in blood-vessels. Out of the blood are produced all the fluids of the body in an entirely direct manner; thus he claims to have found direct passages between the arteries and the salivary ducts in the salivary gland; even the lymph he believes to emanate from the arteries. The purpose of respiration is to give the blood warmth. Haller was keenly interested in the structure of the brain, but the results he attained are not to be compared with those gained by Swedenborg at the same time on purely speculative lines. Haller has only vague ideas on the cerebral cortex; the medulla is the most vital part of the brain, and, in his view, the nerves are filled with a fluid which gives rise to mental impressions. Towards many of the biological points of dispute of his own time Haller tries to adopt a somewhat neutral attitude; as, for instance, in the dispute between the ovists and the animalculists, in which, however, he sided on the whole with the latter, since he held that the *spermium* — "*vermiculus seminalis*," as he calls it — is the origin of man, just as the larva is that of the fly. On the other hand, he describes the follicle, or the egg, as he, like his contemporaries, calls it, as partaking in the production of the embryo. As regards the question of preformation or epigenesis, he is on the side of preformation. On the whole, he gives a conscientious account of such views as he himself does not accept and displays in these works both creditable impartiality and a universal knowledge of literature. He has taken special advantage of the latter quality in his bibliographical works — *Bibliotheca anatomica, botanica,* and *chirurgica* — in which he has compiled information regarding all the literature published till then in various spheres of science. These "*bibliotheca*" are even in modern times of importance to the student of scientific literature and are remarkable for their completeness, though also unfortunately for the mass of misprints which mar their utility.

Haller's reputation

HALLER has been very diversely judged. On the one hand, he has been regarded both by his own age and by posterity as the foremost anatomist and physiologist of his century and as the founder of modern experimental physiology, while on the other, as has so often been the case with scientists of many-sided interests, he has been accused by specialists of unreliability in points of detail. His great service, particularly to the development of physiology, can, however, never with justice be denied; his experimental method and its results are undoubtedly of fundamental significance. As regards his general conception of life, on the other hand, Haller has to a certain extent stood at an old-time view-point, just as in his writings he summarized the results hitherto attained. This to some extent explains how it came about that the immediately succeeding age picked a quarrel with him; thus, Goethe finds fault with him for his views on the limitations of the knowledge of nature, which, it is true, are but little in accordance with natural-philosophical speculation; but above all he fell foul of a contemporary scientist who, starting out from an entirely different standpoint and having different preconceptions, arrived at an entirely opposed fundamental view in regard to science — namely, La Mettrie.

JULIEN OFFROY DE LA METTRIE was born in 1709 at Saint-Malo in Brittany. His father was a wealthy merchant, who had his son brought up to be a priest. He studied theology in Paris and there joined the Jansenistic sect, a movement in the French Church known for the strictness of its rules and ideas, but disfavoured and persecuted by the Government. A physician in his native town, however, succeeded in awaking in the young theologian an interest in his profession, and so it came about that La Mettrie began to study medicine, first in Paris and then at Leyden under Boerhaave. Having passed his examination, he set up in practice for a time in his native town and then became physician to a regiment of guards in Paris; by that time he seemed to have prospects of making a brilliant career, as he was well known both for his successful cures and as a witty and refined man, with social aptitudes. But these high hopes soon had to be abandoned. He had begun his scientific writing by translating into French some of his master Boerhaave's more important works. This was viewed with disfavour by the high-conservative medical faculty in Paris, which had consistently opposed Boerhaave's theories, just as at one time it had opposed those of Vesalius and Harvey. And La Mettrie still further increased their irritation by pub-lishing a number of satirical pamphlets against his opponents. Eventually the latter found an opportunity of taking their revenge; in a work, *L'His-toire naturelle de l'âme,* he had expressed views that were considered to be at variance with the Christian faith. The theologians rushed to battle and La Mettrie's friends advised him to go to Holland until the storm should blow over. At Leyden, however, he printed a new pamphlet, which still further aggravated his position; this was the famous treatise *L'Homme ma-chine,* which was published anonymously, it is true, but which was immedi-ately recognized. Its contents were such that the author could by no means count upon even Dutch tolerance; he had to take precipitate flight and to remain in hiding for a time. His fate would now have been deplorable had not Europe possessed a reigning monarch who was absolutely indifferent to religious problems, but who, on the other hand, was amused by witty companions; this was Frederick II of Prussia. La Mettrie was summoned to Berlin, was appointed lecturer at the royal court, and besides was given an opportunity of practising as a physician. He enjoyed these privileges only for a space of three years; in 1751 he died as the result of an accident. He had always boasted with some pride of his power of enjoying life's pleasures both qualitatively and quantitatively; so at a feast, just to show off, he ate enormous quantities of truffle pasty, immediately fell ill, and died in terrible pain; probably the pasty had contained septic poison. This tragic end, however, still further increased the ill fame caused by his writings;[3] his name has, in fact, been one of the blackest in the whole of the eighteenth century. In many respects, however, he paved the way for ideas which mod-ern biological research has adopted and it is therefore worth while paying some attention to his views.

La Mettrie's polemical works

In his writings La Mettrie shows himself to be a marked oppositionist. It is destructive work that amuses him most, and he likes best to pit his strength against what his contemporaries regarded as the most unshakable foundations both of the knowledge of existence and of the social order and good manners. His polemical writings are sometimes brutally frank, some-times subtly insidious, but he invariably challenges deep-rooted ideals, both scientific and traditional, and is quite prepared to call white black and black white. His love of truth goes just so far as serves his immediate purpose, but undaunted courage we cannot deny him, and he has a firm faith in the

---

[3] In ancient times it was held to be a matter of fact in High-Church quarters that no one can die in peace without the Church's blessing. In this connexion it has been related that Luther hanged himself (a Catholic statement), that Spinoza died under the influence of opium, and Voltaire in a fit of madness. In furtherance of this kind of propaganda La Mettrie's above-described death, which is historically confirmed, was a good find indeed.

subject that he made his own. What his writings seek to create is a general cosmic view based entirely upon "philosophical" — that is to say, natural-scientific — principles, for to him philosophy and natural science are identical. In contrast to the teachings of theology, politics, and the morality that is based on them, he wished to create another ideal of justice and virtue based on "natural" principles, and in contrast to the explanation of life and nature which priests (and "philosophers" dependent upon them) have given in support of ancient authorities, he would set up another explanation, founded upon direct observation of the phenomena of life. He is thus the first to enunciate a purely natural-scientific view of life, and in doing so became the precursor of many similar endeavours in our own time. Herein lies his greatest originality, for in most of his subjects he merely sets forth in detail observations recorded by others, and his writings can hardly be said to possess scientific form in the stricter sense of the term; they are pamphlets published for agitational purposes, often more likely to persuade than to prove. Of these the two which have been cited above won the greatest notoriety; among his other publications there is really only one — entitled *Système d'Épicure* — that is of any great interest.[4] His work on the natural history of the soul, published before he had finally broken with his native country, maintains a somewhat cautious tone and is therefore written in a more scientific form. *L'Homme machine*, again, is nothing but a piece of agitation, and *Système d'Épicure* is a collection of aphoristic contributions to a general knowledge of nature. The view of the functions of the human body on which La Mettrie bases his speculations is by no means a new one; it is the mechanistic theory of the bodily functions, which, founded by Descartes, had been developed by Borelli and Perrault, by Hoffmann and Boerhaave. To their observations La Mettrie has but little to add; the most valuable contribution is an exposition of the independence of the vital functions in the various parts of the body, confirmed by observations of the manifestations of life in detached organs and bodily parts even in the highest animal forms. His theory of fertilization may be worthy of mention; he believes that one single "*sperma*-animal" penetrates each egg and is there further developed into a new individual; he thus belongs to the animalculist party. But it is not the life of the body that interests La Mettrie most; it is the functions of the soul that forms the chief subject of his literary production. With regard to the soul his views are quite clear; it does not exist, or at any rate not in a form that the theologians and suchlike would have it. As a matter of fact, La Mettrie is not quite certain what the soul is;

---

[4] There would be no point in dealing here with La Mettrie's strictly medical writings; of his philosophical treatises, *L'Homme plante* is a development, driven to absurd lengths, of the comparison between the vegetable and animal organs which Linnæus had already drawn up; again, *Les Animaux plus que machines* contains little more than answers to his opponents' accusations.

sometimes he goes so far as expressly to point out that its essence must always remain unknown, while, on the other hand, he is quite convinced as to what it is not: it is not the immortal spirit, distinct from the body, that it is officially declared to be. During a fever La Mettrie had observed how the faculties of the soul within him were affected by the course of his sickness, and in his medical practice he had remarked the same phenomenon in many of his patients. By thus making the influence of the bodily functions upon the intellectual life a subject for investigation and even experiment La Mettrie discovered that field of research which in modern times is termed psycho-physics and which has been so successfully investigated by research-workers with what are, in principle, the same methods as his, though with an entirely different standard of scientific criticism. La Mettrie, in fact, suffered the usual fate of a pioneer in not being able to free himself from the prejudices he attacked.

His general conception of the human soul

He begins by accepting the old division into a vegetative, a sensitive, and a rational soul; he analyses the first two and finds that their functions are dependent upon those of the body, which indeed his predecessors had also taken for granted. He devotes the main part of his investigation into the soul to trying to discover the operation of the sensitive soul; he gives an account of mental impressions and their mechanisms, in the course of which he makes several striking observations, *inter alia* regarding the subjectivity of mental perceptions. He discusses the localization of the mental functions in the brain with extraordinary keen-sightedness and thence goes straight over, with a somewhat daring mental jump, to ideas, which he treats — very naïvely — as bodily entities, the grandeur of which he tries to estimate. After having thus converted ideas in general into bodily phenomena, he discusses in connexion therewith a number of such ideas — memory, imagination, talent, etc. — all of which are to him likewise material, so that finally there is nothing left of the rational and immortal soul that the theologians have made it their mission to cherish. Thus he accumulates a mass of evidence to show that the soul of man is fundamentally the same as that of the animal; he cites examples of animal affection, gratitude, and such feelings, and seeks, on the other hand, to adduce proofs that man possesses animal qualities. He quotes in all seriousness a number of miraculous stories of human beings who have lived like animals in the forests — probably embroidered tales of runaway lunatics — he describes the orang-utan with the entirely human characteristics which were at that time ascribed to that animal, and hopes that it will be possible to teach it to talk by a method of teaching the deaf and dumb to speak which had just been invented.[5] And

---

[5] Why the deaf-and-dumb method should have to be used for an ape which can hear just as well as a man is not explained; probably it was the novelty of the method that made it seem so wonderful and induced the hope of its performing further miracles.

finally he expounds a quite extraordinarily childish theory regarding the "natural" origin of man and the living creatures here on earth.

The "natural" origin of man

STARTING from Buffon's above-mentioned idea of the living particles scattered about in space, out of which all living creatures have arisen, he assumes that similar particles, intended to form human beings, have accumulated in the earth and given rise to a number of human individuals, some defective, others perfect. To the question why the earth no longer produces human beings in this wise he would answer that the earth is now old and weary; in answer to the question how the human babies thus produced eventually developed, it may be supposed that they were brought up by kindly beasts of prey, just as a small child had, it was said, recently been brought up by a she-bear in Poland.

On reading such absurdities one recalls the days of old Empedocles, but there is nothing to indicate that La Mettrie was not serious, as far as he could be serious over anything. There is, it is true, no sign of scientific criticism apparent in speculations such as these — in comparison with them even Buffon's most daring assumptions are temperate and founded on facts — but at least they have their interest as a sign of the times, and the endeavour which finds expression therein points ahead to the "natural-creation stories" of our own day. No one before had ever dared so openly and so rashly to break with the old, officially accepted, traditions, then upheld by the whole authority of the State, and even among La Mettrie's contemporaries there was no one who would have dared to abandon the belief in a God as the Creator of the world and in the immortality of the soul; both were regarded as indispensable bases for even the most liberal-minded morality. But La Mettrie would even reform morals and social life; he desired to create a natural and philosophical system of ethics in place of the official theological and juridical system. Like others of his contemporaries, he believed in man's natural inclination to virtue and happiness and he propounds certain quite justifiable suggestions for reform. He would forbid wearisome memorizing in the schools, maintaining that child education should be based upon the exercise of the natural powers of observation, and he urges the courts of law to differentiate between deeds committed by the mentally deficient and ordinary crimes. The highest aim should be to make the world happy, but the main point of the art of living that he preaches is first and foremost "la volupté" — that is to say, in fact, sexual desire, the satisfaction of which with the greatest possible enjoyment and the least possible risk he discusses in a lengthy treatise, claiming thereby to lead humanity to the height of rational worldly wisdom. This philosophy of licence became, as is well known, widely popular during the latter half of the eighteenth century; in itself it would of course have no concern with this present work had not several of its champions, like La Mettrie, based their philosophy upon arguments derived from natural science, which contributed to bring the latter into double discredit — both as hostile to revealed religion and as promoting all kinds of flippancy and social disorganization. In this the contempt shown for natural science in the age of romanticism, at least in part, finds its explanation. And yet even La Mettrie possessed one feature that is reminiscent of romanticism, or rather of the "*Sturm und Drang*" period: he has left us a characteristic description of himself, in which he talks of his kindly, innocent heart, which never committed any sin, even though his thoughts did so, and he counsels us not to judge his morals by his writings. There are, however, reasons for supposing that his life and his teachings were not inconsistent with one another. He was certainly no paragon of virtue, but he was a child of his age and he had ideas that were in advance of those of that period.

On the whole, we find during the eighteenth century a great number of philosophical speculations widely differing from one another; certain of them broke boldly away from all the old traditional ideas, while others sought to reconcile the old and the new. To the latter type belong in a marked degree those attempts to explain the nature and development of living organisms that were published by Bonnet, who in a remarkable way combines ideas of value for the future with conceptions based on a cosmic theory which had already been abandoned by most thinkers.

CHARLES BONNET was born at Geneva of wealthy parents in 1720. The family had emigrated from France at the time of the persecution of the Huguenots. He studied law and was elected to the council of his native town, but at the same time he evinced a lively interest in natural science and eventually devoted himself entirely to that pursuit. As a pupil of Réaumur he applied himself chiefly to insect biology and in this field carried out work of lasting value. A serious ophthalmic disease, however, soon compelled him to give up making direct observations and all practical work of any kind, so that, being a man of independent means, he spent the rest of his days engaged in purely theoretical speculations in natural science and philosophy. He died in 1793 on his estate in the neighbourhood of Geneva.

Even in his earliest works Bonnet shows himself, apart from his accounts of actual observations, a natural philosopher, and in his last works speculation alone predominates. As a thinker Bonnet is entirely in accord with the Christian point of view, and his writings, by contrast to the freethinking that was so prevalent in his time, acquire a sharply polemical and religiously fervent tone, with the result that sometimes even his purely practical declarations are made in a form that sounds more like those of a lay preacher than a scientist. His writings are extremely difficult for a modern reader to appreciate; one has to search through scores or even hundreds of pages full of enthusiastic praise of the Creator, of effusive outpourings regarding the life of the angels and the existence of the human soul in another life, in order to find amongst it all biological observations of lasting value and shrewd theoretical discussions on natural-scientific problems. Bonnet bases the whole of this speculation, so widely at variance with the spirit of his age, on Leibniz, who, we may remember, strove to reconcile the Christian beliefs with the results of natural science and philosophical thought. Haller's fervently Christian conception of nature also strongly influences Bonnet, who, having sprung from the high-conservative and strictly Calvinistic patrician class of Geneva, was perhaps even in his youth opposed to that tendency to free-thinking which prevailed in most contemporary scientific circles.

Bonnet's discovery of parthenogenesis

BONNET takes his place in the history of biology primarily as the discoverer of parthenogenetic reproduction. For years he studied the reproduction of the aphides, and succeeded in establishing the existence of a number of summer hatches of females who without fertilization propagate by producing live offspring; towards the autumn a new generation arises, this time consisting of males and females, which mate, the females then laying eggs, which hibernate. He also discovered and studied other peculiarities of insect reproduction, as, for instance, the peculiar propagation of the pupiparous flies. Further, Bonnet followed up with great care the study of the phenomena of division and regeneration which Trembley had discovered; he observed a large number of lower, colonizing animals, belonging to the Cœlentera and the Bryozoa, and experimented with them, as also with fresh-water Annelida and common earthworms, observing not only the regeneration which results in normal individuals, but also such as results in malformations — in this respect a precursor in a special field of research, which has been very highly developed in modern times. He studied with great exactness the metamorphosis of insects, endeavouring to discover what changes the parts and organs of the body undergo in the process of evolution from larva, through the pupal stage, to the imago; and at least as far as the intestinal canal is concerned he made, on the whole, correct observations. Again, he studied the adipose tissue and the part it plays during the metamorphosis period of insects as reserve nutriment for the prospective individual. Bonnet also experimented with plants and was one of the first to study their tropisms and growth-movements. The whole of this valuable collection of facts, however, he accumulated to form the basis of his theoretical speculations upon life on the earth — one might even say, in the universe — which were to him the most essential function of science.

His preformation theory

OF Bonnet's scientific theories the best known is his thoroughly worked-out preformation theory. His "incapsulation" theory, according to which every female individual contains within her the "germs" of all the creatures that originate from her, the one generation within the other, and that thus the first female of every species contained within her all the individuals of that species that have ever been produced and that will be produced until the end of time — this theory is really the very foundation on which all his biological speculation was built. He found actual support for it in his observations of the reproduction of the Aphididæ; in the parthenogeneti ally produced, new-born female of the plant-louse he saw the ready-formed rudiments of a new generation, and even the metamorphosing insect shows the imago ready formed beneath the pupal skin. In the plant, on the other hand, the germ and the cotyledons are visible in the seed, and the bud encloses the leaves that are to emanate from it. He therefore considered himself fully justified in seeing in these facts a universal law governing the whole of animate nature; he is strongly opposed to all epigenetic theories and characterizes as legends the observations purporting to show that in the embryo of the chicken certain organs are developed before others. However, it is never made quite clear whether these germs, which thus exist in infinite numbers incapsulated within one another, are to be regarded purely corporeally or whether they are some kind of ideal entities after the manner of Aristotle. Bonnet, as a matter of fact, draws quite a number of his theories from Aristotle, starting from that of the ultimate cause, God or the supreme intelligence, and of the harmony and finality of the universe. At all events, the germs exist not only in the ovaries of the females, but also, in some animals at any rate, scattered all over the body. There is, in fact, no other way by which Bonnet can explain how the bits of a cut-up earthworm are regenerated into new individuals; for the earthworm must, like all animals, be assumed to possess a soul, and the soul is always one and indivisible; if, then, the earthworm is to be regenerated, germs possessing soul-rudiments must lie scattered throughout the body. Indeed, even a separate extremity that is regenerated, as in the crayfish, for instance, must possess a separate germ which is intended to replace it when it is lost, and the same holds good for individual muscles and fibres, which are capable of growing again even in the highest animal forms. — The whole of this germ theory is clearly reminiscent of Leibniz's monad theory and thus has its origin in common with both Buffon's and La Mettrie's doctrines of living particles filling the universe; but while the two latter sceptics utilized the hypothesis to establish a theory of primal creation (spontaneous generation), thereby abandoning the personal creator, the fervently religious Bonnet strongly repudiates all idea of spontaneous generation and gives a number of reasons against it, which are in part valid at the present day — an instance among many of how the scientist's personal conceptions influence his purely scientific theories.

His descent theory

THE preformation theory, however, represents only one side of Bonnet's curious speculative investigations. One idea that occupies his mind quite as much is the thought of the progressive development going on in nature. His firm conviction as to the wisdom of the Creator has made of him an incorrigible optimist; he is absolutely convinced that nature is advancing towards a high goal; he believes that there are heavenly bodies in which this development, which he expects that the earth will eventually experience, has already been attained — in which the stones possess organic structure, the plants are sensible, the animals talk, and men are angels. And just as he expects an advance beyond the present stage, so he believes that this is the result of a process of evolution; the "germs" that are incapsulated within one another in an individual are not alike and never have been; on the contrary, he expressly maintains that if one were to see a horse, a hen, a snake, under the form they had when they first came into existence, they would be unrecognizable. These changes he accounts for by a series of stages of development which the earth has undergone and each of which has been cut short by some vast natural catastrophe, which destroyed all living things, but always spared the germs out of which fresh life-forms arose. The last of these catastrophes was the one that destroyed the earth before the six days of the Creation referred to in the Books of Moses, the historical authenticity of which Bonnet was naturally careful to maintain, but which he interprets somewhat freely, according to the orthodox view. As the result of a coming catastrophe he expects the perfecting of the world's existence, as indicated above, while he assumes from the presence of fossils in the mountains a series of previous epochs of existence with living creatures that did not resemble those now existing and from which these latter have originated.

Just as Bonnet manifestly bases this geological theory on Buffon, so he is the precursor of both Cuvier and Lamarck in the same way; Cuvier's famous catastrophe-theory corresponds too closely with Bonnet's to justify the assertion that the similarity was accidental, while, on the other hand, Bonnet in his express statements regarding the change that takes place in species has forestalled Lamarck's descent speculations, though, it is true, both the biologists mentioned succeeded in elaborating their ideas into a far more perfect whole. But in still another respect Bonnet foreshadows these two great pioneers of biological science: he maintains — again an obvious connexion with Buffon — that nature draws no sharply defined lines between the species, but that all life-forms on the earth pass into one another. He draws up what he calls "*une échelle des êtres naturels*" — a series proceeding from the simple elements through the mineral kingdom, vegetable kingdom, and animal kingdom in a long line right up to man. The transitions in the series are, from the modern point of view, ingenuously chosen: the flying fish provides the transition between fishes and birds; the ostrich, the bat, and the flying squirrel between birds and quadrupeds; the polypus and the sensitive plant between animals and vegetables. But then he also declares that the whole of this division is only approximate and that perhaps the series is not as uniform as he has made it; that perhaps molluscs and insects, lizards and frogs do not follow one another consecutively, but are in reality collateral with one another. Just as the long evolutionary series of Bonnet clearly foreshadows Lamarck's evolutionary theory, so the assumption of parallel evolutionary groups represents a first hint of the type theory that Cuvier founded and whereby he reformed the entire zoological system of classification and rendered possible the approach of the modern descent-theory. And as has already been pointed out, the points of agreement are certainly not accidental; Bonnet enjoyed a great reputation amongst his contemporaries and the immediately succeeding age and was diligently studied. Cuvier in particular has expressed his warm admiration for him and recommended his writings for careful study; and other contemporary biologists certainly knew his works and were to some extent influenced by them.

In the foregoing have been mentioned those theories of Bonnet that have proved to be the most vital for the development of biology, and considerations of space forbid a detailed account of all the shrewd utterances which this imaginative man of genius scattered throughout his writings; for example, his striking criticism of vitalism. In spite of his religious fanaticism he gives a purely mechanical explanation of the bodily functions and cites the pointed objection to the vitalists — mostly Stahl and his school — that "souls" are particularly convenient to have when it is a question of explaining the phenomena of life; they do everything that is asked of them and their non-existence can never be proved. Another time he gives a detailed analysis of the different organs in the same body that are dependent upon one another and shows how a change in one organ must inevitably react upon the others; and on still another occasion he describes his observations regarding different mental impressions — a problem which, as is well known, Goethe made the subject of exhaustive study. Thus Bonnet was a man full of ideas; and though in a great deal he must appear out of accord with our age, yet undoubtedly many of his ideas are nowadays incorporated in the general consciousness.

The experimental biological investigations that Bonnet made the basis of his speculations were continued and considerably widened by LAZZARO SPALLANZANI (1729–99). Born at Reggio, the son of a lawyer, he studied law at Bologna and at the same time took orders. He afterwards devoted himself to natural-scientific studies and became professor of philosophy, first at Modena and later at Pavia. He applied himself to experimental research, particularly in regard to regeneration and fertilization, and left all his predecessors far behind him, in both method and results. In the amphibians, especially salamanders and tritons, he found suitable subjects for the study of regeneration even among the vertebrates, and he made as exhaustive a use of them as was possible under the conditions in which he worked; he studied the re-formation of the tail, extremities, and jaws, and this not merely for the purpose of establishing the fact, but by means of dissection and microscopic investigation he followed the re-forming of the various components of the body: muscles, nerves, and bones. He observed the time that the regeneration lasted and endeavoured to influence the process by means of altering the conditions of food and temperature. He even experimented with the phenomena of fertilization; by filtering the sperm of particular animals he proved that the presence of the spermatozoa was essential if fertilization was to take place; nevertheless he could not be induced to assume a direct influence of these components upon the egg, but believed that the accompanying fluid was the substance that stimulated the egg's development. For he adhered as stubbornly as Bonnet to the preformation theory: he closely studied the development of the frog's egg and followed the formation of the backbone channel, but merely for the purpose of seeking evidence of the entire animal's having been ready-formed in the egg. Eventually he believed that he had discovered incontestable proof thereof, when he saw the frog's egg increasing in size within the body of the mother animal and before it had been fertilized, and as growth is not possible without organs, the larva of the frog must have been ready-made in the egg before fertilization.

Just as Spallanzani was thus convinced that he had found an undeniable argument in favour of the preformation theory, another scientist published a treatise which was to form the basis of a new conception of embryonic development.

CASPAR FRIEDRICH WOLFF was the name of the naturalist who led the science of embryology into fresh paths. He was born in Berlin in 1733, the son of a tailor. He went through a course of medical training at the College of Medicine there and thence proceeded to Halle, where he studied philosophy after the system of Leibniz and his pupil Christian Wolff,[6] and finally,

---

[6] CHRISTIAN WOLFF (1679–1754) was professor of mathematics at Halle, whence he was ejected through the intrigues of the Pietists (Stahl seems to have taken part in the persecution directed against him) and then became professor at Marburg, but later he returned to Halle and there worked as professor of philosophy. Under the general title *Vernünftige Betrachtungen* he published a series of essays covering many different fields of human knowledge, in which he expounds Leibniz's theories in a popularized form, in particular maintaining that everything that happens must possess adequate reason for doing so, because otherwise something might

for his doctor's degree, he published in 1759 the essay which was to make his name famous. Having served for a time as an army surgeon, he applied for and obtained permission to hold lectures in medicine in Berlin, which, however, resulted in his coming into serious conflict with the professors at the Collegium Medicum. Being of a peaceful disposition he was much troubled at this and was delighted when he received a summons to St. Petersburg, where he became an academician and spent the rest of his life carrying on his research work in peace. He died in 1794.

Wolff's generation theory

CASPAR FRIEDRICH WOLFF is one of those who did not win fame until after death. His own age paid little attention to him. Haller, to whom he dedicated his afterwards famous treatise *Theoria generationis*, accepted the honour in a friendly spirit, but paid little attention to the work, as also did other biologists of the period. That Wolff was thus misunderstood by his contemporaries was due mostly to the fact that from the very outset he adopted a course directly opposed to the then prevailing conception of the phenomena of life; he began with a ready-made theoretical program, and the facts he presents are collected for the express purpose of proving his already firmly established convictions. By way of introduction he lays down the plan of his work; by the body's "*generatio*," or, as we should now call it, "evolution," is meant its creation ("*formatio*") in all its parts, and its principle is the force that brings about this creation. The upholders of the doctrine of "predelineation" thus deny, he adds, that any "generation" takes place at all. He starts, therefore, by declaring war on the preformation theory; he does not base his rejection of it on the evidence of the facts he has observed, but on purely theoretical reasons adduced by Christian Wolff's philosophical methods. "He gives a true explanation of generation who derives the parts of the body and their composition from the fixed principles and laws governing them; . . . and he has perfected a theory of generation who has succeeded in tracing the entire ready-formed body from these principles and laws." The principles on which the fresh formation of organism takes place are food and growth; food re-creates the simple components of the organism, while through growth are formed entire parts of the body or fresh bodies. Reproduction is, in fact, brought about by a "weakened growth (*vegetatio languescens*)," whereby the newly-formed seed or embryo is separated from the mother plant or animal and is prevented from growing further in union with the latter. And that which produces all nourishment and growth is, according to Wolff, the "inner force (*vis essentialis*)," a term which he constantly uses to mean the ultimate cause of all that takes place

arise out of nothing, which is impossible. For the rest, he was a clever mathematician and, for his age, a sound botanist, and contributed much towards inculcating an interest in natural science in Germany.

in the organism, the idea of which he himself states that he borrowed from Stahl. For just as the basis of Wolff's research work is the rejection of preformation, so its final object is the abolition of "mechanical medicine" — the theory that holds that the living body should be regarded and treated as a machine, which finds the explanation of the phenomena of life in the form and composition of the bodily parts, or, as it is expressly stated, in anatomical principles. This theory Wolff declares to be a product of the imagination and produces a number of arguments to prove its falseness. Wolff, however, never arrives at any properly worked-out vitalistic theory; after all, he deals with the phenomena of the body along mechanical lines and his "vis essentialis" he does not identify, as Stahl did, with the soul. On the whole, Wolff's theory is vague and inconsistent if we compare it with Stahl's mode of thought, which is certainly hard to apprehend, but is nevertheless in its way loftily conceived. The most serious result of Wolff's philosophical method, however, is that he fancies it capable of explaining practically anything; with a couple of phrases he throws a bridge across even the deepest abysses of natural science; he has a theory ready to hand to explain even such phenomena as those in the face of which modern biology has to be content with merely establishing the fact. In all this he is a precursor of the natural philosophy of romanticism, and it was, in fact, this that eventually procured for his views the honour they deserved.

His cellular theory

WOLFF's treatise deals with the development of both plants and animals in a constant endeavour to find factors common to both. In his opinion, the growth of plants is due to the inner life-force drawing up moisture out of the earth through the roots and into all the various parts of the plants; at the points of growth this moisture is collected in especially large quantities; through evaporation it acquires greater density and forms cuticles, which, through fresh supplies of moisture, assume the form of ampullæ, the walls of which are further thickened by evaporation, and the new ampullæ force themselves in between the earlier ones, whereby the substance of the plant is renewed. The plant's vesicular system is formed through the circulating sap's hollowing out ducts in the vegetable substance, the walls of these ducts being likewise thickened by evaporation. The plant is then formed by these ampullæ and ducts through a system of growth-forms, the abstract and involved details of which it would take too long to follow. As mentioned above, the florescence and germination are caused by a weakened growth — "The adequate reason why within a certain period frondescence ceases and germination begins is a diminution of the supply of alimental sap at the point of growth, as is at once seen from the very definition of growth." This is Wolff's scientific adduction of evidence. Similarly it is proved that the germination and embryonic development consist in a renewed growth induced by the "*perfectum nutrimentum*" with which the pollen provides the seeds; lengthy arguments are brought forward to show why the pollen is and must be the most perfect form of nourishment that exists. Sexual reproduction, then, is nothing more than a renewed growth.

The fundamental principles on which growth proceeds in the vegetable kingdom Wolff discovers in detail in the animal kingdom; in the embryo of the chicken, which is his only subject of investigation in animal embryology, he finds reproduced the same phenomenon of growth; the inner force derives nourishment from the yolk of the embryonic *lamina*, this alimental fluid coagulating, as in the plant, into ampullæ and ducts, the latter here represented by heart and vesicular system. Here, too, the details of the embryonic development, which is described much less fully than that of the plant, are of no particular interest; here again Wolff gives full rein to his speculative imagination at the expense of detailed observation. Such an assertion as that no one has discovered anything with a powerful magnifying-glass that could not equally well have been observed with a lower magnification is sufficient evidence of how his speculations are out of accord with reality. And still worse is his habit of comparing the structure of plants and animals in detail; his comparison of a plant's vessels with the arteries, of its suckers with the veins, rivals in absurdity most of what had hitherto been perpetrated in that sphere — which was by no means little.

And yet it is just through his comparison of plant and animal development that Wolff made his most important contribution to biological history. He was the first to compare the elements of which the plant and the animal are composed, and though the details of this comparison are for the most part incorrect, it was at any rate he before anyone else who pointed out the ampullæ -like structure — in other words, the cell-tissue — that is common to both. He thus carried science a considerable step further along the road marked out by Malpighi and his immediate successors.

His epigenesis doctrine

WOLFF's second service to science is generally said to be his introduction of the doctrine of epigenesis into biology in place of the preformation theory. We have previously found that the epigenesis theory is actually older than the preformation theory; even Aristotle was an epigenetic, and his doctrine was promulgated without contradiction even by Harvey, whereas the first champion of the preformation theory was Swammerdam. It was thus an ancient theory that Wolff adopted, and from the very outset he made it his own on purely theoretical ground; it was only natural, therefore, that his observations should eventually accord with the preconceived ideas. But the progress of science was facilitated by the fact that — whether with preconceptions or not — he saw more correctly in his microscope than his contemporary preformationists; for their part, they considered an embryological study to be in the main superfluous, since everything was ready-formed before, whereas Wolff showed that in this sphere there was still an immense amount to be discovered and investigated, thereby opening up fresh fields of research, in which very successful work was done during the succeeding epoch. It has frequently been said, however, that in this question Wolff was entirely in the right and his opponents in the wrong. This view is utterly at variance with historical facts. When it first arose, the preformation theory not only was fully justified and compatible with the scientific standpoint of the time, but, as has been pointed out above, also constituted a real advance, whereas the epigenesis theory, as Wolff formulated it, certainly shot far beyond the mark. He who denied to the undeveloped egg all trace of organic structure would undoubtedly have found modern ontogenetical research, with its strong emphasis on the orientation of the egg and its various parts and with the maintenance of the immutability of the factors of heredity, highly preformational.

His romantic conception of nature

IT is, however, by no means in his epigenesis theory alone that Wolff shows himself at variance with his age and foreshadows a new era; the whole of his scientific matter and his entire conception of nature differ in a marked degree from those of his contemporaries. He is, as has already been pointed out, a precursor in the course which natural science took at the end of the eighteenth century and which is termed natural philosophy; this course was directed, particularly in Germany, towards an entirely new knowledge of nature, with the utter rejection of both the aims and the means with which natural research had been carried out up to that time. Biological natural-philosophy, however, is only a link in a universal cultural movement of far wider influence, which will be explained later on. But first we must devote some words to the application of the experimental method to botany during the eighteenth century, as well as to one or two anatomical and morpho-logical scientists who were at work during the same period.

In the eighteenth century the science of botany was governed, far more than animal biology, by Linnæanism. During the same period, however, one or two naturalists employed in their study of the vegetable kingdom methods other than the purely systematic; as a rule they worked in obscurity and their results were appreciated only by succeeding ages. One exception to this was the English experimental scientist, STEPHEN HALES, whose work was highly appreciated by his contemporaries; his writings were translated into French by Buffon and into German by Christian Wolff. And he was well worthy of the attention paid to him, for he is without doubt one of the most remarkable biologists of the eighteenth century.

Hales was born in 1679 at Beckesbury in the south of England. He was of good family, and after studying theology at Cambridge he took holy orders in the Church of England. He held various posts and finally became vicar of Teddington, a parish in Middlesex, where he died in 1761. He was known as a zealous priest, who worked for the advancement of his parish from both a moral and a material point of view, and besides found time to devote himself to important philanthropical works, such as the improvement of prison conditions, the administration of charitable societies, and inventions likely to prove of benefit to mankind. By those who knew him personally he was extolled for his kindness and simplicity.

At Cambridge Hales had been attracted to the study of natural science, particularly physics, chemistry, and botany; indeed, during his undergraduate days Cambridge was primarily regarded as Newton's town. He maintained this interest throughout his life; it thus occurred to him to try by way of physics to discover the conditions of the life and growth of plants, an idea that he realized after experimental studies lasting many years. He published his results in the year 1727 under the title of *Vegetable Staticks*. In his ability to organize biological experiments and to draw conclusions therefrom he was excelled by none of his contemporary scientists and by but few of those that have come after him; it has been possible even in modern times to apply his experimental methods with profitable results. In regard to his general cosmic conceptions Hales was, naturally, in conformity with his profession and his age, a pious Christian, but like his master Newton he strove conscientiously to discover the law-bound mechanical processes undergone by the phenomena he investigated; he would never involve himself in hypothetical explanations of the manifestations of life.

Hales's quantitative experiments

WHAT Hales wished to discover by means of his experiments was, first of all, the renewal of substance in plants, both quantitatively and qualitatively. It is above all his quantitative investigations that merited and also won general admiration; he was the first to apply systematically and on a large scale the exact method of physics to animate nature. By watering previously weighed potted plants for a given length of time with a fixed quantity of water, and by weighing the plant daily during that period, he found out its water-consumption; then he measured the leaf- and stem-surface of the plant and calculated therefrom the relation between the surface of the plant and the quantity of water that it absorbed daily. Similarly, by means of measuring and weighing he calculated the quantity of moisture that different plants absorb out of the earth through their roots, as well as the speed with which the sap circulates in the interior of the plant; and finally he proved that plants absorb air through their leaves and stems and not only through their roots, as earlier botanists declared. He was quite specially interested in the problem of the relation of the air to living creatures. He definitely maintains that the air contains component parts which are ab-

sorbed by the plant through the leaves and are converted into solid substances. Likewise he states, though with less certainty, that light penetrates
the leaves and co-operates in the alimental processes in them. In these assertions attempts have been made to find definite proof of Hales's genius, and
there is undeniably in them a brilliant guess at facts that were established
at a later period, but in regard to points of detail his speculations on the
properties of air are undoubtedly far more deficient than his quantitative
experiments. It is true that gas-chemistry had been but little developed in
his time, but it would seem that he scarcely took advantage of what actually
was known; he certainly cites Boyle quite frequently, but he evidently knew
nothing of van Helmont's gas-experiments. To him all gases are "air," both
that which arises from the dry distillation of wood and that which is formed
by treating lime with acid. In such circumstances it was inevitable that the
great trouble he went to in investigating the influence of the air upon vegetable, and even animal, life was to a great extent in vain. What he achieved
as an experimenter, however, is quite enough to ensure for him considerable
fame in the history of biology, and it was to be long before science advanced
beyond his point of view. For this to happen there was required above all
else a reformation of the science of chemistry — which indeed actually took
place at the end of the eighteenth century. This will be described in a following chapter. During the latter part of his life Hales also worked at experiments on animals, particularly in regard to the blood-circulation, and he
displayed in this sphere the same power of arranging experiments and drawing conclusions therefrom that he showed in his botanical investigations.
He measured the blood-pressure in live mammals by introducing into a vein
a tube in which the blood was made to rise; he calculated the speed of the
blood-stream in the veins and capillaries from the volume of the vessels,
the rate of movement of the blood-mass, and the resistance of the walls.
To these investigations he added a quantity of notes on medicinal and hygienic subjects, with particular reference to the injuriousness of alcoholic
liquors, for he was a keen supporter of temperance. This fact gives his
Hæmastaticks, as he called his investigations into the blood, a far more
motley character than his treatise on vegetable physiology; nevertheless,
even these investigations are of some value and he holds a place of honour in
the history of physiology.

Among the plant-physiologists who, after Hales, distinguished themselves during the eighteenth century there are one or two who carried out
important experimental observations regarding plant-reproduction, and who
deserve special mention.

JOSEPH GOTTLIEB KOELREUTER was born at Sulz, Württemberg, in 1733.
We know, on the whole, very little about his life; he seems to have studied
in Berlin and Leipzig and spent some time in St. Petersburg; in 1764 he was

made professor in natural history and curator of the botanical gardens at Karlsruhe. He died in 1806. Before his appointment to the professorship he had already published the first series of notes in which he recorded the results of his experiment with the artificial fertilization of plants. As we have seen, Camerarius was the first to experiment in this field. Linnæus followed in his footsteps, carrying out, it will be remembered, the hybridization of plants, but not being otherwise an experimental naturalist-worker in the true sense.

Koelreuter's experiments in plant-life fertilization

KOELREUTER was the first who exclusively applied himself to experimenting with the cultivation of plants with a view to explaining their fertilization and development. To begin with he investigated the act of fertilization itself; he examined the pollen under the microscope and came to the conclusion that its fertilizing property is due to an oily fluid that it secretes; on the stigma of the pistil he found a similar fluid and concludes therefrom that fertilization consists in a union of these fluids, just as an acid and a base form a salt. Of greater value than these speculations are his careful observations of the method of transmitting the pollen; he is the first to explain clearly that certain flowers are invariably fertilized by insects, and he also pointed out the part played by the wind in the fertilization of other forms. Of greatest interest, however, are his investigations in connexion with hybrid formations, a problem to which he eventually devoted all his attention. In this sphere he paved the way for a field of research that, as is well known, has at the present day attracted the interest of both the scientific world and the public more than most others. To start with, for a number of years he crossed different types of tobacco-plants with one another, afterwards, however, proceeding to other plant genera: pinks, aquilegia, verbascum, and others. Moreover, he was able to vary his experiments and to observe the results thereof; he carefully compared the hybrids with the parent individuals and noted similarities and dissimilarities between them; he mated the hybrids with their parent species and observed the reversion to similarity with the latter; he fertilized the hybrids with one another and obtained results that foreshadowed Mendel's famous observations; he likewise even noted cases which would be regarded at the present day as mutations. However, he naturally did not succeed in utilizing theoretically the results of his experiments; besides, his ideas of the actual essence of fertilization were all too vague — he believed, for instance, that by fertilizing a species with a mixture of its own and foreign pollen it would be possible to obtain a kind of semi-hybrids, which would be somewhat, but not very much unlike the mother species. Further, he mixed his speculations up with certain mystical ideas, particularly in the sphere of alchemy; he expressly compares the change that the characters of the species undergo in hybridization with the conversion of metals effected by the alchemists, and presumes that, just as a vegetable species can, by repeated unilateral crossing, be transformed into another, one day man will learn to convert, by a necessarily gradual process, one metal into another; in further evidence of which he finds a correspondence between the pollen and the sulphur of the alchemists, proved by the fact that pollen can be used as a means of reducing metal oxides, though this is easily explained when the pollen is burned to coal, which has a reducing effect. Moreover, the female sexual product is, in his opinion, "mercurial." It is thus in the sphere of practical experiment that Koelreuter's greatness lies; in this he is a pioneer, and his experiments in crossing were justly taken as a model until Mendel's far more deeply thought-out experiments became known. Koelreuter shared the fate of the latter, the greatest of his successors, in his works' being for a long time entirely neglected; it was not until long after his death that they were rescued from oblivion and accorded the appreciation they deserved.

The same fate of being neglected by contemporary and immediately succeeding ages was suffered by CHRISTIAN CONRAD SPRENGEL, whose investigations into the fertilization of flowers were carried out in association with Koelreuter's. Born in 1750 at Brandenburg, the son of a clergyman, Sprengel studied theology and languages, afterwards devoting himself to the tutor's profession. He was for some years a schoolmaster in Berlin and later became rector at Spandau. After several quarrels with his superiors, his pupils, and their parents he was dismissed with a pension in 1794 and then lived in Berlin in a solitude that increased year by year until he died in 1816. His irascible temperament contributed both to his failure as a teacher and to his subsequent isolation; it was aggravated by the utter lack of understanding with which his contemporaries received the results of the botanical researches that had represented the chief interest of his life. He was unable to print his last botanical writings, with the result that during his last few years he applied himself to philology, apparently with but little success.

Sprengel's experiments on plant-fertilization

THE work which at last brought his name to the knowledge of posterity is his *Das entdeckte Geheimnis der Natur im Bau und in der Befruchtung der Blumen*, published in 1793. Under this somewhat pretentious title he collected a large number of observations in connexion with the florescence of plants, and on them bases a general theory of fertilization in the vegetable kingdom, which in its essentials still holds good today. In conformity with his theological upbringing he was fully convinced of nature's having been preconceived by the wisdom of the Creator down to the minutest detail, and he consequently set about trying to discover for what useful purpose the different parts and properties of the flower were intended. As a result of his inquiries into this subject he found, to begin with, that the flowers' nectaries are always protected from the rain, and further that they are often characterized by special colours, whence he concludes that their object must be to attract insects to the flowers; but then the insects must themselves have some object in their visits, and this he found, as did Koelreuter before him, to be the conveyance of pollen from stamen to pistil. He now studied in detail the relation of the insects to the flowers and noticed that certain flowers are invariably fertilized by special insect forms, others again by several different forms, and that the position of the nectaries in each flower is adapted not only to the flower's general conditions of life, but also to the insects that visit it. Further, he discovered that in a number of bisexual flowers stamen and pistil actually develop during different periods, and that therefore the flower cannot be fertilized by its own pollen, but that pollen is conveyed by the insects from flower to flower. This fact he calls dichogamy, a name which is still used, and he concludes from it that "Nature does not appear to desire that a flower be fertilized by its own pollen." Finally, he explains more lucidly than any of his predecessors the contrast between flowers fertilized by insects and those fertilized by the wind; on this subject, too, he makes many striking observations.

Space forbids a more detailed account of the numerous shrewd and far-reaching observations which Sprengel adduces in support of his theories. Through his work he has laid a lasting foundation for one of the most important sections of vegetable biology, and besides, in regard to insect research, he has pointed out a method of far greater theoretical importance than mere classification and collecting. And so the utter lack of understanding shown for his work by his own age was all the more tragic. The natural philosophers of the Romantic Age deeply despised detailed research work of this kind, and the succeeding generation, which endeavoured to revive the mechanistic conception of nature of the eighteenth century, felt embarrassed by the detailed finality which Sprengel sought and found in the structure and life of the flowers. It was only Darwin's authority that succeeded in rescuing Sprengel from oblivion; in the flowers' and insects' mutual dependence upon one another he found support for his theory of selection and himself carried out investigations in this field, which will be described in a later chapter. Thus Sprengel found redress — tardy but glorious.

CHAPTER XI

THE ANATOMICAL SCIENCE of the eighteenth century appears to be a direct continuation of that of the previous century; no important discoveries of a pioneer character were made, but those fields of research that had been won were well investigated in detail, and this field of inquiry can show names that testify to praiseworthy endeavour, if not so much to brilliant genius. Of these names some of the more representative will be mentioned in the present chapter.

BERNHARD SIEGFRIED ALBINUS was the son of a German physician of repute who, after having studied at Leyden, held various posts in his native country, but eventually returned to Leyden as a professor. Young Bernhard, who was born at Frankfurt an der Oder in 1697, was consequently brought up at Leyden and spent his life there. At the early age of twenty-four he was made professor of anatomy and surgery, and lectured on these subjects and on physiology until his death, in 1770. He was highly esteemed by his contemporaries and honours of many kinds were bestowed upon him. He was, in fact, a thoroughly educated scientist and was gifted in many ways. He was interested in the history of science and published critical editions of the works of the leading anatomists — those of Vesalius, Eustacchi, and Harvey were reprinted by him. His own works were extensive and profound. He studied with great care the bone-structure of the human embryo and its development, and even in the full-grown man it was mostly the bone-structure and the musculature that interested him. He compiled a fine set of engravings illustrating these two organic systems — *Tabulæ sceleti et musculorum corporis humani* — a gigantic work in contents and weight, in which in a series of splendid copperplates, drawn and engraved under his instructions by the famous artist Vandelaar, he reproduces the human bone-structure and musculature in every detail. This work, which cost him a whole fortune, is of its kind still unsurpassed. Besides doing research work Albinus also practised as a doctor, and, thanks to him, Leyden still continued during the eighteenth century to be a centre for anatomical studies.

One of Albinus's most brilliant pupils was JOHANN NATHANAEL LIEBERKÜHN. Born in Berlin in 1711, he was destined by his father for the priesthood, and for several years had to study theology against his will.

After his father's death he took up medicine, studying first in Germany and then under Albinus at Leyden, where he took his degree with a treatise entitled *De valvula coli*. After paying visits to England and France, he settled down in Berlin as a practitioner. He died in 1756. His short life and his extensive practice prevented him from benefiting science as much as he otherwise would certainly have been able to do, yet what he did achieve ensures to him a place in the history of biology. He was above all an excellent technician. He himself made microscopes of splendid workmanship and was able to prepare under the microscope the most minute organic details. Equally remarkable were his injections; preparations made by his own hand are still preserved in the anatomical museum in Berlin. In fact, he used the microscope for the purpose of studying injection-preparations, a thing which had never been done before. His only really important work is his exposition of the structure of the small intestine, in which he describes the Lieberkühnian crypts (called after him), as also those cells existing at the bottom thereof, now called the Panethian cells, whose glandular nature, however, he failed to discover. The whole work bears witness to his technical skill both in injections and in microscopy, and forms a valuable contribution to the development of microscopical anatomy.

Another pupil of Albinus's, who won a far greater reputation in his own age, was PETRUS CAMPER. He was born at Leyden in 1722, studied there, and took degrees in both philosophy and medicine. Having spent a couple of years travelling, he was appointed professor at the academy at Franeker, a small provincial university which, when he first went there, had only four medical students, a number which he succeeded in increasing many times over in a very short time. After five years, however, he obtained a professorship in Amsterdam, and some time later one in Groningen, but he finally gave up teaching and settled at The Hague, where he became a member of the state council and took part in its political life. He died in 1789.

Camper is described as a man of an extremely superior personality, brilliantly gifted, but quick-tempered and despotic. In his own time he was regarded as one of the leading scholars in Europe and attained a splendid position, both socially and financially. His many-sidedness was extraordinary, almost reminiscent of Olof Rudbeck's. Besides carrying out anatomical research in a number of different fields, he was a surgeon and gynæcologist, hygienist, and expert in medical law and veterinary surgery, and in all these spheres he made valuable contributions. He was, besides, an excellent draughtsman and a leading connoisseur of art. He took measurements of the facial angle in human beings of different ages and different races, and in comparison therewith in higher vertebrates, with results of interest both to the history of art and to natural science. This facial angle, which still bears Camper's name, is formed by two lines, the one extending through the opening of the ear and the bottom of the nostril, the other at a tangent to the most protuberant part of the forehead and the chin. When Camper expounded this idea before the Amsterdam Academy of Painting, with a view to giving the artists a more accurate conception of the human form, he little thought that in doing so he was laying the foundations of an entirely new branch of science — modern craniology. In close connexion with this interest in the structure of the human body are his special investigations of the apes, particularly of those resembling man. He had procured as many specimens as he could possibly get of the orang-utan, at that time extremely rare in Europe, and he not only dissected a number of them, but closely studied a live specimen. As a result of especially careful investigations into the musculature of the extremities and the structure of the larynx, he proved conclusively that the animal is unable to walk upright, as La Mettrie and other "philosophers" at that time imagined; nor can it in any form pronounce an articulate language. The philosophers, however, were certainly far too firm in their belief to allow themselves to be convinced by anatomical proofs, all the more so as it could be urged against Camper that he was in all respects, both religious and political, a conservative man.

Camper's anthropological and comparative-anatomical investigations

CAMPER was particularly interested in the anatomical investigation of uncommon and rare animals. He published monographs on the elephant, the rhinoceros, and the reindeer, anatomically useful specimens of which he succeeded in procuring owing to Holland's extensive shipping-communications. Of more general interest than these special researches is his study of the bone-structure of birds, in which he describes for the first time how the bones are filled with air to facilitate flight, and, in connexion therewith, the air-sacs in the body which serve the same purpose. Of immense general interest also are his comparative investigations into the auditory organs of fish, whales, and reptiles, wherein he discusses the reproduction of sound in various media and the ear's adaptability thereto, at the same time making a close study of the different parts of the auditory apparatus. Finally, Camper carried out an anatomical investigation of a highly original kind in his essay "On the Best Form of Shoe," in which, after a detailed description of the bone-structure of the foot, he sharply condemns the unnatural footwear of his time and describes what he considers to be the most rational shape of shoe.

If, then, we find in Camper efforts at comparative anatomy, this is only evidence of his foresight, for as a general rule his contemporary zoologists were content with purely superficial descriptions of types in the Linnæan style. There were a few praiseworthy exceptions, however, among whom John Hunter and Pallas deserve special mention.

JOHN HUNTER (1728–93) was born in a country place in Scotland, the son of a poor farmer. Being orphaned at an early age, he received a very indifferent education, a fact which influenced his whole life. He never even learnt to spell his native language properly, nor at the beginning did he learn any proper profession. At last, at the age of twenty, he went to his elder brother, William, who had become a highly esteemed doctor in London and had been commissioned to examine prospective army-surgeons. John began by assisting at the dissection classes in connexion with this course, but during them he taught himself anatomy, with such success that he was soon able to take over the direction of the entire course. He continued to educate himself, partly under his brother's and partly under other doctors' guidance, finally receiving an appointment as surgeon, attached to the English fleet which sailed to the Spanish main during the Seven Years' War. At the end of the war he settled down as a physician in London, won a reputation as a clever operator, and quickly obtained a remunerative post. He spent all his spare time in anatomical and physiological studies, and as soon as his salary permitted, he bought a house, in which he established a large anatomical museum. On this museum he sacrificed all that he could spare in the way of time and money, so that at the time of his death it was undoubtedly the finest of its kind in existence. He also gave private lectures in anatomy, but he was not a particularly good lecturer. As a practitioner, on the other hand, he was regarded as the best in London in his time. He was universally known as an honest, benevolent, and charitable man, but his personal manners showed his poor education, while his lack of self-control in particular gained him many enemies. In a violent altercation with some of his colleagues he got a stroke of apoplexy, which caused instant death. His museum was taken over by the State and is to this day one of the sights of London in the sphere of natural science. His manuscripts, however, were taken by a brother-in-law, who first plagiarized them for his own benefit and then burnt them in order to destroy all evidence of his plagiarism.

Hunter's work in comparative anatomy

HUNTER's scientific work falls essentially within the sphere of practical medicine; his theoretical researches were always intended as foundations on which to base practical medical work. His famous museum was intended for a similar purpose, but on the broadest lines; he collected all kinds of animals, both higher and lower, dissected them, and experimented with them, setting up the preparations that he made on anatomical principles. Thus he applied for the first time in practice principles of comparative anatomy as a whole, thereby creating a precedent for future research of very great value. Of his writings a treatise on the natural history and diseases of the teeth has been of the utmost value to biology; in it he gives an account of a systematic investigation into the origin and growth of the teeth that is far in advance of any previous work of its kind. In a treatise on inflammation of the blood and bullet-wounds he propounds a curious theory of the blood as a vital principle, which he further developed in a number of lectures on the musculature. He considers the blood to be a kind of primary matter in the body, whence every other bodily substance is derived; all living matter is of a similar nature, so that even the blood of one animal can be transferred into another of a different genus (modern investigations show this to be an error), and life is a kind of independent principle in the body which prevents it from dissolving — a theory reminiscent of Stahl. Though Hunter produced a few solitary brilliant ideas, yet from a theoretical point of view he did not contribute very much to the development of biology; his genius for comparative anatomy was, however, probably greater than anyone else's in his time, and in many respects it has borne fruit in more recent times.

PETER SIMON PALLAS was born in Berlin in the year 1741, the son of a doctor, and studied medicine in his native country, at Göttingen, and at Leyden. At the latter university he got his degree with an essay on intestinal worms. He afterwards spent some years in Holland, working at zoological collections from the tropics, which he described in a series of papers. In 1768 he was summoned by the Russian Government to take part in an important expedition which was being sent to Siberia to explore that country from the point of view of natural history and economics. Pallas spent six years travelling in Siberia, reaching as far as Amur, and he brought home an immense quantity of scientific material, which he worked at in St. Petersburg for a number of years. In 1793 he was sent to explore the Crimean district, which had just then become part of Russia, and he stayed there for a long time, living on an estate which the Empress Catherine II gave him. Finally, however, he moved back to Berlin in order to be in closer touch with the scientific world, and there he died in 1811.

Pallas's work on intestinal worms and on mammals

PALLAS's contribution to the development of biology is particularly many-sided. In his doctor's dissertation he incorporated all the observations he was able to obtain dealing with intestinal worms and he sought to prove that they enter the human body from outside — in his time it was universally assumed that they arose out of "tainted fluids" in the body. In a work on the zoophytes he tries to find out the classification of these animals, their conditions of life, and their relation to animals and plants. He endeavours to prove that the zoophytes form a true transition between animals and plants, following the ancient saying that nature never makes any jumps. He also made a number of interesting observations, both anatomical and biological, on worms and expressly points out how utterly heterogeneous the Linnæan class bearing this name is. Primarily, however, Pallas is a student of vertebrates. In his *Spicilegia zoologica* in particular — a collection of monographs, issued in separate numbers — he describes in detail a number of hitherto unknown higher animals, dealing with their anatomy, morphology, and habits. Among his biological works, however, the place of honour is held by his work on *New Mammal Species from the Rodentia*. In this work he gives an account, with a thoroughness that was quite unprecedented, of the new rodents discovered by him in Russia and Siberia; in it he endeavours to present not merely diagnoses as resulting from his examinations, such as his age was usually content with, but a true general knowledge of the animals described, based on a close study of their exterior, with careful measurements of every part of their body, thorough anatomical investigations and illustrations, and detailed descriptions of the conditions under which the animals lived. The anatomical section is particularly useful and constitutes the best work so far carried out in the investigation of the inner structure of the members of an entire order of animals. Though direct points of comparison do not occur very often in the work, nevertheless the descriptions are so detailed and at the same time so comprehensive that the whole must be regarded as one of the really sound pieces of work that have paved the way for modern comparative anatomy.

At this point we may close our account of the biology of the eighteenth century. Before, however, proceeding to the cultural phenomena — already hinted at above — that represent the basis of the natural science of the nineteenth century, we must take a glance at a radical reform in another sphere of natural science, which contributed towards the creation of modern biology.

CHAPTER XII

THE FIRST BEGINNINGS OF MODERN CHEMISTRY AND ITS INFLUENCE UPON THE DEVELOPMENT OF BIOLOGY

The phlogiston theory

So long as chemical processes had their explanation in the phlogiston theory, it was certainly possible to offer a provisional explanation of a number of phenomena in the sphere of combustion and oxidization, but any deeper insight into the material changes which both animate and inanimate nature daily undergo was of course out of the question. In particular the qualitative side of the chemical process was, as far as this theory went, inexplicable. In spite of this, the theory was stubbornly maintained during the greater part of the eighteenth century, doubtless because so many discoveries had been made under the assumption of its correctness, which the chemists hesitated to interpret anew. For the rest a more accurate knowledge of the process of combustion presupposed a knowledge of the types of gas that play a part therein, and this knowledge was not acquired until the latter half of the eighteenth century. The progress made in this field of inquiry is primarily bound up with three names: the Englishmen Priestley and Cavendish, and the Swede Scheele. Priestley deserves still further mention as a discoverer in the biological sphere; CAVENDISH (1731–1810) is best known as the discoverer of hydrogen, and SCHEELE (1742–86), one of the most brilliant experimental scientists of all time, succeeded in making, in spite of his short life, a large number of chemical discoveries, his treatise *On Air and Fire* becoming especially famous.

JOSEPH PRIESTLEY was born in 1733 of a Free Church family of the artisan class living in the north of England. After studying in his sect's theological training-college he was eventually ordained a minister and served in several parishes, partly in Birmingham. An extreme radical, both in religion and politics, he was a supporter of the French Revolution, which resulted in his being subjected to personal persecution; the mob attacked him in his home, which they pillaged, and he himself escaped with his life and fled to London. As he found no peace there either, he emigrated to America and died there in 1804. Priestley had begun to carry out chemical experiments independently; throughout his life he worked quite unsystematically, heating up and treating with reagents everything that fell into his hands, but as he possessed a great gift for arranging and observing his experiments, he did some wonderful pioneering work. One of the chief results of his observations was his discovery of oxygen, which he found by heating mercury monoxide; further, he experimented successfully with carbonic acid, which brought him into the sphere of vegetable and animal chemistry. He found that rats kept in a volume of air that was confined by water died as a result of the pollution of the air, but by letting green plants stand for a time in that same air, it was so improved that fresh rats were again able to live in it for some time. He found by a series of experiments that the air polluted by the animals' breathing contains carbonic acid, or "fixed air," as he called it. As a theorist Priestley was not particularly original; up to his death he stubbornly maintained the phlogiston theory, which had already been abandoned by most chemists of his age.

The scientist who put the chemistry of combustion on the right road was ANTOINE LAURENT LAVOISIER. He was born in Paris in 1743, the son of a lawyer, and was given an excellent education, special attention being paid to mathematics and natural science. He went in for an official career, however, and in time became a "farmer-general" — that is, a titular member of a body to which the French Government had leased the collection of revenue. This system naturally gave rise to a good deal of abuse, and its officials were not much better tolerated than the publicans of the Jews of old. Lavoisier had never been guilty of fraud, but when the Revolutionary tribunal condemned his colleagues, he was likewise involved in their fall. Condemned for no reason at all, he was guillotined by the Terrorists in 1794. When his services to science were cited as grounds for mercy, the petition was met with the reply: "*La République n'a pas besoin de savants.*"

Lavoisier founds quantitative chemistry

IT has been said of Lavoisier that he never discovered a new substance or a new phenomenon, but that he introduced a new spirit into his science. Even the system of weights and measures on which he based his reform had been used before him by Hales and others, but Lavoisier was the first who, in the study of chemical phenomena, consistently paid attention to the weight conditions and in each chemical process determined their immutability, thereby making of chemistry an exact science in the same way as physics. Thanks to Priestley's discovery of oxygen, he was able to account for combustion and he gave the name of "oxgyen" to the gas which had formerly been called "dephlogisticated air." Likewise, he established the fact of water's being composed of oxygen and hydrogen, the latter discovered by Cavendish. Moreover, he found out that heat is unweighable — a fact which still further explained the process of combustion. He also applied his weighing method to life-phenomena; he shut up animals in a confined volume of air and by means of weighing determined the change brought about by their breathing therein. He established the fact that oxygen is the component of the air which is consumed by respiration and that it is substituted in the lungs for carbonic acid. He saw chemical processes both in respiration and in animal heat, as also in fermentation. His influence on the development of science can scarcely be too highly estimated; through him chemistry was led into entirely new channels; through the discovery that oxygen was a constituent common to a mass of chemical elements, the latter could be viewed from a common standpoint and could be given a nomenclature which in part is still in use today. Moreover, to natural science in general these discoveries meant a complete revolution, as they paved the way for the knowledge of the indestructibility of matter. Lavoisier's association with biology lies, of course, mostly in his knowledge of the respiratory process. It was vegetable physiology in particular that felt the immediate influence of the new advance in chemistry. Two examples of this are given in the following.

JAN INGENHOUSZ was born at Breda, in Holland, in 1730, and studied medicine at Leyden under Albinus. As a medical practitioner he was especially known for his skill in smallpox inoculation — an operation which in those days was not unaccompanied by danger. Persons of high rank came to him to be inoculated, and he was the recipient of distinguished and high marks of appreciation. He died during a journey to England in the year 1799. In the course of a previous visit to England Ingenhousz had learnt of Priestley's above-mentioned attempts to "improve polluted air" by the introduction of live plants, and he resolved to proceed with them in a more extensive and systematic form. And in spite of the fact that his experimental apparatus lacked variety and originality — he immersed different parts of plants in water and collected the gas thus given off by them — he succeeded in establishing a number of facts of fundamental importance for the knowledge of plant life. He found that the production of "dephlogisticated air," which constitutes the plant's role as an air-purifier, is a prerogative of the leaves, and particularly of their under side, and that it is brought about exclusively by the influence of sunlight on the plant, whereas during the night, and even in the shadow by day, a kind of air is produced that is fatal to animal life, and that this air is produced by roots, flowers, and fruit, while these latter, if enclosed in a confined air-space, render it impossible for a light to burn in it. Ingenhousz also carried out quantitative investigations, though of a somewhat primitive nature. In this field both he and every one of his contemporaries were far outrivalled by a younger scientist, who, it is true, had the inestimable advantage of being able to avail himself of Lavoisier's new methods.

NICOLAS THÉODORE DE SAUSSURE was born at Geneva in 1767. His father was a scientist of repute and was interested in botany, but his real *métier* was geology. The son also eventually became professor of geology, afterwards entering the Genevan representative council and making a great reputation both as a scientist and as a public citizen. He died in 1845. He was both a chemist and a physicist, but he is chiefly known for his work on vegetable physiology, spending years in the investigation of the subject and finally publishing his results in 1804. The greatest service performed by this work lies in the fact that here for the first time the quantitative method of chemical research, as founded by Lavoisier, together with its results, were systematically applied to living subjects of investigation. This opened up for Saussure entirely new possibilities for methodically organizing his experiments that his predecessors never possessed. He enclosed plants and parts of plants in a quantity of air which had been previously weighed and carefully analysed, and after having let them live there under different conditions, in light and in darkness, he investigated the changes in the composition of the air which their manifestations of life had brought about. He thus established the quantitative relation between the amount of carbonic acid absorbed by the plant in light and the quantity of oxygen simultaneously given off by it. In the same way he found out the quantity of oxygen absorbed by a plant at night, and also the quantity of water consumed in association with the absorption of carbonic acid that is required for the growth of the plant. While the plant was thus found to derive the quantitatively most considerable portion of its nourishment from the air, Saussure on the other hand established the indispensability of the mineral constituents which it drew from the earth, and which he determined by careful analyses of the ashes of the plants investigated. Finally, he also found out that the percentage of nitrogen that the plants possess is primarily absorbed in the form of ammoniac associations. On the other hand, Saussure was wrong in thinking, in contrast to Ingenhousz, that the green colour of the leaves is not essential to their vitality — a misconception (based on the existence of red leaves in certain varieties) that, owing to his authority, was long associated with that line of research.

But while Lavoisier's new method was thus immediately applied to biology with a large measure of success, the more speculatively inclined scientists were led by it to make bold guesses — as is usually so with new discoveries. In the romantic natural philosophy we shall find ideas which were awakened to life by the great revolution in chemistry.

CHAPTER XIII

CRITICAL PHILOSOPHY AND ROMANTIC CONCEPTIONS OF NATURE

1. Kant and his Immediate Successors

Materialism and spiritualism in the eighteenth century

THE TRANSITION PERIOD between the eighteenth century and the succeeding era is characterized by the violent political and social convulsions beginning with the French Revolution in 1789 and ending with the fall of Napoleon in 1815. During this period came into being the modern social system, which, based on the claim of the private citizen to be allowed both to determine his own actions and to take part in the administration of the State, is sharply contrasted with that of the preceding age, with the State possessing unlimited authority in all matters, both secular and spiritual. But even from a purely scientific point of view the beginning of the nineteenth century involved a radical revolution, which had been long preparing, like the political revolution, throughout the centuries. In the eighteenth century's conceptions of nature and life, the two tendencies described in the foregoing — the mechanical and the mystical-spiritualistic — appear in deep contrast to one another. Out of the former, which has its origin in the natural philosophy and natural-scientific research of the seventeenth century, and which, like its predecessors, seeks to explain natural phenomena on purely mechanical lines, there develops towards the close of the eighteenth century — during the so-called "Era of Enlightenment" — a general materialism of the kind that we have seen in La Mettrie: a conception of life expressing itself partly in a dogmatically formulated theory of existence as a play of exclusively material forces, and partly, in the ethical sphere, in a doctrine of a state of blessedness common to all mankind, based ultimately on the liberty to enjoy life independent of traditional rules of conduct. This doctrine, which assumed its best-known and most popular form in HOLBACH's work *Système de la nature*, is remarkable for its readiness to answer every conceivable question in accordance with the formula, laid down once and for all, that, provided the mechanical explanation of nature is maintained, the most daring constructions of thought and the weakest verbal subtleties may pass as complete scientific evidence. Intellectual superficiality and banal hedonistic morality thus became marks

LINNÆUS

CHARLES BONNET

of the enlightened philosophy and contributed in succeeding generations towards concealing its services in the political and the social sphere; the philosophers of enlightenment have striven unceasingly for humanity and tolerance in the life of the State, and in that respect their activities have left a deep impression on the social life of our own day. Parallel with the philosophy of enlightenment, however, there developed another, entirely contrasted, conception of nature, the precursors of which had been Paracelsus and van Helmont, and which, possessing in Stahl, Swedenborg, and Caspar Friedrich Wolff its scientifically most important representatives, appears throughout the eighteenth century under various forms; a view of life which sees in natural phenomena an expression for the operations of spiritual powers, whereas, according to its tenets, the mechanical explanation of nature admits of only a superficial observation of what takes place, without any insight into that inherent connexion in existence which the spiritual powers imply. This attempt to regard nature as a living entity, to look for connexions in phenomena where, when viewed superficially, none are apparent, has constituted this tendency's greatest service, besides which the freedom of mechanical principles, in many cases, admitted of greater liberty in the interpretation of special phenomena, as Wolff's embryological and Sprengel's botanical investigations proved. The weakness of this spiritualistic view of nature has lain in the frequent desire to solve by mystical formulæ problems the solution of which would have required observation and deep thought, and, generally speaking, in its tendency to degenerate into meaningless phrases. As, moreover, this natural mysticism was associated with moral and religious speculations and was upheld by specially founded mystic communities, there was thereby created that extremely unsound "secret wisdom" that under various names and forms spread with incredible rapidity at the close of the eighteenth century, in spite of protests and ridicule on the part of the adherents of enlightenment.

Kant and his philosophy

BESIDES these two directions of thought, which offered, at least in their more extreme forms, but slender possibilities for the further advancement of science, there appears towards the close of the eighteenth century a new system of thought which really gave the scientific activities of the next century their peculiar character — namely, critical philosophy. Its founder was EMMANUEL KANT (1724–1804), whose life's work has undoubtedly represented the greatest contribution to the history of human thought since Socrates and Plato, and for this reason his work merits attention even as concerning the history of biology. Kant was born, lived, and died at Königsberg, in Prussia, where he was professor of philosophy and applied himself entirely to his work as a thinker and teacher. In his youth he had studied, besides philosophy, certain exact sciences, chiefly physics and mathematics, and throughout his life he retained his interest in natural research, not least in biology. His first papers, in fact, dealt with mechanical and cosmological problems; the best known of these is his *Allgemeine Naturgeschichte und Theorie des Himmels*, in which he tried to set up a mechanical theory concerning the origin of the universe. On this subject Swedenborg and Buffon had been his precursors; Buffon, an account of whose cosmological theory has been given above, seems to have been his chief source of inspiration. In contrast to the latter, Kant believes that the planetary system has evolved from a collection of dust particles which moved in space and eventually became concentrated. This theory, the details of which need not be recounted here, all the more so as it has often been referred to, testifies to Kant's efforts to find a mechanical explanation of existence. Towards the end of the work, however, he becomes involved, doubtless under the influence of Swedenborg, in fantastic speculations about life on other heavenly bodies; he believes that on the more distant planets, Jupiter and Saturn, there are beings of a higher order of intelligence than that of man — this because the inhabitants of the more distant planets must be made of lighter material in order that the less intense solar heat there may set them in motion; but the lighter the corporeal matter the greater the intelligence, while heavy bodily fibres and dense, "sluggishly cooking" fluids result in inferior abilities. Strangest of all, he cites Newton's calculations in support of this theory, which might more naturally have originated from the earliest Greek philosophers. Kant, however, soon rose above these fantasies; in a paper published ten years later entitled *Träume eines Geistersehers* he settles with Swedenborg, as indeed with all metaphysical speculations upon the relation between spirit and matter. He ironically examines all the old theories about the location of the soul — now existing everywhere in the body, now located in a small section of the brain — and finally proves the impossibility of determining how the soul influences the body or whether spiritual beings can exist without material space; reason is as little able to decide this question as it is to determine how anything can be a cause or can possess a force — which are all matters that can only be determined by experience; and alleged experiences of single individuals, such as Swedenborg's visions, cannot form the basis of a law of experience for the very reason that they are isolated cases. He ends by pointing out that there certainly are many things that we do not understand, but there is also a very great deal that we do not need to understand. We must be quite clear as to what is necessary for us to know and what in that respect can and must be dispensed with.

Kant, having thus exposed the futility of the old metaphysical speculations, spent more than ten years in trying to find out the limitations and conditions of the human capacity for knowledge in general. The result of these researches he recorded in his *Kritik der reinen Vernunft*, published in

1781 — one of the most epoch-making works in the whole history of human development. Kant's purpose — which, in fact, he at least partly achieved — is to lay the foundations of a new philosophy, to meet not only all the needs of human life in the way of knowledge, but also its moral and religious aims. The many different points of view from which he examines the workings of the human mind, as well as the laws he lays down therefor, cannot of course be recounted here. Chief in importance for the future advancement of natural science is his attempt to determine what justification natural science has for assuming the truth of the knowledge of nature which it expounds. Kant first of all discusses the ideas of space and time and finds that they are not grounded in experience, but in human nature itself; all experience, on the contrary, is based on our having the ideas of space and time that we have. And the same part that time and space play in our views, the idea of causes plays in our understanding. The knowledge we gain by experience is a knowledge of the phenomena that appear to us owing to our organization's being what it is. What those things that cause the phenomena are like in themselves we can never know for certain. Natural science is thus a knowledge of reality such as we observe it, not a knowledge of reality as it actually is. Natural laws are based on our own capacity for knowledge and are binding on us because this capacity has certain fundamental qualities that are the same for all men. Natural science is thus fully justified in drawing its conclusions in the world of experience; on the other hand, it can never give any enlightenment as to the intrinsic meaning of things — that is, what is not phenomenon — nor indeed does it need to do so for the purpose of its physical explanations; but even if, say, some influences from the immaterial world were to arise, it should pass them over and base its explanations upon what the senses are able to reveal and what is reconcilable in accordance with the laws of experience, with our actual observations. On the other hand, all things on which the experience of the senses can give us no knowledge, such as what the soul, the world, God, actually are in themselves, fall outside any rational knowledge. Of these things, then, we can know nothing — we can maintain neither their existence nor their non-existence. But for that very reason we are able, if our feelings require it, to take them for granted; we are justified in believing in God, in the immortality of the soul, and in the free will, and reason has no right to reject any such belief as irrational. These things are, in fact, a part of practical reason — that sense of duty and right which Kant is firmly convinced is inherent in everyone; that which says, not *why* we act in this or in that way, but *how* we *should* act in order to obey the dictates of conscience within us. — Kant himself, in spite of his keen criticism of the life of the human soul, was an ideally minded personality throughout — an enthusiast over such questions as human justice and social equality, who hoped for universal peace in the future. The highest feeling he knew he used to express in the following words: "The starry heavens above me, the sense of duty within me." These words are actually carved on his gravestone.

His influence

THROUGH his critical philosophy Kant has played an important part in human cultural development in general, and not least in scientific development. Thanks to his criticism, biology was freed from the question, which had so often arisen and yet had never been solved, of the relation between soul and body; biological research had, as its exclusive mission, to explain the material course of the phenomena of life, while the investigation of the spiritual side of the soul-life became the function of the science of psychology, employing entirely different methods. But in other respects, too, Kant's critical philosophy exercised an influence upon the development of biology in the century that followed; many of its leading biologists have been keen supporters of Kant; for instance, Johannes Müller, to mention only one of the most eminent. But Kant's practical criticism of reason has also indirectly affected the development of natural science. He thereby established that reason can neither prove nor disprove man's personal ideas of faith and conscience, so that any attempt to influence what the individual holds in high esteem and deep reverence is both unjustifiable and irrational, whether it is done in the name of the Church or in that of science. His principle, just and reasonable though it is, has nevertheless found it difficult to gain a hearing; ever since then, and indeed up to the present day, there have been disputes between "faith and knowledge," brought about not least by the fact that the ecclesiastical authorities claim that their doctrines shall be accepted in their entirety as objectively true. The Roman Catholic Church in particular has banned Kant and his philosophy. But even his contemporaries found it difficult to reconcile themselves to the strict self-control that Kant enjoins upon human thought; that nothing was to be known of "things in themselves" annoyed both the old philosophers of enlightenment, who found Kant's thoughts difficult to grasp and oversubtle, and also the champions of the mystical-romantic class, which strove after a uniform, comprehensive view of existence. In particular, thinkers in the latter direction, while adopting certain of Kant's principles, thought to bring human knowledge beyond the contrast between personal consciousness and the "Ding an sich"; as a matter of fact, the whole of the beginning of the nineteenth century was full of efforts of this kind, which have left their mark on every science, and indeed on the whole of human culture during this period of history. And so far as they influenced the development of biology, they will be briefly touched upon in the following pages.

JOHANN GOTTFRIED HERDER (1744–1803) was a fellow-countryman and disciple of Kant's. He was ordained a minister and held a living for a time in Riga, afterwards spending some years travelling in Europe, and finally, on Goethe's recommendation, becoming court chaplain at Weimar. At the same time enthusiastic and irascible, he had great difficulty in getting on with people; at times even he and Goethe would be on bad terms, but they would soon become reconciled again. As a poet and student of folk-lore Herder has contributed much to literary history. Pronounced romanticist as he was, he sought earnestly for a uniform conception of existence; in these efforts Spinoza was his principal master — he rescued the latter's writings from oblivion and was an ardent supporter of the more mystical views contained therein, whereas Kant's criticism attracted him but little. In his principal work, *Ideen zur Philosophie der Geschichte der Menschheit*, Herder tries to prove how one and the same spirit dominates the whole of nature; all living beings have been created according to one common plan; their various characteristics correspond to their peculiar functions in life, which finally reaches full perfection in man. In the whole of this conception of the course of life Herder is a precursor of the romantic philosophy, which left such a deep impression even on biological history.

JOHANN GOTTLIEB FICHTE (1762–1814) is generally regarded as the first of the purely romantic philosophers. The son of poor parents, he suffered many hardships before becoming a professor, first at Jena, where, on account of his strictly moral principles, he came into conflict with both professors and students and was finally dismissed for "atheism"; and then in Berlin, where he worked hard for the elevation of morals and of the national spirit under the oppression of Napoleon's rule. His philosophy, too, is mostly concerned with ethics; he is of only indirect importance in biological history, as having been the teacher of Schelling, the founder of natural philosophy. Fichte bases his philosophical speculation on Kant, but he also felt the influence of Spinoza. Kant thought that our consciousness gives us the idea that we have of a thing, whereas the thing itself is unknown to us. Fichte also starts from the idea of consciousness, but denies the existence of the thing in itself: he believes that the consciousness or the ego, "*das Ich*," as he calls it, is the only true thing existing; through its operation it then gives rise to existence apart from itself — "the ego places the non-ego," runs the oft-quoted phrase, which primarily refers to the creative work which the moral will of man performs, for the moral will is man's true ego and the central point in the whole of Fichte's extremely abstract and involved speculations. But besides the individual ego, Fichte assumes an "absolute ego" — a kind of world-soul, which can be attained by man only through "intellectual intuition" — a kind of mystical impulse on the model of Spinoza. It was Schelling who further developed his idea, making it one of the foundations of his natural philosophy.

FRIEDRICH WILHELM JOSEPH SCHELLING was born at Leonberg in

Württemberg in 1775. He was the son of a clergyman and from childhood was destined for the same calling. He developed early; after a brilliant career at school he matriculated at the age of fifteen and in his twentieth year had taken his doctor's degree in both philosophy and theology. His first philosophical studies had dealt with Spinoza, Kant, and Fichte. He afterwards spent a couple of years as a private tutor at Leipzig and there studied natural science, chiefly chemistry and physics. At the age of twenty-two he published a work which at once brought him fame — *Ideen zu einer Philosophie der Natur*. Thanks to this work, he was appointed assistant professor at Jena — Goethe, who was much interested in the book, had recommended him to the Saxe-Weimar Government for the post — and there he came in contact with a circle of men and women of genius with pronounced romantic views on science and art. There was one who exercised special influence on him — Caroline Michaelis, a gifted and energetic woman, who, although she was twelve years older than he and had had a somewhat adventurous past, became his wife and highly influenced his work as an author. After her death, in 1809, Schelling's influence actually came to an end. Six years before, however, he had already left Jena, where, through his extraordinary insolence, he had acquired many enemies — with one or two of these he entered into a dispute that ended by their all being condemned for libel. After this he was for a time professor at Würzburg. He then spent a long time in Munich as secretary to the Academy, but he was finally summoned to Berlin (in 1841) for the purpose of using his romantic philosophy to counteract the increasing radicalism. In spite of the support of the Government, however, he was utterly defeated; his enemies published his lectures with insolent comments, with the result that he withdrew altogether from public life. He died in 1854. His character was conspicuous for ostentatious egotism and uncurbed violence by the side of a devoted faith in the doctrines he expounded. Early successes spoiled him, and when later he was confronted by opponents who did not allow themselves to be frightened by his overbearing manners and scornful polemics, his creative power vanished entirely. The work that brought him fame was completed before his thirtieth year; the half-century that he lived after that added nothing to his renown.

As a thinker Schelling based his ideas on Spinoza and Fichte. With Kant, on the other hand, he had but little sympathy; the doctrine of the strict limitation of the capacity of human reason that Kant taught was really the direct opposite of what Schelling desired and thought himself able to produce. His relations with Fichte, however, were at first those of a loyal pupil, though he later broke entirely with him. It was from this master of his that Schelling borrowed the principle of the ego as the basis of everything, both in the spiritual and in the material world. The greatest influence on Schelling, however, was exercised by Spinoza, with his doctrine of spirit and matter as different forms of one and the same "substance" and with his principle, derived therefrom, of the validity of the laws of human reason even in nature. When afterwards, in Leipzig, Schelling became acquainted with natural science, chiefly with chemistry, which was then making great strides, there awakened in him a desire to create, like Spinoza, one common system of thought embracing the whole of existence, which was to prove the connexion between the worlds of nature and of the spirit, in that the world of nature would be derived from that of the spirit, and, vice versa, the world of the spirit from that of nature. The latter became the aim of Schelling's natural philosophy proper, which in one place he terms " *Spinozismus der Physik.*"

Schelling's natural philosophy

FROM this a new natural science was to arise, which was not only to observe individual phenomena and from them derive certain universal principles, but which would actually understand the fundamental forces that cause all that happens in nature. Thus it was a program of natural research directly opposed to that developed theoretically by Bacon and practically by Galileo, which, indeed, research has followed since then. Nevertheless, Schelling expresses the deepest contempt for this natural research; in one place he calls Bacon, Newton, and Boyle the bane of natural science, and Lavoisier's chemistry is treated with no less disdain. It is natural enough that the so-called Spinozism which Schelling would put in its place should become a mere dogmatic system of thought; moreover, as he was entirely lacking in patience and consistency in matters of detail, his theory became vague and fragmentary. In view of the great influence it exercised on the development of biology, however, an attempt must be made to describe it.

In a paper entitled *Darstellung meines Systems*, which Schelling, after the manner of Spinoza, wrote in the form of a series of statements and proofs — though unfortunately entirely without that strictly binding logic which characterizes every sentence of the great Jewish thinker — he describes first of all the "absolute reason" as "*eine totale Indifferenz des Subjektiven und Objektiven,*" which is to be attained by thinking of reason while being fully abstracted from one's thinking self. This is indeed ultimately the same as the mystical view with which Spinoza concludes and with which Schelling thus, strikingly enough, begins. Outside this reason there is nothing, and in it is everything. The supreme law governing the existence of reason is the law of identity — that is, $A = A$. "*Die absolute Identität kann nicht unendlich sich selbst erkennen, ohne sich als Subjekt und Objekt unendlich zu setzen. Dieser Satz ist durch sich selbst klar.*" Thus arises the contrast between subject and object, by which Kant, as we know, meant the consciousness that conceives and the thing which is conceived, and which in Schelling means about the same. Further on, the absolute identity is said to correspond to the universe, whereupon the subject and object are denoted by $A = B$; finally matter corresponds to A = $\overset{+}{B}$, because in matter the objective predominates. Then the absolute indentity is identified with light, the opportunity being here taken to sneer at Newton for his spectral investigation and to compliment Goethe upon his optical theories. From matter, on the other hand, are derived gravity, cohesion, and magnetism: "*Die Materie im ganzen ist als ein unendlicher Magnet anzusehen*," and "*Der Magnetismus ist bedingend der Gestaltung*," with the result that "*alle Körper sind blosse Metamorphosen des Eisens*." Electricity and magnetism are indentified according to a derivation that reasons of space compel us to pass over, as also the derivation of heat from the foregoing. Finally, organism is derived from the absolute identity. As an example of Schelling's biological speculation may be cited one of the paragraphs in the work *in extenso* (the spacing is Schelling's):

„ *Der potenzierteste positive Pol der Erde ist das Gehirn der Tiere, und unter diesen des Menschen. Denn da das Gesetz der Metamorphose nicht nur in Ansehung des Ganzen der Organisation, sondern auch in Ansehung der einzelnen gilt, das Tier aber der positive (Stickstoff) Pol der allgemeinen Metamorphose ist, so wird im Tier selbst wieder das höchste Produkt der Metamorphose der vollkommenste, d.h. potenzierteste Pol sein. Nun ist aber (wie bekannt) das Gehirn das höchste Produkt u.s.w. Also etc.*

„ *Anmerkung 1. Der Beweis dieses Satzes ist freilich nicht aus den chemischen Analysen zu führen, aus Gründen, welche künftig allgemein werden eingesehen werden. . . .*

„ *Anmerkung 2. Das Bestreben der Metamorphose im Tierreich geht, wie aus dem bisherigen leicht zu schliessen ist, notwendig durchgängig auf die reinste und potenzierteste Darstellung des Stickstoffs. — Dieses geschieht in dem gebildeten Tier fortwährend durch den Prozess der Assimilation, der Respiration, welche bloss dazu dient, den Kohlenstoff vom Blut loszureissen; ruhiger und nicht mehr in einem stetigen ununterbrochenen Prozess, gleichsam als ob die Natur über sich schon zu Ruhe gekommen wäre, durch die sogenannte willkürliche Bewegung. — Das erste ruhende Tier stellt die bereits ganz aus sich selbst herausgekommene Erde dar; mit der vollkommensten Gehirn- und Nervenmasse aber ist ihr Innerstes entfaltet und das Reinste, das die Erde der Sonne gleichsam als Opfer darbringen kann.*

„ *Zusatz 1. Das Geschlecht ist die Wurzel des Tieres. Die Blüte das Gehirn der Pflanzen.*

„ *Zusatz 2. Wie die Pflanze in der Blüte sich schliesst, so die ganze Erde im Gehirn des Menschen, welches die höchste Blüte der ganzen organischen Metamorphose ist.*"

This quotation may suffice. If the reader desires more he is referred to the original, the 159 paragraphs of which are, some slightly less, others somewhat more, absurd than the one quoted. Quite out of our subject is Schelling's transcendental philosophy, which resolves itself into a glorification of art, as being the identity of the conscious and the unconscious, and in which subject he certainly felt far more at home than in natural science.

The influence of his system

BUT even in regard to natural philosophy it would be entirely unhistorical to dismiss Schelling as simply and solely a half-witted fool, as is so often done. This is at once inadmissible, owing to the extraordinary influence he exercised on his own age. And it must be admitted that among all the identifications and derivations that constitute his system, there are, besides much madness, a number of really brilliant ideas, which, although expressed as mere fancies, nevertheless undoubtedly exerted an influence upon the future development of science. Thus we may at least suppose that Schelling's comparison of electricity and magnetism was not without its influence upon Örsted, who along experimental lines discovered electromagnetism and who in his youth was a great admirer of Schelling. It should also be noted that Schelling had a keen eye for the physiological contrast between plants and animals, which lies in the former's oxygen-production and the latter's oxygen-resorption; the significance of this contrast for the general economy of nature he has realized and expressed quite clearly, though, it is true, he draws the odd conclusion that the plant has no life, for its arises merely through the development of the life principle and possesses only the semblance of life *"im Moment dieses negativen Prozesses."* — The whole of the extraordinary thought-system which he built up finds its explanation partly in the vast possibilities which the new gas-chemistry had just then opened up for research and speculation — even in our own time hopes of the solution of the riddle of life have more than once been placed upon important discoveries in the field of natural science — partly in the change from criticism to dogmatic philosophy which Fichte had brought about with his theory of the ego as the origin of all things and which was in complete harmony with the romantic tone that was peculiar to this epoch, particularly in Germany. People dreamt of a uniform conception of existence, they looked for spiritual forces in nature, they had grown accustomed to the mystical dreams that were propagated by a number of secret brotherhoods, and all these vain strivings Schelling met with his explanation of existence as an "absolute identity," an explanation that was no more dogmatic, indeed, but certainly more poetic, than La Mettrie's and his successors' materialism, which had constituted the natural philosophy of the previous generation. What, after all, makes Schelling's natural philosophy useless from the point of view of natural science is its absolute lack of practical value; if the object of natural science is to extend and consolidate man's dominion over nature — and that has indeed been its aim ever since the days of Aristotle and Hippocrates — then certainly most of Schelling's efforts have been in

vain, however much genius he put into his work. But even as a purely speculative thought-system it suffers from serious defects — inconsistency, daring
conclusion, and lack of cohesion. All this, however, Schelling took quite
lightly; he was indeed a genius and an artist, and these factors work, according to his theory, half unconsciously and without being worried by the
pedantry of the man in the street. It was these very faults, however, that soon
proved his undoing; it was just in the purely theoretical sphere that his
philosophy was out-distanced by another system, the Hegelian, which was
equally abstract and unreal, but far more consistently thought out, and besides, from the scientific point of view, it had the undeniable advantage of
not involving nature in its speculations.

GEORG WILHELM FRIEDRICH HEGEL (1770–1831) was a fellow-countryman and school-friend of Schelling's and, though older, was at first under the
influence of his precocious friend. However, he eventually worked out for
himself a theory of his own and in his first independent work published a
severe criticism of Schelling's theory of the absolute, which is described as
the "simplicity of the emptiness of knowledge; a night in which all the cows
are black." Hegel eventually became professor in Berlin and founded a well-
attended school, which he subjected to strong discipline. What impressed
his pupils, and indeed his entire age, was, besides his commanding personality, the splendid consistency that he developed in his system of thought.
His dialectical method, however, according to which every idea has its
opposite, both of which are afterwards brought together and combined into
one larger idea, has no concern with our subject, especially as Hegel and his
disciples expressed the deepest contempt for nature and its study — which
gradually resulted in natural scientists' turning the tables by generally regarding all that is meant by philosophy as empty prattle about empty
fancies. On the other hand, Hegel performed a great service to the study of
history in insisting upon the necessity of ascertaining not only the events
that took place, but also the spiritual movements that brought them about.
A similar position to that of Hegel in Germany was held in Upsala by KRIS-
TOFER JAKOB BOSTRÖM (1797–1866), who for half a century governed that
university and made of Linnæus's ancient seat of learning a centre of abstract
speculation.

But though Schelling was thus worsted in the theoretical sphere, his
natural philosophy survived as a general theory of life with the support of a
whole generation of contemporary scientists. The cause of this strange phenomenon must partly be sought in the fact that there was no other equally
comprehensive explanation of nature available, and some such explanation
was an absolute essential of existence at that time. But there were many contributing causes thereto, including the fact that Schelling's natural philosophy was embraced by a man who was regarded by his age as an authority in

every sphere of culture — namely, Goethe, the poet and universal genius. As is well known, he has been an influence even in the field of biology and it is this side of his work that will be described in the following section.

2. Goethe

JOHANN WOLFGANG GOETHE was born in 1749 of wealthy middle-class parents at Frankfurt am Main. He studied jurisprudence, first at Leipzig, then at Strassburg, and after passing his law examinations practised for a time as a lawyer and at the same time acquired a reputation as a poet. In 1775 he came to the court at Weimar, which was interested in literature, and there, thanks to his brilliant intellectual and personal advantages, obtained an eminent position — not only as poet and organizer of the pleasures of the court, but also as an official he held the highest appointment in the little Saxon capital. For a long time he held the reins of government with success as Minister of State to the principality of Saxe-Weimar. In 1786 he made a journey to Italy, which lasted two years and which proved of decisive importance in his life, especially in regard to his scientific work. Having returned home, he gradually withdrew from public life and devoted himself whole-heartedly to poetry and science. Active and possessing his full intellectual powers to the last, he attained a great age, dying in 1832.

Even as a child Goethe had evinced a lively interest in nature; he examined flowers and carried out experiments in electricity and magnetism. In his poems, too, there was conspicuous from the very first a keen interest in nature — a gift of observing and describing its life in its different phases, which greatly contributed to his fame. During his student days he received varied impressions from the extremely chequered intellectual life prevailing in Germany at that time; he became acquainted with French materialism, which seemed to him dry and unanimated; on the other hand, he engrossed himself in mystical literature, studying the writings of Paracelsus, van Helmont, and Swedenborg, which made a somewhat deep impression on him and influenced his poetry. In Strassburg he made the acquaintance of Herder and, as he himself declares, his association with him increased his interest for the study both of nature and of human development. Like Herder, Goethe admired Spinoza and sought in him a basis for the unity between spirit and nature that he desired to find in life.

Goethe's anatomical researches

AT Weimar Goethe's interest in the natural sciences was increased through his intercourse with scientists at the University of Jena and through periodical collaboration with Herder. While the latter was putting the finishing touches to his above-mentioned *Ideen,* Goethe was studying anatomy at

Jena. Herder, it will be remembered, was endeavouring to find one common type for the form and functions of human beings and animals. During the latter half of the eighteenth century, moreover, a dispute had been going on with regard to the relation of man to the apes, about which we heard through La Mettrie and Camper; the former, indeed, had sought to prove that the orang-utan was a kind of human being, which it should be possible to civilize, and the indignation the theory aroused in the Christian-conservative people had found expression in violent polemics, while Camper, through his paper on the anatomy of the orang-utan, referred to above, gave support to those who maintained the dignity of man. Goethe, who as a young man was somewhat averse to religion, to which his famous poem *Prometheus* in particular testifies, entered into the dispute on the side of the materialists. Camper had asserted that in the facial skeleton of the orang-utan there is a suture which, starting from the nasal cavity, extends on either side as far as the space between the corner tooth and the foremost front tooth; this suture does not exist in man, in contrast to the apes and other mammals. In consequence of this Goethe wrote a short treatise in which he maintains that the intermaxillary bone, which terminates in the said suture, is found also in man — an assertion based chiefly on the existence of sutures which in the gum and above it separate the bone in question from the upper jaw and adjoining bones. Goethe also described it as existing in certain other mammals in which this bone had not previously been found. The treatise was sent in 1784 to Camper, who expressed courteous thanks for it and specially complimented Goethe on having established the existence of the bone in the walrus. In regard to the discovery in man, on the other hand, Camper had no remarks to offer, and that for sound reasons; as a matter of fact, the bone had been known ever since Vesalius's days and had been described in man, in whom in the embryonic stage it is clearly separated, while in full-grown individuals its outer suture disappears. This difference between man and the ape existed just as Camper had pointed out, and for obvious reasons Goethe had not been able to disprove it. The fact that he imagined he had "discovered" the intermaxillary bone in man was no doubt due to the accident that some text-books of that time treated the incompletely separated bone in the full-grown man as if it were one with the maxilla superior. The claim to this discovery has on Goethe's authority even reappeared in literary histories and is believed by the public, unjustifiable though it is. Goethe's pamphlet on the question remained for the time unprinted, presumably owing to lack of encouragement on the part of the specialists; at any rate, there were no financial obstacles standing in the way of its publication.

Goethe, however, continued his anatomical experiments and finally published the theoretical views at which he arrived, in a paper entitled

Erster Entwurf einer allgemeinen Einleitung in die vergleichende Anatomie, dated 1795. He starts with the principle that natural science is on the whole based on comparison — a principle upon which Aristotle had already laid great emphasis. As the standard of comparison Goethe sets up an ideal type, with which the anatomical details in each animal form are to be compared. Thus one should at once be able to interpret an anatomical detail in an individual by comparison with the ideal type. Goethe offers no detailed description of what he imagines this type to be like, and indeed it would have been difficult to conceive one. Herder's above-mentioned speculations on an ideal type have obviously influenced Goethe here far more than the already existing comparative anatomy such as Buffon, Daubenton, and Camper practised. Even in this theory Goethe indulges in wild philosophical fancies, as when he states that the tail of mammals "*als eine Andeutung der Unendlichkeit organischer Existenzen angesehen werden kann*," or when he says of the body of the snake that it is "*gleichsam unendlich*," because it does not need to expend matter and force on extremities. This paper also remained in manuscript form for the time being.

His metamorphosis of plants

BEFORE this, however, Goethe had published the treatise that is generally acknowledged to be his principal contribution in the field of natural science — namely, his *Versuch, die Metamorphose der Pflanzen zu erklären*, which was printed in 1790. The gist of it is, briefly, that the leaves of plants gradually develop through "metamorphosis," which first gives rise to cotyledons, then to the stem-leaves, and finally to the flower-leaves: food-leaves and petals, stamen and pistils. This metamorphosis can partly be "regular" or "progressive," "*welche sich von den ersten Samenblättern bis zur letzten Ausbildung der Frucht immer stufenweise wirksam bemerken lässt, und durch Umwandlung einer Gestalt in die andere, gleichsam auf einer geistigen Leiter zu jenem Gipfel der Natur, der Fortpflanzung durch zwei Geschlechter hinaufsteigt.*" Irregular metamorphosis is one of nature's retrograde steps: "*wie sie dort mit unwiderstehlichem Trieb und kräftiger Anstrengung die Blumen bildet und zu den Werken der Liebe rüstet, so erschlafft sie hier gleichsam, und lässt unentschlossen ihr Geschöpf in einem unentschiedenen, weichen, unseren Augen oft gefälligen, aber innerlich unkräftigen und unwirksamen Zustande.*" Here he refers to double flowers, whose stamens are converted into petals. Then he goes through the different leaf-forms: the seed-lobes are thick because they are filled with raw material, while the leaves of the stem, and still more of the flower, become finer and finer on account of the fact that only finer saps penetrate into them. Another idea that, besides the saps of various degrees of tenuity, plays a conspicuous part in Goethe's vegetable physiology is "*Anastomosis*," by which he apparently means the intercommunication between various parts of plants; the idea, however, remains obscure and is certainly not made any clearer by the fact of fertilization's being called "*eine geistige Anastomose.*" Even florescence is caused by "*geistige Kräfte,*" since the operation of these forces preponderates over the raw saps that form the leaves. And, finally, Goethe develops a theory of germination, according to which the seeds and the leaf-buds are compared. Goethe himself admits that his metamorphosis work does not contain any really original observations; the metamorphosis theory itself occurs in Linnæus's *Philosophia botanica*, in which, under the heading "*Metamorphosis vegetabilis,*" the bud, the leaf, and the flower are analysed and the leaves in their various transformations identified. Goethe, who admits that he had read that work, nevertheless claims to have "discovered metamorphosis"; by this he cannot reasonably mean anything else than its philosophical side — the theory of the ideal type, according to which the leaves are transformed. The "*Geistige*" that so frequently recurs in the treatise on metamorphosis is explained by the fact that it was this that Goethe considered to be the essential, as also did Herder, of whose theory the plant-metamorphosis doctrine is most reminiscent. For it is romantic philosophy from beginning to end; it bears no resemblance whatever to modern natural research.

His theory of colours

IT was in the course of his journey to Italy, as a result of the impression made upon him by the southern vegetation, that Goethe first had the idea of his metamorphosis theory. During the same journey he had also studied, in company with some artists in Rome, the laws of colour-combination and its effect upon the sight. Not content with the results obtained in this respect, he resolved upon his return home to devote himself to the study of colours from the physical point of view as well. He procured a prism and with its aid studied a number of light and colour phenomena. These he described with great lucidity and accuracy in a work entitled *Beiträge zur Optik*, published in 1791. He had, however, made one or two observations — that the centre of a large white surface viewed through a prism remains white and that a black line on a white ground is resolved by the prism into colours — which he considered it impossible to explain by the Newtonian laws of optics. It is true, some physicists in the profession who read his book explained the phenomena to him in the light of Newton's theory, but Goethe does not appear to have been much edified by it. Then Schelling took up the question. As mentioned above, to him light represented the "absolute identity," and he enthusiastically hailed Goethe as a liberator from Newton's detestable spectral theory. The poet, who was extremely sensitive to applause as well as to criticism, was thereby entirely won over to the new natural philosophy and felt encouraged to go on with his optical investigations in the hope of creating a new "colour theory" in place of Newton's. After years of preparation he finally (in 1808) published his *Farbenlehre* — the greatest of his scientific works and the one that he himself valued most highly. The theory of colour that he develops in this work agrees entirely with Schelling's polarity theory. All colour-effect is derived from a "primal phenomenon" — namely, the contrast between light and darkness; between these two stands as a connecting link "*das Trübe.*" When pure light is broken up by a prism, it is disturbed by the action of the glass, and from this arise the spectral colours. That these colours arise through the disturbance of the light Goethe tries to prove, *inter alia*, by the fact that the sun, when viewed through a darkened glass, appears red. Newton's view that the pure white light actually arises as a result of the combination of the various colours in the spectrum puts Goethe into a furious passion; he goes through Newton's optics point by point and provides them with marginal notes which are as irrational in content as they are scurrilous in tone. The coarsest expressions in the vocabulary denoting stupidity and dishonesty are lavished on page after page of the work. Goethe has here — to his own discredit — adopted his admirer Schelling's polemical vocabulary, while his general attitude towards Newton displays in a deplorable manner the narrow limitations of even a universal genius. Goethe deserves no place in the history of optics.

Nevertheless, Goethe was not entirely wrong when he considered the colour theory to be his best natural-scientific work. In fact, it contains a section in which Goethe's finest gifts as an observer of nature are given full play as nowhere else — namely, the chapter on "physiological colours." In this chapter, as well as here and there in other parts of the work, are a large number of observations of subjective colour-perceptions, recorded with all the exactness of a scientist and with the keen insight of an artist. These detailed observations concerning colour harmony, colour contrasts, complementary colours, and other optico-physiological phenomena, attracted great attention even among his contemporaries; they resulted in continued research by scientists possessing professional knowledge of quite a different type from that of their model, and even in our own day, when mental physiology has become a specialized science, they have won justifiable recognition. In no field has Goethe so nearly approached the spirit of exact natural research as he has here, and that, too, in spite of the false theory for the sake of which this subsequent research work was carried out.

During the remaining years of Goethe's life the colour theory absorbed most of his scientific interest; in fact, in his old age he ended by valuing it above all his poetic works. This was manifestly due to the fact that as a poet he considered himself neglected; his neo-romantic protégés, Schelling's friends, had come to dominate public opinion, and though they always treated him with courtesy, they wounded his feelings by placing their own quite mediocre leaders on a level with him. On the other hand, they loudly praised his scientific speculations; there grew up quite a school of scientific students of natural philosophy who looked up to Goethe as a prophet —

all these being factors conducive to continued scientific production. The colour theory was reprinted and amplified; the anatomical writings of his youth were published and largely added to, the ideal primal type and Schelling's polarity theory reappearing in several variations. In these works he tries to find a primal form for the anatomical details, just as he did a primal phenomenon in optics; he invented the word "morphology," knowledge of form, and it still survives in modern science, although, it is true, in an entirely different meaning from that which Goethe originally gave it. Of these works may be mentioned an article on the six vertebræ composing the cranium. Ten years before, Oken had expounded a similar theory, which will be referred to later on; he has thus the prior right to the idea and declared, moreover, that he mentioned it to Goethe in the course of conversation, which, however, the latter emphatically denied. It is scarcely possible now to find out exactly what happened; besides, it is of not very great interest nowadays, as the whole theory is out of date.

Spiral theory

THE article *"Über die Spiraltendenz" der Vegetation* belongs to Goethe's last years. Both in its idea and in its method this article is one of the most eccentric imaginative creations of the romantic philosophy, but for this very reason it aroused great enthusiasm amongst the supporters of that tendency, whilst those who hoped to see in Goethe a modern natural scientist passed it over in complete silence. According to this article,[1] the plant is composed of two indissolubly connected "tendencies": the vertical, which represents the eternal essence, and the spiral, which represents the nourishing, the cultivating, the reproductive. The latter tendency, naturally represented by the spiral vessels, is given a number of utterly incomprehensible definitions: *"das Spiralsystem ist abschliessend, den Abschluss befördernd. Und zwar auf gesetzliche, vollendete Weise. Sodann aber auch auf ungesetzliche, voreilende und vernichtende Weise."* The aquatic plant Vallisneria, in particular, the male flower of which grows straight, while the stalk of the female flower after fertilization contracts into a spiral, is analysed in connexion therewith, the result being that as a general rule the vertical represents the male in the plant, and the spiral the female, which is confirmed by the ancient metaphor of the tree and the vine-tendril which winds itself round it, as a symbol for the masculine and the feminine in life. With this glimpse into the innermost soul of existence Goethe concludes the "spiral" article, which was written six months before he died, so that after all it is the poet in the old philosopher Goethe that has the last word, which is only right, as the need for a deeper and wider poetic view of nature was undoubtedly the true reason for his coming to grips with the study of nature.

---

[1] Here, too, Oken had previously dealt with the question; in his natural philosophy there is a fantastic exposition of the spiral vessels in plants.

His influence

GOETHE's posthumous reputation as a natural scientist has varied generation after generation. From the exact scientists of his own age he met with but little encouragement — optical science in particular, which was just in his time making brilliant progress in the hands of Fresnel, Wollaston, Brewster, and others, naturally left him far behind — while, on the other hand, the entire school of natural philosophy saw in him an inimitable master. And with good reason, for as a matter of fact he did more than any other to preserve the reputation of natural philosophy. When, however, this tendency was finally abandoned and mercilessly given up to ridicule, Goethe was accorded far more indulgent treatment; his great authority as a poet and man of culture exempted him from harsh treatment at the hands of scientific critics. Goethe's morphological speculation received a new lease of life through the coming of Haeckel; for reasons that will be explained later on, Haeckel expressed a boundless admiration for Goethe and regarded him as one of the foremost precursors of Darwinism. On his authority both the general public and literary history have since willingly accorded Goethe, who in other respects has left so many marks on the cultural development of our time, the further honour of being a natural scientist in the modern sense of the term. Nevertheless, his biological writings have certainly been more admired at a distance than read in the original, a fact that has no doubt contributed in the long run towards concealing their true quality.[2] Goethe was no exact scientist, but a romantic natural philosopher; in that capacity, however, he has also exercised an influence, which must not be underestimated; his psycho-physiological observations and speculations have formed the basis on which men like Johannes Müller and Purkinje have built up their work, and although Goethe may have had no eye for comparative morphology in the modern sense, yet many an eminent anatomist of later date has been induced by Goethe's ideas to devote himself to a comparative study of form that has proved of benefit to science. Goethe takes his place in the history of biology as a stimulating force; his influence was, it is true, both good and bad, but by no means inconsiderable.

---

[2] So far as is known, no separate edition of Goethe's scientific writings has existed until recent times; those desirous of studying them have had to have recourse to the somewhat expensive editions of his *Sämtliche Werke*. An edition of these writings was published some years ago by R. STEINER, who, as is well known, made Goethe's conception of nature the basis of his "anthroposophical" theory of existence, which does not in the least accord with modern natural science.

CHAPTER XIV

NATURAL-PHILOSOPHICAL BIOLOGY

1. Germany and Scandinavia

Character of the natural philosophy of the time

THE DIRECTION taken by natural-philosophical thought that has been described in the foregoing has played a very important part in the cultural development of the world and not least in the science of biology. There were, of course, during the natural-philosophical period a large number of scientists who were not at all, or only very slightly, affected by the speculative tendencies of natural philosophy, while others, it is true, embraced its tenets either temporarily or permanently, but at the same time carried out research work in exact natural science with lasting results. The work of these scientists will be recorded later on; in the present chapter we shall devote our attention to a group of scientists who applied themselves entirely to a speculative explanation of nature and sought to incorporate in it all the known facts about nature that they considered necessary and attainable, or who, at any rate, in a more or less pronounced way gave themselves out as champions of such views. It was in Germany and Scandinavia in particular that these faithful disciples of Schelling and the other idealist philosophers won for a time extraordinary success and managed to present to the public, and particularly to the universities, their master's theory, with amendments of their own, as the only true natural science. The causes of this phenomenon, which must seem strange to us, as it was to earlier generations, were manifold. The universal cultural tendencies that favoured romanticism in general — disappointment at the failure of the efforts to win liberty under the revolution and weariness after the great wars of independence — naturally also played an important part in the development now under discussion; again, the interest in mysticism that spread far and wide at the close of the eighteenth century and was cultivated by the numerous brotherhoods, undoubtedly had a great influence. The possession of some form of knowledge that is unattainable for the majority has always been an attractive prospect for human egotism — now it was possible for the professor of philosophy or natural science at the university to present to his hearers a theory which at any rate had the advantage of being incomprehensible to the uninitiated;

LORENZ OKEN

LAMARCK

when, moreover, according to its tenets, the elect possessed the right to be hailed as geniuses, it is easy to realize the enthusiasm which the new wisdom evoked. The result was that there developed at the universities such student insolence as had never existed since the days of scholasticism; and it survived for a long time, especially at the academies situated in the provincial towns, where neither masters nor pupils had much contact with practical life. This very academic isolation explains to a certain extent how it was that these theories, so out of accord with reality, could survive for so long, and explains the fact that they were localized in Germany and Scandinavia, while in western Europe, with its more lively and practical activities, speculation at least adopted more dispassionate forms.

One of the most notable and influential personalities in German natural philosophy was LORENZ OKEN (1779–1851). He was of south German peasant stock, his family name being really Ockenfuss, and was brought up in poverty, though he managed to obtain a school education and afterwards studied medicine, eventually becoming a doctor, in 1804. Medicine, however, was not of very great interest to him; at an early age he had formulated a natural philosophy of his own. After having maintained himself under severe privations at several universities, he was in 1807 appointed assistant professor at Jena, where he published as his inaugural address his paper on the subject of the cranium's being composed of several vertebræ. This resulted in his falling into disfavour with Goethe, which caused him considerable unpleasantness, all the more so as he was of a passionate nature and found it difficult to exercise discretion in his behaviour. Being an ardent German patriot, moreover, he was enthusiastic for his country's unity and was consequently suspected by the authorities in the reaction after the War of Independence, for which he zealously agitated. At last, in 1819, he was forced to resign, although he had the support of the entire University; he was for a time without an appointment, but afterwards became a professor at Munich; there too, however, he was unable to get on with the authorities, so that he was glad to accept a post in Zurich in 1832. He carried on his work there, respected and esteemed, for the rest of his life.

Oken's activities were many-sided and his influence upon the development of culture considerable. For many years he published the journal *Isis* — the name is characteristic of his half-mystical philosophy — which became a focus for the scientific life of Germany; with great impartiality it accepted papers by scientists of different camps; the discussion of problems was encouraged, and prizes offered for solutions, with the object of promoting scientific research. Oken took the initiative in another idea which has proved of value to the future of science; he organized meetings of scientists for the purpose of exchanging views and encouraging sociability. Thus it was he who originated those gatherings that are so much appreciated in our own day and which he himself stimulated and developed by his lively personality and his keen interest in the whole domain of thought. Finally, by his writings he promoted also an interest in the study of nature; his *Allgemeine Naturge-schichte für alle Stände* is a compilation of a very high standard of excellence, based on comprehensive material, which has widely increased the knowledge of and interest in the study of nature.

Oken's natural philosophy

OKEN's own contributions to exact natural science are, on the other hand, of but little importance. In his youth, before going to Jena, he carried out an investigation into the development of the intestine in the embryo, which contains a number of sound observations, though his conclusions were partly drawn from principles that were not very successfully thought out. Oken, in fact, considered himself above all a natural philosopher and set great store by his best work, *Lehrbuch der Naturphilosophie*, which he rewrote twice. He was, however, not very learned as a philosopher; his conclusions were as fantastic as Schelling's, but had not even the latter's small degree of formal consistency and logic. Nor did he possess Goethe's poetical imagination, and his speculations were therefore as grotesque as they were irrational. In particular, the first part of his work, which, strikingly enough, is called "*Theosophie*," is simply extraordinary; its first sentence runs: "*Die höchste mathematische Idee oder das Grundprincip aller Mathematik ist das Zero = 0.*" Then we learn that God and the world $= O + -$, while God alone or the primal idea $= O$, and space is $O = + O -$. When we come to the living creatures, the whole is certainly somewhat closer to facts; organic life is derived from a primal slime, which is described as "*oxydierter, gewässerter Kohlenstoff*," and which had its origin in the sea, whence all life comes. Life is formed of three "entelechiæ": magnetism, chemism, and respiration. In regard to plants, Oken, like Goethe later on, speculates upon the spiral ducts; to Oken they are "*das Lichtsystem in der Pflanze.*" The parts of the plant correspond to the four elements, the root being the earth-organ, the stem the water-organ, the leaf the air-organ, and the flower the fire-organ. With regard to the animal kingdom, all animal life is derived from a follicle; there are four consecutive formations thereof — the point-, the ball-, the fibre-, and the cell-formation. The organs of animals constitute special systems, first "*pflanzliche*" (namely, intestine, gills, veins); then "*thierige*" (bony, muscular, and nervous systems). Moreover, the animal is composed of "*Hirntier*" and "*Geschlechtstier*," both of which possess organs that correspond to one another, as, for instance, lung and bladder, mouth and rectum, thorax and pelvis. The animal kingdom in its entirety is regarded as one mighty animal, the various parts of which correspond to different animal forms; the lowest animals have only intestine, as the polypi; then come such as have intestine and skin — snails and insects; finally, those having in-

testines, skin, and flesh — the mammals. Further quotations would be superfluous. We find in the above a great deal of ancient mysticism, such as the
mysticism of numbers — recurring groups of three and four — the comparing
of the animal kingdom to a great body, reminiscent of Swedenborg's speculations, and, finally, the strange idea of figures at the beginning of his work. At
the same time we find here some occasional idea that recalls biological theories of our own day, as, for instance, the follicle-shaped primal animal, the
idea of the sea as the origin of animal life. Oken was undoubtedly a man of
ideas, many of which might be of value to the future; his unbridled imagination, however, made him a warning to the succeeding generation and an ill-
directed example of what the results of natural-philosophical speculation
may be.

CHRISTIAN GOTTFRIED DANIEL NEES VON ESENBECK (1776–1858) forms a
parallel to Oken in the sphere of botany. From the south of Germany, like
Oken, he was the son of a public official; he studied medicine at Jena, where
he was won over to Schelling's philosophy and came into contact with
Goethe. Having completed his studies, he settled down on an estate that he
had inherited and there worked as a private scholar until the year 1818, when
he was appointed professor of botany at the newly-founded University of
Bonn. He established the botanical institute and gardens there and wrote a
number of books on both botany and natural philosophy. Later, having
been appointed professor at Breslau, he began by doing some successful work.
In old age, however, the romantic natural philosopher became ultra-radical;
he took part in the labour movement, zealously supported ideas for the reform of Christianity, and worked hard in theory and practice for free marriage without State co-operation; the end of which was chicanery, dismissal,
and death in poverty. The Labour Union at Breslau, whose chairman he was,
followed him to the grave.

As a classifier of plants, Nees von Esenbeck has acquired a distinguished
name; he has won special fame for his tropical floras, dealing with the phan-
erogamous plants of the Cape and of Brazil; his works on the cryptogams,
on lichens and hepaticæ, algæ and fungi were also at one time highly thought
of. He himself, however, set greater store by his natural-philosophical speculations. In his *Lehrbuch der Botanik*, which he dedicated to Goethe, he carried
the latter's metamorphosis theory to the uttermost extreme. To him the
leaf is a kind of symbol for the plant as a whole; the entire vegetable world
is to him one mighty leaf, just as the animal world was to Oken one mighty
animal. Even in the vegetable world the number three plays an important
and mystical role and is the basis for a good deal of play upon words. Polarity
in the style of Schelling recurs once more; the fungi represent the north and
the plants the south, the animals midnight and man noon, while the chemical components of plants are dealt with just as arbitrarily. The colours in

the vegetable kingdom are of course treated in accordance with Goethe's theory of colour. Even the spiral vessels give rise to a number of speculations, although Goethe's theory had not yet been published. The spiral theory was, in fact, developed later by a large number of botanists who, like Nees von Esenbeck, could be quite rational collectors and systematists, but who at the same time gave play to a wild and reckless imagination on the subjects of the spiral and polarity, till at last, in this sphere also, exact research claimed its due.

One of the last survivors of Germany's natural philosophers is worthy of mention — CARL GUSTAV CARUS (1789–1869). Born in Leipzig, he became professor of comparative anatomy there in 1811 and afterwards professor of gynæcology and court physician at Dresden. As a doctor he was an eminent specialist and, besides, a man of unusually varied interests; a personal friend of Goethe's, he had a truly artistic nature and was himself a good painter and writer on art, as well as a comparative anatomist and natural philosopher. He experimented in comparative osteology, insect anatomy, and zootomy. His comparative anatomy (of 1828) stands to a certain extent on the border-line between the contemporary and a more modern conception of that science. Carus, for instance, no longer goes in for the plus and minus signs with which his predecessors wasted their time without in the least solving their problems; to him, indeed, nature is an expression of an idea, and life is a flux, and the three-grouping recurs here and there; but he is at any rate able to describe an organ or a system of organs without at once becoming involved in sheer incomprehensibilities. He sets up an animal system arranged in circles, one inside the other, with the protozoa outermost and man innermost, and with not very successful descriptions, but, on the other hand, he gives a comparative account of the nervous system throughout the animal kingdom that is arranged clearly and in an exact form throughout. In 1861, as an old man, Carus summarized his ideas in a work entitled *Natur und Idee*, which certainly strikes a curious note, considering the advances that natural science had already made by that time. Here we find ether regarded as the primal substance and the essence of all chemical elements; after which we are told that "*die Urhandlung des Äthers ist Leben.*" The nervous system is the central force in the animal kingdom, like the primal fire and the electrical principle in the earth; the universe is an infinite sphere whose centre point is everywhere and whose periphery is purely ideal. Again, animals arise out of a sphere, the ovum, and develop through new spheres' being added to the original; the senses are drawn in the corners of a pentagram inscribed in a circle. When this work was published, Darwin's theory of selection had been known for two years. Thus the last representative of natural philosophy in its most extreme form survived up to modern times.

The neo-romanticist natural philosophy was brought to Scandinavia by HENRIK STEFFENS, who, although he has played little part in the development of biology, nevertheless deserves mention here owing to his importance in cultural history. Born at Stavanger, in Norway, he studied in Copenhagen and Kiel and after a period of wandering came to Jena, where he became an enthusiastic admirer of Schelling, a friend of Oken, and a natural philosopher heart and soul. He returned to Copenhagen in 1802 and for a year or two lectured at the University, but as he obtained no permanent post there, he accepted an appointment in Germany, where he remained for the rest of his life. During the years he spent in Denmark he exercised great influence by his enthusiastic promulgation of natural philosophy, although his exaggerations aroused doubts in the minds of the Danes, which were not perceived by his less critical friends in Germany. His principal work on natural philosophy dealt with the internal natural history of the earth; in it he seeks to prove, *inter alia*, that the various strata of the earth are sections of a galvanic element. Of importance to the history of biology was his theory of the origin of the circular coral islands; he believed that they grew up on the edge of volcanic craters in the ocean, and this theory was accepted as true by many, until Darwin disproved it by his well-known investigations into the subject.

In Sweden natural philosophy was embraced by the famous CARL ADOLF AGARDH (1785–1859), known as one of Sweden's most many-sided geniuses. He was a native of Scania, matriculated at Lund, and eventually became lecturer in mathematics and professor of botany and economics at that University. He ended by being Bishop of Karlstad, after having won renown as a botanist, mathematician, national economist, priest, and politician. Only his sphere of activity as the first belongs to this narrative. At Lund Agardh became acquainted with the Linnæan system of plant classification, and in the course of journeys in Germany he came to know Schelling and natural philosophy. His most lasting fame he has won as one of the founders of algæ classification; much of the system that he created still exists today. He has also made valuable contributions in connexion with plant classification as a whole; in particular he was one of the first to pronounce against the sharp difference that had hitherto been held to exist between phanerogams and cryptogams. His general views with regard to animate nature he has collected in a handbook of botany published in the years 1828–32, the first part of which he dedicated to Schelling. This first part, entitled *"Organography,"* contains also traces of the influence of natural philosophy; nevertheless Agardh displays a degree of caution in speculation that is in favourable contrast to the rashness of his German master. Thus he at once declares on the first page that natural objects cannot be exactly defined on a logical basis; he believes that we have to content ourselves with establishing in each what is the most common phenomenon or the most usual form, without venturing to lay down rules having no exceptions. And when he quotes Nees von Esenbeck's above-mentioned comparison between natural objects and the points of the compass, he does so with the — certainly mild — reservation that "this method of philosophizing is beautiful but obscure." On the other hand, he can hit upon such eccentric ideas as that the human hands may represent leaves between which the head sits like a bud; and the metamorphosis theory likewise leads to a number of extremely bold comparisons between the stages of development in plants and animals. But there is observable throughout a keenness of observation in points of detail which proves that the teachings and example of Linnæus had not lost their influence in his own country. The second section, on vegetable biology, treats of the manifestations of life in plants and is on the whole more exact than the former section.

Another important representative of natural philosophy in Scandinavia was ISRAEL HWASSER (1790–1860). He was the son of a priest at Älvkarleby, and after being educated at home he matriculated at Upsala, where in 1812 he took the degree of doctor of medicine. Five years later he became professor of medicine at the academy of Åbo, where he exercised considerable influence. The medical education there, which had fallen into decline, was improved by him in regard to both the number of students and the standard of the knowledge imparted, while he himself succeeded in gathering about him friends and pupils who took part in his idealistic labours. In 1830 he applied for and obtained a professorship in Upsala, where he afterwards worked until his death. The reason for his transfer was that he wished to counteract the scheme just then being proposed for removing the medical school to the Carolinian Institute in Stockholm, which, he argued, was at variance with the principle of the connexion of ideas in scientific education. But he maintained his interest in Finland throughout his life. He was in everything an ideally minded personality who in speech and writing as well as in private company was a zealous supporter of patriotism, loyalty, and clean morals and in this respect exerted great influence on the young people in the University. His scientific activities he desired also to place entirely at the service of moral ideals. He was, it is true, a natural philosopher, but he did not approve Schelling's speculations, especially the attempt to construct nature out of an idea; on this attempt he passes the weighty, and on the whole correct, judgment: "It was a scientific extravagance pushed to extremes, which in the minds of some of those who took part in it seemed to have been fostered and supported by a pride nearly akin to madness." On the other hand, he was a great admirer of Sydenham and still more so of the French anatomist Bichat, who will be described later on. His whole conception of life in nature is characterized by his deeply ethical aims. He is dissatisfied with those who, starting from the lower, would try to understand the higher, whether they do so by means of theoretical construction or practical investigation. In general he despises any dealings with matter as such and denies its indestructibility. To him life is a magnificent process of ethical refining; to his mind the development of the individual represents the selfish element, while reproduction, which entails self-sacrifice for the welfare of the race, is the highest element in the organism. He is zealous also for an ideal conception of love and marriage and for that reason disapproves of Goethe's sensual love-poetry and therewith of all that Goethe did. His theory of disease as self-destruction in the individual is in accordance with this, his conception of life. Nevertheless, he by no means disapproved of practical medical education, though he himself had little to do with it, owing to bodily clumsiness, and, as he himself declared, consequent laziness. As will have been seen from the above, his biological theory was no less inconsistent with true nature than Schelling's, but at any rate it had the advantage of assuming as its chief mission the improvement of morals, which Schelling's certainly did not. It was in any case Hwasser's personality that exercised the best influence; it is mostly for this that he is remembered today.

In Finland Hwasser's natural philosophy was maintained after his death by his faithful friend and pupil IMMANUEL ILMONI (1797–1856), who shared his ideas and warmly defended them. With the passing of these two men natural philosophy disappears from the universities of the North, where, as a matter of fact, it had never made the same progress that it had done in Germany; biology was mostly carried on throughout the natural-philosophical period on Linnæn principles; moreover, men like Berzelius and Anders Retzius were working for exact natural research, and natural philosophy had no rivals in Scandinavia to compete with them.

2. England and France

Cultural development in western Europe

NATURAL philosophy has by no means played the same part in the two lands of culture in western Europe as it did in Germany. The reason for this may be sought ultimately in the national character of their peoples, Englishmen and Frenchmen always showing themselves less speculative and more inclined to direct their energies towards practical aims than the Germans. And indeed practical functions were far more attainable in those uniformly governed and well-organized western-European kingdoms than in the divided and politically disillusioned country of Germany. The reaction against the opinions of the eighteenth century sought and found its expression, both in England and France, in politics, both of Church and State, and in literature, while science was able to continue its work undisturbed by any serious revaluations of the old standards. Speculative tendencies had indeed existed in these countries at an earlier period — Buffon is, of course, the most brilliant example — and the theories of the true natural-philosophical age often appear as continuations of those tendencies, while, on the other hand, they serve as the transition between the past and exact science in the nineteenth century. Nevertheless, even during this period there arose scientists whose speculations had more in common with the natural philosophy so far described, and who to a certain extent actually had direct points of contact with it. It is now proposed to give one or two examples of speculations of this kind, while such scientists as seem to stand in a more direct relation to modern biology will be discussed at the beginning of the next section of this work.

In England natural-philosophical speculation has been a familiar practice from early times. Many of its pioneers have combined a wealth of original ideas and theories for explaining natural phenomena with somewhat unsystematic methods of thought and experiment. More or less gifted authors of this type there were in abundance, especially in the numerous circles of private scholars in England. In this category may be included ERASMUS DARWIN (1731–1802), whose speculations caused a sensation in his day, not only in England, but also on the Continent. Born at Nottingham of an old stock, he devoted himself to medicine, studied in Cambridge and Edinburgh, and finally practised as a doctor at Lichfield. He is described as very original, vigorous and somewhat coarse-grained, honest and straightforward; besides, he was a keen worker and well thought of in his profession, kind towards the poor, and an ardent supporter of temperance. He had many children; one of his sons was the father of Charles Darwin, and a daughter became the mother of Francis Galton, the student of heredity. Apart from his own profession, Erasmus Darwin was an indefatigable author and wrote a great number of papers for the Royal Society and also published one or two collections of poems, with which he himself was highly satisfied, but which fell a victim to the ridicule of his contemporaries and the neglect of posterity. The work, however, which alone has made his name memorable is his *Zoonomia*, an attempt to find out the laws of organic life, which was published in 1794 and was translated into several European languages.[1] It excited a good deal of attention at the time; the German natural philosophers in particular have quoted it with appreciation. In modern times, however, it would certainly have caused but little notice had not the author been the grandfather of Charles Darwin. At first sight it gives the impression of being a most extraordinary conglomeration of diverse notes on scientific

---

[1] The author has not succeeded in coming across the original of this work, but has been compelled to have recourse to Brandis's German translation.

and medical subjects; but on closer inspection we find that it deals with a number of problems that engaged the minds of the author's contemporaries, although from a curious point of view.

Erasmus Darwin's natural philosophy

THE work begins with the assertion that spirit and matter are the foundations of nature; then life is defined as being due to a special force, which, after Haller, is called irritability and by aid of which all life-phenomena are explained in a somewhat peculiar manner. All manifestations of life, both physical and psychical, are due to the contraction of fibres, which is induced by irritation; by "idea" the author understands a contraction of the fibres that form the direct sensory organs. La Mettrie himself could hardly have expressed himself more materialistically, but Erasmus Darwin is by no means a materialist as the term was understood by his own age; true, he refers to the sceptic Hume's inquiries into cause and effect, but at the same time certifies his invincible faith in the Bible, quoting verses of the Psalms on the wisdom of the Creator and citing the words of Moses in the Book of Genesis touching the creation of Eve from Adam's rib as a proof of his own theory of reproduction. This latter theory, more than anything else that Erasmus Darwin wrote, attracted great attention, which indeed it undoubtedly deserved. To a certain extent it is reminiscent of Caspar Friedrich Wolff's theory of evolution — that is to say, in so far as it is markedly epigenetic. Whether the author had recourse to Wolff is not clear from his work — possibly he had on second-hand information. But, above all, his theory is pronouncedly animalculistic; just as the whole basis of his theory of life rests on the assumption of "irritable" fibres being the basic substance of all living things, so the origin of the embryo is a "filament," which is derived from the father and to which the mother only gives nourishment. As a result of this latter the embryo grows, by no means, however, as a result of the development of ready-formed rudiments, but through the addition of fresh matter. The author seeks to disprove the preformation theory by arguments both serious and facetious; Bonnet's incapsulation theory in particular seems to him exceedingly ludicrous; the dimensions of the infinite number of embryos contained one inside the other remind him of how St. Anthony was tempted by twenty thousand devils, all dancing on a pin-point. In further confirmation of his epigenesis theory he declares that the male "filament" which gives rise to the new individual is manifestly influenced by the nourishment which the mother provides; any resemblance between mother and child is due thereto, as is particularly shown in bastards. And still further: the conditions under which the parents live clearly influence the character of their offspring, as is proved by the new varieties obtained from domestic animals; organs which the animals need are produced by irritation in the parts of the body which form them, and they are afterwards inherited by their progeny.

Thus stags have got horns, and cocks spurs, while fighting for their mates. In fact, one is justified in assuming that all living creatures, different from one another as they now are, nevertheless originate from one and the same "primal filament," whose offspring have become changed as a result of different conditions of life, this being confirmed by the existence of transition between all animal and vegetable forms, both higher and lower. This theory undoubtedly sounds to a certain extent "Darwinian," but what differentiates the grandfather's doctrine from the grandson's is the problem each set out to solve: Erasmus Darwin really had no interest in the origin of species; with him it was a question of obtaining as strong evidence as possible for the epigenesis theory, which was a marked feature of his speculations. Nor indeed was it this side of his work that evoked contemporary interest; it was rather his speculations on the subject of the life-force and, further, his theory of irritability and his observations of sense-impressions, which he made with a view to confirming the latter theory and which in a certain degree foreshadow Goethe's. All this afforded special interest to the German natural philosophers, who not infrequently refer to his writings. He was entirely forgotten by the succeeding generation; in fact, it was not until after Charles Darwin had become world-famous that interest in Erasmus revived, when an attempt was made to see resemblances between his speculations and those of his grandson. Some resemblance there certainly is, but it is undeniable that the originator of the theory of selection had worked with entirely different qualifications from his grandfather's in order to produce a universal theory of the evolution of life.

In France biological speculation during the latter half of the eighteenth century was essentially governed by Buffon's ideas. He and his friend Daubenton had, it will be remembered, carried out comparative investigations into the anatomy of various animals, especially the bone-structure of mammals. During the following epoch also French scientists were keenly occupied in investigations of this kind, and natural philosophy, to which such investigations were at that time referred, thereby assumed a more defined and practical character than in Germany. Moreover, its exponents stand out far more clearly as the precursors of modern biology and deserve to be discussed in connexion therewith. In particular, he who is regarded as the foremost natural philosopher that France has produced — Lamarck — seems, in view of the great influence he has exercised on modern research, to be most worthy of record amongst its pioneers. A natural philosopher who, on the other hand, is far more closely associated with the German speculation, and who was actually connected with it, was Geoffroy Saint-Hilaire. He may therefore suitably be described in this context.

ÉTIENNE GEOFFROY SAINT-HILAIRE was born at Étampes, near Paris, in the year 1772, the son of a public official. His father had him educated for the priesthood and actually procured him a benefice, but at the same time allowed him to follow his bent for natural studies. Highly gifted, impulsive, and passionate, young Geoffroy became deeply engrossed in the study of chemistry, crystallography, and anatomy. During the Revolution he distinguished himself by rescuing a number of priests from death during the massacres in September 1792 at the risk of his own life. In spite of this he was appointed in the following year by the Revolutionary Government professor of zoology at a newly-founded educational establishment and at once became known for his brilliant energies and success. Cuvier, who was at that time still quite unknown, was promoted to another professorship under his recommendation. Eventually Cuvier was to rise above the head of his patron, but for the time they collaborated with success in the sphere of comparative anatomy. When Bonaparte made his famous expedition to Egypt, Geoffroy accompanied him as zoologist and succeeded in making there a number of splendid collections, which he later, thanks to his resolute action, prevented from falling into the hands of the English. The result was that, upon returning to Paris, he won still further honours. He won less glory in an expedition that he made to Portugal, in whose museums he brought together "collections" at the command of Napoleon on behalf of the French State. His later years are mainly characterized by an increasing rivalry and enmity with Cuvier; they were very largely contrasts to one another. In his old age he became blind and finally also paralytic. He died in 1844.

Geoffroy's comparative anatomy

THE comparative anatomy introduced by Buffon and Daubenton was enthusiastically embraced by Geoffroy and Cuvier. The development to which this science attained, chiefly thanks to Cuvier, who made it one of the most important foundations of modern biology, will be described in the following section. Nor, indeed, was Geoffroy's contribution towards the progress of this science without its significance, but at quite an early age there developed in him a fancy for imaginative speculation, which justifies his being placed in the category of theorizing natural philosophers. It has been thought possible to trace the influence of German natural philosophy, especially Schelling's,[2] in this tendency of his, but as it was not until later that Schelling's writings were translated into French, one can hardly suppose that there was any direct influence from that quarter. These fantastic speculations were certainly characteristic of the age; he had predecessors in this respect even in French literature; one need only recall the name of Bonnet. The main idea in Geoffroy's philosophy is the existence of a common fundamental type, beginning with the organization of all the vertebrate animals, and then for the entire animal kingdom in general. He worked principally at the anatomy,

---

[2] According to Kohlbrugge in his essay on Goethe; see the Bibliography at the end of the book.

particularly the bone-structure, of the vertebrates, and here he has really
worked out a number of ideas which foreshadow results that have been
gained by modern comparative anatomy. He thus derives the auditory bone
of mammals from the cranial bone in fishes — doubtless from the opercula
and not from the bones now regarded as the starting-point. He derives the
cartilage of the larynx from the fishes' branchial arches — as has indeed
been, at least partially, done in present-day comparative anatomy. But here
at once his unbridled imagination manifests itself in the utter lack of detailed
criticism and ability to limit his speculative field; thus, for instance, he finds
a sternum in fishes, and he derives the annular cartilages of the trachea from
the gill-arches, just as he delights in making direct comparisons between
fishes and mammals in general. (In passing, he makes the truly natural-
philosophical assertion that the auditory apparatus of birds is the finest
there is, which is proved by the fact that they are so musical.) He indulges
in his wildest flights of fancy, however, when he compares vertebrates and
invertebrates. To his mind, insects and crustaceans are composed of vertebræ, in which both apophyses and ribs can be distinguished — the joints
are vertebræ, and the extremities ribs. The shells of the tortoise and the snail
are compared, and the ink-fish is a vertebrate animal with a duplication of
the back. One realizes that this application of the theory of a common fundamental type for the animal kingdom must have impressed Goethe. In the
dispute that eventually broke out between Geoffroy and Cuvier — which
will be described later — Goethe loyally supported Geoffroy. When Cuvier
died, Geoffroy was left free to promulgate his own theories, which were
also adopted, in part at least, by his son, Isidore, who was likewise an eminent biologist. The more critical comparative anatomy, besides eradicating
the worst exaggerations, eventually acknowledged the wealth of ideas and
the, in many respects, productive thoughts that were to be found in Geoffroy Saint-Hilaire.

The natural-philosophical school of thought which we have endeavoured to describe above has had a deep influence on the development of biology. Its extravagances cannot, of course, be regarded as other than features
tending to retard the sound progress of science; time has also helped to put
them out of mind, or at worst they have been recalled only for the purpose
of ridiculing the weaknesses of an older generation. The service it has rendered to humanity lies in the lively interest for the study of nature which
it evoked in the scientists of its era — an interest in striving to find law-
bound phenomena in existence. Otherwise its age certainly specialized in
speculation upon abstract ideas, as Hegel and his school would have it; but
the fact that during this period the study of nature did not disappear altogether nor degenerate into a mere handicraft is at any rate due in no small
measure to natural philosophy. Many of its ideas, indeed, recur, in a more
or less revised form, in the biology of the nineteenth century, an account
of which will be given in the next section of this work.

PART THREE

MODERN BIOLOGY

BIOLOGY DURING THE FIRST HALF
OF THE NINETEENTH CENTURY

CHAPTER I

FROM NATURAL PHILOSOPHY TO MODERN BIOLOGY

1. The Predecessors of Comparative Anatomy

IN THE HISTORY of biology the nineteenth century will undoubtedly be always regarded as one of the most important epochs. With regard to the value of the discoveries that were then made, that period can certainly vie with the most brilliant of the periods that preceded it, and if we consider the reputation which biology enjoyed in the world of culture of the time, it was an unrivalled epoch. It is primarily the latter half of this century, after the appearance of Darwin, that witnessed the greatest advance that biology has ever been able to record, particularly in regard to the volume of its discoveries. But this advance was being prepared for during the immediately preceding decades in the splendid development that took place in most branches of research — a development which in its turn was evolved during the preceding eras out of events that have previously been recorded in this work. Thus, in the biological science of the nineteenth century we find elements derived from the exact scientific research that during the two preceding centuries had sought for a mechanical explanation of the phenomena in animate nature, but there were also features from the speculative natural philosophy that endeavoured, by means of purely theoretical systems of thought, to solve those problems of existence which exact scientific research had found itself compelled to leave unexplained. As has already been described in the last few chapters of the previous section, this natural philosophy chiefly flourished in Germany and Scandinavia, but also existed in England and France; it exercised a decisive influence upon the cultural development of that period, and, as we shall see later on, far beyond it. During the age when natural philosophy flourished, however, exact scientific research was by no means dead; it only worked the better in peace, and, moreover, there were of course scientists who, though convinced supporters of natural philosophy, nevertheless carried out exact investigations on special subjects. A few examples of this exact natural research during the age of natural philosophy deserve to be quoted here as representing a transition to modern biology, which developed during the succeeding epoch.

In France, as we have already found, comparative anatomy had been considerably developed through the work of Buffon and Daubenton. But descriptive anatomy had also had a brilliant representative during that same century in the Dane Jacob Benignus Winslöw, a relative of Steno, who, like the latter, became a Catholic and was fully naturalized abroad, but who, in contrast to his predecessor, enjoyed an unusually long life of activity. He died in 1760 at the age of ninety-one. His description of the human anatomy was especially complete, particularly as regards the topographical section, and he made the medical faculty in Paris, where he was professor, an important centre of anatomical study. He and his immediate pupils were, however, outshone by a man who was able to develop even the human anatomy along comparative lines on the model of Daubenton and Camper.

Félix Vicq d'Azyr was born in 1748 at Valogne, in Normandy. He was the son of a physician, and after being educated at school, he chose his father's career and studied in Paris with such success that only eight years after entering the profession he was able to give lectures there. He had no academical career, however. He was passed over when a vacancy was filled in the professorship of anatomy at the Jardin des Plantes in 1774, and again upon the appointment of a successor to Buffon. Instead, he was sent by the Board to study and stamp out serious epidemics in certain provincial parts of France, and he wrote some valuable accounts of them. In the field of veterinary science also he made some important contributions. Besides this work, he held private courses in anatomy, which were very popular, and he also collaborated in the founding of the Royal Society of Medicine in Paris, of which he became permanent secretary; in that capacity he composed a number of brilliant epitaphs upon past distinguished physicians, on account of which he was appointed Buffon's successor in the French Academy. At last he was made personal physician to the King, but when the Revolution broke out, shortly afterwards, this post of honour was a cause of much trouble and danger. His health, which had already given way under stress of work, now broke down; finally, as a result of attending the famous feast of the Supreme Being, under compulsion, he contracted a chill and died a few days later.

Vicq d'Azyr's career was thus a short one, and, moreover, his energies were divided as a result of the practical work he had to carry out in order to make a living. On these practical activities, indeed, he expended much labour, and his works on epidemics, veterinary surgery, and organizational problems in practical medicine are fairly numerous. But in spite of all this he found time for carrying out serious theoretical research in the sphere of anatomy and physiology, and though the results to a large extent exist only in the form of brief accounts published in academical proceedings, nevertheless he has thereby contributed largely to the development of biology. Of one important work that he started on anatomy he managed to publish only the first part, wherein he lays down the principles on which he considers that the study of anatomy should be pursued.

Vicq d'Azyr's classification of the functions of the organism

FOR this purpose he takes as his starting-point, firstly comparative anatomy, as created by Daubenton, and secondly Haller's physiological theories and experiments. He begins by discussing the ancient division of natural objects into three kingdoms and finds that the essential difference lies between animate beings and inanimate things; plants and animals possess common properties that stones and minerals lack. In connexion therewith he strongly rejects the old comparison — which is sometimes repeated even in modern times — between the growth of the organism and that of the crystal; he points out the mathematically regular shape and homogeneous structure of the crystal as contrasted with the rounded forms and variously constituted systems of organisms, but above all he emphasizes the organisms' definitely characterized functions as a peculiarity of life. These functions he divides into the following categories: (1) digestion, (2) nutrition, (3) circulation, (4) respiration, (5) secretion, (6) ossification, (7) generation, (8) irritability, (9) sensibility. The existence of the various functions, together with their respective organs, is then examined in the different life-forms; regarding digestion, it is stated that man, the quadrupeds, whales, birds, and crustaceans possess one or more stomachal cavities clearly distinct from the œsophagus and the intestine; oviparous quadrupeds, snakes, selachians, and osseans have a stomach in the form of a single extension; insects, worms, and zoophytes have only one intestinal tube, and plants no digestive canal — the classification is noticeable as being more reminiscent of Aristotle than of Linnæus. With regard to generation, a distinction is made between viviparous, oviparous, and gemmate reproduction. In regard to irritability, he differentiates between insect larvæ, worms, and polypi, which have an entirely contractile or muscular body; vertebrate animals, whose muscles cover the skeleton; insects and crustaceans, in which the skeleton covers the muscles; and plants, which possess no free movements. It would be possible, of course, to adduce weighty detailed objections to this system and its various categories; it gives evidence, however, of a careful study and a penetrating analysis of the phenomena of life. And Vicq d'Azyr undeniably possesses a keen eye for certain manifestations and functions of life — far keener indeed than any of his predecessors and many of his successors.

In particular he keenly criticizes the current theories of the essence of life; on the subject of vitalism, as maintained by Stahl's successors, he holds that, while it is true that a number of phenomena exist only in living creatures, nothing is gained by referring to the soul as their cause; they should rather be regarded as physical phenomena and studied through observation and experiment, but not ascribed to a principle "whereby thought retires in the belief that everything is done, when in reality everything remains to be done." This criticism he extends to several current hypotheses; thus, he rejects the assumption of a fluid controlling the impulses in the nervous system, for by this particularization of a little-known function a number of illusions have been created, while such an expression as "nervous force" would be far more applicable to the actual knowledge we possess of the phenomenon. Similarly he criticizes the theory current at the time, which originated in Buffon, that the different parts of the embryo are derived from corresponding parts in the parents — a theory that even Darwin afterwards entertained. In disproof of it Vicq d'Azyr adduces the fact that two parents who have lost one and the same part of the body nevertheless produce normal offspring. In this point he has thus foreshadowed views that are expressed by modern students of heredity.

As his teacher in scientific criticism Vicq d'Azyr mentions the philosopher of enlightenment CONDILLAC (1715–80), who enjoyed a great reputation in his time and who made a special study of the relation of sense-perception to the consciousness — he asserts that the consciousness is composed of what the sense-impressions communicate from the outside world — and in connexion therewith he maintained that in science words should exactly convey the ideas they are intended to denote. By a careful study of his writings Vicq d'Azyr undoubtedly learnt to realize the necessity for clear ideas and unambiguous terms in natural science as much as in anything else.

Vicq d'Azyr's comparative anatomy

VICQ D'AZYR's influence has been felt not only on account of this criticism, valuable as it is, but in a still greater degree as a result of his studies in comparative anatomy, which were unfortunately fragmentary; the principles on which he worked, however, he summarized in the form of a program for a course of lectures in anatomy and physiology. The subject is first of all divided up according to the nine life-functions referred to above. Under the heading "ossification" he first deals in descriptive form with the bone-structure and its articulation and forms of connexion, then by way of comparison the individual bones in different animal forms, and further a number of physiological experiments in connexion with the growth and regeneration of bones, and finally he gives an account of the chemical composition of osseous tissue. Under the heading "irritability" is discussed the muscular system, first descriptively, then comparatively, as regards both conformation and finer structure, the vascular and nervous ramifications, and, finally, ex-

CUVIER

MARIE FRANÇOIS XAVIER BICHAT

perimentally. In the last-mentioned respect are observed both the contraction of individual muscles when the nerves that are connected with them are irritated, and the different forms of movement in man and animals — a subject in which Borelli and Perrault are cited as precursors. Similarly, under the heading "sensibility" the nervous system is dealt with, followed by the other functions of the body. The most remarkable point of this exposition is the detailed comparison made between the same organs in different animals — a study in which Vicq d'Azyr certainly finds support in the preliminary works of Daubenton; but the former without the least doubt carries through the comparative program far more thoroughly than the latter.

Its valuable contributions to science

IN points of detail Vicq d'Azyr's investigations contain many contributions of immense value, and also a wealth of ideas of great significance for the future. He paid special attention to the comparative anatomy of the mammals. He continued and widened the comparison of the bodily structure of man and the apes which Camper had initiated, arriving, too, at the same results as the latter (see Part II, p. 260); on this subject he made a special study of the musculature of the extremities, and in general closely compared the extremities throughout the mammalian class. Further, he made investigations into the teeth of the entire vertebrate class; he points out the difference between teeth fixed in dental sacs and provided with vascular and nervous systems, and those that are fixed on the jaw-bone; he observes the dissimilarity in the number and structure of teeth in mammals of different structure and habits; he draws attention to the pointed teeth of the beast of prey, the knobby teeth of omnivorous animals, and the enamel-coated teeth of herbivorous animals; he notes the presence and absence of various kinds of teeth in different animals. He points out the correlation existing between different organs in animals; a certain shape of tooth presupposes a certain type of structure in the extremities and the digestive canal, because all its bodily parts are adapted to the animal's way of living. He also shows how these different characteristics give every animal a special role to play in the great struggle that is constantly going on in nature between the various life-forms. The weakest point of his comparative investigations is the comparison between vertebrates and invertebrates; although he is far more cautious than Geoffroy Saint-Hilaire was later, he nevertheless has no true eye for the difference between the organs in the important main groups in the animal kingdom, and, on the whole, it was certainly fortunate for him that he did not find time to extend his studies to the invertebrate animals. He even includes the vegetable world in his comparative studies, sometimes without much success, as when he compares the symmetry in pinnate plants with that of animals, and sometimes with an insight that looks far ahead into the future, as when he crosses white and red tulips and finds that the descendants are white, red, and intermediate. Finally, it is worth mentioning that as a descriptive anatomist he produced a copiously illustrated work on the brain and its nervous system — a splendid work for his period and in its extent the most considerable of all his productions. Thus, in more than one respect Vicq d'Azyr has left his mark on the history of biology; he will be especially remembered as a pioneer in the sphere of comparative anatomy.

In Germany at the beginning of the nineteenth century, biology was dominated by the romantic natural philosophy that was described at the close of Part II, in which it was also pointed out that during that period there were working by the side of the natural philosophers a number of scientists who pursued their inquiries by exact methods, thereby upholding the traditions of the preceding era and at the same time paving the way for the succeeding age's magnificent progress in the field of biology. Of these exact scientists working during a period given over to fantastic dreams some of the most prominent merit description here.

JOHANN FRIEDRICH BLUMENBACH was born at Gotha in 1752; his father was a schoolmaster, and his mother, to whose memory he dedicated a special work, was a good and gifted woman. Even as a child he was interested in natural science; one of his keenest delights was putting together skeletons out of bones that he collected. He studied first at Jena, and then at Göttingen, where he took his degree with a dissertation on the human races, which brought him immediate fame and procured for him, as early as in the year 1776, the professorship in anatomy at that university. He carried on his work as a teacher for nearly sixty years, during which he led a quiet life, interrupted only by a few collecting-expeditions. At last, in 1835, he resigned, and died in 1840. In his old age he was a very original character; he was regarded as one of the sights of Göttingen, and he was always quite willing to show himself, especially to distinguished visitors. As an author, too, Blumenbach is peculiar; his style is heavy and full of long periods, occasionally lightened by dry humour, which is always inoffensive, though now and then not in the best of taste according to modern standards. It is also said that his lectures were interspersed with witty remarks, which recurred year after year in a given context to the delight of generations of undergraduates. And Blumenbach had innumerable pupils; he had the ability both to gather round him and to train scientific experts, and not a few scientists of European reputation derived their knowledge from him. As an author of text-books and manuals he was very fine for the age in which he lived, and, generally speaking, he has contributed very largely towards stimulating his countrymen's interest in the study of nature, which, indeed, was to attain, during the generation that immediately followed his own, unexpected heights in his country; so that he has honourably deserved the title of "*Magister Germaniæ,*" which he enjoyed even in his lifetime.

Blumenbach's comparative anthropology

As the chief service that Blumenbach has rendered to science is generally quoted the fact that he introduced into Germany the study of comparative anatomy — and this at an earlier date than Cuvier introduced it into France. There can be no doubt that the two collaborators in research Buffon and Daubenton have the prior claim to the introduction of comparative anatomy, but it is certain also that Blumenbach was essentially a comparative anatomist and that he brought that science up to a high state of development. There is especially one branch of it in which he is a pioneer — namely, anthropology. Here, it is true, Buffon, with his descriptive and statistical method, and Camper, with his studies of the facial angle, had paved the way, but Blumenbach was the first who systematically worked at the subject, thereby laying the foundations on which all subsequent research has carried on its constructive work. He instituted a collection of skulls, skeletons, and illustrations of human beings of as many different races as he could procure, and he methodically studied the peculiar characteristics of the material he thus got together or was able to borrow from other museums. The result was a close comparison of the characteristics, both external and internal, of different human types, and on that basis a division of mankind into races. A similar attempt had indeed been made before, such as Buffon's, for instance, but Blumenbach's was the first that really proved successful, and his five races — Caucasian, Mongolian, Ethiopian, American, and Malayan — have been the foundation on which all subsequent racial divisions have been based, just as his postulate that the races are varieties of one and the same species is also regarded as true, in spite of isolated attempts to create several species. His characteristic descriptions of the cranium, with accompanying illustrations, have certainly been more recently improved upon by Anders Retzius, Virchow, Broca, and others, but Blumenbach's nevertheless form the groundwork on which his successors have built.

Besides the study of races, Blumenbach paid special attention to the question of determining the characteristics wherein man differs from the other mammals, and particularly from the manlike apes, whose anatomy he closely studied. Like Camper, he strongly maintained that man is fundamentally unlike the apes; it was he who divided the Linnæan order Primates into two — Bimana for the human and Quadrumana for the apes. And he collected as much anatomical, morphological, and psychological evidence as he could in proof of this. Many of these detailed anatomical characteristics are without doubt correctly observed, whereas others are the result of anatomical misconceptions, such as the statement that the apes have four hands, while man has two. This point of view was accepted, however, by the biologists of the succeeding age, and in actual fact Blumenbach was the originator of most of the reasons which eventually were to be adduced against

Darwinism by its conservative opponents. In other points, too, these champions of man's high dignity could be satisfied with Blumenbach's views; he believes, for instance, in species' having been created as one pair of each — at least as far as man is concerned — and he holds the view that the Caucasian race was the original out of which the others were created later by "degeneration," due to climatic and economic conditions. During the fighting days of Darwinism, therefore, old Blumenbach received but little gratitude at the hands of the champions of progress, but a later age must indubitably do him the justice of acknowledging him as the founder of comparative anthropology.

Blumenbach, however, extended his research and educational work to other spheres; actually, like Linnæus, he has dealt with all the three kingdoms of nature. His botanical knowledge is based entirely on Linnæus, whose system he uses in its entirety; in mineralogy, too, his activities were not very remarkable. As a zoologist he is likewise limited; he deals with the invertebrates in a summary fashion and has but little new to tell of them. The vertebrates, on the other hand, he studied carefully from the comparative point of view, with special reference to the mammals; he discusses the latter's anatomy in detail, principally their bone-structure, and his work on this subject is worthy to be compared with Daubenton's and Vicq d'Azyr's. In contrast to the last-named he divides his comparative anatomy according to organs and not according to physiological functions, which gives to the entire work a far more modern character.

His vitalism

HE was also, however, keenly interested in physiological problems and believed that he had created something essentially new in this field; in a treatise entitled *Über den Bildungstrieb* he expounds a theory of reproduction and embryonic development, which he afterwards advances repeatedly in various connexions. In this work he first of all gives an account of earlier theories of evolution; the preformation theory is criticized and rejected, with a certain degree of satire, his principal target being, as usual, Bonnet's incapsulating theory; the spermatozoa are declared to be parasites, and finally the epigenesis theory is advanced as the true explanation of the phenomenon of evolution. With C. F. Wolff, he holds that the prospective individual is evolved out of a completely indifferentiated mass. Blumenbach, however, rejects his predecessor's theory of "*vis essentialis*" (see Part II, p. 249), holding instead that the development is caused by a special "formative force," which is displayed not only in the development of the embryo, but also in all kinds of growth, regeneration, and reproduction in animate beings. This formative force — *nisus formativus* — must not, however, be confused with other "life-forces," such as irritability and sensibility, but it operates with those forces in order to maintain life. Blumenbach specially points out that these "forces" are merely expressions by which to denote phenomena, the cause of which we do not know, but the effects of which we can observe. By the side of this, however, he speaks also of the body-mechanism. Thus we find that Blumenbach did not create, like Stahl, any elaborated vitalistic thought-system, and his speculation in general is not particularly logical. His service lies rather in the field of comparative observation than in that of speculation.

Blumenbach's contemporary and equal in scientific reputation was SAM-UEL THOMAS SÖMMERRING (1755–1830). He was born in the Polish town of Thorn, though of a German family; his father was town physician, and the son was destined early for the medical profession. He carried out his elementary studies at Göttingen, where Blumenbach was one of his younger teachers; he afterwards studied anatomy, in Holland under Camper and in England under Hunter. His youth recalls that of Swammerdam, in so far as his father desired to see him early established in practice and stubbornly opposed his going in for expensive scientific studies; but the young man refused to give in; through his personal ability and the mediation of friends he managed to secure extended help towards his studies away from home and thus continued his scientific work in difficult economic circumstances, until he was able to earn not only a good reputation, but also his daily bread. He held professorships, first at Kassel, then at Mainz and Munich; but during that period spent some years as a practitioner at Frankfurt am Main, where he married and found his true home. There, too, he spent the last ten years of his life in peace and happiness, surrounded by friends and continuing his scientific work until the end.

In his general conception of nature Sömmerring was to a certain extent influenced by the speculations in mystical natural philosophy in which his age indulged. At Kassel he entered the Rosicrucian Brotherhood and in it carried on both alchemy and spiritualism, although he afterwards realized his delusion in both respects. At Mainz, where his activities were most productive, he applied himself exclusively to anatomy and carried out in this field a number of valuable investigations on special subjects. Like Albinus, whom he chose as his model, he employed a clever draughtsman and with his assistance published several excellent compilations, one on the human body in its entirety — which was never completed, but the published sections of which are remarkable for their clear method of presentation and their sound descriptions — as well as a number of special investigations, such as a treatise on monstrosities of various kinds, which he spent some years in collecting, and, further, a comparative study of the visual and auditory organs in different races of mankind, and finally some investigations on various subjects and of varying value.

Sömmerring's work on the brain

His curious work *Über das Organ der Seele*, which he dedicated to Kant, is a combination of anatomical inquiry and philosophical speculation. In it he gives a detailed description of the brain and its nerves, illustrated with splendid engravings. His account of the origin of the nerve-stems in particular is admirable, considering the age in which it was written, and he has rendered a still greater service to science by treating, for the first time, the sympathetic nervous system as a pair of nerves independent of the central nerve system, "which pair is in indirect, but not direct, connexion with the brain and spinal cord." The whole of this study of the brain, however, forms the basis of a highly fantastic speculation upon the brain as the organ of the soul, or, to be more exact, upon the location of the "*sensorium commune*," which in the German is translated as "*das gemeinschaftliche Empfindungsort.*" By this is meant that part of the brain in which the sense-impressions converge and co-operate. Ideas of this kind in regard to the localization of the soul in the brain had indeed long been current; Descartes adopted for this purpose the *glandula pinealis*, Perrault the *medulla oblongata;* Swedenborg alone was guided on the right path by his brilliant intuition when he drew attention to the pyramid-cells of the cerebrum. Sömmerring tries to prove that all the cerebral nerves open into the central cavity of the brain and that in connexion therewith the cerebral fluid is the organ of consciousness; the only point that worries him is: "*Kann eine Flüssigkeit animirt sein?*" which, however, he answers in the affirmative on arguments derived from the Bible, Aristotle, and modern writings. This assertion, which in our own day, when protoplasm and its derivatives have so often had to serve as wholly or at least partially fluid, should not be regarded as utterly absurd, nevertheless aroused grave doubts in the minds of Sömmerring's contemporaries, just as his philosophical argumentation in general shocked the students of natural science; his good friend Goethe wrote him a letter in which, with reflective and observant criticism, which he unhappily did not always employ in his own writings, he warns him against letting philosophical speculations interfere in scientific investigations. And Sömmerring in actual fact learnt wisdom from the opposition he met with; his later works are in the main based on exact natural science; his reputation as one of the leading anatomists of his age was greatly enhanced by them and has been confirmed by posterity.

There was one scientist, a contemporary of Sömmerring's, who approached far more nearly to the modern conception of the structure of the brain and nervous system, but through his own fault he managed to acquire a somewhat doubtful reputation; this was the "phrenologist," FRANZ JOSEPH GALL (1758–1828). Born at Baden, he went as a medical student to Vienna, became a doctor, and carried on a medical practice there. At the same time he began to interest himself in the study of the brain's structure and its manifestations of life. He promulgated his ideas on the subject both in writings and in public lectures; when these were prohibited as being "materialistic," he left Vienna (in 1805) and wandered about Germany for a couple of years, accompanied by his friend and pupil SPURZHEIM, everywhere demonstrating his ideas and wherever he went attracting the attention of the public, which was occasionally flattering, but often quite the opposite. Eventually he settled down in Paris, became naturalized, and lived on a practice which had its peculiar features — for instance, he kept strictly secret the composition of the medicines he prescribed — and which brought him into strained relations with other doctors. He was also on bad terms with scientific specialists; universities and academies closed their doors to him. The public became all the more interested in his doctrines, which were promulgated after his death by many, mostly dilettanti, who brought his theories into utter discredit, so that ultimately they were entirely forgotten.

Gall's theory of brain and nerves

NEVERTHELESS, Gall has exercised an undeniable influence even upon serious science. For he was without doubt one of the most brilliant brain-anatomists of his age, and the ideas he produced on the subject have proved of great significance for the development of that branch of science. In his exposition of the nervous system he does not start from the brain, as his contemporaries did, but from the simple nerve-fibre, which he considers to be the simplest type of nerve; it is found even in worms, and out of it "nature has evolved" all the higher nerve-forms: the ganglia as the junctions of several nerve-fibres, and the spinal cord, which consists of a series of ganglia drawn through by a mass of nerve-fibres and connected by means of cross-fibres. Through the spinal cord the nerves lead up to the brain, where they end in the cortex-substance which represents the brain's "ganglion"; in this latter are combined the functions of the nervous system, particularly in the folds of the cerebrum; this, indeed, is the reason why, the more highly the great brain is developed, the greater is the intelligence. In the cortex of the brain are situated the different intellectual qualities of man; these qualities are due to hereditary tendencies and together form the soul, which is thus not confined to any particular spot in the brain, as earlier anatomists had declared. What is new and of value to the future in this nerve theory is, first of all, the emphasis he lays on the significance of the nerve-tracts; further, and above all, the placing of the soul-functions in the cortex of the great brain; and, finally, the assumption of hereditary intellectual tendencies. In particular, the idea that the cortex of the great brain is the organ of intelligence has been fully verified. It is not easy to determine how far Gall, who undoubtedly possessed a thorough knowledge of the details of cerebral anatomy, himself established this fact, or how much he borrowed from his predecessors. It is at any rate peculiar to him that he cites among authorities on this subject even

Swedenborg, whom he must thus have studied, and who may well have been able to inspire him with ideas pointing in that direction; at all events, he has not adopted his predecessor's theory of the essential part played by the pyramid-cells in the work of the brain; on the contrary, he overlooks them and believes that the cerebral cortex is composed of matted nerve-fibres. Gall's theory nervertheless represents a great advance towards the modern standpoint and has undoubtedly exercised considerable influence on the development of cerebral research, in spite of contemporary opposition. This is also the case with Gall's assumption of hereditary intellectual tendencies, which represented a definite advance in face of the naïve belief of the philosophers of enlightenment that all men possess like tendencies to virtue and genius, which require only proper education in order to be able to develop. Unfortunately Gall went to the most ridiculous extremes in developing his theory; he sought and discovered in the brain organs for all kinds of intellectual and moral qualities, for genius and beauty, love and piety, and even for stealing and murder. And he went so far as to imagine that he could discern these very qualities in the irregularities on the surface of the skull, since he believed the skull to be exactly fashioned after the brain. This study, which he called "cranioscopy," rapidly degenerated into sheer humbug; in particular, the discovery of the "bumps of genius," which were supposed to denote special talent, brought much profit to quacks and rogues. This expression has survived in modern phraseology as the best-known relic of Gall's activities, which has served to conceal the really sound work that he accomplished in biological science.

Another scientist who was closely connected with natural philosophy was JOHANN CHRISTIAN REIL (1759–1813). Son of a clergyman of East Friesland, he studied medicine at Göttingen and Halle, practised for some years in his home district, was then appointed professor of internal medicine at Halle, and at the same time became town physician there. When the University of Berlin was founded, he was elected a professor, but resigned his post upon the outbreak of the War of Independence against Napoleon and volunteered as an army surgeon. When acting in that capacity he fell a victim to the typhus epidemic that raged during the war.

Reil's life-theory

REIL's influence has been both many-sided and important. He was highly esteemed by his contemporaries; among Scandinavian doctors Israel Hwasser in particular studied and admired him. As a practitioner he enjoyed a wide field of activities; he recorded in a compendious work all the knowledge that his age possessed of fevers and their treatment. Still more influential, however, was his work in the sphere of psychiatry, which he radically reformed; he effected improvements in the appalling conditions prevailing in the lunatic asylums and insisted upon the elevation of psychiatry to the position of

an independent branch of study at the universities. All his practical endeavours, however, he preferred to base upon a careful study of the functions of
the body; for this purpose he started in the year 1796 the journal *Archiv für
Physiologie*, which under various names and editorial conditions has survived up to the present day. In an essay in this journal he expounded a
general biological theory that exercised great influence on his own time and
is therefore worthy of reference. This essay, *"Von der Lebenskraft,"* contains,
like many others written in that age of pioneers, a number of ideas fruitful
both for the contemporary world and for posterity, side by side with a mass
of uncritical and fantastic nonsense. After a philosophical introduction
touching the terms "matter," "phenomenon," and "idea," Reil criticizes
the vitalistic speculation of preceding ages and declares, with an obvious
reference to Stahl, that phenomena in the animal kingdom cannot emanate
from an immaterial soul, because assumptions as to supernatural influences
explain nothing. Rather, the basis of all phenomena in the animal body that
are not ideas must be sought only in corporeal matter and in "the form and
composition" in its various constituent parts; to matter's different "composition" in muscles, nerves, and bones are due the different properties and
functions of those parts. Reil has here learnt not only from Stahl, but also
from the animal chemistry that started in connexion with Lavoisier and had
been developed in his own time. Unfortunately, however, he does not by
any means come up to the standard already reached by his contemporaries
in this respect; thus, he did not realize the relation between the interchange
of gas in plants and animals, a fact which Schelling, for instance, during the
same period realized well enough to be able to ascribe fundamental importance to it (Part II, p. 277), and he shares the ancient popular belief that
the grain of seed in the earth and the still unbrooded egg are "dead" and acquire life by being provided with warmth and other fine components of life.
Herein lies the weakness of Reil's speculation, and, generally speaking, both
chemistry and philosophy lead him into somewhat strange paths. Having
thus started by defining the idea of force in nature and the relation between
phenomena and the properties in matter which produce them, he goes on to
explain the life-force in animate creatures as being the relation between
more individualized phenomena and a special kind of matter, wherein a
differentiation is made between vegetative force in plants, animal force in
animals, and reason in man. Growth in inanimate and animate nature is
declared to be of an identical character, so that animal growth is at once
termed *"tierische Kristallisation."* But besides these fanciful ideas Reil succeeds in producing a definition of the term "organ" itself that undoubtedly
represents a real advance. Starting from the elder Darwin's theory of fibre
as a basic component in the animal organism (Part II, p. 295), he describes
various categories of fibres: cell-tissue fibres, bone-, muscle-, and nerve-fibres.

Out of these are formed more complex organs, such as nerves, bones, ligaments, cartilages, muscular mass; and finally there are formed of these components in various proportions the higher organs — namely, intestines, sensory organs, and musculature. Here we undoubtedly catch a glimpse of the idea of tissue, which Bichat afterwards developed, independently of Reil and more universally and radically than the latter. There is no doubt, however, that Reil also helped the succeeding generation, in Germany especially, to define the terms on which anatomical science has since developed.

2. Humboldt

In this connexion there is also worthy of mention a scientist who, like those described in the foregoing, belongs both to the history of natural philosophy and to that of exact natural science, but whose fame far outshone the rest and who is universally looked upon as one of the greatest personalities in the whole range of science: ALEXANDER VON HUMBOLDT. He was born at Berlin in 1769 of a distinguished and wealthy family; his father was chamberlain at the court, his mother came from a French family, who had gone into exile for their Protestant faith. Having studied at the University of Göttingen and at the mining academy at Freiberg, he entered the service of the Prussian Mining Department and worked there for some years, until an ample inheritance placed him in a position of being able to devote himself to natural science without having to earn his living. After preliminary studies and travelling in Europe, he equipped at his own expense in the year 1799 a journey of exploration to South America, which region he traversed in various directions and explored so thoroughly that he was called the second discoverer of America. After five years out there he returned home with rich collections, which it took many years to work up. For this purpose he spent a long time in Paris and there published an extensive account of his expedition, which made him world-famous. He spent all his fortune on the journey and its description, but the King of Prussia indemnified him by presenting him with a well-paid post as chamberlain; he rejected offers of university appointments. In 1827 he settled in Berlin and there spent the rest of his days, except for a short expedition to Russia and Siberia. In close contact with the royal family, yet retaining the liberal ideas of his youth, respected as one of the great men of science and highly esteemed for his lovable personality, he lived to a great age, working incessantly at different branches of science, though certainly towards the end with diminished energies. He died in 1859.

Humboldt was an unusually highly gifted personality, artistically as well as scientifically, and he has exercised an extraordinarily varied influence upon the development of natural science, although he did not rise to the highest levels in any particular sphere. As a scientific explorer he is without a rival and he raised geography to the rank of a science. Climatology especially owes its fundamental principles to him: thus, the method of indicating on the map by means of isothermal lines places having a similar annual temperature was invented by him. He devoted many years of methodical study to terrestrial magnetism, and the magnetic-meteorological observatories which are now established throughout the globe have him to thank for their existence. As a geologist he deserves especially well of science, owing to his studies of the problem of Vulcanism; he established the fact that the volcanoes exist grouped in ranges along cracks in the earth's crust. But he was also highly interested in biological problems.

Humboldt's idea of life-force

In his youth he expounded a theory of life as a whole in the form — characteristic of the man himself and of his age — of a mythological story entitled *Die Lebenskraft oder der rhodische Genius*. The gist of it is that life is maintained by a force that prevents the elements of which the body is composed from obeying the laws of affinity that hold good in inorganic nature.[1] In his old age, however, he abandoned this fantastic theory and in his later writings utters a warning against any kind of speculating upon the life-force. He displayed greater exactitude, however, in his investigations, published shortly before his South American expedition, into the influence of electricity upon muscles and nerves, which he carried out partly with himself as subject, and which, together with a number of natural-philosophical illusions, contain ideas that have been utilized in research work of a later period in connexion with electrical phenomena in the animal kingdom.

His vegetable geography

The greatest service rendered to biology by Humboldt, however, was his creation of vegetable geography. Even as early as in Linnæus we found a lively interest and a keen eye for the life-habits of plants. Linnæus's investigations into the question of the habitat and distribution of plants (Part II, p. 215) were, however, based entirely on his classification system. Humboldt's interest in plant life, on the other hand, is at the very outset of quite a different nature. As is the part of a natural philosopher, he takes as his starting-point life in its entirety, examines its various manifestations, and finally dwells on the special advantages which soil and climatic conditions offer to the vegetable world in different latitudes. He puts the question: How is the shape of plants affected by these conditions of life? And he searches for the connexion between the impression made by the landscape

---

[1] The idea contained in this story is without doubt directly or indirectly influenced by Stahl's previously mentioned theory of the soul as the force that prevents the chemical components of the body from disintegrating (see Part II, p. 181).

on the observer and the shape of the plants that dominate the landscape. Thus he produces semi-artistic, semi-scientific pictures of vegetation in different latitudes; he declares that each latitude possesses its own characteristic natural physiognomy, and he finally differentiates between certain vegetable types, not according to systematic characters, but according to the impression the observer receives of their form as a whole. He distinguishes sixteen of these landscape-forming vegetable types, though he states that their number could certainly be increased. Among these types may be mentioned: the palm type, the banana shape, the heather type, the cactus type, the orchid type, the fir type, grasses, ferns, lilies. The whole of this conception of plant life and this grouping of its individual components according to common conditions of life, instead of according to the nomenclature of species, represent a new idea; it is true that here Humboldt has learnt something, as he himself acknowledges, from Buffon, as well as from a number of earlier describers of landscapes, but out of these ideas and as the result of his own observations he created a new field for research, which was cultivated and extended at a later period with great success.

His cosmos

DURING the last decades of his life Humboldt devoted himself to formulating a universal cosmology, which was intended to reproduce every imaginable conception of and all the known facts about the universe: the purely scientific, the historical, and the artistic. This gigantic work, the execution of which was far beyond the powers of one single man, he called *Kosmos;* its first part was published in his seventy-fifth year and a final part of the unfinished work came out after his death. Never has any natural scientist of modern times conceived a plan on a grander scale, and though its execution is naturally both fragmentary and defective, the work nevertheless contains a vast amount of valuable material in the way of facts and is, besides, like all Humboldt's work, unequalled in style. Romantic natural philosophy's idea of a uniform conception of nature has received in Humboldt's *Kosmos* its most glorious memorial; it seems almost symbolical that its creator should have died in the same year as that in which Darwin published his work on the origin of species; the modern theory of evolution stepped in where natural philosophy ended.

3.   Lamarck

JEAN BAPTISTE PIERRE ANTOINE DE MONET, usually styled Chevalier de LAMARCK, was born in Picardy, in northern France, in 1744, one of the youngest of a large and poor noble family. At an early age he was sent to a Jesuit school with a view to eventually securing a comfortable living as a priest.

From the outset this prospect failed to have the least attraction for him, but as long as his father lived, he had to obey him. When he was seventeen, however, his father died, and he inherited a sum just sufficient to enable him to buy a nag; on this he rode away and joined the French Army, which at that time was in the field during the Seven Years' War. On the day after he enlisted, a battle took place, in which his company suffered severely, losing all its officers and non-commissioned officers, whereupon Lamarck, with his one day's war-experience, collected the survivors and held out at his post until help arrived. This deed was rewarded with a lieutenant's commission, but his promotion went no further; he was sent to Toulon on garrison duty, and on the conclusion of peace he resigned his commission for reasons of ill health and was granted a small pension. He now had to look about him for a fresh means of livelihood, and for this purpose betook himself to Paris; there he remained for the next fifteen years as a literary hack, living in a garret in the Quartier Latin just the kind of Bohemian life that has so often been described in novels. During these difficult years, however, there developed in Lamarck an ever-increasing love of natural science, particularly botany; even during his garrison life on the shores of the Mediterranean the abundant and wonderful flora of that coast had deeply interested him, and this love of knowledge grew apace in Paris, where in those days the interest in animate nature was kept alive by Buffon. It was he, too, who paved the way for the scientific success of the penniless writer; he became interested in a flora of France that Lamarck had written and procured his admittance to the Academy of Science. Further, Lamarck was commissioned to travel through several European countries as companion to Buffon's young son, and he finally became an assistant in the botanical department of the natural-history museum. It was during the Revolution, however, that Lamarck first obtained a secure position; the National Convention, which wanted to reform everything, instituted a number of professorships, including two in zoology. As no more suitable candidates could be found, the one chair was offered to the botanist Lamarck, and the other to Geoffroy Saint-Hilaire, who had till then been mostly occupied with mineralogy. These two improvised zoologists shared between them the duty of lecturing, Geoffroy undertaking the vertebrates, and Lamarck the invertebrates. Thus, at the age of fifty Lamarck started research work in the field in which he was eventually to win fame as a pioneer. The rest of his life passed in assiduous work in the career he entered so late; retiring and modest as he was, he sought no outward honours, nor did he win any; he remained throughout his life in poor circumstances, especially at the end, having lost by unsuccessful speculation what little capital he had saved. He suffered, too, from domestic troubles more than most people; he was married four times and lived to see all his unions dissolved by death, while of his seven children the majority also died prematurely. Two daughters, who devoted themselves entirely to administering to him, were his one consolation in his old age; with their aid he was able to carry on his work unremittingly to the end, although he was blind during the last years of his life. He died in 1829, and a year later the last part of the work that had occupied him up to the last was published.

Just as the strangeness of Lamarck's fate is unique in the annals of biology — a discharged lieutenant without any scientific grounding, who from being a Bohemian literary hack works himself up to lasting fame as a scientist and who at the age of fifty becomes professor in a subject that he had never studied before — so his posthumous reputation has likewise been unique. By his contemporaries he was mainly looked upon as a systematist, and, as we shall find later on, he certainly did accomplish valuable and sound work as one. But besides that he published a number of works on evolutionary history based upon speculation; these attracted little attention, however, either in his own day or in the immediately succeeding period. They were neglected by the natural-philosophical school for reasons that will be explained later, and were regarded by the subsequent representatives of exact research as fantastic speculations. It was not until after the launching of the modern theory of the origin of species that Lamarck came into his own. Haeckel in particular, who searched everywhere for precursors of that theory, the promulgation of which he made his mission in life, referred to Lamarck as a pioneer of modern natural research, and there followed in his footsteps a whole group of scientists who saw in Lamarck's theories the basis for a correct view of evolution in nature. During the last few decades this so-called neo-Lamarckian school has, it is true, fallen off considerably in both numbers and influence, but Lamarck himself is still counted one of the pioneers of modern biology.

Lamarck's multifarious works

THE cause of these varying opinions lies essentially in the very character of Lamarck's scientific productions. As will be seen from the above account of his life, he was a self-taught man, without any systematic scientific training, with the result that his production to a large extent bears the mark of dilettantism — many-sided interests, vagueness of both thought and expression, daringly brilliant ideas side by side with foolish fancies. His earlier works especially — up to about the close of the century — are extremely multifarious as to contents, as well as of unequal value. Besides a number of partly still valuable botanical writings, he wrote numerous works on meteorology and geology, as well as a collection of essays with the striking title of *Mémoires de physique et d'histoire naturelle, établis sur des bases de raisonnement indépendantes de toute théorie.* Towards the close of his life, however, he concentrated entirely upon zoology and in that field produced his best works. His enthusiastic admirers have as a rule passed over his earlier speculations in silence; yet without a knowledge of them it is impossible to gain any idea of Lamarck's scientific development, all the more so as throughout his life he firmly adhered in all essentials to the views he held in his youth.

In the above-mentioned work, *Mémoires de physique*, Lamarck endeavoured to form a general theory of existence, a combination of physics, chemistry, and physiology. This theory represents a continuous attack upon what he calls "pneumatic chemistry" — that is, Lavoisier's quantitative method (Part II, ch. xii). For Lavoisier himself Lamarck has nothing but praise, his polemics being invariably objective and honest, but on the composition of things he has ideas that are entirely his own. Lavoisier had conceived combustion as a process of oxidization; Lamarck finds this explanation absurd — the idea of oxygen's being an essential component of both water and air is in his opinion utterly irrational; no chemist has ever seen it and nobody has been able to prove its actual existence. And equally irrational is the theory of chemical affinity as a cause of chemical associations between the elements: "It is not compatible with reason and is therefore impossible." As essential components of nature Lamarck assumes the four known elements — fire, air, water, and earth — and adds a fifth, light. The purest earth is — rock-crystal. The chemical associations are not at all bound together by any affinity; rather, they strive to disintegrate into their simple components. What creates chemical associations on the earth is exclusively life; all inorganic associations that exist — rocks, minerals, metals — are disintegrated remains of living beings. Lamarck sets up an evolutionary series that is unique of its kind, beginning with blood, bile, urine, bone-substance, snail-shell, and proceeding to increasingly greater "disintegrations" through shell-lime, marble, gypsum, to precious stones, metals, and lastly "simple" rock-crystal. The problem of what life really is is of course a question that largely occupies the mind of Lamarck and is discussed by him with great particularity. The essential factor in life he finds to be motion; an animate being is composed of various parts which affect one another and are kept in motion partly by mutual influence and partly by influence from without, and it undergoes constant change in consequence of this motion. Life itself is motion and nothing else — that is, a purely mechanical phenomenon. The essential components in the living body are partly solid (fibres and membranes), partly liquid (blood, lymph, and other special "fluids," of which more later on). Of the functions of life, secretion within the organism is an expression for the afore-mentioned efforts made by the chemical associations to disintegrate; nutrition counteracts these efforts by providing the living being with fresh substances, a difference being made between the power of a plant to form out of simple alimental substances complex bodies, and the dependence of animals upon these same complex products for their nourishment. — In these and other phenomena, in both animate and inanimate nature, there is, according to Lamarck, incorporated as an essential component fire, which penetrates the whole of existence; it is a "fluid," which appears under various modifications, as heat, as electric and magnetic fluid, and in living beings still further specialized. It produces colour-perceptions, sound — Lamarck denies that the air conveys the sound, for a cannon-shot is heard at a distance better with the ear to the ground than in the air — and, further, chemical changes, the various kinds of which it would take too long to enumerate here. The serious offences against contemporary physical and chemical knowledge of which Lamarck is guilty in this work will have been sufficiently illustrated by the above. And it would be a waste of time to trace the sources out of which he created these wild fancies; he himself, indeed, asserts that they are "independent of any theory" and they are characterized from beginning to end by sheer dilettantism. Fortunately, however, Lamarck did not retain this standpoint always; although more than fifty years old when he published his *Mémoires*, he managed to escape out of the helpless maze of thought that they involve and to create works which have kept his memory alive even up to the most recent times — a spiritual test of strength indeed, which is almost without its counterpart in the history of science. That this was so is not due to his having acquired any essentially better knowledge of physics and chemistry than others,[2] but to his having applied himself whole-heartedly to zoology. In this field, thanks to his long experience as a lecturer and a museum-worker, he had gained a many-sided knowledge of form, whereon he was able to base a system of thought that was not only original, but also truly scientific, as regards both form and substance.

The result of Lamarck's theoretical speculations in the sphere of biology — he it was, in fact, who created the word "biology" — is recorded in three separate works: *Recherches sur l'organisation des corps vivants*, of 1802; *Philosophie zoologique*, of 1809; and the introduction to his great work *Histoire naturelle des animaux sans vertèbres* (1815–22). The first of these presents in short and concise form the theory of the development of life that made Lamarck famous. It has been completely overshadowed, however, in the history of biology by *Philosophie zoologique*, which is the one work of Lamarck that is regarded as a classic and which has in more recent times been frequently reprinted and translated into many languages. It is really an expansion of the previous work, full of repetitions and containing a number of additions, which in many instances, but not in all, are improvements. In the third of these works the author once more recapitulates the theory in summary form, as he entertained it towards the close of his life.

[2] In his latest works, it is true, he acknowledges the existence of oxygen, a fact of which he had apparently been convinced by some chemist, but on the whole he maintains the old standpoint.

His life-theory: life is motion

LAMARCK begins his work *Recherches* with a protest against that dry systematization that is content with differentiating as many species as possible without troubling to make a comprehensive survey of the connexion between the life-forms in nature. Rather, he would start by regarding life in its entirety, and he thereupon finds, in accordance with his conception referred to above, that the most essential quality of life is motion. All that occurs in life is motion; through it the organism strives to develop and to specialize the organs; motion is also the absorption of nutriment, whereby the individual, during its days of physical power, compensates for the losses caused by excretion, whereas during the later period of life excretion becomes superior to the power of absorbing nutriment, so that eventually death results; it is through motion that development proceeds in every living being, the fluids of the body making their way through the surrounding solid parts, with the result that in these latter are formed organs which assume various functions, and canals which convey nourishment to them.[3] Thus is gradually formed not only the individual, but also, step by step, all living beings of various types, while the qualities that have been developed in the individual life-forms are transferred by reproduction to the descendants. On this basis it is possible to place all living beings in one series, beginning with the lowest and ending with the highest. It is more instructive, however, to examine the organization of animals in the opposite direction, in that, if we start from the highest forms, we can follow the "*dégradation*" that appears in the series, one organ after another becoming changed, simplified, and finally disappearing. The mammals are naturally the highest; they are the only creatures that really produce their young alive; they possess milk-secretion, independent lungs, and complete diaphragm. The birds come lower than the mammals, for they lay eggs, their lungs are fixed, and they have no diaphragm. Below these two warm-blooded animal groups come the reptiles, owing to their cold blood and incompletely formed heart and lungs, which latter are in certain forms represented during earlier stages by gills (the batrachians, as is well known, were still at that time grouped together with the reptiles); further, the two pairs of extremities in these animals gradually disappear, wherefore the snakes, which possess no extremities, are the lowest of the order of reptiles. Upwards, again, the transition between reptiles and birds is formed by the Chelonia (tortoises), while the then newly-discovered duck-billed platypus assumes the same role between birds and mammals. The fishes, on the other hand, are lower than the reptiles, for they have entirely lost lungs and extremities; that is to say, their fins are not real extremities. With the transition from the fishes downwards the backbone and the inner skeleton disappear from the

---

[3] This theory recalls a similar one of Caspar Friedrich Wolff's (Part II, p. 250), but it is uncertain whether Lamarck knew his works — at any rate, he never quotes them.

animal kingdom. Of the invertebrates, the molluscs stand highest, for they have gills like the fishes and possess brain, nerves, and single-chambered heart. Next to them come the Annelida, which Lamarck, after Cuvier, distinguishes from the worms, and which he has named; they likewise breathe by means of gills, sometimes visible, sometimes concealed in the skin; moreover, they possess a nervous system, a vascular system with red blood, and a pair of extensions thereof corresponding to the heart. These are followed by the crustaceans, also possessing gills and heart, but after them these latter organs disappear from the animal kingdom. The spiders come next; because they have a concentrated respiratory system and emerge from the egg in the same form as they retain afterwards, they are above the insects, which possess scattered tracheæ and undergo metamorphosis. With these animals, in Lamarck's view, sexual reproduction disappears from the animal kingdom. Thus, the worms, which follow next in the series, are reproduced by gemmation; as a matter of fact, they may possess a nervous system and tracheæ. With them disappear visual organs and nervous system from the animal kingdom. The next class is the Radiata, another systematic creation of Lamarck's; these animals lack visual organs, but possess organs of generation — though sexless — and are thereby distinguished from the polypi, which possess no organs at all.

Lamarck having thus classified the animals in a series on a basis of the absence or presence of certain principal organs, he goes on to state that the sequence thus formed does not refer to the separate animal individuals, but to the great masses of animals that form one entire class; within such a class it is possible that, owing to dissimilarities in less essential organs, ramifications may take place in various directions, but the above arrangement, which has been made in the animal classes on the basis of the structure of the most vital organs, is presented with such certainty that "no enlightened natural scientist will be able to produce another." It shows how, the higher we come in the series, the greater becomes the specialization in the organs, while the lower we go, the simpler we find the organs becoming and the wider their functions. On this ever-increasing specialization of the organs Lamarck now bases his theory of how the various life-forms have arisen, a theory which at the very outset he formulates as follows: "It is not the organs — that is to say, the form and character of the animal's bodily parts — that have given rise to its habits and peculiar properties, but, on the contrary, it is its habits and manner of life and the conditions in which its ancestors lived that has in the course of time fashioned its bodily form, its organs, and its qualities." He seeks to prove this basic argument by innumerable examples: moles and blind mice have lost their sight as a result of living underground for several generations, the ant-bear its teeth through swallowing its food whole; waders have acquired long legs and long neck through stretching those parts of the body in their search for food on the shores, swimming birds their webbed feet through stretching out their toes during their movements in the water. In connexion herewith he declares that if a number of children were to be deprived at birth of their left eye and they were allowed to have children by one another, there would eventually arise after a few generations a one-eyed race of men. But it is not merely influences of this detailed kind that re-create the life-forms; but also the effect of geographical conditions in general; climate, humidity, abundance or scarcity of food, have in the long run had a transforming influence upon animals as a whole and have produced new organs or caused the old ones to disappear. This is rendered possible owing to the fact that there is an infinity of time at the disposal of evolution; "Time has no limits," says Lamarck explicitly. Thus life becomes purely and simply a mechanical process; all its manifestations are motion and nothing else. It is true, existence has a Supreme Originator, for whose name Lamarck always slows respect, but His greatness lies in the fact that He has created nature in such a way that it has developed its profuse multiplicity without interference from without. This development out of given qualifications Lamarck does not succeed in establishing, however; "nature" appears constantly as a creative power and is spoken of in terms suggestive of a personality; this is especially so in *Philosophie zoologique*, from which a few examples will be cited further on in this chapter.

His theory of life-fluid

In the continuation of *Recherches* Lamarck further develops his mechanical theory of life, which to him is really the main problem, in which his evolution theory is only one detail among many. To his mind, life itself is the condition in all the parts of the body that makes their organic movements possible. This condition consists in the existence of "*l'orgasme vital*," a state of tension, a "*tonus*," which maintains the molecules in the soft parts of the body in a definite position and which by increasing and diminishing enables the organs to contract and expand. The cause of this tension is a fluid that is secreted by the blood and thence absorbed by all the organs of the body, but is particularly concentrated in the nervous system. This fluid is really a peculiar variety of fire, related to heat and electricity, a "*feu éthéré*." It is transmitted on fertilization from the male sexual product to the embryo, which derives life from it — immediately in mammals, but not until later in the bird's egg, which only receives life by brooding. But this same fluid exists scattered everywhere throughout nature, so that everywhere, and especially in hot countries, with their humid climate, there takes place a spontaneous production of life. Lamarck asserts that this spontaneous generation under the influence of heat, light, and electricity goes on incessantly, the lowest animal forms — and even plant forms — being continually reproduced out of inanimate matter; he declares it to be probable that the fresh-water polypi freeze to death every winter and spontaneously generate again every spring. According to his idea, the most primitive living creatures consist of a mass of gelatinous substance, which absorbs nourishment through pores on its surface. Out of these nature gradually evolves a special organ for the admission of food; first there arises as a result of the movements of the animal a small depression in which the food can easily collect; through the pressure exercised by the food, this slight hollow expands into a sack-like cavity, which similarly becomes in process of time still further extended; thus arose the polypus's digestive canal. The next important life-property which nature developed was reproduction; this consists in reality of a growth over and above the normal dimensions; a division must therefore take place, and actually does so in the lowest animals, the Infusoria, which never die of old age, but divide themselves in two when they have attained a certain size. Through the division's not being uniform, gemmation arises, which is the manner of propagation characteristic of the polypi; when this is repeated, one particular area becomes specialized for the purpose, and thus originated the internal gemmation by means of which the Radiata propagate. Through further evolution in this direction there arose the eggs, being incomplete buds which, in order that they may develop further, require to be influenced by the male sexual product.

The evolution of man

HERE Lamarck interrupts his exposition of the origin of the most vital organs and proceeds direct to a consideration of the evolution of man. Like Camper, he emphasizes the differences between the anatomical structure of man and of the higher apes, but all the same he maintains, in conformity with his view that all properties are evolved by exercise, that both the physical and the intellectual superiority of man has been achieved through his having in the course of ages exercised his faculties to an ever-increasing perfection, while, on the other hand, the higher apes can also be trained to attain a high standard of intelligence and a finer character. But there is still a vast difference between Lamarck's ideas of human development and La Mettrie's and his contemporaries' enthusiasm over the intellectual similarity between primitive man and the higher animals, Lamarck maintaining that it has been given to but few men in the whole course of the ages to achieve real intelligence, whereas the majority have remained in a state of bestial ignorance; they have prayed to beasts and perpetrated acts of the wildest folly; even where a nation has attained the highest culture, this has been due to the work of a few highly gifted persons, while the majority of their countrymen have indulged in the maddest aberrations. Lamarck is without doubt thinking here of the Reign of Terror during the Revolution, which he had witnessed at close quarters. It was probably these memories of human degradation that deprived him of his taste for inquiry into the characteristics of primitive man and the man-apes. Possibly, too, fears of censure on the part of the Napoleonic Government had their influence in this respect.

After a brief discussion of the term "species" in the vegetable and animal kingdoms in relation to the mineral kingdom, wherein he states that the mineral differs from plants and animals in not possessing individuality, and further emphasizes the influence of environment upon the development of species, adducing such examples as the deep-water and shore form of *Ranunculus aquaticus*, Lamarck proceeds to record his views on the nervous system and its functions. As he deals with the same subject more fully in his subsequent works, we may postpone our account of his views on these questions and here close our résumé of his *Recherches* — the work that displays his genius and his limitations more clearly, perhaps, than any other.

In his *Philosophie zoologique* Lamarck discusses once more his theory of the development of life in nature. By way of introduction he examines the question of how much is human invention and how much is nature's own law in natural science, and he comes to the conclusion, with Buffon, that all systematic classifications are arbitrary products of human thought; in nature there are only individuals, which can certainly be placed in groups in respect of certain characteristics, but the lines between which are always arbitrarily drawn. As regards the problem of evolution itself, he adopts the same plan as in *Recherches;* he first describes the *"dégradation"* throughout the animal kingdom, and then expounds the theory as to how the organs, and therewith the animal forms themselves, have developed by habit and way of living. He further insists upon the importance of the essential organs for purposes of development in contrast to the non-essential, citing the old instances of how organs develop, to which reference has been made above, and a number of new ones besides — sometimes quite absurd ideas, such as that the males of the Ruminantia have acquired horns through the blood having gone to their heads in the mating-season. In regard to the general conditions under which life has developed, he holds that the earth has evolved continuously and not as a result of catastrophes, as Buffon, and after him Cuvier, maintained, and also that no animal species have died out, except those that man himself has eradicated, but that the fossil species that are not found at the present time have been transformed into now existing forms. With increased emphasis and with his criticism directed especially against Cuvier, he seeks to prove that all animal classes are derived from one another and should therefore be arranged in a line and not parallel or "reticularly"; nevertheless, it is permitted for the genera in each class to form ramifications from a common primal form. Further, in a supplement to the work he extends this ramification theory to the classes in the Vertebrata, in that the birds are derived from the tortoises and the mammals from the crocodiles.

Reform of the animal system

THEN follows a review of the animal system, which is one of the most brilliant features in the whole of Lamarck's work. Here he draws up the invertebrate system that, except for one or two alterations, has held good ever since. He distinguishes the Infusoria from the Polypi, and the Cirripedia from the Mollusca, and thus gets ten invertebrate classes: Infusoria, Polypi, Radiata, Vermes, Insecta, Arachnida, Crustacea, Annelida, Cirripedia, Mollusca. Of these classes the Radiata are now divided into two: Cœlenterata and Echinodermata; the Polypi have been grouped with the Cœlenterata and the Cirripedia with the crayfish. A number of fresh divisions have certainly been made in modern times, but at any rate Lamarck created a system that, in comparison with Linnæus's invertebrate grouping, represents an extraordinary advance; and it is all the more to Lamarck's honour that he so generously acknowledges his predecessor, whom he calls one of the greatest scientists that have ever existed. But Lamarck was not only a natural philosopher, he was also an expert on form, and as such he was bound to realize the value of the preliminary work carried out by Linnæus, although he did not accept his hard and fast rules governing species. In this connexion Lamarck describes the difficulties with which the systematist is overwhelmed as a result of the aggravated chaos in scientific nomenclature; to cure this evil he recommends that the nomenclature be fixed by international agreement, and this has actually been done, though not until quite recently.

After reviewing the system of the vertebrate animals, which he has borrowed from another zoologist, DUMÉRIL, and is therefore not to be compared in point of interest with the invertebrate system, Lamarck once more takes up the question of the origin of man. He says that the centre of gravity in a man standing erect is situated far in advance of the vertebræ, so that muscular effort is required to hold himself upright, which indicates an origin from quadruped animals. He drafts a theory as to man's descent from the anthropoid apes, but adds that this might have been so if man had not a different origin from the animals. He has evidently not dared to draw the obvious conclusion from his theory, but has taken refuge behind a reservation, similar to that made by Descartes in his hypothetical views on the creation. Lamarck apparently feared that Napoleon would not have felt flattered by a genealogy based on the orang-utan.

Theoretical speculation on life

MORE than half the *Philosophie zoologique*, however, is taken up with purely theoretical speculations on life and its manifestations, and in this sphere Lamarck again shows his weaker side almost as much as he does in physics and chemistry. Although in the foregoing he constantly makes nature appear as a creative power, he defines it, in his introduction to the speculative section of the work, in the following manner: "Nature — that word that is so often pronounced as if it referred to a particular being — should not appear to us as anything else than the comprehension of things, embracing: (1) all physical bodies that exist, (2) the general and particular laws which direct the changes in the condition and position of these bodies, and (3) the motion that is current in different forms among them, eternally maintained and renewed, infinitely varying in the products it creates . . . . '' But he is so little capable of adhering to this view that only a few pages further on he is able to say: "Every step which Nature takes when making her direct creations consists in organizing into cellular tissue the minute masses of viscous or mucous substances that she finds at her disposal under favourable circumstances.''[4] A personal god could not have acted more personally. And Lamarck's belief in creative nature is as dogmatic as was Linnæus's belief in God. He develops afresh his old statement that life is nothing but motion, and that motion is produced by this wonderful and ubiquitous ethereal fire, which is to Lamarck what the soul was to Stahl; we can apply to the one as to the other the saying of Bonnet, that it performs anything that one requires of it and its non-existence can never be proved. On this basis Lamarck creates an extremely curious psychological theory. To his mind the soul-life is a purely mechanical process, which is dependent for its nature upon those organs that the animal in question possesses; animals that lack muscles and nerves have practically no sense-impressions; they are "apathetic," they move only as a result of influences from outside, through the ethereal fire's penetrating them and stimulating them. Animals having a nervous system certainly receive sensible impressions, but they react to them purely schematically and are incapable of combining the impressions as a guide for their actions; animals that possess a brain can retain the sense-impressions they receive and combine them to form ideas as a guide for their actions. Lamarck's way of explaining all the manifestations of the human soul-life — sense-impressions, ideas, and moral conceptions — with the aid of the ubiquitous and universally applicable ethereal-electrical fluid, is in itself of but little interest; in this he associates himself with thinkers of the eighteenth century: Locke, Condillac, and in particular the physician CABANIS (1757–1808), all of whom taught that ideas are exclusively based on sense-impressions; the last-named, the most pronounced materialist of them all, was, however, a far more trained, and therefore also a more cautious, thinker than Lamarck, who blindly relied on his fluid, by means of which he explained everything, while his predecessors were content to analyse certain definite phenomena in the soul-life.

Lamarck managed to complete one or two further important works in his old age: the above-mentioned lengthy systematic survey of the Invertebrata

---

[4] *Philosophie zoologique*, ed. cit., Part I, pp. 349, 362.

and a work on fossil Mollusca, which is worthy to be associated with Cuvier's contemporary works on extinct vertebrate animals. In the introduction to his former work he deals for the last time with his theory of evolution. Nature is here represented with greater emphasis than in his earlier works as a creative force. "Nature is . . . an intermediary between God and the various parts of the physical universe for the fulfilling of the divine will." And "Nature has given to animal life the power of progressively consummating the organization and of developing and gradually perfecting it." It is thus an inner striving after perfection that, besides the influence of environment, has here been the cause of evolution. This striving after evolution, which is also hinted at in his earlier writings, became, as we shall see, a stumbling-block for Darwin, which evoked his opposition to Lamarck's theory. It now remains to examine the hypotheses on which this wealth of scientific production rests and the influence it had.

Influence of Buffon and Bonnet on Lamarck

THE scientist by whom Lamarck as well as other biologists in France at that period was undoubtedly most influenced was Buffon. We recognize this influence in Lamarck's emphatic assertion that only individuals exist in reality, while the categories of the classification system are products of the mind, as also in the whole of his general conception of life as one vast evolutionary process of a purely physical character;[5] even the very idea of evolution as a result of habits of life and environment we find developed in Buffon, who cites in proof thereof the featherless face of the rook and the padded feet of the camel. If we compare these two scientists, we find that Buffon is without doubt superior as a thinker; he realizes the difference between hypothesis and fact, as he is aware of the limitations of natural science — things for which Lamarck has absolutely no mind. On the other hand, Lamarck is decidedly superior in his knowledge of form and has a far keener eye for classification, which is certainly not exclusively due to the fact that he was acquainted with a greater number of forms than his predecessor.

But Lamarck also learnt a good deal from Bonnet, as indeed he expressly acknowledges. The classification of the animal kingdom in one single series was adopted by him from this source, as also the actual French expression for it — "*échelle*" (scale). The idea of animals' "degeneration" through the loss of certain organs is, however, reminiscent of Vicq d'Azyr. And,

---

[5] Among Lamarck's precursors it is also customary to mention BENOÎT DE MAILLET (1656–1738), for a long time French consul in Egypt and the author of a work on natural philosophy published under the name of Telliamed (the anagram of his surname), wherein is described in an extremely fantastic manner how the entire earth was once covered by the sea, and the ancestors of all existent land-animals were aquatic animals, which gradually became accustomed to living on land. It is, however, difficult to determine what influence this work, which was treated with contempt by Voltaire and was speedily forgotten, may have had upon Lamarck.

finally, Cuvier, with whom he was so often in controversy, has undoubtedly exercised great influence upon him through his earlier writings; we have already seen how Lamarck gained from them important ideas in regard to classification, but in connexion also with problems of evolutionary history he has undoubtedly felt the influence of his younger rival — a point that will be more closely dealt with in connexion with the latter's own work. But though we can thus trace outside influences in Lamarck's speculations, it is nevertheless to his lasting credit that he formulated and elaborated as he did the idea of origin. In that respect he is truly a pioneer of modern biology. It is true that the theory of the heredity of qualities acquired through the influence of environment has not stood its ground in face of modern exact research in this field, and also that the actual method of working out the idea leaves very much to be desired. Thus, for instance, one would have expected that when he so definitely differentiates between essential and non-essential organs and qualities, he would have tried, following his own method, to trace the development of the essential organs (heart, lungs, backbone), instead of dilating upon such details as the legs and feet of waders and swimming birds. But the idea itself is nevertheless conceived and elaborated not only with splendid consistency, but also with a keen eye for peculiarities in the interrelation of living forms, which left all his predecessors far behind. And indeed details can be instanced which are highly original, as, for example, the theory of the origin of the digestive canal through invagination. It is in all probability here that Haeckel, the originator of the "gastræa" theory and an enthusiastic admirer of Lamarck, obtained the idea for his hypothesis, which has proved so valuable to modern embryology. The superficiality of Lamarck's psychological speculation has already been pointed out, but this quality he shared with many of his predecessors and contemporaries; he is here a child of the era of enlightenment; the Supreme Originator who was at one time creative, but afterwards inactive, and also the subsequently omnipotent nature, are ideas that, since the days of Voltaire, often recurred in the works of scientists of that epoch, while even the mechanical soul-theory constantly occurs, better or worse expounded in the writings of that period. In the purely systematical sphere, on the other hand, Lamarck is, as previously pointed out, one of the foremost of all time, and perhaps he has made in this sphere his most lasting, although not his most brilliant, contribution to the development of biology.

Importance of Lamarck's life-work

WE have already suggested that Lamarck's reputation in his own age was based entirely on his work as a systematist. That he did not receive recognition as a natural philosopher was essentially due to the fact that his materialistic conception of nature, which originated during the previous century, was already out of date when he came on the scene; the two scientists who entirely controlled the development of biology at that time, and who will be discussed in the next chapters — namely, Cuvier and Bichat — both embraced an entirely different theory from Lamarck's as to the intrinsic nature of life, nor, indeed, did the latter's speculation really possess that force and consistency which would have enabled it to hold its own against more modern directions of thought. Neither, on the other hand, did contemporary natural philosophers pay any attention to this thought-system, which, even from its own point of view, became rapidly out of date and, besides, was not particularly complete in form; to one such as Schelling and his school it must indeed have been an abomination, if only on account of its connexion with the hated seventeenth-century materialism, and even Goethe had much the same motive for leaving it at its worth. The day arrived, however, when the mechanical conception of life again came into its own and when Lamarck's *Philosophie zoologique* underwent a brilliant revival. The account of this revival is reserved for a future chapter; but this much may be pointed out here, that Lamarck's greatest admirer in modern times, Haeckel, not only adopted his theory of evolution, but also a considerable amount of both good and bad out of his materialistic psychology — this, too, having thus exercised some influence up to our own day.

But though Lamarck's ideas were thus to have a future, the biological research of his own age was being directed by a man with an entirely different conception of nature and its phenomena, a man who possessed to a rare degree a conception of those problems which at that time most urgently required solution, and who was, moreover, capable of dealing with these problems in a manner that redounded to the lasting benefit of science. This scientist was Cuvier, one of the foremost of those who laid the foundations of biology in the modern sense of the word.

CHAPTER II

CUVIER

GEORGES LÉOPOLD CHRÉTIEN FRÉDÉRIC DAGOBERT CUVIER was born in 1769 at Montbéliard, a small town not far from Basel, which, although entirely French, belonged at that time to the Duchy of Württemberg. He came from a French Huguenot family, which had at one time sought refuge from religious persecution at home; his father, however, had been an officer in French service, but in his old age had returned to his native town, where he married and lived on a small pension given him by the French Government. At an early age young Georges displayed brilliant intellectual gifts; he passed through the local school with honours and during his time there became acquainted with Buffon's writings, which he diligently studied. The poverty of his family, however, threatened to prevent him from continuing his education, when a chance opportunity procured him free entry into the Karlsschule at Stuttgart. This one-time famous educational establishment was originally a military academy, but had been extended by the reigning Duke Karl into a college providing for the training of Civil Service officials as well. The school was renowned for its excellent staff of teachers and at the same time feared for the severe military discipline exercised there under the personal supervision of the despotic Prince. Schiller, the German poet of liberty, had been one of its first pupils, but had escaped from the insufferable constraint by flight, and others had followed his example. Cuvier, on the other hand, who was not only naturally gifted, but also possessed a sense of discipline, got on well there; although upon first entering the academy he had no knowledge of German, he soon became one of the best pupils in the class for the science of State finances, which he entered because natural science was most widely taught there for the benefit of aspiring argicultural and forestry employees. The teacher of biology here was KARL FRIEDRICH KIELMAYER (1765–1844), one of the most extraordinary of German biologists, afterwards professor at Tübingen, a man who allowed none of the courses of lectures that he gave during a long life to be printed, though they were highly thought of, copies of them being made and eagerly studied. He appears to have been a speculative natural scientist, who had been influenced by Herder's ideas of a common primal type for all living creatures and their several organs, and who consequently strongly recommended the study of comparative anatomy. Cuvier received a thorough grounding at his hands and gained from him many valuable ideas, which indeed he gratefully acknowledged throughout his life. Having successfully passed out of the school at the age of eighteen he returned home; he could not afford to work his way up as an unsalaried official in the Civil Service, so he had to accept the post of tutor in a Protestant family in Normandy. Here on the Channel coast he found an entirely new animal world, which he at once began to study with keen interest; in his spare time he dissected all the fishes he came across and compared their structure, and with even greater enthusiasm took up the study of the innumerable lower animal forms that the ebb tide left stranded on the shore — molluscs, worms, and starfish. In Linnæus's *Systema Naturæ*, which was the examination text-book of the time, these creatures were not thoroughly dealt with; even Aristotle had at one time displayed greater interest in marine animals, and in his writings Cuvier found not only records of their life, but also ideas suggesting ways of comparing their different structure. He drew everything that he studied, for he had learnt to be a clever draughtsman. Some of these pictures, which were submitted through an aquaintance to Geoffroy Saint-Hilaire, then newly-elected professor in Paris, proved of momentous importance for Cuvier's future. He was summoned to Paris and within a short time was appointed professor of comparative anatomy, although he had never dissected a human body — an appointment similar to that of Geoffroy and Lamarck the year before. Thus his fortune was made and new promotions and honours followed in rapid succession, more than space allows us to enumerate. Cuvier stood especially high in Napoleon's favour; contemporary with the Emperor in regard both to the year of his birth and to the period when he first became eminent, he possessed something of the latter's genius for organization; his energy was inexhaustible, he could discharge many duties at the same time without neglecting a single detail, he was full of ideas touching problems of organization, and he also possessed a theoretical knowledge of statecraft which he had acquired during his school period at Stuttgart. Thus he became "*inspecteur général*" in the department of education and carried out his duties in that post, at the same time attending to his professorship and his science, so successfully that under his leadership the educational system in France was thoroughly reformed and a number of new universities founded, both in France and in its extensive subject countries, Italy and Holland. When Napoleon fell, Cuvier became an indispensable authority in the spheres of science and education; in spite of the Catholic reaction that succeeded the Bourbon's regime, he, a Protestant, was allowed to retain his appointments and received still further promotions, becoming a baron and minister for Protestant ecclesiastical affairs. Throughout this period he was wise enough to maintain his political independence, and after the July revolution he rose still higher, becoming a peer of France.

By that time, however, his days were numbered; he died of cholera during the first epidemic that ravaged Europe, in 1832. His wife survived him, but all his children had died before him.

As a personality Cuvier has been very differently judged, both by his contemporaries and by subsequent generations. It may be taken for granted that one who served Napoleon with such great success was himself something of a despot, and he certainly did not escape the personal hatred that is always the lot of such men. Bitter accusations have been made against him even in modern times, but their truth is contradicted by the reputation he enjoyed amongst his contemporaries. Better evidence of his true character is provided by the unfailing dignity with which he carried on his controversy against Geoffroy Saint-Hilaire, as well as by his widely attested kindliness and helpfulness towards younger scientists. In his political views he was conservative, though not one of the servile type; since the appearance of the origin-of-species theory he has been accused even of scientific conservatism on account of his having maintained the immutability of species; in this respect he must naturally be judged according to the standards of his time and, viewed from this standpoint, his opposition to Lamarck's theories is easily explained. As to his vital importance for the development of biology, however, there can be no two opinions; a survey of his most important work will confirm this.

Cuvier's comparative anatomy

When Cuvier set out to deal with comparative anatomy on scientific and educational lines, he started from a point directly opposed to his predecessors' line of advance. All of these had been medical men: Daubenton and Vicq d'Azyr as well as Camper and Blumenbach; to them man was the primary object, with which all other living creatures were compared. Cuvier, however, had begun by studying marine animals: fishes, molluscs, and worms. Upon coming to Paris he carried out a number of valuable investigations, in the style of Camper, on special subjects, such as the orang-utan, the rhinoceros, and the lemur, and later on, the Vertebrata became his chief object of investigation. He believed his mission in life to be the creation of a general comparative anatomy; he worked for it throughout his life and in his other writings often referred to the forthcoming work, but it was never completed. In preparation for it he published his lectures on comparative anatomy, written down by his pupil Duméril. The system of thought that he elaborated in these lectures was adopted in several treatises on special subjects: fishes, molluscs, and fossil vertebrates. Finally he published a systematic work, Règne animal, based on the same principle. As a result of these works he became, as W. Leche says, "the founder of modern comparative zoology. He became so not through bringing to light a large number of fresh facts, but rather through having introduced a new method." Much the same judgment has been expressed concerning another of the great pioneers whom France has given to scientific research — Lavoisier.[1] This ability to create new scientific values by way of method seems to be inherent in the French nation, with its keen and critical faculty.

The new method that Cuvier thus introduced was comparative anatomy in the modern sense of the term. True, as he himself admits, he was not without precursors — the two collaborators Buffon and Daubenton, and also Camper, Vicq d'Azyr, and Blumenbach, had made weighty contributions to the branch of science in question. But Cuvier's great contribution is his consistent and far-reaching application of the comparative method, whereby he actually created an entirely new view of the connexion of causes in nature, in respect of both the construction of the separate individual and the mutual relation of the various animal forms. In this sphere he has perhaps learnt most from Aristotle, whom he resembled in his power of discovering and comparing formal qualities of fundamental importance for the conception of life in nature.

In his first more important work, the above-mentioned *Leçons sur l'anatomie comparée*, which came out in the years 1799–1805, Cuvier still to a certain extent holds to the old point of view, the influence of his predecessors, chiefly Daubenton and Vicq d'Azyr, being clearly apparent. But what at once strikes one on reading this work of Cuvier's youth is the clarity and soberness of thought that dominate his whole conception, particularly in the purely theoretical problems. All speculation upon the innermost essence of existence is carefully avoided; he frankly acknowledges the powerlessness of the human capacity for thought in this sphere and the worthlessness of those systems of thought that earlier and contemporary natural philosophy had created in order to fill the gaps in our knowledge of nature. In this Cuvier stands out in sharp contrast to such scientists as Buffon, Bonnet, and Lamarck, to say nothing of the German natural philosophers of his age. This tendency to criticism was undoubtedly innate in Cuvier; it was certainly stimulated by the study of Kant, whom he quotes in one place, just as, on the whole, his acquaintance, initiated in Stuttgart, with the German world of thought contributed towards broadening his field of vision beyond what was customary in his countrymen at that time. The consideration of the problem of life with which the work referred to starts is thus introduced by the emphatic declaration that life in its innermost essence is and must remain a riddle: "a word that the untrained mind is ready to regard as an expression for a special principle, although actually it can never denote anything but the summary of the phenomena that have given rise to its formation." Then follows a description of these phenomena, which recalls that which Hum-

---

[1] See Part II, p. 265.

boldt gives in his early work mentioned above: we observe a state that hinders the ordinary physical and chemical forces in their efforts to dissolve the body into its simple components: this is called life, and its maintenance requires a constant renewal of chemical components; fresh components are absorbed by the body at the same time as others already existing in it are given off. Finally this process ceases, whereupon death ensues, accompanied by that dissolution of the components of the body which life had prevented. And this life can be produced only by previous life, but the problem of the production of life is as much beyond our grasp as is that of life itself. Cuvier therefore does not accept the principle of spontaneous generation. Next, he gives an account of the various components of the body — their composition and function. His account shows that he had mastered contemporary chemistry, as developed by Lavoisier and his successors; he emphasizes the part played by oxygen in respiration, which is expressly compared with a process of combustion; he enumerates the simple components of the body — carbon, hydrogen, oxygen, and nitrogen — pointing out the importance of the last-named element in the animal organism as opposed to the vegetable organism. It is hardly necessary to lay stress on the vast difference between this substantiated and critical exposition of life-phenomena and their basic structure, and Lamarck's fantastic speculations upon life. The one predecessor whom Cuvier most recalls in this early work of his is without doubt Vicq d'Azyr. But in contrast to him there stands out at once Cuvier's originality, chiefly in his conception of the object of anatomy and the consequent arrangement of the details of his exposition; whereas in Vicq d'Azyr the function of the organs is the essential, and the basis on which the work rests is therefore physiological, Cuvier thrusts the form of the organs into the foreground. He holds that respiration, whose role in the renewal of substance is the same throughout the entire animal kingdom, is performed within the separate animal classes by means of organs which are so unlike one another that no comparison is possible between them. And this is also the case with the organs of motion.

His correlation theory

ON the other hand, in similar animals there takes place a co-operation between the organs that makes them, as far as regards their form, entirely dependent upon one another; the correlation between the separate organs in the same body, which Vicq d'Azyr had already described in its main features, is studied in detail by Cuvier and to him represents the very basis of his conception both of animals' habits of life in nature and of their systematic classification. He points out that a carnivorous animal, while having a digestive canal intended to absorb this kind of food, must also possess sharp teeth for tearing the meat, jaws adapted to these teeth, claws for clutching its prey, power of rapid motion, and good visual organs; a beast of prey thus never has hoofs or flat molars, for they suit only herbivorous animals. A practised naturalist should thus be able to determine from the shape of one single, suitably selected part of the body the whole of the animal's structure, habits, and place in the system. And in this system, therefore, only such animals should be grouped together as fully conform to one another, at least in the organs that are most essential to life. The creation of a system based entirely upon such conformity in the organs henceforth became one of the missions in life that Cuvier never let out of sight. For the time being, however, he contented himself with a system of grouping that differs from the old only in that the vertebrates and the invertebrates are distinguished from one another, and also that the lowest animals are grouped together under the name of Zoophyta — a name of which Lamarck strongly disapproved. Otherwise Cuvier retained the variously composed class Vermes, and he also made his anatomical comparisons cover the entire animal kingdom all at once. In doing so, however, he is at the very outset careful not to extend the comparisons in detail beyond what he can vouch for — in sharp contrast to the audacity of both Vicq d'Azyr and Lamarck, not to speak of Geoffroy Saint-Hilaire.

His studies of fossils

DURING the succeeding period Cuvier applied his method to the special investigations into fishes and molluscs that have been previously mentioned, but principally to mammals. Within this class he soon found himself engaged in a special field of research, the study of fossil forms. As is well known, Paris is situated in the centre of a calcareous district, in which the stone used for building-material is particularly rich in fossils. These had already attracted Buffon, for purposes of both observation and speculation; it was on the basis of material gathered from this and other districts that he formed his theory of the evolution of the earth and of the creatures living on it (Part II, p. 224). Cuvier, however, was the first to apply himself to a systematic exploration of the richly fossiliferous Paris area; with the assistance of his friend BROGNIART, he organized systematic excavations, in the course of which the location of the fossils was closely observed and the animal remains scattered about in each place were noted as carefully as possible. After this Cuvier began to apply his correlation theory to fossils; for every single bone that was discovered he searched in the neighbourhood for such bones as appeared from their structure to belong to the first one found, if the resultant skeleton nevertheless remained incomplete, he drew his conclusions from the structure of the available bones as to the habits of the animal, and from them again as to the structure of the bones that were missing; from the bone-structure it was afterwards possible to determine the construction of the soft parts. The accuracy of the method was still further ensured by the extinct animal's skeleton being regularly compared in detail

with the corresponding bones of closely related existing animals. Through
this method of reconstruction, which he expounded in his famous work
Recherches sur les ossemens fossiles (1812), Cuvier created the science of palæontology in the modern sense. And at the same time he largely reformed
the system of zoological classification by introducing fossil animals into it;
by account's being taken of the extinct animal forms the investigations into
the problem of affinity in the modern animal world have been far more firmly
substantiated and placed on a sounder basis than had been possible before,
and, moreover, they have led to results that to the systematists of earlier
times would have been utterly inconceivable. Of Cuvier's own investigations
in this field his comparative study of the order of elephant in particular has
won high commendation; he has here shown in the most convincing way
what results his new method is capable of giving. He begins by examining
the difference between the Indian and the African elephant, which were formerly grouped as a single species, but which, as he proves by comparison of
their teeth and bone-structure, are two widely different species; moreover,
he has established the fact that the extinct mammoth, of which he secured
as many remains as he possibly could, is in reality more closely related to
the Indian elephant than the latter is to the African. And, finally, he compares with existing elephants a number of other extinct types, which had
either been known before and described by Buffon, or else were in the form
of newly-discovered remains; among these fossils there are some from America that possess knobby molars, which warrants their being formed into a
new genus, Mastodon; the members of this genus must, however, be regarded as true elephants, for their heavy head postulates a short neck, and
this again, as well as the long legs, show that the animal must have possessed a trunk, while from the knobby molars it may be concluded that its
food was similar to that of the hippopotamus. Generally speaking, the Pach-
ydermata especially interested Cuvier; he studied their existing forms:
rhinoceros, hippopotamus, and tapir, in comparison with their extinct ancestors; of these he described a number of new genera, Palæotherium, Dino-
therium, etc. Even the small Hyrax he removed, for anatomical reasons,
from the rodents, with which it had previously been associated, to the proximity of the elephants — one of the most daring applications of his comparative-anatomical method. The first detailed descriptions of the American
giant sloth likewise originate from him. He also carried out some rather
sporadic studies of extinct birds and reptiles, which are of considerable value.

Cuvier as a geologist

THESE investigations into the existence and relationship of extinct animal
forms, however, brought Cuvier, as they had formerly brought Buffon, face
to face with the question: What changes have taken place in the character
of the earth's surface that have caused the dissimilarity between the animal

world of the past and that of the present? He has tried to give the answer in a survey of the process of the earth's development through the ages, which forms the introduction to his *Recherches* and which became his best-known and his most discussed work. Herein, basing his argument on material derived from the finest observations of earlier times, supplemented by his own, he seeks to prove that the changes in the character of the animal world have been caused by great catastrophes undergone by the earth's surface in prehistoric times. He at once takes it for granted that these changes had the character of violent catastrophes; that they were violent he considers to be established by the fact that stratifications which, judging from the nature of the fossils, have demonstrably taken place in the sea, are now found on the one hand elevated to enormous heights and on the other hand overthrown and inverted. That all this took place with great rapidity is obvious to his mind, not only from the sharp lines of demarcation shown by the various strata, but also from the fact that many of them contain such extraordinarily numerous animal remains that it can only be assumed that they died a sudden death as the result of upheavals which obliterated all life for the time being. The assumption of such catastrophic changes on the earth's surface also affords, in Cuvier's opinion, the best explanation as to why the animal species of ancient times have disappeared and been succeeded by new and entirely different forms. And as a further confirmation of this assumption he adduces the fact that most nations possess legends which tell of a mighty catastrophe, a flood that drowned all living creatures, and in which undoubtedly the mammoth and the other great land-animals living in Europe in earlier times perished.

His catastrophe theory

It is this universally known "catastrophe theory" that without doubt brings out both Cuvier's strength and his weakness as a natural-scientific thinker. He does not, however, deserve any very severe censure for the actual theory of these vast volcanic upheavals, with their resultant inundations; the geological material available for observation was still somewhat scanty and was, moreover, as far as French research was concerned, largely gathered from the Alps, with their greatly subverted formations, which even to this day are difficult to interpret, and which are peculiarly likely to induce a belief in violent upheavals. But there undoubtedly existed in Cuvier a very pronounced tendency to pursue the theories he had once set up to their uttermost conclusions — a tendency which may well be attributed to his marked aptitude for the formal side of science. Thus, he expressly declares that each stratum has its definite fossil species, which are characteristic of it and do not exist elsewhere; the catastrophes that took place entirely eradicated all then existent species; never has a species survived from one period to the next, so that species found in the form of fossils cannot be the same as those living now; the fossil remains of lions, bears, elephants, belong to species other than those at present existing; fossilized human remains do not exist, the human bones that have been declared to be such have become mixed up with fossil finds by accident. In this connexion he maintains, in opposition to Lamarck, that the periods of geological development have not by any means been going on for an indefinite time, but, on the contrary, during a fairly limited space of time, and that therefore the assumption that species change through habits and environment is unwarranted. If change of species were conceivable, it would be possible, he thinks, to come across transitions between extinct and now existing animal forms, but there are none. The immutability of species is to Cuvier's mind an absolute fact; he has not a trace of Linnæus's hesitation, which he expressed in his old age, in face of the difficulty of drawing a line of demarcation between the species; according to Cuvier's definition, species consist of "those individuals that originate from one another or from common parents and those which resemble them as much as one another." In this definition no mention is made of the creation of the species, which, it will be remembered, Linnæus took as his starting-point, but which, on the whole, Cuvier does not discuss at all. The assertion that so often occurs in literature that, in his view, life has been created anew after each catastrophe is utterly incorrect; on the contrary, he points out that isolated parts of the earth may have been spared on each occasion when it was laid waste, and that living creatures have propagated their species anew from these oases, which indeed he expressly applies to the human race. But as a rule Cuvier is not particularly interested in what might conceivably have happened; he adheres to what he considers to be definitely proved, leaving hypotheses to the "metaphysician." Nor is it true, as has also been stated of him, that he allowed religious beliefs to invade the realm of science; he certainly embraced with conviction the tenets of the Protestant Church, whose guardian he eventually became, but in his scientific arguments these doctrines play no part whatever; as a matter of fact Lamarck refers to the Creator far more often than Cuvier. It is true that the latter cites the First Book of Moses in support of his flood theory, but Chaldean and Egyptian documents are quoted at the same time and with exactly the same authority; and to ascribe historical authenticity to popular legends was an illusion shared at that time by most professional historians.

His Règne animal

In his work *Le Règne animal, distribué après son organisation*, which was published in 1817, Cuvier develops his ideas further. In the foreword he enters a strong protest against those who would arrange all living creatures in one series and declares that such a method is unforgivable. He emphatically denies that mammals, which come last in the system, are the lowest, or that the last mammal in the series is more perfect than the first bird. Here Cuvier has certainly laid his finger on one of the weakest points of the whole series-theory, as expounded by Bonnet and Lamarck, and has undeniably fore-stalled the conception of the relativity of the degree of evolution as held in modern times. And as an application of this doctrine of his he presents his famous type-grouping system. According to this, the animal kingdom is divided into four main groups, Vertebrata, Mollusca, Articulata, and Radiata. Within each of these groups there is a special "ground-plan" for the construction of the life-forms — a plan that appears modified in various ways in the different systematic categories within the type. Thus, the animals within the same type may be compared with one another, but there is no comparison between the ground-plans of the different types. This type theory is Cuvier's greatest contribution in the sphere of systematization and represents the farthest advance in animal classification since Linnæus; in fact, it represents, although in a somewhat modified form, the basis of all subsequent animal classification, and it is thanks to it that modern biology has been able to lay firmer foundations for the theory of descent than Lamarck succeeded in doing with his uniform evolutional series. But this is certainly due to the fact that in modern times it has been possible to compare ground-plan and organic structure even in animals belonging to different types. Here Cuvier was far too reluctant, as indeed he was in the application of his geological theory, to draw his conclusions from the observations on which he based his system.

Besides the account of the type theory, the work in question also contains a number of observations on general scientific problems, and here, as everywhere, Cuvier maintains the strictly critical attitude which to him was one of the essentials of life. He is a master in not giving utterance to more than he can stand for, and sometimes it is only in a roundabout way that one can guess his train of thought. Thus, he repeats his above-mentioned principle regarding life's quality of counteracting the manifestations of chemical affinity in the elements that form the body, and he adds that it would be irrational to assume that the force which acts in that way has a chemical nature. But he enunciates no definite vitalistic theory. With equal caution he expresses himself in regard to fertilization; how the embryo arises we cannot tell, we can only study its subsequent development. Similarly, the essence of the soul-life is a mystery; materialism is an arbitrary hypothesis, "so much the more so as philosophy cannot offer any direct proof of the true existence of matter." Here Cuvier has undoubtedly learnt from Kant; on the other hand, his analysis of the influence of sense-impressions upon the brain seems rather to have been influenced by Condillac and his school. However, the knowledge with which Cuvier applies the theories of the new chemistry to zoology represents a remarkable advance; in this respect he has laid the foundations upon which subsequent research has built further.

His controversy with Geoffroy Saint-Hilaire

IN the writings which Cuvier produced during the last years of his life there stands out with increasing distinctness his clear, though narrow, conception of the interrelation of animal types. This is especially conspicuous in his controversy with Geoffroy Saint-Hilaire, which attracted much attention in his own time and has been keenly debated by later generations, up to the present day. These two had indeed been friends from youth and had for a long time loyally collaborated. Gradually, however, their ways parted. Cuvier insisted more and more upon the truth of his principles regarding the immutability of species and the incomparability of types, while Geoffroy became more and more deeply engrossed in the study of the comparison of organs in different animal forms and speculations inferred therefrom upon the question of one uniform type of life. Cuvier did not like personal controversy; his objections to views of which he did not approve he invariably made without personal remarks and clothed in a sometimes rather haughty, but always courteous, style. While, then, Geoffroy for years propounded his fantastic comparisons between the segments of the Articulata and the vertebræ, tortoise-shell, and mussel-shell, which have been referred to in the foregoing, Cuvier never directly opposed these, to him, absurd ideas, but, on the other hand, formulated with increasing distinctness his own theories and his arguments against all that contradicted them. At last, however, came the inevitable clash, in the year 1830. Geoffroy had submitted to the Academy of Science a paper written by two younger scientists, containing a detailed comparison between ink-fish and vertebrates: the ink-fish was regarded as a vertebrate animal reflexed in the middle, with the anal opening pressed on to the head, possessing a diaphragm, cartilages corresponding to the cranial bones, and in general most of the organs peculiar to a vertebrate animal. The essay contained a direct attack on Cuvier, though this passage was struck out when sent to the press; but it was read before the Academy, and therefore called for a reply. This produced from Cuvier a courteous but sharp criticism dealing with the whole of Geoffroy's natural-scientific speculation; by illustrating side by side the organs of an ink-fish and of a vertebrate animal in the reflexed position, which it had been claimed constituted the likeness between them, he demonstrated the fundamental difference between the organs common to both, both in structural detail and position, showing, moreover, that many organs existing in the one form do not occur at all in the other. But, besides this, Cuvier rejected the entire fundamental principle on which Geoffroy based his research, at the same time emphasizing the latter's brilliant services as an exponent of the comparative anatomy of vertebrates. He made special reference to the similarities between bones of different vertebrates during the embryonic stage that Geoffroy had established, maintaining that the method employed in Geoffroy's investigations was by no means new, but originated in Aristotle, and that Geoffroy's talk of a uniform plan for the structure of the entire animal kingdom was mere empty words without any real meaning and without any equivalent in nature. This reply greatly offended Geoffroy, whereupon there started one of those long-drawn-out controversies so common in scientific history, when two persons of utterly different temperament fall foul of one another, and when the longer it lasts, the more unprofitable it becomes. Strikingly enough, Geoffroy at once desisted from maintaining the comparison between ink-fish and vertebrates; instead, he transferred the whole discussion to the sphere of the vertebrates. Similarly, he replaced the expression "*unité de plan*," to which Cuvier had objected, by the phrase "*théorie des analogues*," but at the same time emphatically declared that this theory was entirely new; for while the old comparative anatomy concerned itself merely with the form and function of an organ, the new theory took for comparison all the parts of which an organ was composed. As an instance of this he cited the hyoid bone in mammals, which he found to be composed of different parts in different animals, and also the opercular bones in fishes. Here Geoffroy was clearly referring to what we nowadays call homology — the likeness that exists in the evolutional history of certain organs, which warrants comparison in a manner different from what the mere functions of these organs would justify. But unfortunately he was too vague in his speculations to be able to give them plausible form; in the subsequent discussions before the Academy, Cuvier pointed out a great number of errors of detail even in Geoffroy's comparisons of the hyoid bones in the vertebrates, not to mention his idea that this bone occurred in crayfish. Further Geoffroy had a weakness for general philosophical speculation that must have seemed utterly absurd to his sober-minded opponent. In the introduction to a book in which he collected his contributions to the discussion, there occurs the following passage, which, like Schelling's, must be quoted in the original: *Pour cet ordre des considérations il n'est plus d'animaux divers. Un seul fait les domine, c'est comme un seul être qui apparaît. Il est, il réside dans l'Animalité; être abstrait, qui est tangible par nos sens sous des figures diverses.*" Such an expression of views was quite in Goethe's style — he, too, as is well known, took part in the dispute as a warm supporter of Geoffroy; he considered that the latter's cause was the cause of natural philosophy itself, and in this he was certainly right. For if there existed in Geoffroy's speculations advanced ideas of the greatest value even for modern comparative anatomy, they were nevertheless an expression for that same romantic natural philosophy, that same striving after an ideal unity in existence, which was then prevalent in Germany and which, in fact, Geoffroy persistently claimed in his favour. Cuvier, on the other hand, displaying his somewhat narrow range of vision, claimed the object of science to be an exact knowledge of natural phenomena. It is not worth while following the discussion between these two any further; it developed into a repetition of old arguments and a more and more stubborn adherence to statements once uttered.

The results of the dispute

NEVERTHELESS, one more point in connexion with this dispute must be noted. Among the assertions that Cuvier made and that Geoffroy at the very outset quotes with disapprobation, there is one which deserves attention, not only because it shows up Cuvier's limitations, but mainly because it emphasizes the contrast between the origin-of-species theories of earlier times and that of our own day. Cuvier says of the ink-fish: "They have not resulted from the development of other animals, nor has their own development produced any animal higher than themselves." In face of the modern theory of evolution the former of these statements is undoubtedly untrue, whereas the latter is correct. To Lamarck and Geoffroy both statements were equally untrue and they became even more excited over the latter than over the former. For what they were particularly in search of was just that connecting link between the highest form in each class and the lowest type in the next one — ink-fish and fishes, tortoises and birds, to name two examples. On this rock the earlier theories of origin regularly suffered shipwreck. The fact that the modern historian of evolution has learnt instead to search backwards in the evolutional series, in order to find among more primitive forms primary types for the separate, highly specialized groups, has been rendered possible owing to modern zoology's having accepted Cuvier's type theory, which avoids direct comparison between highly developed life-forms of different types — but thanks also to embryology, which Geoffroy endeavoured, though without success, to make the basis of his theory of comparison between organs of the same kind in different animal forms. It happens then, that both the parties to the dispute of 1830 have contributed to the creation of the modern view of natural evolution.

CHAPTER III

BICHAT AND HIS TISSUE THEORY

IN THE IMMEDIATELY PRECEDING SECTION of this work one chapter (chapter v) was devoted to giving an account of the two mutually opposed ideas as to the nature of life that were prevalent during the early part of the eighteenth century: the mechanistic, which conceived the phenomena of life from the purely mechanical point of view, and the vitalistic, which, represented by Stahl and his pupils, saw in the soul the real entity of life and regarded the body as existing for and through the soul. Curiously enough, Stahl's doctrine, the most markedly vitalistic of them all, won support particularly in France, where it was preserved and further developed by the medical faculty at Montpellier. It was especially Stahl's idea of the complex chemical composition of the body and the easy decomposability of its constituent parts, and the peculiar structure of them characteristic of different beings, that was developed by the Montpellier school. On the other hand, these scientists paid less attention to Stahl's speculations on the soul itself; rather, it was life, the life-force, that was believed to be the binding force that prevents the chemical components of the body from disintegrating. We have seen Stahl's theory recur in this form both in Humboldt and in Cuvier. In actual fact the sharp distinction between mechanism and vitalism was to a certain extent removed towards the close of the eighteenth century; the progress of chemistry made it necessary to consider other functions in the body besides the purely motive phenomena — a fact that even the most convinced mechanists eventually had to realize; while, on the other hand, a number of active natural forces were discovered — primarily the electric and the magnetic — of which earlier ages knew nothing and in face of which biology — whether vitalistic or mechanistic — was bound to adopt a definite attitude. As examples of the influence of these new discoveries may be mentioned, on one hand, GALVANI's experiments with electrical phenomena in the organism, which were continued by Humboldt and others, and on the other MESMER's investigations into "animal magnetism," or what we should nowadays call hypnotic phenomena. As a result of all these complications, that age's conception of life-phenomena became a mere groping in the dark; it was only after the discovery of the law of the indestructibility of energy that biology also gained a fresh basis on which to build, as a result of which it became possible to form a fresh mechanical conception of life-phenomena.

Among the scientists of the Montpellier school who disputed the prevailing mechanistic theory of life, which was maintained chiefly on the authority of Boerhaave, may be mentioned THÉOPHILE DE BORDEU (1722–76). The son of a doctor living in the south of France, he settled down in practice, after taking his degree, first in his home district and later in Paris. He wrote an account of his views on life-phenomena in a work entitled *On the Glands*. Contemporary physiologists of the mechanistic school of thought believed that glandular secretion was due simply to the mechanical pressure of adjacent muscles. Through a series of careful experiments and extensive investigations based thereon, Bordeu proves that mechanical compression cannot produce glandular secretion. This is due rather to the direct influence of the nerves leading to the glands. Through this nervous influence the supply of blood to the gland is increased and by means of a purely mechanical arrangement — Bordeu believes that he found openings capable of expanding or closing through the influence of the nerves — the follicles of that gland absorb out of the blood such fluids as are characteristic of the secretion. This individual power of the gland to absorb fluids that are suitable to it Bordeu names "sensation" and he ascribes to each organ in the body a special power of self-operation, a "tact," as he calls it; the stomach absorbs certain substances, and reacts against others by the process of vomiting; the eye has its special reaction against the outside world, and likewise the ear. Life proceeds as the result of co-operation between the individual action of all the organs. The brain and the nervous system control this co-operation; their action is expressed in the alternate contraction and expansion of their mass. Although Bordeu evinces great admiration for Stahl, it is nevertheless with extreme caution that he expresses any opinion on the question of the soul's relation to the body, just as in general he avoids entering into more abstract regions of thought.

We find, on the other hand, speculations of a markedly natural-philosophical character in a somewhat later pupil of the Montpellier school, PAUL JOSEPH BARTHEZ (1734–1806). He was first of all a practitioner, then professor at Montpellier, and finally chancellor of its university. Being of a pugnacious and irascible nature, he became involved in many a dispute, especially after he had taken sides with the aristocracy during the Revolution. Having been deprived of his post, he lived for a time as a private individual. He published his theoretical opinions in a work bearing the striking title of *Science de l'homme*. By way of introduction he gives an analysis of causality, which he afterwards examines with special reference to the cause of life. Barthez finds the ultimate cause of life to be inexplicable and considers that neither Boerhaave's nor Stahl's theories are satisfactory hypotheses or of any use to medical science; in their place he assumes a special "life-principle" as the foundation for the vital manifestations of all living creatures. In man this principle does not coincide with the conscious psychical life; it is rather a kind of general force, which contains within itself irritability, sensibility, and the other vital phenomena described by physiologists of that period. This speculative tendency, it appears, follows a line of thought exactly opposed to that which Bordeu pursued; he tried to regard the vital manifestations of the different organs as isolated phenomena, while Barthez sought above all for one common principle that would hold good for all manifestations of life.

These and other members of the Montpellier school during the eighteenth century would, however, scarcely deserve mention beyond the borders of France had not their work formed the basis on which Bichat built further. MARIE FRANÇOIS XAVIER BICHAT was born in 1771 at Jura, in eastern France. His father was a doctor of repute, and the son was from early youth destined for the same profession; after finishing school he studied surgery at a hospital at Lyons, but when that city was destroyed during the wars of the Revolution, he betook himself to Paris. There he found a patron in the surgeon Desault, with whom he worked both as a surgeon and as an anatomist. After the death of his benefactor, in 1795, he spent a couple of years editing his writings; in return he found in Desault's widow a maternal friend and a practical adviser. Bichat displayed throughout his short life an enthusiasm for science that has hardly ever been equalled; although he lived through the most exciting events of the French Revolution that occurred in his immediate neighbourhood, he was nevertheless able to devote himself entirely to his anatomical works. He spent his days and quite often his nights in the anatomical theatre, in order not to waste time. Nor did he care much about promotion; in 1797, however, he began to give lectures and four years later was appointed to a professorial chair, without having applied for it. Moreover, he carried on very intensive work as an author and took part wholeheartedly in the life of the medical faculty. In the spring of 1802, however, he was attacked by a malignant fever — whether as a result of septic poisoning or whether owing to some other infection is apparently not known — and he died in spite of the utmost care of his friends and colleagues; at the time of his death he had not yet reached his thirty-first year.

Bichat's character is described by his contemporaries as mild, modest, and unselfish. His writings testify to a general knowledge that was surprisingly extensive for such a young and extremely busy man, and yet his is the work by no means of an unpractical bookworm, but of one who took a deep interest in life and also observed a great deal in his fellow men. His works were written during the last four years of his life; in the early maturity of his creative powers he resembled Linnæus, as also in the fact that his genius was primarily formal and systematic. Bichat has introduced a new system into the science of anatomy and it is in this fact that his chief greatness lies.

In his writings Bichat shows himself above all a medical man; the functions of the body are invariably described in close relation to its morbid changes and to the manner in which they should be treated. Pathological anatomy engages his interest quite as much as normal anatomy, and postmortem examinations formed a considerable part of his practical work. He studied the various parts of the body in both its healthy and its diseased state, employing a number of different methods for the purpose; besides dissection he mentions drying, cooking, and maceration, as well as treatment with acids, alkalis, and alcohol. On the other hand, he did not use a microscope, for he thought that this only gave rise to fallacies and delusions. And yet it is as the founder of a science of microscopy that he won his highest fame. Another peculiar fact about him was that he despised the illustration as a means of reproducing the results of research; in his view, all representations, even plastic, illustrate only in an imperfect and misleading manner the facts which the research-worker wishes to convey. His writings do not contain a single illustration.

Bichat's conception of life

BICHAT's conception of life has always been regarded as vitalistic. Indeed, his theoretical fundamental view is unquestionably reminiscent of Stahl; life, says he, is "the sum of the functions that resist death." It is a far cry, however, between Bichat's so-called vitalism and Stahl's; the latter's theory of the soul as the ultimate end and conservator of the body Bichat strongly denies. Stahl, he declares, had realized the incompatibility between physical laws and animal functions, but because the soul was everything to him in explaining the functions of life, he failed to discover the laws of life. With equal emphasis, however, Bichat rejects Boerhaave's theory that life should be regarded as a purely mechanical process. "The true essence of life is unknown; it can only be studied through the phenomena it manifests"; and among these phenomena the most conspicuous is that previously mentioned — that it resists the influence of those forces which strive to disintegrate the body and which achieve their object as soon as life has departed.[1] As is well known, Stahl laid special stress upon the complexity of the body's chemical composition and its consequent easy decomposability as being something essential to life; this truth was appreciated by Bichat more than by any of his predecessors and was further developed on the basis of the epochmaking discoveries in chemistry in his own age. The primary lesson he learnt from Stahl, however, is the importance that different structural conditions have for the functions of the organism; in fact, the theory of structure represents Bichat's greatest contribution to the development of biology; it forms one of the corner-stones on which our conception of life and its manifestations rests.

---

[1] This definition recalls Humboldt's early ideas referred to above and may certainly be derived from the same source.

His classification of tissues

ACCORDING to Bichat's classification, the body is built up of tissues, which may be grouped in systems — for example, the bone system, the cartilage system, the muscle system. An organ is composed of several systems (e.g., the stomach, the lungs, the brain); several organs form an apparatus (e.g., the respiratory apparatus, the digestive apparatus). The knowledge of the tissue systems forms what Bichat calls "general anatomy," which he discusses in an important work;[2] upon this knowledge should be based the theory of organs or, as he calls it, descriptive anatomy. Bichat claims that this method of research and investigation is new, for, as he adds with justifiable self-appreciation, general anatomy hardly existed before he produced his works on the subject.

The tissues, Bichat declares, are the true conservators of the life of the body. He distinguishes between twenty-one different kinds of tissues — namely, (1) cellular (closely corresponding to what is now called retiform connective tissue); (2) the nervous tissue of animal life; (3) the nervous tissue of organic life; (4) arterial; (5) venous; (6) the tissue of exhalation; (7) absorbent; (8) bone-tissue; (9) medullary tissue (in the bones); (10) cartilaginous; (11) fibrous; (12) fibrocartilaginous; (13) animal musculature; (14) organic musculature; (15) mucous tissue; (16) serous; (17) synovial; (18) glandular; (19) dermoid; (20) epidermoid (dermis and epidermis); (21) capillary tissue. These tissues, however, are by no means alike everywhere; rather, they invariably possess the power to adapt themselves to the organs in which they are incorporated. The tissues are the true conservators of life; not each individual organ, as Bordeu asserted, but each individual tissue has individual life. Therefore diseases, in so far as they attack individual organs, are localized in their tissues; in abdominal catarrh it is the mucous membrane that is affected and not the muscles in the abdominal wall; in inflammation in the brain it is in most cases the cerebral membrane that is the seat of the disease. "If we would study a bodily function, we must consider the organ which performs that function from a general point of view, but if we would become acquainted with the vital qualities of the organ, we must disintegrate it" — that is, into the tissues of which it is formed. The tissues are thus the vehicles of life; in maintaining this view Bichat definitely dissociates himself from a number of earlier and contemporary scientists, who considered the fluids to be the true vital elements of the body.[3] But the vital qualities are not identical with the actual structure, for this still remains after life has departed; not even the fluids of the body are the same after death, and if we analyse them chemically we find only an equivalent

---

[2] The *Traité des membranes*, cited in the bibliography, may be regarded as a preliminary study to this work.

[3] See, for instance, Sömmerring's above-mentioned theory on the cerebral fluids.

to the anatomical nature of the dead body. Life consists rather in certain qualities, which the living tissues possess and which are not found in inanimate nature. Here Bichat takes as his starting-point Haller's previously mentioned theories of irritability and sensibility and he develops them further, expressing great appreciation of Haller, who, to his mind, had a far more correct view of life than Stahl. According to Bichat, sensibility is the characteristic quality of the nervous system; the muscular system displays a quality that he calls contractility; this has different characteristics in different organs and should not be confused with the tensibility that the tissues possess independently of life. But life manifests itself not only in these qualities, but in still another phenomenon, unknown in inanimate nature; this is called sympathy and expresses itself in the effect that the vital functions of the various organs have upon one another in conditions of sickness and health. Bichat made serious attempts to ascertain the nature of these vital phenomena by experimenting with living organs under various conditions. Thus he tried to analyse especially muscular contractility and distinguishes several categories thereof — namely, he holds that the muscle comes into action: (1) as the result of impulses from the brain received through the nerves — that is, normal contractility; it ceases if the nerve is severed; (2) through chemical or physical influences — that is, organic and sensible contractility or irritability; it ceases if the muscle is deadened (e.g., by opium); (3) through the fluids which the vascular systems convey to the muscle and which distend its minutest parts — that is, passive contractility or tonicity; it ceases as a result of death; (4) finally, the muscle contracts on being severed — that is, the contractility of tissue itself, which only ceases as a result of putrefaction. Of sensibility he distinguishes two categories — namely, "organic," which consists in the power of receiving an impression, and "animal," which not only receives the impression, but conveys it farther to a common centre and is thus a higher category of the previous one.

Organic and animal life

THE contrasted ideas, organic and animal, frequently referred to above, play an important part in Bichat's explanation of life. "Organic" are vegetable life and the unconscious life of animals; "animal" are the functions in animals that are controlled by the will of the individual and are consequently the more developed the higher the life is. Even in modern times one sometimes differentiates between animal organs, among which are included especially the nervous system and the motive organs, and the vegetative, among which are included the digestive, circulatory, respiratory, and excretal organs. Bichat, however, carries this differentiation to most absurd extremes when he consistently speaks of "the two lives," the animal and the organic, and assures us that the former's organs are always symmetrical and the latter's unsymmetrical, much labour being spent on trying to prove that lungs and kidneys are in reality unpaired. Similarly, he endeavours to show that the animal functions are always "harmonious," while the organic are "discordant," by which is meant that it makes no difference if one lung functions more or less than the other, whereas dissimilarity in the visual or auditory organs causes serious disturbances; the lack of a gift for music Bichat considers to be due to the fact that the ears possess different powers of hearing. He does not include the sexual organs in either category, because they serve the genus and not the individual. Of the psychical qualities, the intelligence belongs to animal life, while the passions are derived from organic life, from disturbances in the digestion and the blood-circulation. The community is thus only a development of animal life, while the passions have brought about all human disasters — revolutions and reigns of terror. In all this Bichat shows an inclination for sophistry, which not infrequently accompanies a highly developed genius for the purely formal. Several others of his systematic divisions are also by no means wholly successful. At all events, if only for the new system that he introduced into anatomical science, Bichat must be counted as one of the greatest pioneers of that science that have ever lived. Considerations of space forbid a more detailed account of his thorough exposition of the different tissue systems which he gives in his general anatomy, as also of his application of it to the theory of organs in his descriptive anatomy. His work contains no histology in the modern sense, but this is only natural, as he refuses to learn from microscopical observation; on the contrary, he dismisses with some compassion Leeuwenhoek's attempts at determining the form and size of muscular fibrillæ; the true nature of muscular fibre is unknown, and that is all there is to be said about it. He is far more interested in the chemical composition of the tissues, as far as it was possible to ascertain it in those days, and in their condition under processes of drying and maceration. It is at any rate the topography of the tissues that chiefly engages his attention; their finer quality did not concern him; for instance, Malpighi certainly knew more about the structure of the brain than he did. Bichat's greatness, then, lies in his having so convincingly proved the quality of the tissues as fundamental constituents of the body and its functions. He thereby placed the study of the phenomena of life on a definite basis, the value of which is best realized if we compare his tissue theory with the fantastic ideas of a "nervous fluid" and "microscopical life-units," in which the works of even the most brilliant biologists of the immediately preceding epoch abound. Even the terms "sensibility" and "contractility," which were invented by him, have been incorporated in modern terminology. And although his ideas of the application of physics and chemistry to biology must appear primitive to a modern reader, still, he had the eye of a genius for essentials in the contrast between animate and inanimate matter, which many a modern biologist has lacked. That he so strongly maintained this difference was certainly well justified as a reaction against those clumsy mechanical theories of life which were then being propounded by Lamarck and others. Nor indeed can it be said that Bichat prostituted his contractility idea to metaphysical or mystical speculations. His mind was, in fact, trained through studying the sceptical philosophers of the eighteenth century — he quotes both Condillac and Cabanis — and their criticism formed a sound counterbalance to the bold ideas which he learnt from Stahl and his school. Bichat knew to a fair degree how to retain the best of what he learnt from his predecessors and how to establish on the basis of the knowledge thus gained a consummate field of research of his own.

CHAPTER IV

THE FIRST HALF of the eighteenth century shows in general a lively development in the sphere of biology. The splendid progress made simultaneously in physics and chemistry created innumerable fresh problems also in the biological sciences; voyages of geographical exploration, which were made to hitherto unknown lands on the precedent of Humboldt, resulted in new material for investigation, which broadened existing ideas and broke down old systematical barriers — we need only mention such animals as the duck-billed platypus and the lung-fish in order to show clearly the importance of these discoveries — and, finally, the vast technical and economic progress of that epoch awakened an interest in the study of nature, which also proved of benefit to biology. Among the technical inventions that belong to this period may first of all be mentioned the improvement in the construction of the microscope, which alone has given mankind a knowledge of a whole series of hitherto unknown life-forms; the economic progress, again, rendered possible the instituting of collections such as earlier times had never dreamt of, as well as the carrying out of costly experiments on a large scale. As a result of all these circumstances, of which many keen scientists took full advantage, biology achieved more and more brilliant results as the years went by, with the consequent enhancement of its general cultural reputation — in spite of the indignant protests and the scornful rejection of the idealistic philosophers.

Progress of comparative anatomy

ONE branch of biological research which, more than any other, made rapid strides during the period under discussion was that of descriptive and comparative anatomy. Cuvier, its most brilliant precursor, had many pupils in different countries, both direct and indirect, who carried on the numerous ideas he brought out, and besides these there were many others whose research work resulted in valuable contributions towards the progress of science. A number of the most representative of these scientists will be dealt with in the present chapter.

CARL ASMUND RUDOLPHI was born in 1771 in Stockholm, of German parents. He studied medicine at Greifswald, at the German university of Swedish Pomerania, becoming professor in anatomy first there and afterwards in Berlin, where he worked until his death, in 1832. He founded the Berlin zoological museum, now one of the finest in the world, and did much extremely successful educational work; he was highly esteemed by pupils and friends on account of his zeal for science and his noble, almost supersensitive temperament. He could never get himself to perform vivisections and once declared that even the prospect of world-wide fame would not induce him to possess the insensitiveness of a Brunner.[1] At the same time he was severely critical towards others as well as towards himself and laboured hard for the abolition of the mysticism that natural philosophy had introduced into biology, so that his writings, in spite of a number of inaccuracies, give on the whole an impression of solid reality and of being far more modern than those of many of his famous contemporaries.

Rudolphi's work on parasites

IT is in three particular branches of biology that Rudolphi has made valuable contributions: parasitic research, comparative anatomy, and physiology. In the first-named he is a pioneer; his work *Entozoorum historia naturalis* has so considerably widened the knowledge of intestinal worms which Pallas founded that all subsequent research has been based on it; this work is the result of investigations into numerous germ-carrying animals and gives detailed accounts of the appearance and conditions of life of the parasites existing in them. Through this work the number of known species of intestinal parasites has tripled. But while Pallas believed that the parasites or their eggs enter the host from outside, Rudolphi is convinced that they are produced by diseased processes inside the hosts — a false idea, which is all the more curious because otherwise he most emphatically denies the possibility of spontaneous generation. In these circumstances it is natural that his account of the evolutional history of the intestinal parasites should be the weakest part of his work and far inferior to the masterly description that he gives of the different forms.

In a collection of short essays Rudolphi has recorded a number of valuable investigations in comparative anatomy. Of special interest are his comparative microscopical studies of the intestinal villi in different vertebrates. He gives an account in this work of a large number of different forms of these appendices, thereby increasing not only the knowledge of the tissue theory as created by Bichat, but also the use of the microscope. This investigation therefore deserves to be remembered as one of the first in the sphere of comparative histology. Another useful work of his was the study of the cerebral cavities, wherein he attacked Sömmerring's above-mentioned theory of the cerebral fluid's being the intellectual organ and his belief in connexion therewith that the nerve-fibres end in these cavities. Rudolphi considers that the

---

[1] JOHAN CONRAD BRUNNER (1653–1737), whose name is preserved in the glands of the duodenum called after him, made some well-known experiments with the extirpation of the pancreas of live dogs.

entire brain is the "intellectual organ" and maintains, in contrast to the attempts of earlier times to localize the soul, that such a complex phenomenon may very well require a complex organ as its foundation. On the other hand, Rudolphi's attempt at a new systematic grouping of the animal kingdom, into nerveless, single-nerved, and double-nerved, was not very successful and was forgotten long ago — a scheme, in fact, which had already been rejected by his contemporaries and which had to give way to Cuvier's better and more natural basis of classification.

His text-book on physiology

RUDOLPHI's most important work, however, was his *Grundriss der Physiologie*, which occupied his old age and was still unfinished at his death. This work best displays his great knowledge, founded on his own experiences and his wide reading, as well as his critical faculty and elevated mode of thought. Physiology, he says, is "the doctrine of the human organism." An organism without life is unthinkable; when the one is created, the second must be present; a dead body is not an organism, but only the remains of one. The classifying science of physiology is therefore anthropology: here Rudolphi, like Blumenbach, strongly insists upon man's dissimilarity from the apes, but considers in opposition to the latter that the human genus should be divided into species, not races. In this connexion he declares that human beings cannot have originated from one single pair — a sentiment which, during that reactionary period, it certainly required some courage to express. The chapter on anatomy that follows this section is one of the most brilliant parts of the work; the clear and concise manner in which he expounds the composition of the human body was unrivalled at the time; as compared with Bichat's tissue theory it represents a great advance, on account of both its simple and concise grouping of the tissues and its sound criticism; here we find no fantastic theories of life, no absurd speculations on symmetry, but a clear and sober account of the different parts of the body, which is mostly consistent with modern conceptions. In regard to the essence of life, Rudolphi associates himself most closely with Reil's theory of life as bound up with the form and mixture of matter, while Oken's and Schelling's extravagant ideas are utterly repudiated. Likewise, he rejects Stahl's theory of the soul as a cause of bodily phenomena: "*Das Dasein oder das Hinzutreten eines Geistes oder einer Seele zum Körper erklärt uns das Leben nicht im geringsten.*" On the other hand, he strongly emphasizes the importance of the chemical processes for the vital functions, in this associating himself with Berzelius's animal chemistry, which, thanks to his childhood's having been spent in Sweden, Rudolphi was able to read in the original. His account of the functions of the nervous system and the sensory organs is an extremely careful piece of work, which sharply criticizes all the mystical nonsense that was prevalent at that time: animal magnetism, interpretation of dreams, divining-rods, and the like. Occasionally, however, his criticism goes beyond the mark; he expresses doubt not only of Gall's theory of the nerves' leading to the grey matter of the brain, but also of the existence of sensory and motor nerves. The caution with which he treats current theories is at any rate attractive. Space forbids a more detailed account of his exposition of the digestive organs, respiration, and the musculature; these organs and their functions are described with the same thoroughness and care that marked his previous chapter. The whole work testifies, even in its incomplete form, to the strides that exact research had already made during the first decades of the nineteenth century, as regards both methods and results, foreshadowing the immense successes of subsequent periods.

Contemporary with Rudolphi there was working in Germany a scientist who made important contributions in the sphere of exact biology, although in his theoretical conceptions he maintained the point of view of the natural philosophy of the time. JOHANN FRIEDRICH MECKEL was born at Halle in 1781. Both his father and his grandfather had been professors in anatomy there; both had by their keenness and insight improved the anatomical collection existing there — especially the father, who had even declared in his will that his skeleton was to be mounted and set up in the museum. Young Meckel followed in their footsteps; having matriculated and taken his degree at Halle, he worked for a year or two with Cuvier and afterwards made a tour through Europe. At the age of twenty-five he was appointed professor in his native town, where he worked until his death, in 1833. During his lifetime he exercised great influence both as a teacher and as a research-worker, and not least as a result of his having undertaken the publication of Reil's above-mentioned *Archiv*, in which he developed his own ideas and gave accounts of a number of special investigations. These partly fall within the sphere of descriptive and comparative anatomy, and partly they are purely speculative and philosophical. In the former sphere Meckel proved a worthy pupil of Cuvier, while in the theoretical sphere he was undoubtedly influenced by Geoffroy Saint-Hilaire. The name of "the German Cuvier," which his contemporaries gave him, thus only partially corresponds to his point of view; what made him most worthy of the title was the work he did by exhortation and example to introduce into Germany the study of comparative anatomy, which in course of time was to reach its highest development in that very country. Amongst his contemporaries, at any rate, there was not one, with the exception of Cuvier, who had mastered the anatomy of all animal forms, both higher and lower, so thoroughly as he, though his most important investigations he carried out in the field of vertebrate anatomy.

Meckel's system of comparative anatomy

OF these specialized investigations of Meckel's the most exhaustive are his anatomical monographs on the ornithorynchus and the cassowary, but besides these he has made considerable contributions in several other spheres of anatomy: investigations into the development of the nervous system and the intestinal tube in the embryonic stage, into the structure of the brain of birds, into the intestinal villi, which reasons of space make it impossible to discuss in detail. Finally, in his later years he collected the results of his researches in a large work entitled *System der vergleichenden Anatomie*, which, like Rudolphi's *Physiology*, was never finished. The first part of it forms a summary of Meckel's theoretical speculations; the following sections deal with the structure of the individual organs.

If we turn from reading Rudolphi's *Physiology* to Meckel's general anatomy, the first impression will undoubtedly be that we have taken a long step backwards in time — from a critical and almost modern method of presentation back to romantic natural philosophy. The very foreword contains the statement that the "*Bildungsgesetze*" which govern the animal kingdom may be grouped under two main principles, "multiplicity" and "unity," the latter also being termed "reduction." And the exposition of these "formative laws" is introduced with the assertion that the form of the animal may be regarded either in itself and with reference to the physical force which is its origin, or with reference to the purpose intended to be served thereby and the creative spiritual force that forms the basis thereof. This at once is far more reminiscent of Schelling than of modern natural science, and yet we constantly come across proofs of the author's many-sided and radical knowledge of anatomy and of his genius for combining acquired facts. By the formative law of multiplicity is meant all those qualities which distinguish the life-forms from one another, and herein are included not only the characters that differentiate the species, genera, and higher groups, but also the qualities of the individual organs in the same and in different animals and the changes in them, such as are brought about by age, habits of life, and heredity. Under this heading, then, comes descriptive anatomy, while the "reduction" law embraces comparative anatomy, or, as Meckel says, the proofs that all formations in the animal kingdom are variations of one single type — that is, the same idea that Geoffroy and Goethe tried to develop.

Law of multiplicity

UNDER the law of multiplicity is described, to start with, the body's formation of tissues — a chapter in which Meckel does not compare well with Rudolphi in clarity and conciseness. As the fundamental substance for all the parts of the body he gives a solid matter, shaped like minute globes, which lie embedded in a fluid; these are clearly visible in the lower animals and in the embryos of higher animals, while in the latter themselves the fluid is coagulated and in conjunction with the globes forms fibres, membranes, and tissues. In this speculation Meckel is without doubt influenced by Caspar Friedrich Wolff, whom he held in high esteem and whose writings he translated into German. As regards the system, Meckel to a certain extent adopts an attitude of indecision; on the one hand he has to accept Cuvier's types, but on the other the whole object is to prove the possibility of one single primal type; consequently Meckel rejects the definite line that both Cuvier and Lamarck draw between vetebrates and invertebrates, and as an intermediary form between them he places the ink-fish, whose carapace is declared to be the rudiment of a backbone. Further, Meckel, like Lamarck, believes in a common spontaneous generation whereby a number of lower life-forms arise in various parts of the world and thus increase the number of existing species. In this, as in his general idea of the origin of life-forms, Meckel inclines towards Lamarck, his indebtedness to whom he acknowledges when quoting him. Each of them sought to produce a "history of natural creation," and it must be admitted that on this subject Meckel was able to derive advantage both from the work of his predecessors and from his own thorough knowledge of anatomy. Meckel's theory of origin thus contains many interesting and suggestive ideas of importance for the future of science, but it certainly contains also masses of weird fancies and ridiculous conclusions. What distinguishes Meckel's theory from Lamarck's — and even from Darwin's — is the fact that he does not assume one single cause of evolution, but a number of causes, and his exposition herein lacks the easy comprehensibility that characterizes both his predecessor's and his successor's work, which, indeed, explains why it is that he failed to win the same degree of popularity that they did. Among the causes of evolution, it is true, Meckel, like Lamarck, lays great stress on the influence of habit and environment, or, as he expresses it, the formative influence of mechanical forces. In this connexion he quotes stories of how bobtailed equine and canine races have been created as a result of the tails of the animals' ancestors having been docked, and in the same way mechanical pressure in the course of ages has produced the numerous interlacings and various divisions of the digestive canal, as also other changes in the internal organs. And he ascribes similar transforming force to light, heat, and electricity; in particular, the electrolysis of fluids, which was then newly discovered, leads him to indulge in fantastic speculations upon the effect of this force on the development of life-forms. Moreover, he drags into his theory of the formation of species the entire category of mostly unknown phenomena that give rise to malformations; thus, he cites the old belief that mothers can give birth to malformed children after "getting a fright," as at least a plausible cause of the appearance of new life-forms, and he finally mentions hybridization as an important cause of the arising of fresh species. To this factor in the evolution of life, however, which, as is well known, has excited special attention in modern times, he attributes utterly irrational effects: he believes in old tales of a cross between a cat and a hare, a cock and a duck. There is indeed far better justification for his expressly ascribing the abundance of species in the insect class to hybridization.

Law of unity

ALL the speculations just referred to, Meckel includes in the sphere of the law of multiplicity. As already mentioned, under the law of reduction come the proofs of the unity of the life-type. Here Meckel displays the whole of his extensive knowledge of anatomy, both good and bad; here he presents his most brilliant ideas and also indulges in the most ridiculous absurdities. The latter necessarily result from the false preconception of one single life-type and evolutional series, the same fatal constructions of thought, in fact, which led Lamarck and Geoffroy to such wild delusions. And it may be said that herein Meckel outrivals them to the same extent as his anatomical knowledge is more extensive than theirs. It is hardly worth while going further with him along these erratic courses; as when he compares the shell of the tortoise and of insects, or the papillæ on the tongue of cats and of snails, or when he likens the double malformations occurring in man, now generally called Siamese twins, to a colony of polypi. It would be better to ponder over the numerous ideas of great value for the future to which he gives expression in this connexion. Among these ideas that have been generally adopted in modern times may be mentioned his view that the lungs of the land vertebrates are derived from the air-bladders of fishes, his comparison between the male and the female sexual organs and his derivation of their several parts from an indifferent embryonic stage, his comparison of the segmentation of worms and articulates with metamerism in the vertebrates, and, above all, his foreshadowing of Haeckel's biogenetical organic law, when he declares that each higher animal during its embryonic development passes through the same forms as those that are lower in the evolutional series possess when fully formed. This theory of "the development of the special organism in accordance with the same laws as the entire animal series" he supports by a number of very ill-founded arguments — for instance, he makes the human liver undergo a crayfish and a mollusc stage — but also on reasoning which has been accepted by modern supporters of the theory. This doctrine, which had already been sweepingly rejected by Rudolphi, has certainly been very widely debated in our own day, but at all events it has had a highly stimulating effect upon research; its subsequent fate will be discussed later on. The theoretical conception that Meckel thus formed he applies in detail in the special part of his work wherein, with a many-sided and at the same time thorough knowledge of his subject, which but few had so far emulated, he describes the structure of the organic systems throughout the animal kingdom. With extreme thoroughness he discusses and compares the bone-structure in the vertebrates, describing especially the bones of fishes in great detail and making new discoveries in this latter field; the musculature and digestive canal, the respiratory and circulatory systems, are also carefully described. This radical detailed knowledge of Meckel's has had a considerable influence on the development of biological science. Moreover, his general conception of the phenomena of life has likewise made a deep impression. His theory of evolution deserves to be mentioned by the side of Lamarck's; he is at all events worthy to be named by the many who at the zenith of Darwinism sought for "pre-Darwinists," instead of Goethe, who never discussed the problem of species. He undoubtedly had a great influence also on biological research in Germany during the succeeding era; the very word "*Bildungsgesetz*" sounds quite familiar to anyone who has studied Haeckel's works for instance, in which the word "'law" occurs so often in passages where "hypothesis" or "theory" should have been written instead. Nor can there be any doubt that his penchant for bold comparisons and derivations has not been without its influence on the modern school, which has made the derivation of the organs of the higher animals from corresponding less developed forms the chief aim of biology. This morphogenetical school has largely applied Meckel's ideas, although employing an entirely different standard of criticism, so that Meckel, who so essentially belongs to romantic natural philosophy, stands as an intermediary between this school of research and that which included a man like Gegenbaur amongst its notable members.

Comparative anatomy in France

WHILE thus the exact biological method in Germany only gradually succeeded in getting rid of its connexion with natural philosophy, the same method in France had no difficulty in triumphing over the more speculative method of natural research represented by Lamarck and Geoffroy; it was Cuvier and Bichat who with their pioneer work determined the direction in which the scientists of the next generation were to follow with fair unanimity. Thanks, then, to these precursors, France acquired before other European countries a science of life-phenomena free from the romantic infusions of speculative philosophy, but, on the other hand, this science eventually became extremely conservative, and when the theory of the origin of species appeared, French research repudiated it with greater vehemence than any other. Of these French biologists from the beginning of the century some practised a purely experimental method; these will be dealt with below. As a direct pupil of Cuvier and Bichat, however, Blainville is worthy of mention, for he furthered the cause of comparative anatomy long and successfully.

HENRI MARIE DUCROTAY DE BLAINVILLE was born in 1777 at Arques, in Normandy, of noble family. He was brought up at a monastic school, and when it was closed down during the Revolution, he went to Paris, where he first of all applied himself to painting, apparently with but little enthusiasm or success. A mere chance — a lecture by Cuvier that he happened to go to hear — aroused his interest in biology; at about the age of thirty he became a pupil of Cuvier's, quickly obtained, through his recommendation, a post as assistant, and eventually, after having held certain other appointments, became his master's successor. By that time, however, the relations between them had long been severed, for the pupil's quick-tempered and unreasonable disposition could not reconcile itself to the calm considerateness of the master. Blainville himself, however, was a splendid teacher, attracting large audiences and devoting himself with indefatigable energy to his teaching and research work up to a ripe old age. He died in 1850.

Blainville was a biologist with many and varied interests. Amongst his works the *Manuel d'actinologie et de zoophytologie* is worthy of mention, in which he gives the results of his thorough investigations into the lowest animal forms, and further an *Osteographie*, which deserves to be coupled with Cuvier's investigations into the present-day and fossilized vertebrate animals. His theoretical conception of biology he has recorded in three works: *De l'organisation. des animaux*, *Cours de physiologie*, and *Histoire des sciences de l'organisme*. In these he presents a view of life-phenomena that is in many respects original and has proved of value for the future. The first work is introduced with a survey of the objects and methods of comparative anatomy. First of all an account is given of vegetable and animal chemistry, wherein, curiously enough, the universally accepted contrasts between the alimental process of plants and animals are considered doubtful. Then the animal is characterized as a "combination of certain organs, which give rise to certain forces — *inter alia*, a digestive and a motive force — assuming a definite form and influencing external surroundings in a definite manner." As methods of getting to know the structure of the animal are adduced observation, experiment, and a logical mode of thought, after which are named certain pioneers in this sphere, among whom one seeks in vain for the name of Cuvier — a characteristic touch, showing the pupil's bitter feelings towards his master.

Blainville's theory of cellular tissue

NEXT, an account is given of the structure of the animal body — this forming one of the most brilliant sections of the work. Here Blainville declares with a confidence such as was never shown previously that the cellular tissue is the fundamental substance of the animal organism, the element which is formed earliest and out of which all the organs are evolved. Of this tissue it is said that it represents the finest and most extensive element in the animal body and that it is formed of thin membranes, which cross one another, so that cystic interstices arise. As modifications of this tissue are mentioned the skin, the mucous membranes, connective tissue, bone, vascular systems, and finally — the most complex of all — muscle and nerves. Regarded from the modern point of view, Blainville's conception of the cellular tissue as the basis of the animal body is certainly imperfect, but when compared with the cellular-tissue theories of his predecessors, Caspar Friedrich Wolff and Bichat, it represents an undoubted advance; it is one of many examples of how knowledge in a given sphere progresses, as it were, fumblingly from generation to generation until, finally, the decisive word has been spoken. In regard to the conception of tissues and the part they play in the formation of the organs, Blainville otherwise associates himself closely with Bichat, and he likewise adopts the latter's theory of organic and animal life, which, however, he employs with considerably greater moderation than its creator. For the rest, he believes a living body to be a kind of chemical workshop wherein fresh molecules are constantly being conveyed and old ones removed, where the combination is never fixed, but always, so to speak, "*in nisu,*" resulting in constant motion and heat. This view of life is certainly not vitalistic, but Blainville nevertheless emphasizes in what follows the contrast between "general" and "vital" forces, both unknown as to their real nature, but the former far more measurable than the latter; both operate in the living body and life's intensity is dependent upon the ascendancy of the life-forces over the general forces. In this sphere, then, Blainville wavers somewhat between divergent principles, and on the whole he has been counted amongst the vitalists. The two primary qualities of life are, according to him, "*composition*" and "*décomposition*"; in the former is included the absorption of nourishment, in the latter not only excretion, but also reproduction. Among the alimental organs are also counted the organs of motion and, in general, everything that moves the external bodily form, to which Blainville ascribes a fundamental importance for all knowledge of animal life. His system rests entirely on this basis and thereby acquires a somewhat artificial character; nevertheless, it has done considerable service, chiefly in the fact that here for the first time a definite line of demarcation is drawn between amphibians and reptiles, which all subsequent research has confirmed. For the purposes of his special presentation of comparative anatomy Blainville prefers to go from the higher form to the lower, his arguments in justification of which are somewhat reminiscent of Lamarck's "*dégradation*" theory. In other respects, too, his treatment of comparative anatomy is based on a form of theoretical speculation that renders the actual method of presenting his subject highly artificial; it would, however, take too long to go more deeply into these questions. Undoubtedly Blainville's works contain, besides much that is absurd, a number of ideas of immense value, both in detail and as a whole. Among these may be specially mentioned the importance he attaches to the stages of embryologic development as a basis of comparison between the animal forms, a principle that, as is well known, has since proved of fundamental importance to comparative anatomy. For this fact science has to thank a number of works in the sphere of embryology that were brought out during the period now under discussion. To this sphere, therefore, we shall now proceed.

CHAPTER V

Ovists and animalculists

Tнᴇ ᴍᴏѕᴛ ɪᴍᴩᴏʀᴛᴀɴᴛ ꜰᴇᴀᴛᴜʀᴇѕ of the earlier history of embryology
have already been referred to in previous sections of this work — how
even Hippocrates had observed the development of the hen's egg, how
Aristotle studied the embryology of various animals, how Fabricius,
Harvey, Malpighi, and C. F. Wolff each in turn made valuable contributions
to the knowledge of the development of the embryo, especially in the hen's
egg, which had remained throughout the most easily available object of
investigation, but also in a number of other animals, chiefly, of course,
mammals. These inquiries were naturally much influenced by the speculations
on the process of development that succeeded one another during different
epochs; in this respect, the "preformation" theory, which prevailed for a
time, had a most unfavourable effect, seeing that its champions, for obvious
reasons, cared but little for practical observations of the embryonic development — everything having been ready-formed from the beginning, there
was, of course, no need for observation. This explains why the eighteenth
century, during which the preformation theory held sway, proved so barren
in embryological observations; instead of investigating, scientists wasted
their time in profitless speculations and controversies between ovists and
animalculists. Some of the latter certainly reached the height of absurdity
when they saw in the spermatozoa the true agents of reproduction, with the
consequence that they succeeded in distinguishing under the microscope in
every human *spermium*, with the aid of their imaginations, a complete
miniature human being with all the limbs ready formed. It was not until
the close of that century that embryology received a fresh impetus; C. F.
Wolff made a beginning with his, certainly exaggerated, epigenesis theory
and his embryological observations based thereon; Cuvier, who was interested in all biological problems, also made weighty contributions to this
subject; Blainville has just been mentioned as a promoter of embryological
research; nevertheless, science has mainly to thank certain German scientists
for its most important progress in this direction, progress which, in fact,
gave rise to an essentially new view of life-phenomena. As has often happened with pregnant problems in the history of science, this, too, was dealt
with simultaneously by several observers, each of whom contributed his

portion towards its solution. In the following we shall deal with Pander, who investigated the germ layers in the embryo of the hen; Rathke, who discovered the branchial slits in the embryo and the circulation in conjunction therewith; and, in another connexion, Purkinje, who discovered the germ-cell in the hen's egg. The first place among the creators of modern embryology, however, is held by von Baer, one of the great personalities in the field of research in the nineteenth century.

KARL ERNST VON BAER was born in 1792 on the Piep estate in Esthonia, the son of a landowner belonging to the German nobility of that country. Upon leaving school at Reval he matriculated at the recently founded University of Dorpat, where he applied himself to medicine. He continued his studies in branches of that subject in Vienna, but from there, realizing that he was not made for a doctor, he proceeded to Würzburg in order to be trained as a theoretical scientist. The teacher of anatomy in that university at the time was IGNAZ DÖLLINGER (1770–1841), a disciple of Schelling's, who, combining his master's passion for philosophical speculation with a radical knowledge of anatomy, especially interested himself and his pupils in problems of evolutional history. It was here that von Baer's research took the course in which he was eventually to go further than any of his contemporaries. After completing his studies he was appointed professor at Königsberg and carried out his principal investigations at that place. In 1834 he accepted an invitation to become an academician at St. Petersburg. There his activities won a brilliant success, and honours were lavished upon him accordingly. The scientific works of his old age, however, do not possess the same importance as those of his early years. This is due primarily to the fact that he to a great extent divided his interests; upon official request he undertook several journeys to different parts of the Russian Empire and as a result became interested in a number of different problems—anthropology, ethnography, archæology, and even etymology. Having been allowed to resign his post, he settled at Dorpat and died there in 1876. He was honoured by his countrymen in many ways; the Esthonian nobility published at their own expense a splendid *édition de luxe* of the autobiography that he had written in his old age, and after his death a bronze statue was raised to him at Dorpat.

Von Baer discovers the egg of mammals

THERE can be no doubt that von Baer won his greatest fame through the embryological works written in his youth. He published the results of these in a brochure entitled *De ovi mammalium genesi*, which came out in 1827, and a larger work, *Über Entwicklungsgeschichte der Tiere*, of the years 1828 and 1837. In the first-mentioned treatise he describes the most important of the discoveries he made in this field — namely, the egg of mammals in the ovary. Apart from the vague ideas of earlier scientists on this subject, de Graaf (Part II, p. 172) was the first to explain at all the conditions obtaining at the earliest stages of development of mammals. He described the follicles named after him in the ovary and believed these to be eggs; when later he discovered eggs in the uterus of a rabbit in a later stage of growth, he supposed that these had been moved thither from the ovary for their further development; he met with an insoluble difficulty, however, in the fact that the further advanced eggs in the uterus were smaller than the follicles, and, moreover, the latter proved to be not very constant, wherefore Haller, who carefully investigated the matter, assumed that the egg was formed out of the follicular fluid through coagulation. By carefully following the development of the egg in dogs, von Baer learnt to know its later stages, afterwards tracing its origin back by investigating a series of animals approaching nearer and nearer to the fertilization stage. Here he found the egg to be a minute yellowish cell inside the follicle, after which he was able to continue the study of its progressive development.

His pioneer work on evolution

BESIDES these studies in mammal embryology von Baer devoted himself to that ancient classical object of study in evolutional history — the hen's egg. He followed its evolution with the utmost care and published the results in the first section of the above-mentioned work *Über Entwicklungsgeschichte*, which, besides, summarizes all the then existing knowledge of the subject, thereby becoming a pioneer work on which all subsequent research has had to be based. The latter half of the work is a survey of the embryonic development of all the vertebrates. Through this book von Baer has created modern embryology, not only as an independent field of research, but also as an important branch of comparative anatomy and a means of proving the affinity of different animal forms. In the embryo of the chicken von Baer discovered the spinal cord, which he identified by comparison with the cord of the selachians. He also showed in its proper light Rathke's discovery of the gill-slits and gill-arches in the embryo. Further, he has explained the cause of the amnion formation — a discovery comparable with the foregoing — and has also given concise accounts of the development of the uro-genital apparatus of the formation of the lungs, of the various stages of development of the digestive canal and the nervous system. And finally he makes his proved ideas a basis for a general evolutional theory, which, it is true, contains a mass of natural-philosophical notions, but on the other hand gives a clear survey of the connexion in evolution and far excels all previous theoretical representations, although, since it is ignorant of the part played by the cells in the generative process of the organism, it cannot be called modern. For the process of fertilization is thus substituted a vague hypothesis in which the idea of a growth over and above the individual plays a conspicuous part: *"Zuerst wird die Möglichkeit eines neuen Tieres durch unmittelbares Wachsthum des mütterlichen Körpers gegeben. Es bleibt aber nur Teil. Durch die Befruchtung wird aus dem Teile ein Ganzes.*" This definition of fertilization is pure metaphysics, here closely akin to Aristotle's. In opposition to Wolff's one-sided epigenesis theory, von Baer declares that in reality no new formation takes place in the egg, but only transformation in the direction of increasing specialization. True, this theory is also based on purely speculative grounds — that the idea of the producing animal-form controls the development of the embryo — but it at any rate leads to a result that has been accepted even in modern times. Further, against Meckel's theory that during the embryonic stage higher animals pass through the form of lower animals, von Baer makes one particularly striking criticism; he maintains that no lower animals exist that really resemble the embryonic stages of the higher animals, but, on the other hand, the embryo of a higher animal and that of a lower animal resemble one another more closely than do the fully developed animals. The tissues of the embryo are less differentiated than those of the animal itself and are therefore more like the tissues in lower animals; but a fish embryo is from the very outset, and always remains, a fish, just as every vertebrate animal's embryo is from the beginning a vertebrate animal. Since, then, evolution involves a differentiation, the principle holds good that "the more dissimilar two animal forms are, the further we have to go back in evolutional history to find an agreement." The common primal form for all animals is the simple cell-form, the form of the egg and of the first embryonic stage. Starting from these considerations, von Baer emphatically rejects the Bonnet-Lamarckian theory of a uniform chain of development in the animal kingdom and instead associates himself with Cuvier's type theory, which he further develops. He maintains that a series of animals can in respect of the development of one organ be progressive, while another organ in the same animal series is regressive, and that one animal within a lower type may attain to a very high development in comparison with another form which comes low down within a higher type — the bee and the fish are cited, with the intelligence as the standard of comparison. Each organ, therefore, should be judged not only according to its definitive form, but also with reference to its evolutional history; the different animal types often possess organs having the same function, but an entirely different origin. He predicts that a comparative investigation of the different organs in the animal kingdom on this basis will prove of great importance for the future, and this prediction has of course been fulfilled.

His natural philosophy

SIDE by side with these progressive ideas, and often curiously interwoven with them, we find in von Baer a wealth of ideas of manifestly natural-philosophical origin which must certainly seem highly grotesque to the modern mind, but which nevertheless have undoubtedly had some influence on the biological speculation that was to come. A number of these ideas he had adopted from contemporary natural philosophers, as for instance Oken's theory of the head's being composed of vertebræ and of the jaws' having the qualities of ribs. Others again he has certainly invented himself — for example, the theory that the vertebrate animals are composed of a number of tubes lying inside one another in the shape of the figure 8. He reaches the extreme heights of Schellingism with a scheme he works out, according to which the three tubes lying within one another are each divided into a positive and a negative half; the epidermis, the muscles, and the nervous membrane are thus given a plus sign, while the cutis, the bones, and the nerve-fibres are denoted by a minus sign. Strangely enough, one comes across fancies of this kind in many of the eminent biologists of that period; some have already been mentioned, others will be discussed later on. It would be quite irrational, however, to accept these confessions of the weakness of the period for more than what they are; they are certainly striking from the point of view of cultural history, but their significance, whether for the activities of the scientists named or for their contribution to the general development of science, should at any rate not be exaggerated.

MARTIN HEINRICH RATHKE may claim an eminent place by the side of von Baer among the pioneers of embryology. He was born in 1793 at Danzig of a wealthy burgher family. He studied at Göttingen under Blumenbach, practised for a time as a doctor in his native town, was invited to Dorpat in 1829 as professor in physiology and thence, as von Baer's successor, to Königsberg, where he worked until his death, in 1860. Being personally a lovable character, keenly active on behalf of his science, constantly seeking to increase his knowledge by research work at home and abroad, he was universally esteemed by his colleagues and pupils.

Rathke's work as a biologist was many-sided and important. Among his earliest works was an article published in a journal, "On the Development of the Respiratory Organs in Birds and Mammals," which in point of value may be compared with von Baer's above-mentioned embryological treatises. It has already been pointed out that Rathke discovered the gill-slits in the embryo of birds and mammals, as well as the ramification of blood-vessels connected therewith. He further compared them with those of the fishes and followed their later development — how the gill-slits disappear and how the blood-vessels adapt themselves to the lungs, which are developed from an expansion of the front part of the digestive canal. He has also described and compared the development of the air-sacs of birds and the larynx of birds and mammals. In another work he gives an account of the so-called Wolffian bodies discovered by him, which he characterizes as "head kidneys" (pronephros), and which for a time performed the function of excretal organs, only to disappear later according as the true kidneys developed, while their efferent ducts in certain animals serve as part of the sexual organs.

In regard to his theoretical conception of these conditions, Rathke accepts without reservation Meckel's theory of the lower animal forms that the higher animals pass through during their embryonic life. A more independent theory is developed by him in a treatise *Über die rückschreitende Metamorphose der Tiere*, in which he records the results arrived at during his research work in comparative embryology. Although full of difficult abstract ideas, this article presents a view — original for the time at which it was written — of a hitherto neglected phenomenon in animal life. The phrase *"rückschreitende Metamorphose"* is characteristic of the age. Rathke makes a cautious reservation against the confusion of his ideas of metamorphosis with those of Goethe; in the form in which it is presented here, it has in view the regressive development which certain organs undergo during their embryonic and early life and which concludes with their total disappearance or survival as rudiments. Rathke cited examples of phenomena of this kind from the entire animal kingdom, but he supports his argument mainly on examples taken from the vertebrates, as for instance the gills and tail of the tadpole, the Wolffian organs, etc. He declares that such organs are either dissolved or are reabsorbed by the rest of the body, or else are knocked off and disappear; the latter occurs if they are horny and lack blood-vessels, the former if they possess blood-vessels by which their substance can be absorbed and made use of in the body. And it always happens, he declares, that such a disappearance of one organ is succeeded by the development of another which takes its place, as for instance the lungs of the frog, which develop according as the gills disappear, or the kidneys in the bird and mammal embryo, which take the place of the Wolffian bodies. Only an entirely altered mode of living during more advanced stages of development can bring about the total loss of previously existing organs, as happens in the parasitic crustaceans. It will at once be realized that here Rathke has shed light on one of the most important problems of modern biology.

Rathke's marine-zoological studies

BUT Rathke's activities were not merely confined to embryology; he is also one of those who have opened up for biological research the vast field that the seas have to offer. Cuvier was the first in modern times to draw the attention of science in this direction. Rathke, who was born and grew up in a seaport and who in the course of his travels — in Scandinavia amongst other countries — had his own interest in this field of research stimulated hereby, contributed largely towards awakening it in his countrymen. In this respect he gained much from the acquaintance he made in Bergen with the then priest MICHAEL SARS (1805–69), another of the pioneers of marine research, who presented him with many valuable animal-forms and gave him much information. Of Rathke's work in this sphere may be mentioned his careful description of the lancet-fish — that extremely primitive vertebrate animal

which has been so widely studied in modern times and which, not long before Rathke's period, had been thought to be a worm of a mollusc. Rathke's
monograph was the first detailed anatomical description of the animal and
was written with the same thoroughness that characterized his embryological investigations. His abundant and valuable writings further contain several
monographs on crustaceans, both independent and parasitic, molluscs and
worms, as also a number of monographs on the vertebrates — for example,
on the lemming — which are worthy of mention as examples of Rathke's
extensive and radical research-work.

The third in order of the above-mentioned pioneers of embryology was
HEINRICH CHRISTIAN PANDER (1794–1865). Born at Riga, the son of a wealthy
banker, he was able to give undivided attention to scientific work, which
had attracted him from an early age. He studied at Dorpat, Berlin, and Göttingen and afterwards, having made the acquaintance of von Baer, at Würzburg in the latter's company. There he carried out his pioneer work on the
development of the embryo of the chick, the results of which, thanks to
his great wealth, he was able to publish in a very fine edition. In 1826 he
became an academician at St. Petersburg, but the very next year he resigned
and lived for some time as a landowner in the neighbourhood of Riga. In
1842 he entered the Russian Mining Board, after which he published only
works on geology.

Discovery of the germinal layers of the hen's embryo

PANDER's above-mentioned treatises on the development of the hen's egg,
which were published as early as 1817, were the fruits of work carried out
under the guidance of Döllinger and with the collaboration of von Baer.
They thus represent to a certain extent the basis on which the latter scientist worked further, although there is no doubt that while they were being
written, the younger of the two friends came under the influence of the elder.
As Pander's greatest service should be recorded the fact that, taking as his
starting-point the preliminary work of Malpighi and C. F. Wolff, he distinguishes the different layers out of which the organs of the chicken embryo
are built up. These layers, which, following Wolff, he names "*Blätter*" — a
relic of the latter's attempts to compare plants and animals anatomically —
were afterwards further investigated by von Baer and have since then formed
the foundations of modern embryology. In his presentation of the continued
course of development of the embryo, however, Pander is not to be compared
with his above-mentioned contemporaries and he was unable to follow up
the promising ideas that he had produced in his early work. A work *Vergleichende Osteologie*, which he published as an *édition de luxe* in collaboration
with the artist d'Alton during his visit to Germany, and which attracted
the attention of Goethe, has not the same interest as his embryological
works, and upon his return home he divided his genius between a number

of small tasks, the results of which have not attracted the attention of posterity. There was created, however, by the scientists described above, a new branch of comparative morphology, which proved of the greatest importance for the development of biology in general, in that it made possible a far more universal and extensive study of the organ in living creatures than had been conceivable before, embracing not only the present characteristics of the organs, but also their evolutionary history, thus proving not only a morphological, but also a morphogenetical subject of research. From this period we can also consider that the advent of comparative anatomy in the modern sense dates, and its development during the succeeding decades, especially in Germany, was splendid. It was this line of research that really dominated biological science in that country during the greater part of the nineteenth century. But it was certainly not merely the embryological discoveries that produced this fresh impetus. In other spheres, too, there opened up for biological research, as a result of new methods and new facts, vistas of an extent hitherto unknown. In particular, there were two methods, already previously known, it is true, but not sufficiently appreciated by the immediately preceding generation, which were adopted at this period with renewed interest and considerable improvements, and which produced results that fundamentally reformed the views on life-phenomena — namely, the experimental method and microscopy.

CHAPTER VI

THE DEVELOPMENT OF EXPERIMENTAL RESEARCH AND ITS
APPLICATION TO COMPARATIVE BIOLOGY

The development of organic chemistry

IN THE PREVIOUS SECTION it has been pointed out that during the eighteenth century the experimental method was applied with great success both in animal and in vegetable biology; names such as Haller and Spallanzani, Hales and Ingenhousz are sufficient proof of this. During the reign of romantic natural philosophy, conditions were different; the representatives of that school, who imagined that they could solve all the riddles of existence by speculation, deeply scorned experiment, which they considered led to nothing but fruitless artifice. Indeed, the physiological works which saw the light during this epoch are for the most part purely speculative or else devoted to morphological problems. Gradually, however, reason came into its own even in this sphere; the immense success which the experimental method brought to contemporary physics and chemistry induced attempts at applying that method also to biology. And this all the more so as during the immediately preceding period eminent scientists had begun to apply themselves with considerable success to the study of the chemical composition of living organisms. A glance at the development that had taken place in that branch of chemistry may therefore not be out of place in this connexion.

CARL WILHELM SCHEELE, mentioned in the previous section as a pioneer in gas chemistry, is worthy to be called the founder also of animal and vegetable chemistry. German in origin and upbringing — he was born at Stralsund in 1742 — in his youth he adopted the profession of apothecary in Sweden, finally settling at Köping, a small town, where he died in 1786. During a brief life spent in very poor circumstances he managed to carry out unusually fruitful research work. As one part of his work it may be mentioned that he subjected to a more thorough chemical revision than anyone had done before him elements from the animal and vegetable kingdom; among the results he obtained was the discovery of lactic acid, cyanuric acid, hydrocyanic acid, and uric acid, and, further, glycerine, citric acid, and malic acid, not to mention other equally important elements. Lavoisier, it will be remembered, also studied phenomena in the animal and vegetable kingdoms. A successor to him was ANTOINE FRANÇOIS DE FOURCROY (1755–1809).

ALEXANDER VON HUMBOLDT

JOHANN WOLFGANG GOETHE

Brought up in poverty, he received financial aid for his medical studies from Vicq d'Azyr and when still a young man became a professor of chemistry. He was a zealous supporter of the Revolution, was a member of the famous committee of public safety, and eventually became director-general of instruction. In a handbook that he wrote on chemistry, *Philosophie chimique*, as well as in a number of other writings, he deals exhaustively with animal chemistry; the so-called "chemical philosophy" excited considerable attention and was translated into several languages. In this work an account is given of the chemical composition of plants and animals. The essential difference between substances derived from the vegetable and the animal kingdom is claimed to be the latter's azotic content. Vegetable elements are divided into sixteen separate substances, including gum, sugar, fatty and fugitive oils, resin, etc. The elements derived from the animal kingdom form three groups: albumen, lime, and fibrin; characteristic of both the vegetable and animal kingdoms are fermentative processes of various kinds, whereof is described the fermenting of wine and vinegar, and putrefaction. Besides this grouping of the components of living creatures, Fourcroy has given us the results of a large number of valuable special investigations into animal substances: milk, blood, gall, serum, and others. His services have been fully acknowledged by Berzelius, who was the great discoverer in this sphere as much as he was in chemistry in general.

JÖNS JAKOB BERZELIUS was born in 1779 at Väversunda in Östergötland, Sweden, the son of a poor priest. Being left an orphan at an early age, he had to carry out his studies under severe privations, first of all earning his living by private teaching and practising as an apothecary, and later, having taken the degree of bachelor of medicine, in medical practice. In 1809 he became a doctor of medicine and obtained an appointment as doctor at the College of Surgery in Stockholm (out of which, thanks mainly to his activities, the Carolinian Institute was eventually founded). Here he had a laboratory in which he was able to apply himself to those chemical investigations which in a few years were to bring him world-wide fame. An unrivalled capacity for work made it possible for him, apart from his research and literary work, to devote himself to a large number of public duties — thus, he became permanent secretary to the Academy of Science, whose annual report he used to write in a masterly style — and he was also the recipient of innumerable honours both at home and abroad. Ill health weakened his powers during the last years of his life. He died in 1848.

Berzelius's animal chemistry

BERZELIUS's work as the creator of the science of chemistry in general is universally known, and falls, moreover, outside the scope of this history. In actual fact, he mastered the whole of chemistry as no one else has ever done since his time, and he created something new in all the spheres in which he worked. One of his most important contributions, however, is his investigation of the substances that are produced by life on the earth. In his *Lectures on Animal Chemistry*, published during the years 1806–8, he expounds his conception of life-phenomena in general and, besides, records a number of fresh facts in the sphere of organic analysis, which he later still further augmented. These purely chemical investigations into the composition of blood, gall, milk, bone, fat, and many other elements, really belong to the science of chemistry rather than to biology; although they exercised a fundamental influence on the knowledge of life and the functions of the living body during the succeeding period, a detailed account of their results would hardly be in place here, but the views on the phenomena of life which Berzelius formed as a result of these investigations have naturally played an important part, both in his own time and in the age that followed; an account of them is therefore of interest, quite apart from the interest always excited by ideas expressed by one of the great pioneering personalities of the world.

As sources from which he derived his view of the phenomena of life Berzelius mentions first of all the works of Bichat and Fourcroy; in particular, the former's explanation of the tissues of the body and their functions formed the basis of his entire conception of life-phenomena. Reil, too, contributed largely thereto, while Fourcroy's role was primarily that of a precursor in the purely chemical sphere. What chiefly distinguishes Berzelius's general conception of nature from the conceptions of his predecessors is his severe criticism of and opposition to any kind of "hypothesis-mongering." While contemporary natural philosophy created brilliant thought-systems, Berzelius introduces his "animal chemistry" with the words: "I have everywhere sought to avoid hypothesis, and where I have at any time ventured to make insignificant guesses, they are all of such a nature that they will soon be decided by experience. I prefer to say: 'This is entirely unknown to us,' than to try by means of a number of probabilities to gloze over a gap in our knowledge."

Berzelius rejects vitalism

In conformity with this principle he rejects the vitalistic theories of his age: "Life does not lie in any extraneous essence deposited in an organic or living body; its origin must be sought in the common fundamental forces of primal elements, and this is a necessary consequence of the condition wherein the elements of the living body are combined." And further on he says: "All, therefore, that we explain with the words '*own vitality*' is entirely unexplained and it is an illusion if they are given any other meaning than that of a still unknown mechanico-chemical process." Even the functions of the soul are explained in the same way: "Unreasonable as it may seem . . . nevertheless, our judgment, our memory, our reflections, as well as other functions of the brain, are organic chemical processes, as well as, for instance, those of the abdomen, the intestines, the lungs, the glands, etc.; but here chemistry rises to a higher plane, where our spiritual research can never reach her.'' Even La Mettrie himself has never expressed himself more clearly, and though he adds in a note on materialism that ''it is not in accord with our hope and our practical innate feeling regarding the immortality of soul,'' yet, in view of what has been said above, this makes but little impression. It is obvious that a man like Israel Hwasser must have felt antipathy for the man and the university whence such words originated.

When, later, it becomes a question of applying in detail the mechanical theory in life, Berzelius, like so many of his predecessors and successors, gives way to the temptation to simplify too much, and in spite of his honest endeavours he is unable to free himself from speculative construction. Under the heading ''Principle Components of the Animal Body'' he declares that ''the phenomena of animal life are divided into two systems, the *nerves* and the *other organs*. Life is really placed in the former, and through them the animal lives for the moment. The latter, on the other hand, promote those conditions whereby the animal is to live in the immediate future.'' The nerve system thus represents the essential difference between organic and inorganic nature; the plants, therefore, must also possess nerves, although they are still unknown. This contrast between the nervous system as the real conservator of life and the other organs, which are called instrumental organs, is just as unnatural as Bichat's theory of the two lives, which Berzelius criticizes, and the idea that life cannot exist except in a nervous system is still more unfortunate. The functions of the nervous system are, according to Berzelius, unknown, and he utters a serious warning against adducing electrical and other forces to explain what one does not understand. He has only vague ideas of the structure of the brain and the nerves; he takes no account of Malpighi's microscopical discoveries and Swedenborg's application of them. Moreover, he makes a number of unfelicitous assertions regarding the vascular system; he believes that the capillaries open out between the organs and that the latter grow up as a result of a stratification of solid matter around the opening — ''a kind of crystallization.'' In this connexion it may be mentioned that he believes in the spontaneous generation of lower animals. On the other hand, his description of fertilization is clear and definite; he frankly acknowledges the inability of science at that time to explain the process, and his ideas on the development of the egg are extraordinarily clear, considering they were arrived at twenty years before von Baer's epoch-making discovery.

Berzelius, therefore, cannot be said to have been a very deep thinker in the sphere of theoretical natural science, as was Galileo, for example, but his honest and modest acknowledgment of the limitations of natural science and his opposition to any kind of hypotheses doubtless to a great extent cleared the atmosphere in a generation that had been befogged by the fantastic ideas of natural-philosophical speculation, all the more so as he certainly possessed all the authority to which a discoverer of the highest rank in the world of natural phenomena can lay claim. His pupils were numerous and brilliant; the most gifted chemists of the following generation received their training from him and undoubtedly they disseminated the master's aim to try to establish the actual phenomena in nature without any "explanations by means of *qualitates fere occultæ*."

Experiments on live animals

THE experimental research of which an account has been given above concerned the chemical composition of the various organs. But the functions of the organs as well — their vital manifestations, each separately and in collaboration — were during this period the subject of radical experimental investigations. In this field of research Haller was a pioneer; in his footsteps there followed an increasing number of scientists who sought by means of experiment on live animals — that is, vivisections — to ascertain the course of events in animal life, both in the isolated organs and in groups thereof, to an ever-increasing extent. These experiments led to the discovery that the actual phenomena of life were themselves bound by laws to an extent hitherto undreamt of; it was found that they could be made the subject of exact research just like the chemical and physical processes in inanimate nature. As pioneers in this sphere French and English scientists were conspicuous; in Germany, where formerly natural philosophy and afterwards comparative morphology predominated, the representatives of experimental research were also comparative morphologists, at least those of the older generation, wherefore it is often hard to decide to which category this or that scientist rightfully belongs.

CHARLES BELL was born in 1774 in the neighbourhood of Edinburgh, the son of a country parson. Having studied in circumstances of great poverty and taken his degree at Edinburgh University, he came as a doctor to London, where he rapidly gained a great reputation, becoming professor of surgery and curator of the Hunter Museum, mentioned in a previous section of this book. In his later years he returned to his native country as professor of anatomy and died there in 1842. He enjoyed a universal reputation as a clever doctor, a distinguished university tutor, and a warmly religious personality; he was the recipient of an extraordinary number of honours. As a scientific author he was very productive; he published a text-book on general anatomy that gained a wide reputation, and also a large number of papers on special subjects. In several of these latter his Christian piety is a marked feature, particularly in an essay on the structure and function of the hand, which represents from beginning to end a hymn of praise to the wisdom, power,

and love of the Creator. What chiefly made his name famous to later generations, however, was his original investigations in the sphere of nerve-
physiology. Even Galen had been aware that some nerves had to do with
motion, while others received sense-impressions, and since then the nervous
system had been investigated by innumerable scientists with increasing accuracy in the matter of detail. But there still remained the question of how
the nerves running from the spinal marrow can act as intermediaries between not only motive impulses, but also sense-impressions; to this question
no answer had been given or else only very unsatisfactory ones. Bell recorded
his experiment on this subject in a brief paper entitled *Idea of a New Anatomy of the Brain*. He had a few copies of this printed in 1811 and presented
them to his friends. In this work he describes how he severed the posterior
root of a medullary nerve without causing any muscular contraction, whereas
the act of touching the anterior root caused convulsions in the muscles. From
this he concludes that the medullary nerves have a double function, due to
their double roots. The idea, however, is hinted at rather than followed up,
and indeed this small brochure contains several similar suggestions — regarding the specific mental energies, a problem which J. Müller afterwards
examined thoroughly, as well as ideas on the localizations in the great brain
and the connexion between them, an inquiry which at the time held wide
possibilities. During the next decade, however, Bell did not follow the line
he had opened up, with the result that others got in advance of him, especially Magendie in Paris. Nevertheless, Bell's work received high praise
later on, which it deserves without a doubt.

FRANÇOIS MAGENDIE was born in 1785 at Bordeaux, where his father
was a surgeon. He went to Paris to study medicine, and after studying in
great poverty and anxiety and having successfully passed his examinations, he
became prosector at the anatomical institute, then a hospital doctor, and
finally a professor at the Collège de France, where he worked with immense
success as a lecturer, gathering a number of distinguished pupils around him.
He systematically originated the method of ascertaining the vital phenomena
by operations performed on live animals, thereby exciting both admiration
and disgust amongst his contemporaries; the sensitive Rudolphi mentions
his experiments with horror, and his notorious ruthlessness has even gone
down to posterity. Perhaps this has been exaggerated owing to his manners
towards his fellow men; his manners were rough and his self-esteem and
scorn for the opinions of others often reached outrageous heights. But if
he was hard on others, he did not spare himself; when for the first time
cholera raged in Paris, he voluntarily undertook the task of ministering to
the sick among the poorest of the population, defying both the risk of infection and the fanaticism of the populace, who believed that the doctors
had caused the disease by poisoning the drinking-water. Magendie was

active both as a doctor and as a research-worker until the time of his death, which took place in 1855.

Magendie's criticism of contemporary vitalism

EVEN in his earliest writings Magendie appeared as a keen opponent of Bichat's vitalism, and the whole of his subsequent work turns on his insistence on the possibility and the necessity of applying to the phenomena of life the laws that hold good in physics and chemistry, and, in connexion therewith, the experimental method that brought these sciences such success. But he also possessed from the very outset a keen eye for the limitations of that method; he repeatedly declares that it is not possible to explain all life-phenomena merely as physical and chemical processes. The life-manifestations of the nervous system in particular are called by him "vital" and are excepted from the mode of thought that he applies to other life-processes. These vital phenomena are to his mind inexplicable for the very reason that the physical-chemical principles cannot be applied to them; simply to invent on their account a number of theories of a speculative kind he considers to be harmful; he hopes rather that in future the exact method will be applicable also to as much as possible of this field of research, for that method alone, he says, can produce results of lasting value. He would take as the basis of his research the method that was created by Galileo and perfected by Newton — "to observe and to question nature by means of experiment." Among his predecessors in the biological sphere he names first of all Borelli, whose previously described investigations into the mechanism of animal movements he cites with admiration and pursued still further. He applies the same mechanical idea especially to the respiration and the circulation of the blood, at the same time endeavouring to take as full advantage as possible of the progress made by chemistry during his age. In doing so he came into constant disagreement with Bichat, whose theory of the independent life-manifestations of the various organs he desired to replace as far as possible by purely mechanical processes such as could be confirmed by experiment or observation. He considers hypotheses in general to be useless; facts alone have any scientific value and what cannot be explained with their aid must for the time being remain unexplained. This scepticism, which he pursues with absolute consistency, undoubtedly proved a useful counterbalance to the unbridled speculation of preceding periods. True, even his criticism could sometimes lead him astray, as when he accepts Spallanzani's assertion that the spermatozoa play no part in fertilization, and yet doubts von Baer's discovery of the egg in mammals; but on the whole his conception of nature is both sound and keen-sighted. It rests, too, on a broad basis; although it is the vital manifestations of the human body that he studied most carefully, nevertheless he makes constant reference to other animal forms and he supplements his text-book on physiology with a systematic survey of the entire animal kingdom.

Magendie's greatest service to biology, however, is not on the theoretical side, useful though his criticism of his contemporaries' hypothetical ideas was; his most valuable positive contribution was without doubt the experimental technique that he created. In working it out he took advantage of experiences gained from methods of physics and chemistry as well as from those of surgery and internal medicine, thereby originating an experimental procedure that even to this day forms the basis of the method of research in physiology. He employs it in a number of important processes in the higher animal life, especially in connexion with the phenomena of circulation and resorption. To both these studies he successfully applied his mechanical system of thought; to him the circulatory apparatus, as also the respiratory system, was a mechanism, the operation of which should be calculable, and indeed was partly calculated by him. With regard to resorption, he took advantage of DUTROCHET's then recent discovery of the osmotic processes. Magendie's investigations into the nervous processes, however, brought him the greatest fame; independently of Bell he took up the problem of the roots of the medullary nerve, studying it both experimentally and theoretically, with far greater attention to detail than his predecessor. He actually claimed as his own the discovery of the physiology of the sensory and motor nerve-roots, and if by this is meant thorough investigation into the subject, he is no doubt justified in his claim. Bell, however, appeared once more and maintained his old claims; this led, as usual, to a not very edifying controversy between Magendie and his predecessor. As a matter of fact, the contrast between them, as far as regards their general conceptions of nature, was as wide as it could possibly be — Bell looking to the glory of God in his scientific results, and Magendie refusing to accept any other explanation of nature than the purely mechanical. In actual fact, both have performed considerable services in the sphere mentioned — Bell in having been the first to determine the bearing of the problem and to establish the different functions of the two nerve-roots in the medulla, and Magendie in having dealt with the problem experimentally throughout and established the fact in all its details. Bell is said to have been deterred by the painfulness of the experiment from pursuing it beyond establishing the above-mentioned fact in the case of one live subject; Magendie on the other hand, who had no qualms on that score, carried out the investigation from as many sides as possible.

Magendie gathered round him several distinguished pupils. Among these may be mentioned MARIE JEAN PIERRE FLOURENS (1794–1867), who continued his master's work in the sphere of nerve-physiology, besides making valuable contributions to the knowledge of the function of the skin and several other organic systems. Chief among French physiologists, however, must be named CLAUDE BERNARD. He was born at Saint-Julien, near the Rhone, in

1813, of poor peasant parents, but through the kindness of a priest was given an opportunity of studying. For a time he served as an apothecary's apprentice at Lyons, afterwards trying his hand at literature, but eventually he devoted himself to medical studies, which he completed in Paris under great privations. He was saved from the necessity of having to seek a living as a country doctor by Magendie, who discovered his brilliant genius and made him his assistant. After holding a number of other posts he became his master's successor, but in his old age exchanged that appointment for a professorship at the Jardin des Plantes. He had to carry out his experiments for a great number of years in chilly and damp premises, with the result that he contracted an illness that prevented him from doing any practical work for about ten years. Instead he spent this period of his life in literary work on subjects in the theoretical sphere, his writings being very highly thought of. Finally he succumbed to his illness in 1878. During the last years of his life he enjoyed a brilliant reputation; he was the recipient of many high distinctions, both at home and abroad, and his funeral was undertaken at the expense of the French Government. A competent judge (Chr. Lovén) declared at his death that the greatest physiologist of the age had passed away, and subsequent generations have not challenged that judgment. And he was no less great as a personality; he was of a warm-hearted and modest nature, and at the same time a brilliant writer and an eloquent speaker. His experiments were carried out in less brutal fashion than Magendie's, but they were as deeply thought out and, if possible, even richer in results than the latter's.

Bernard's theoretical conception

As will have been seen from the above, Bernard's research work comprises not only a series of experimental investigations, but also, during the latter years of his life, a collection of theoretical speculations upon the phenomena of life. He had already formulated his theoretical conceptions in their main features in his early youth, however, and throughout his life worked for the creation of a completely elaborated theory of life. From the beginning he rejects as emphatically as Magendie the vitalism of the Bichat-Cuvier school, though he is not content, like his predecessor, with a general attitude of scepticism, but endeavours to analyse the problem of what life really is. Here he arrives at the conclusion that it is not possible to define what life is, but only to analyse its manifestations — that is, Galileo's principle. He groups the manifestations of life under the following headings: "*Organisation, Génération, Nutrition, Évolution.*" Of these he finds the last to be both the most characteristic of life and the most difficult to explain from the purely mechanical point of view; the development, out of an egg, of an individual, all of whose parts, both large and small, are produced in regular sequence and in definite likeness to its parents' — it is that, he thinks, which mostly distinguishes living from dead matter. But, all the same, it is not to be supposed that living creatures and their organs are, as Bichat imagined, independent for their functions of the laws of inorganic nature; on the contrary, heat and cold, electricity and chemical reagents exercise a law-bound influence on them just as they do on dead matter. But vital phenomena definitely differ from the processes of inorganic change on account of their constant alternation of regeneration and dissolution, of building up and breaking down. This relation between the vital phenomena and the general physio-chemical conditions that govern them Bernard calls "determinism," a term that he would substitute for "vitalism" and "materialism," both of which he rejects. For in contrast to Magendie he considers hypotheses and theories useful to science; they possess, it is true, little real value, but they are never-theless inevitable, "for in every science it is impossible to proceed from a known fact to an unknown fact without the aid of an abstract idea or the-ory." And yet the general view of life is the business of the research-worker himself; "no one asks whether Harvey or Haller were spiritualists or mate-rialists; we only know that they were great physiologists, and it is their observations and experiments that have been handed down to posterity." Bernard thus sought to compromise between the sheer unimaginative estab-lishing of facts, and speculation that in its efforts to create a general theory of existence loses sight of those fundamental realities which form the vital conditions of all natural science.

His investigations into nutrition

BERNARD's fame, however, does not by any means rest primarily on the theo-retical view of life that he propounded. It is as a practical pioneer in the sphere of experimental biology that he has acquired so great a name. His investigations have especially aimed at following up the process of nutri-tion and metabolism in the animal body, and the result he attained created in many respects an entirely new conception of them. In particular, he showed clearly for the first time the function of the liver in the process of digestion; he established the percentage of sugar in the liver and studied the conditions under which this secretion takes place. Likewise, entirely new light was thrown by him on the part played by the liver in the body's economy in general; the liver is characterized by him as "*un véritable labora-toire vital.*" Whereas after the discovery of the lymphatic system the liver was considered to be merely an organ for the preparation of bile, Bernard found that a number of substances from the intestine are conveyed through the cystic vein and are transformed there. Thus, thanks to him, the knowl-edge of the absorption of nourishment in the digestive canal was placed on a new basis. In connexion with this subject Bernard studied the production of sugar in the human body and in animal bodies in general. He established the fact that a stab in the *medulla oblongata* of an animal causes diabetes — a fact that now bears his name and laid the foundations of our knowledge of that disease. Further, Bernard found out the function of the pancreatic juice in the process of digestion, investigated the function of the vasomotor nerves and the problem of heat-production in animals, and, finally, carried out a great deal of important work in the sphere of pathology — for instance, in regard to the effect of poisons — all contributions of the greatest significance to the development of biology.

Whereas in France, then, the experimental method as applied to biology was used for the purpose of finding out purely physical and chemical phenomena in living creatures, in Germany the same method had a somewhat different application; to begin with, it had to serve the purposes of the purely speculative philosophy that was still exercising a dominating influence at the time and was later on practised in connexion with comparative anatomy, being aided by the use of the microscope. This co-operation had brilliant results; a new direction was given to biology, which placed Germany in the first rank among the centres of research in that science. We shall now proceed to give an account of the most important of the representatives of this school.

JOHANNES EVANGELISTA PURKINJE was born in 1787 at Lobkowitz, in Bohemia, of Czech parents. His father, who was a bailiff on an estate, died early, but through his mother's efforts the boy became a pupil at a theological college, where he learnt German and general school-subjects, and for three years devoted himself to theology; shortly before he was to be ordained, however, he relinquished this career and began studying philosophy and medicine at Prague. His dissertation was on the subject of sight and was influenced by Goethe's *Farbenlehre*, with the result that it attracted the interest of the poet; through the latter's influence Purkinje, who had sought in vain to procure a situation in his own country, was invited by the Prussian Government to become professor in physiology at Breslau in the year 1823. The faculty had recommended another for the appointment, so that from the beginning Purkinje found himself in a difficult position, which was still further complicated by the fact that he was not a good lecturer, probably owing to his having a poor ear for German. For many years he worked and struggled to get a physiological institute of his own, and eventually, having overcome the opposition of his superiors and colleagues, especially his lifelong enemy the professor in anatomy, he was able to open the institute in 1840 — the first of its kind in Germany, and very modestly equipped. Hitherto Purkinje had had to carry out his experiments in his own home, the comforts of which he had for many years sacrificed to the ends he desired to achieve. From here emerged a number of pioneer works in various spheres of biology, performed by himself and the, in part, very distinguished pupils he had gathered around him. After the completion of the institute Purkinje's interest in science began to wane; instead he applied himself more and more to his country's culture and politics. Even in Breslau he appeared as an author in the Czech language, translating the poems of Schiller and Goethe into his native tongue. Having been called to Prague in 1850, he devoted himself heart and soul to the national cause; he now spelt his name Jan Purkyně; he worked hard for the founding of a purely Czech university and was a member of the Young Czech party in the Bohemian Diet. National and foreign honours were showered upon him; and, worshipped by his countrymen, though at the same time hated by the Germans in his country, he laboured indefatigably to a great old age. He died in 1869.

Purkinje's discoveries

PURKINJE is one of the great geniuses in the field of biological discovery; a great number of facts of the highest value to our knowledge of life have become known through him. On the other hand, he never systematically and thoroughly investigated any particular field of inquiry, nor did he discuss theoretical problems. Even his greatest work, his investigations into the physiology of the senses, is really only a collection of different experiments and observations without any connexion other than the organ with which they deal. His experiments on sight-physiology, the ideas for which, as mentioned above, he obtained from Goethe's *Farbenlehre,* and which, in fact, are dedicated to the poet, represent a work of fundamental importance in their sphere. They were carried out with extraordinarily well-trained and keen powers of observation and a corresponding gift for experiment. Visual sensations induced by mechanical influence, by galvanic current, by various kinds of light-impressions, are described and analysed; especially well known are the chapters "*Indirektes Sehen*" and "*Wahre und scheinbare Bewegungen in der Gesichtssphäre*"; famous, too, is the venous figure named after him, which is caused by the oblique illumination of the eye. As a microscopist Purkinje has likewise made remarkable discoveries, among which should be mentioned the germinal vesicle in the chick, discovered two years before von Baer found the mammalian egg; and, further, the spiral apertures of the sweat-glands and the structure of cartilage. Of peculiar interest are the thorough investigations into the existence of the cilia in the animal kingdom; formerly these hairs were known only in protozoa and molluscs. Purkinje discovered them and their movement in the oviduct and respiratory duct in vertebrates — he established the movement's independence of any extraneous force, although, not yet being aware of the nature of the cell, he was unable to adduce the latter's autonomous life as a cause of the phenomenon. There is still one more of Purkinje's discoveries that deserves mention here — namely, the axis cylinders of the nerves, and the large ramified cells in the cerebellum, which bear his name. As a physiological chemist he became known for his investigations into the effect of rennet on the digestive process. There are, indeed, still several of his discoveries that might well be referred to here if space allowed.

Contemporary with Purkinje there was working in Germany a scientist who, in many respects, might be called his personal antithesis, but who who was his equal in importance for science.

JOHANNES PETER MÜLLER was born in 1801 at Coblenz, on the Rhine, the son of a shoemaker. He was of a well-to-do family and was able to indulge his passion for study. After a brilliant career at school he went to Bonn, where he settled down to the study of medicine. After having taken his doctor's degree he spent three terms in Berlin, where he was welcomed with paternal kindness by Rudolphi and received impressions that proved a decisive factor in his further development. Having returned to Bonn, he became, first, lecturer and afterwards, in 1830, professor at that university. When Rudolphi died and the question of his successor arose, Müller submitted a letter to the Prussian Minister of Education in which he applied for the appointment, at the same time drawing up an ambitious program for his future work. He was accordingly appointed and held the professorship until his death, in 1858. Both as a teacher and as a scientist he worked with unique success; the circle of pupils he gathered around him has few parallels in the history of science, as regards both results and the fame to which many of them attained. Müller began by devoting himself to experimental and microscopical research; it was he who introduced experimental physiology into Germany, and his services to microscopy were of no small value. During his later years he applied himself chiefly to comparative anatomy and evolution, and in connexion therewith to marine research, which had first been taken up by Rathke. For this latter purpose he visited both the Mediterranean and the Scandinavian coasts, everywhere enriching biology with his valuable observations. The violent exertions demanded by this many-sided activity had, however, told upon not only his bodily powers, but also his mind; anxieties of a practical nature also further weakened his health. He was University Warden during the years of the Revolution of 1848 and, being a conservative, came into repeated conflict with the revolutionary-minded students. Some years later he was in a serious shipwreck, which cost the life of one of his friends. These events seemed to have proved too much for his powers. Ever since his youth his personality had been a curious combination of nervous unrest, proud egotism, and deep melancholy; the last gained the upper hand according as his worries increased and his powers declined. One morning he was found dead in his bed without sickness's having intervened; a common rumour, which was never contradicted, declares that in despair he laid violent hands upon himself.

Johannes Müller's scientific career may be said to be typical of that of contemporary German biology in general — it begins in natural philosophy and ends with exact research, first physiological and then comparative anatomical. From childhood Müller possessed a mobile and imaginative temperament; he had a tendency to hallucinations, which he later on studied from a scientific point of view, and the education he received at Bonn was well adapted to develop this over-imaginative side of him. Among his tutors, it is worth noting, were Nees von Esenbeck, the fantastic botanist, who has previously been described, and the Schellingian Brandis. His dissertation for his doctor's degree, *On the Relations of Numbers in connexion with the Movements of Animals*, is also entirely in the spirit of Oken;[1] here one may read, amongst other things, that "bending and stretching are the two poles of life, the former resembling the closed bud, the latter the opened but withered flower; in both night prevails, but between them moves life." In his old age Müller is said to have destroyed all the copies of this fantastic production that he could lay hands on. His visit to Rudolphi distinctly cooled his ardour for extravaganzas of this sort, while the reading of Berzelius is said to have had an even deeper influence upon him in this respect. Before he entirely abandoned the natural-philosophical school, however, Müller published his investigations into subjective sense-perceptions, which were undoubtedly the finest work on natural science produced by German romantic philosophy. Like Purkinje, who in this subject was his predecessor, Müller takes as his starting-point Goethe's colour-theory. The physical qualities of light do not interest him at all and he accepts the theory of light's "primal phenomena (*Urphänomen*)" although he is not blind to its weaknesses; what attracted him to Goethe is the latter's observations on the subjectivity of the sense-perceptions; taking these as his starting-point, Müller builds up with ample material derived from personal observations his general theory of the specific forms of energy of the sensory organs. He establishes the fact that every sensory organ reacts in its own special way towards every kind of irritation; for instance, the eye through light-impressions reacts just as much to blows and electric current as to daylight; on the other hand, different organs of sense react each in its own way to the same irritation; thus, to irritation caused by electricity the eye responds through light-impressions, the ear through sound, the tongue through taste; and finally each sensory organ can express its individual reaction to impressions from within, in which are produced "imaginary sense-phenomena," or what would nowadays be called hallucinations. Through these facts Müller has laid the foundations of experimental sense-physiology, which has been so diligently studied in modern times; thanks to his extraordinary powers of observation and clearness of thought, he succeeded in doing so in spite of the natural-philosophical principles on which his research was based. We

---

[1] A German résumé of this work is included in the 1822 number of Oken's *Isis*.

find here, for instance, a number of statements quite in the spirit of Goethe's speculations. The experimental method is scornfully rejected — Magendie's experiments in particular are the object of adverse criticism — and the search for "divine life in nature" is highly commended. The function of physiology is said to be to comprehend the phenomena of life, not from the point of view of experience, but from that of the idea of life. This again is a proof that a keen observer can create fresh values in spite of a weak theoretical standpoint. Müller's propensity for observing the life-manifestation of his own senses was really unique, but the danger that always attends such self-introspection threatened him no less; his nervous system was shattered by his "fantastic sense-observations" and he fell into a state of melancholy bordering on insanity. Rest and careful tending restored him, it is true, but he gave up for ever these "subjective" researches and therewith also most of the natural philosophy upon which they were based. We may say that this mental disease involved the downfall of natural philosophy in Germany.

J. Müller's experiments on sensory and motor nerves

MÜLLER, indeed, never abandoned his idealistic view of life, but his natural research was now based on the principles laid down by Rudolphi — a comparative study of the phenomena of life based on the knowledge of their organs in different animal forms. He thus took up for renewed investigation the Bell-Magendie experiments on the sensory and motor nerve-roots and he succeeded in finding a more suitable subject for investigation than his predecessors; they had experimented on dogs and rabbits, while Müller had recourse to frogs, which are of a more enduring nature and can therefore lend themselves to more careful observation. Müller in fact essentially widened the knowledge of these important phenomena. His reinvestigations have thrown a special light on reflex movements. Another field of study that particularly interested him was the embryonic development of the sexual organs, in which he considerably widened the field discovered by Rathke and von Baer; he also threw considerable light on the knowledge of the evolution of the mesonephros or middle kidney. Further, he made important observations in regard to the glandular systems of the higher animals; in particular, he definitely determined the glands' character of closed tubes without connexion with the blood-vessels. He embodied the whole of his knowledge on this subject in his *Handbuch der Physiologie des Menschen*, the work which contains the clearest exposition of his general biological views and which became the authoritative source of the contemporary conception of life-phenomena, which held good up to the advent of Darwinism.

Müller introduces his physiology with some general observations on the essence of life, which show how deep was the influence that natural philosophy still had on him, even after he had broken away from it. He is a vitalist, and a much more positive one than Bichat, for instance, who really only maintained that the life-process was inexplicable by chemical and physical methods. Müller, on the other hand, definitely declares that there is a special "organic creative force" that is the essential condition of life. He points out the resemblance between his theory and that of Stahl, but with this difference, that Stahl considered the conscious soul to be the condition of life, whereas Müller holds that the consciousness is something apart from the organic creative force; the latter belongs to all living beings, while the consciousness, "which does not create any organic products, but only ideas," is found only in the higher animals. "This rational creative force manifests itself in each animal in accordance with a strict law, which the nature of every animal requires"; it exists in the embryo before its parts are present and it produces these parts. "*Der Keim ist das Ganze*, Potentia; *bei der Entwicklung des Keimes entstehen die integrierenden Teile desselben* Actu." Here, apparently, Aristoteleanism recurs word for word, and it is still more conspicuous in Müller's constant declaration that the organization of the living being is governed by finality. "*Die organischen Körper unterscheiden sich nicht bloss von den unorganischen durch die Art ihrer Zusammensetzung aus Elementen, sondern die beständige Tätigkeit, welche in der lebenden organischen Materie wirkt, schafft auch in den Gesetzen eines vernünftigen Planes mit Zweckmässigkeit, indem die Teile zum Zwecke eines Ganzen angeordnet werden, und dies ist gerade, was den Organismus auszeichnet.*" This gives the gist of Müller's biological views; among the details it may further be pointed out that he emphatically maintained the immutability of both species and genera, as well as of other higher systematical categories in the animal and vegetable kingdom; and again that he holds the same epigenesis theory as C. F. Wolff, whom he greatly admired, that he believes with Rudolphi that intestinal worms are produced by spontaneous generation, and lastly that he considers the spontaneous generation of the Infusoria can be neither proved nor disproved.

His vitalism

IT need hardly be specially pointed out that this organic creative force is a product of natural-philosophical thought; likewise, it will at once be realized that it in no way helps to explain the course and connexion of the vital functions. We might apply to it Galileo's above-quoted words on the omnipotence of God as a ground for natural phenomena: that one can derive from it anything whatsoever because it is based on no kind of necessity. And as a characteristic consequence of this mode of thought results the idea of finality as a law governing organic evolution. Müller's strong insistence upon the complete finality of the organisms is no doubt connected with his often expressed religious respect for nature, but he arrives at no explanation of nature in that direction; science indeed has always striven to give to its conclusions the character of laws of necessity, but where the domination of necessity is established, there is no room for finality; no one has commended the finality of mathematical conclusions, however useful they may have been to science. Müller's ideas, however, deserve all possible attention; there is no doubt that he largely created the standard of thought which prevailed in biological circles up to the appearance of the origin-of-species theory, and from which the opposition to that theory largely recruited its forces. At any rate, this legacy of Müller's from the age of romantic natural philosophy certainly had its influence on successive generations; it even crept into biological theories fairly effectively, although in a roundabout way, during the greatest days of Darwinism.

The influence of Müller's physiological text-book has been all the greater because its special section contains information of great value based on the results of his own original research-work; here we find a very careful and exhaustive account of the law of specific mental energies to which we have previously referred, and here are explained in a manner unexcelled by his age the functions of the nervous system; here his above-mentioned investigations on that subject are summarily described. Further he declares that the ganglion-cells of the brain perform the latter's functions, and he explains the connexion between them. Specially noteworthy in this respect as a summary of his results is the chapter on "*Mechanik des Nervenprinzips*," wherein his keen powers of observation and combination, undisturbed by any philosophical adjuncts, are very conspicuous. His exposition of the alimental and vascular systems, as well as of the sexual organs, is very fine, although somewhat brief.

After the "text-book" had been completed, Müller gave up physiology. According to his own statement, he shared Rudolphi's dislike of experimenting on live animals, as practised by Magendie and his school, and his physiological works were actually based very largely on comparative anatomical observations. He clearly realized that physiology could not be carried any further in this way and he consequently went over entirely to comparative anatomy, which at that time had very large fields of inquiry still unexploited. Müller made a particularly happy choice when he devoted himself to investigations into the structure of the lowest Vertebrata. Among his works on this subject may be mentioned his monograph on the lancet-fish, which exhaustively supplements Rathke's previously mentioned work on that animal. But in connexion with this group special mention should be made of his monumental work on the skeleton-system, muscles, and nerves of the Myxinoidei, on which he spent nearly ten years. He took up for study this subject of the most primitive group in the order of Cyclostomi because, as he says, the boundary forms in a class are the most interesting in that they lose a good deal of the character of the class and thereby show us the type of the class in its most simple form. The work contains a detailed description, exemplary in its accuracy, of the said organic systems in the *Myxine glutinosa* and its African related types; taking that as a starting-point he makes a detailed comparison of the skeleton, muscular and nervous systems of all Vertebrata. It is not only the accuracy of this work that has made it a standard for the future; but the method itself — starting from special inquiry, comparing the results with the conditions existing in the related types of the subject and thus throwing light on the form-connexion in a wider or narrower group of living types — was imitated during an entire period and is to this very day by no means exhausted. It would take too long to examine in detail the result recorded by Müller in this work; certain it is that his successors have had little to add to the material he investigated, while the comparative section is also of immense value, although naturally some of its conclusions have since been disproved. Thus, Müller adopts the theory of the cranium's being formed of vertebræ; though he deals with the subject more cautiously than either Oken or Goethe, his conclusions are at any rate too far-fetched to be acceptable in modern times.

His marine research work

THE result of Müller's occupying himself with these marine animals was that he took up with increasing interest marine research work; through his holiday trips to Heligoland, the coasts of Scandinavia, and the Mediterranean he was irresistibly attracted to the study of the life of marine animals, which had been so very little investigated before. In this field, as also in that of anatomy, he became a pioneer. A long series of extremely important discoveries in the sphere of marine biology is due to him; chief among these should be mentioned a great number of the larval forms of worms, Echinodermata, and molluscs, the evolution of which was found out partly by him and partly by others who followed his example later on; further, the discovery of that curious parasitical mollusc, the Entoconcha, whose origin in the host, a holothurian, he was nevertheless unable to discover, and furthermore a number of interesting observations on the life and evolution of fishes. He is thus not only a pioneer in marine zoology but also one of the greatest in that field that the world has ever seen. The idea of special stations for the study of this type of life was vigorously promoted by him, while at the same time he originated a good deal of the methodology applied in work on the subject. If we add that Müller also followed in Cuvier's footsteps as a palæozoologist with great credit, we shall have given a picture, however incomplete, of one of the most prolific scientific achievements that the history of biology has to record.

Müller was also very distinguished as a teacher. Few biologists, if any, have succeeded in gathering around them so many incipient scientists of the highest rank. One of the most eminent of these has declared that the master never taught dogmas, but only his own method. The pupils had themselves to form their own ideas; only the method and the results achieved were common to all. This explains to a certain extent how it was that so many scientists of independent and original thought could be trained in this school, men such as Schwann and Virchow, Henle, Remak, Kölliker, Du Bois-Reymond and Helmholtz; the same circumstance may also explain the widely differing lines of research upon which they entered, but it naturally confirms also the extraordinary many-sidedness of the master himself. Thus, microscopy and cytology as well as experimental physiology in its most strictly limited sense were here developed side by side. We shall now consider, to begin with, the development of the two first-mentioned branches of research.

CHAPTER VII

MICROSCOPY AND CYTOLOGY

Improvement of the microscope

A s HAS BEEN PREVIOUSLY POINTED OUT, microscopical research had a pe- ✔
riod of brilliant success in the seventeenth century, the age of Malpighi and Leeuwenhoek. Afterwards, however, this method made
no further advance for more than a hundred years; the eighteenth century ⌐
certainly produced some microscopists of importance, such as, for instance,
Lieberkühn, but on the whole little was achieved during this period with
the aid of magnifying apparatus. The reason for this was that the aforesaid
scientists of the seventeenth century and their contemporaries did all that
could be done with the instruments at their disposal; microscopes were and
remained imperfect, and improvements were a long time in coming. The most
serious difficulty lay in the chromatic aberration of the lenses; a colourless
object seen under the microscope would shimmer with all the colours of the
rainbow, a fact which naturally gave rise to countless misinterpretations of
the objects investigated. To procure achromatic glass, free from this fault,
was a task that occupied many scientists at that time; Newton himself declared the problem to be insoluble. Eventually a Swede, SAMUEL KLINGEN-
STIERNA (1698–1765), professor of physics at Upsala, succeeded in working
out how the achromatic glass should be made, and under his instructions an
English mechanician, DOLLOND, constructed the first achromatic lenses. It
was some time, however, before the invention could be utilized for microscopical purposes. Among those who in the beginning of the nineteenth century
constructed microscopes with achromatic lenses may be mentioned the
Frenchman CHEVALIER and the Italian AMICI; the latter's microscopes in
particular were very fine, and there soon arose in every country microscope-
makers who produced gradually perfected instruments. The year 1827 is
named as that in which Amici demonstrated his first achromatic lens-sys-
tem and during the thirties the biological institutions, at least the more important ones, were able to obtain specimens of these improved microscopes.
It was at the beginning of that decade also that microscopical biology first
showed any notable advance, and after that the great discoveries in this field
followed one another in rapid succession.

There were two spheres in which the pioneers of the new method were
induced to try their strength; on the one hand, the structure of the higher

animals and plants and the problem of their fundamental constituents, and on the other hand that world of minute, independently living creatures that the new instruments made it possible for the eye to see — in collections of water, in infusions on parts of plants (hence Infusoria), and indeed everywhere in nature. As a result of these investigations there arises an entirely new conception of the composition of organisms — cytology, or the knowledge of cells. An attempt to show the development of this branch of knowledge in summary form offers certain special difficulties; as Richard Hertwig strikingly remarks: "The way was paved for the reform of the cell theory through discoveries made in very different spheres and not until late in time concentrated in a focus." We must therefore give a brief summary of these various discoveries, though it should be mentioned in this connexion that many important steps were taken in this field by people who otherwise exercised little or no influence upon scientific progress. For the sake of brevity we must confine ourselves to discussing only the most important achievements and personalities in the history of cytological research.

Works on plant-tissues

In the beginning of the eighteen-thirties Bichat's tissue theory was still accepted in zoology, though more or less modified by various investigators. In botany it was different. Since the days of Malpighi and Grew it had been known that the wood of plants is composed of cells — minute chambers having more or less thick walls. It was a matter of dispute whether a number of other elements in the plant, especially spiral vessels and bast, were compact or cellular. The first who attempted to compare the composition of animal and plant was C. F. Wolff; he believed, it will be remembered, that the construction of each represents a mass of cell-shaped forms. Among later scientists Blainville produced a theory, mentioned in the foregoing, that the animal organism is composed of cells, but this theory was not very clearly developed and therefore won but little acceptance. The knowledge of cells, however, made steady, if slow, progress, the botanists still leading the way. To start with, it was a question of deciding whether all the parts of the plant consist of cells, and this led to lively discussion. Among those who contributed to its solution may be mentioned CHARLES FRANÇOIS MIRBEL (1776–1854), professor of botany at the Jardin des Plantes, and LUDOLF CHRISTIAN TREVIRANUS (1779–1864), professor at Bonn, where he succeeded Nees von Esenbeck. Mirbel especially examined the cell-structure in certain mosses, making valuable contributions to the subject, besides which he resolutely maintained the cell's quality as a basis for all structures in the vegetable kingdom. Treviranus, on the other hand, performed a signal service in the observations he made in regard to the regular movements of the cellular contents in a number of vegetable forms; moreover, he observed that the spiral vessels in plants originate in cells which become stratified upon one another and lose their intermediate walls. The scientist who is generally mentioned as the creator of modern plant-cytology, is, however, HUGO MOHL. He was born in 1805 at Stuttgart, of a brilliantly gifted family in the Government service. He became a doctor of medicine and a professor, first of physiology at Berne, then — in 1835 — of botany at Tübingen, where he remained until his death, in 1872. His life was typical of the modest and reserved man of science; being unmarried, he spent his days in the laboratory, and his evenings, after the manner of his countrymen, at a "*Stammtisch*" with a few friends. Even his research work has the same quiet character; accurate observation of phenomena, a great capacity for placing known facts in their proper light, extremely conscientious examination of his own ideas, and praiseworthy consideration for those of others. He was decidedly against philosophical speculations and he never produced any summary of his own field of research; his writings consist of a large number of short papers. The valuable results that he achieved, however, have been acknowledged by both his own and succeeding periods. During his life he received many honours, including that of being raised to the nobility with the name of von Mohl, and after his death his reputation was still further enhanced.

Mohl's work on cell-reproduction

AMONG Mohl's works should be mentioned, to begin with, his observations of cell-reproduction. Before his time, and even later, opinions differed on this point. He upheld clearly and convincingly that the cells in algæ and even higher plants arise through partition-walls being formed between previously existing cells. These partition-walls he investigated and described with great accuracy. We must, however, leave his and his contemporaries' detailed researches in this sphere, however influential they may have been in the development of vegetable anatomy; it need only be mentioned here that Mohl established the cellular structure in spiral vessels, bast, bark, and other components of plants, a point that had formerly been much debated. Further, he carefully investigated the process of development in spores of various cryptogams, finding therein both a confirmation and an extension of his theory of cellular division. In another connexion we shall make reference to some of his further important contributions to this subject. For the rest, he was also an expert optician; a work which was unique at the time and which is still worth reading was his *Micrographie*, a text-book on microscopy and microtechnique.

Discovery of the cell-nucleus

AMONG the investigators who contributed to the development of cell research, which was particularly active at this period, may further be mentioned the English botanist ROBERT BROWN, a scientist of many parts, whose work will be described in another connexion. Here it need only be pointed out that it was he who, in 1831, published the discovery that to the contents of every cell there belongs, as an essential component, an "areola" or, as he also calls it, a nucleus; this cell-component he discovered in the epidermis of the Orchidaceæ and later he established its existence in a great number of other plant-cells. It was, however, reserved to other investigators to discover its true significance.

Cytological research was given a new direction by MATTHIAS JACOB SCHLEIDEN, one of the strangest scientific personalities of his age. He was born in Hamburg in 1804, the son of an eminent doctor. He began by studying jurisprudence, became a doctor of law, and took up a practice as a barrister in his native town. He had, however, but little success as a pleader, a fact that increased his naturally melancholy disposition. Finally, in a fit of despondency he shot himself in the forehead, but without the result he intended; he recovered and then resolved to devote himself to natural science. He became doctor of both philosophy and medicine, gained a great reputation by his writings, and in 1850 became professor of botany at Jena. After twelve years, however, he resigned; a professorship at Dorpat, to which he was appointed shortly afterwards, he relinquished within the year and after that led a life of wandering, with brief sojourns in various German towns, which lasted till his death, in 1881. The life he led fully testifies to a soul without balance, and this is reflected in more ways than one in his scientific work.

The work that at once brought Schleiden fame was an essay in Müller's archives of the year 1838 entitled "*Beiträge zur Phytogenesis.*" The question he propounds is: How does the cell arise? Here Schleiden takes as his starting-point Brown's above-mentioned discovery of the cell-nucleus, and his service to science lies in the fact that he was able to appreciate its fundamental importance, which Brown himself failed to do. From the nucleus, or, as he calls it, the cytoblast, Schleiden sought to reconstruct the course of development of the cell, and he made a very happy choice when he selected for the purpose the embryonic cell as his starting-point. He made a special study of the embryo-sac in different phanerogams, carefully examining the nuclei in the cells in question, and discovered in them the formation that is now termed the nucleolus or nucleal body. This discovery, however, led him to continue the investigation along the wrong lines; he thought he had discovered that the nucleolus is first formed through an accumulation of granulate mucus in the uniform content of the embryo-sac and he believes it to consist of gum; around this element is afterwards stratified the rest of the nucleus, and not until the latter is complete is there formed on its surface a small vesicle that grows outwards until it encloses the entire nucleus; the walls of the vesicle thicken, and thereby the cell becomes complete. According to Schleiden, during the further development of the cell the nucleus is in most cases dissolved — a statement that of course does not accord with the facts. As will be seen, the whole of this cell-formation theory is quite out of keeping with the truth, and this is still further emphasized in the eyes of a modern reader by the fact that Schleiden uses a number of romantic-philosophical terms: expressions such as *"potenzierte Zellen,"* *"edlere Säfte,"* and other similar terms are clearly reminiscent of Goethe. What made this paper so original ✗ is its insistence upon the independence of the cell; the plant is presented for the first time as a community of cells, a *"Polypstock,"* as it is expressly called, and it was from this standpoint that future investigators started who with a finer critical sense made use of the idea that Schleiden had produced.

Schleiden's text-book on botany

THERE is still one more important work from the hand of Schleiden that is worthy of mention — his *Grundzüge der wissenschaftlichen Botanik,* which was published in 1842 and at the time created an extraordinary sensation, criticism being both favourable and unfavourable. Really in its way it is a pioneering achievement; it implies a fundamental agreement both with the purely systematic botanical training that had hitherto been in vogue, and with the natural-philosophical conception of the phenomena of life. In a lengthy "methodological introduction" Schleiden propounds his general conception of nature, which represents the most interesting part of the work. It shows that he held a well-thought-out philosophical view of nature, acquired under the guidance of JACOB FRIEDRICH FRIES, professor of philosophy at Jena, and one of the few thinkers who during the age of romantic speculation maintained an interest in Kant's mode of thought. Following him, Schleiden declares that the aim of natural science is "to relate all physical theories to purely mathematical grounds of explanation." With this ideal of exact research before him he tries to convert botany into a comparative investigation of life-forms and life-manifestations, with special reference to the evolutional phenomena in the vegetable kingdom. As the cause of all that happens in nature, both animate and inanimate, he assumes one and the same "form-building force"; on the other hand, he strongly denies the existence of any special life-force, and, as had often been done before, he refers the growth of the crystal and the organ to the same category of phenomena. In spite of this he is definitely opposed to the idea of spontaneous generation of the higher animals and even rejects Meckel's "biogenetical principles." In a purely philosophical connexion he maintains, with Kant, the contrast between subject and object, and consequently also between spiritual and material entities. Schelling's and Hegel's theories on the unity of spirit and matter he dismisses with scorn. His "free-thinking" brought him into dispute with the theologians; at that period the latter were monists, following Hegel, while dualism was upheld by their opponents among the biologists; in Haeckel's time, it will be remembered, just the contrary was the case, which fact indicates that it was really the contrast between personalities that was the essential point.

In the special section of the handbook Schleiden gives an explanation of the cytology, morphology, and physiology of plants, after much the same plan as has been followed in similar works since then. This method of presentation really bears the stamp of genius; the actual contents, however, offer nothing essentially new from the point of view of that age; with constant and often extremely abusive criticism of the botanists of the time — Brown and Mohl are the only ones who are let off lightly — he presents a summary of the facts already known. Concerning the formation of the cell he propounds his old theory, although by that time it had lost much of the validity it formerly possessed. In fact, another scientist had entered this field of research with an entirely new idea that eventually directed its further line of development.

THEODOR SCHWANN was born in 1810 in a small town in Rhenish Prussia, where his father had a book-shop. He studied under his fellow-countryman J. Müller, at both Bonn and Berlin; having taken his doctor's degree, he became his master's assistant. In 1839 he was called to the chair of anatomy at the Roman Catholic University of Louvain, and some years later to Liége, where he worked until shortly before his death, in 1882. He was of a gentle and reserved disposition; he avoided polemics and therefore accepted none of the professorships that were offered to him at German universities — he did not like the way the German histologists quarrelled, he said — and throughout his life he remained a devout Catholic; thus, he was in everything a contrast to Schleiden, with whom nevertheless he was on friendly terms. His scientific activities fall entirely within the period during which he worked with Müller; it apparently needed his master's will-power to spur his easy-going and peaceful nature on to any exertion. As a professor he published only some few text-books and summaries. His teaching was always conscientiously carried out.

Schwann's work on cell-structure

SCHWANN's research work during his Berlin period was both many-sided and important. His doctor's dissertation dealt with the respiration of the embryo of the chick; he discovered the ferment of gastric juice, to which he gave the name "pepsin"; he studied Infusoria and experimented with fermentative phenomena, which led him to deny spontaneous generation and to declare that fermentation and putrefaction are caused by organisms. All these works, however, fall into the shade beside that by which he established his fame as one of the pioneers of biology — the work published in 1839 entitled: *Mikroskopische Untersuchungen über die Übereinstimmung in der Struktur und dem Wachstum der Tiere und Pflanzen*. He here takes as his starting-point Schleiden's above-mentioned cell-formation theory, which he accepts in its entirety and expands into a general theory of the basis and origin of life-phenomena. By way of introduction he points out the fundamental difference between animal and plant, which the biology of preceding ages had realized in the fact that animals possess a vascular system, which plants lack; the plants' "*gefässloses Wachstum*" was accounted for by the cell-structure or, as it was then called, the plant's composition of independent units. Now, Schwann had discovered in the notochord of tadpoles cells provided with nuclei, similar to the plant-cells, and both there and in the embryonic cartilage he believed he saw a process of cell-reproduction such as Schleiden had described. This induced him to look for cells in all the tissues of the animal body, and by examining these in the embryonic stage and afterwards following their development he succeeded in establishing the fact of cell-structure even in tissues that in a state of full growth show little or no trace of any such structure. It is not difficult to realize the great influence that this discovery was to have on the tissue theory, and what follows will make it still clearer. Of still greater significance for the future, however, was his general cell-theory, according to which, as he says, "one common principle of evolution is laid down for the most highly differentiated elementary parts of the organisms, and this principle of evolution is the cell-formation." This conception of the cell as a general unit of life and as a common basis for the vital phenomena in both the animal and the vegetable kingdom was immediately and universally accepted; so self-evident did its truth seem to be that it met with hardly any opposition, and in fact became the foundation on which since then both animal and vegetable biology have developed. It is thanks to this theory that the present age has been able to work out its conception of life-phenomena as a connected whole; without Schwann, Darwinism would hardly have been victorious.

In its details, however, Schwann's cell theory is very primitive; he not only embraces Schleiden's belief in a free cell-formation out of moisture, but takes it further. Out of moisture is concentrated, first the nucleolus, then the nucleus, and finally the cell; this process is explicitly compared with crystallization, and the whole concludes with reflections as to whether the hollow form of the cell might not be accounted for by the "*Imbibitions-fähigkeit*" of its component parts — in modern terminology, its colloidal qualities; according to him, then, the cell-formation would be a kind of crystallization in non-crystalline elements. For the essential part of the cell is, in Schwann's view, its hollowness; in its essence it is a space surrounded by walls; its content is a moisture, which runs out if the wall is damaged, while the nucleus is a transitory formation, which disappears in later stages of development. These views were in their essentials corrected in the immediately succeeding future. For the rest, Schwann made his cell-formation theory the basis of a general theory of life, which proved to be considerably more materialistic than that of his master, J. Müller; as a devout Christian he believed in the world's serving a purpose given it by the Creator, but in contrast to Müller he found no further or greater finality in living nature than existed in inanimate nature; the same purely mechanical forces shape both the cell and the crystal.

Further development of the cell theory

THE cell-theory which has just been described, and which has always been called the Schleiden-Schwann theory, after its founders, was adopted and at once followed up by other investigators, while, as mentioned above, the two pioneers withdrew from the field. The most important contributions made during the next few years were those of Mohl, who published a series of new observations regarding the role of the cell in the vegetable kingdom. In these brief but weighty papers he analyses the different components of the cell. To him the cell is still "a vesicle formed of a fixed membrane and containing a moisture"; the character of the membrane is the essential thing, and the shape, consistency, and interrelation of the cellular walls are described before anything else. Moreover, an account is also given of the contents of the cell: the "viscid moisture" that forms its fundamental constituent is carefully described; its currents, which were discovered by the Italian CORTI and rediscovered by Treviranus, are depicted in detail in various plant-forms — inter alia in those, since then, classical objects of demonstration, the Tradescantia hairs — similarly, the evolution of the cell-content is followed through its different stages of growth, and the secondary formations that accompany it — vacuoles, chlorophyll- and starch-granules — are described. The fundamental substance in the cell Mohl calls protoplasm; he thereby establishes the fact that the cell-content is an element by itself and not merely "slime" of some indeterminate kind, as Schleiden supposed. The name, which in spite of its clumsiness has come into permanent use, is, as a matter of fact, based on the false assumption that all the component parts of the cell, even (and above all) the nucleus, originate in this element, the "primal slime." The nucleus is described by Mohl in greater detail than by his predecessors; true, it is still stated to be, as mentioned above, a derivative of protoplasm come into being through an accumulation of a granulate substance in young cells and disappearing in the older ones, but the grossly mechanical precipitation-theory that Schleiden and Schwann held was accepted with reserve. And, above all, division is mentioned as being the normal method of cell-reproduction; independent cell-formation is confined to the embryo-sac alone. In regard also to the alimental physiology of the cell, Mohl offers some interesting observations, but they must be passed over here.

At the same time, valuable contributions to the evolution of the cell were made by KARL NÄGELI, an investigator whose far-reaching activities will be described later on. In an essay on the pollen-formation in the phanerogams, particularly in the Liliaceæ, he describes the cell-divisions (in 1842) with such care and reliability as had never been done before; even the division of the nucleus was observed with great accuracy. His "transitory cytoblasts" are chromosomes, although, with the inferior means at his disposal at that time, he was unable either to follow the course of development to the end or to interpret it aright.

While, then, plant-cytology was making rapid progress, cell research in the animal kingdom was by no means unproductive. Among those who collaborated in the working up of this field of research it is only possible to name a few of the most influential: to begin with, some of Johannes Müller's pupils, Henle, Reichert, Remak, and Kölliker.

JACOB HENLE was born at Fürth, near Nuremberg, in 1809, the son of a Jewish merchant who later, with his entire family, adopted Christianity. He studied at Bonn under Müller, afterwards becoming the latter's prosector in anatomy at Berlin. There, however, he became the victim of political persecution; he was a liberal and a member of the Burschenschaft, with the consequence that, like so many other youths at that time, he was arrested by the scarified Prussian police and after lengthy law-court proceedings was condemned for treason. His scientific reputation, however, saved him from further rigorous treatment; Humboldt, among others, interceded for him, with the result that he was pardoned, but he received no further appointment from the Prussian Government. In 1840 he accepted a professorship at Zurich, somewhat later one at Heidelberg, and finally, in 1852, one at Göttingen, where he worked until his death, in 1885.

Under Müller's leadership Henle worked both as an anatomist and as a biologist in the invertebrate field; afterwards he also devoted himself to pathology. His activities as a student of cell-life are associated with a number of special essays, and also with his *Allgemeine Anatomie*, an excellent work for its period. Among his contributions in the sphere of invertebrate research his discovery of the hair-sac mites is universally known. Best of all his specialized work, however, is his investigation of the histology of the intestinal epithelium; it was he who discovered the cylindrical epithelial cells and explained the existence of the pavement and columnar epithelium in the various parts of the intestinal canal. He also carefully studied the ciliated epithelium and its distribution and it was he who created the term "epithelium." In connexion with the intestinal mucous membrane, he investigated the chyle vessels with great care, particularly with reference to their terminal ramifications, which had hitherto been misinterpreted.

Henle's *General Anatomy* is the first histological handbook based entirely upon cytology and undoubtedly the most original since the days of Bichat. It begins with a chapter on animal chemistry, viewed from a contemporary standpoint, and goes on to describe the part played by the cell as a primary formation. His cell-theory is on the whole that of the Schleiden-Schwann school, which has previously been mentioned: the nucleus formed through an accumulation of a granulate substance, the cell formed round the nucleus and consisting of membrane, nucleus, and fluid content. Cell-division is denied as far as the animal kingdom is concerned; the cell-formation is rather compared with the emulsive phenomena that arise when oil and albumen are shaken together — an attempt at an explanation which, as is well known, has been the subject of endless variations in modern time. On the other hand, Schwann's comparisons between cell-formation and crystallization are not accepted. Henle adopts a decidedly critical attitude in regard to speculations on the primary vital phenomena. "Explaining a physiological fact means tracing its necessity from physical and chemical natural laws. It is true, even these laws offer no explanation as to the ultimate grounds, but they make it possible to combine a mass of details under one point of view." On the life-force theory adopted universally by his contemporaries he passes the following striking judgment: "The life-force is formally as good an explanation as the force of gravity, but it is one force the more and this is at variance with our striving after unity."

Henle thereupon proceeds to give an account of the tissues, and, of these, first of all the epithelial system, which indeed was best mastered and is very well expounded. A number of other details are also excellently explained, especially the vascular musculature, which is here for the first time satisfactorily dealt with. In regard to the division of tissues, Henle is, of course, far in advance of Bichat, but even his system is, from the modern point of view, difficult of comprehension; in particular, the category nowadays called connective tissue is split up into a mass of sub-headings which are often somewhat unhappily formulated. Another weak chapter is that on the glandular system, as indeed Henle himself admits, referring to the paucity of the investigations that have been made in that sphere. But, on the whole, Henle's general anatomy deserves the judgment passed on it by a later histologist who declared that it laid the foundation of modern histology and on that account will survive.

KARL BOGISLAUS REICHERT was born in 1811 in a provincial town in East Prussia, where his father was mayor. He studied at Königsberg under von Baer and in Berlin under Müller, was called to a professorial chair first at Dorpat, then at Breslau, and finally in Berlin, where after Müller's death, when the latter's professorship was divided, he took over the professorship of anatomy, which he retained until his death, in 1883. He began his activities as a comparative anatomist with a valuable work on the development of the gill-arches in the Vertebrata and another equally eminent work on the embryonic formation of the frog's head. He then devoted himself to cytology in the spirit of Schwann and applied the latter's theories to the evolution of frog's spawn, not, it is true, without falling into the misconception prevalent at the time — he believed that the separate granules in the yolk of the egg are independent cells — but nevertheless with great accuracy in his observations of the consecutive stages of development, resulting in his establishing the cell character of the products of division, out of which the embryo is formed. Even the two afterwards oft-recurring expressions *"Bildungs-"* and *"Nahrungsdotter"* originate from him. One or two works on the evolution of the tadpole likewise contain sound observations; one of them contains a number of general reflections on the subject of organic formation by means of invagination, which in a certain degree foreshadowed Haeckel's gastræa theory. Reichert's greatest contribution, however, lies in his study of the evolution of the connective substance; he introduced this term to imply a number of connecting-tissue elements of different structure and has based it upon arguments from evolutional history. In his old age Reichert was completely isolated; he refused to accept the new protoplasm theories, and still more the origin-of-species theory, and he made no attempt to hide his disgust when these ideas prevailed. In particular, he attacked with great vehemence Haeckel's theory of the germ layers being homologous throughout the animal kingdom; instead, he maintained the independent origin of the separate organs. While he was scorned by Haeckel and his contemporaries, Reichert has to a certain extent been justified by the results of modern research, whereon we shall have more to say in a later chapter.

ROBERT REMAK was born at Posen in 1815. Like Henle, he was of Jewish extraction, but in contrast to him held to the faith of his fathers. After studying under Müller he became his assistant lecturer, eventually being given the honorary title of professor, though never holding a post as ordinary professor. He made a living by carrying on a medical practice, and this gradually diverted him from a scientific career. He died in 1865. His contribution in the field of cell research is concerned partly with neurology and partly with embryology. Thus, he discovered and described the sympathetic nerve fibres called after him, and he established the fact that in the embryonic life the nerves are constructed in the form of fibres which grow out from the nerve-cells. He is specially worthy of remembrance, however, for the determined opposition he made to Schwann's theory of free cell-formation; he studied the evolution of frogs' eggs and thereby proved that the egg is a cell that divides itself up into new cells and that this division starts from the nuclei; he does not accept any cell-formation by accumulation in a formless matter. Further he drew a comparison between the embryonic development in the egg of the frog and in that of a bird: it was he who invented the terms "holoblastic" and "meroblastic," which are still used for these two types of egg. He distinguishes three germinal layers, which he believes to be common to the embryonic development of all vertebrates and which give rise, the outermost to the nervous system, the middlemost to the musculature, and the innermost to the intestinal tube — all observations that have been confirmed by modern research. In his later years Remak paid special attention to the study of electrotherapy, making in this sphere a valuable contribution, which, however, does not belong to the history of biology.

RUDOLF ALBERT KÖLLIKER was born in 1817 at Zurich, the son of a wealthy merchant. He studied zoology in his native town under the aged Oken and afterwards went to Berlin, where, under the guidance of Müller and Henle, he was initiated into their method of research. When Henle went to Zurich, Kölliker became his prosector, but in 1847 he was invited to become professor at Würzburg. There he gave lectures up to 1902, when he resigned; he died three years later. He remained a Swiss subject all through his life. He was one of the foremost teachers of his age; many of the most eminent biologists of the succeeding generation belonged to his school.

Kölliker's research activities lasted as long as his educational career. He was active far into his ninth decade and was successful to the end; his later work therefore belongs to the following epoch. As a research-worker he was above all a microscopist; the connecting link in his work was formed by the microscopical method, which he employed in a great number of fields of research, everywhere with immense success, although no discovery or idea of supreme importance attaches to his name. He gave a splendid summary of contemporary knowledge on this subject in his *Handbuch der Gewebelehre des Menschen*, published in the year 1852, which deserves to be called the first modern histology. Its purely external form has been repeated, with the necessary modifications required by the progress of science, in innumerable text-books on this subject. Kölliker here expounds with impartiality and far-sightedness the contemporary cell and tissue doctrine; he does not, indeed, entirely deny free cell-formation, but he limits its existence as much as possible. The cells themselves he considers to be constructed of elemental parts: granular and vesicular formations, to which he ascribes a certain degree of independence in growth and development — an idea which was later adopted by many others. He strongly insists upon the importance of the role played by the nucleus in the life of the cell, in its multiplication by division and its other vital manifestations. We must pass over his classification of tissues; he did not adopt Reichert's connective-substance category, but otherwise his classification is clear and concise and his elucidation of the structure and vital manifestations of the different tissues is full of original observations and excellent in its form. Kölliker produced another splendid text-book in his *Entwicklungsgeschichte des Menschen und der höheren Tiere* (1861), a summary in clear and comprehensive form of the embryological knowledge of the time.

Kölliker's investigations

OF Kölliker's numerous original investigations it is possible to quote here only a few of the most important, in so far as they come within the period now being dealt with. Especially noteworthy is his investigation into the spermatozoa (1841), in which he proves that they are not parasites, but a true sexual product. Further, his fine monograph *Entwicklungsgeschichte der Cephalopoden* (1844), which he worked out in the course of a visit to Naples and which contains an account of egg-division and embryonic development in those animals, to which account subsequent research has had but little to add. Of great importance, too, was his study of the smooth musculature, the elements of which he definitely isolated for the first time, describing them as single-celled fibrillæ; their distribution in the different organs of man and the mammals he elucidated with unprecedented completeness. Henle, indeed, had established the musculature of the blood-vessels, but it was Kölliker who explained its character in detail. In the sphere of neurology he also made valuable discoveries; thus, he proved convincingly that the nerve-fibres are connected with processes of the ganglion-cells, thereby making important contributions to the knowledge of their structure. If we add that Kölliker investigated with valuable results certain unicellular animals, as, for instance, the gregarines, we shall have given some idea of his extraordinarily many-sided research work.

There is one scientist who is worthy of mention by the side of Kölliker — namely, FRANZ LEYDIG (1821–1905), who was a native of Württemberg and who was professor at Bonn from 1875 to 1895. As a cytologist he was remarkable for his investigations into the invertebrates. He, too, published a *Lehrbuch der Histologie*, which, remarkably enough, pays as much attention to the tissues of the invertebrate animals as to those of the vertebrates, thereby laying the foundations of comparative histology, which has since been so extensively developed.

Leydig's classification of tissues is more in accordance with the modern method than that of Kölliker; thus he groups under the heading "connective substance" not only connective tissue and cartilage, but also bone tissue, which Reichert still kept separate. His presentation of the life and development of the cell is likewise more modern than Kölliker's, but Leydig's book was published four years later, and during that period cytology made great strides year by year. A good deal of Leydig's own pioneering research-work is recorded in this treatise; his detailed studies of the structure of the insects, especially their digestive, glandular, and sensory organs, should be mentioned first of all. Leydig's other extremely conscientious microscopical investigations into worms and molluscs, as well as vertebrates, belong to the specialized literature on those subjects; no histological specialist can afford to neglect them, but considerations of space forbid any further reference to them here.

Cell research entered upon a new phase through the work of RUDOLF LUDWIG CARL VIRCHOW. He was born in Pomerania in 1821, the son of a country shopkeeper, and after finishing school applied himself to medicine. He was a pupil of J. Müller and after completing his studies he became assistant at the Charité Hospital in Berlin, rapidly acquiring a reputation on account of his writings on pathology and his *Archiv für pathologische Anatomie und Physiologie*, which he founded in 1847 and edited until his death. He was sent by the Government to be a medical officer in an industrial district in Silesia, where a serious epidemic of typhus had broken out; in the report on his mission he represented social distress in the district as being the true cause of the disease in such terms as created resentment in high bureaucratic circles. When, moreover, during the revolutionary year 1848 he joined the opposition, he was dismissed from his post. He then moved to Würzburg, where he became professor in pathological anatomy and developed such brilliant activities in the spheres of research and education that his school soon rivalled that of his master, Müller. The Prussian Government recalled him in 1856, and from that date until shortly before his death he was one of the most brilliant personalities at the University of Berlin. He died in 1902 as the result of an accident. He remained throughout his life faithful to his liberal ideas; as a member of the Prussian Diet and the German Parliament he indefatigably supported the cause of liberalism and thereby came into constant conflict with Bismarck and the adherents of that statesman. Virchow naturally had no chance against such an antagonist, and his purely political activities were unproductive. On the other hand, his influence on the public health services in Germany was extraordinarily effective; it was largely due to him that the German medical system became a model for other countries. His energy sufficed for all claims made upon it, from the reform of the sanitary system in Berlin to the organizing of the medical corps during the War of 1870. Above all, however, the care of the sick in Berlin stands as a monument to his organizing genius.

Virchow's cellular pathology

As a research-worker Virchow was really a pathologist; it was diseases and their causes that was the chief object of his investigations. This led him to the problem of the cells as fundamental constituents of the organism both in health and sickness, and in the middle of the eighteen-fifties he laid the foundations of his "cellular pathology": a theory of the cells as the true causes of disease. When, therefore, a decade later, bacteriology began to make headway, he refused to accept its results. An important work that he published on tumours was never completed, and he subsequently devoted himself, apart from politics, mostly to anthropology and archæology. In these spheres also he achieved much that is of value, not least on account of his initiative — the great Museum für Völkerkunde in Berlin, for instance, was founded by him — but this work is by no means to be compared in importance with the products of his youth.

Virchow's cellular pathological-theory has been of great importance ✓
to the development of biology, owing to the fact that he established, as
no one else had done before him, the cell's character as an independent life- ✗
unit. He denies any form of spontaneous generation, whether within the
organism or without in nature. Just as it is impossible for an ascaris worm
to arise out of intestinal slime or an infusorian out of decaying matter, so
it is not permitted in the physiological or pathological tissue-theory for a
cell to be constructed "*aus irgend einer unzelligen Substanz.*" And he continues:
"*Wo eine Zelle entsteht, da muss eine Zelle vorausgegangen sein, ebenso wie das Tier
nur aus dem Tiere, die Pflanze nur aus der Pflanze entstehen kann.*" It is this principle of cell multiplication, and thereby also of the cell's role in the organ- ✗
ism as a whole, that represents Virchow's great contribution to the history
of biology. He himself applied his principle mostly to the sphere of pathology, in which he created with its aid a new theory of the origin not only
of tumours and other new growths, but also of purulent bodies. Otherwise,
his conception of the cell was in no way original; he mentions as its neces- —
sary components the membrane and the nucleus, and considers its other
"fluid" contents to be less essential. In his general conception of the vital
phenomena Virchow is to a certain extent undecided; on the one hand, he
declares that there is a special life-force, that life is not a mechanical result
of the molecular forces of the bodily parts, while, on the other hand, he
holds that this life-force is probably of mechanical origin. Throughout his
life Virchow was of a very pugnacious disposition and used to defend his
views with great vehemence; with Haeckel in particular — once his own
pupil — he entered into violent controversies, not only on scientific, but
also on social questions, which both disputants were strongly inclined to
confuse with one another. But these disputes belong to the next era.

The modern conception of the life and component parts of the cell was
founded by Max Schultze, a man who, in spite of a short life, made a lasting name in the history of biology. MAX JOHANN SIGISMUND SCHULTZE was
born at Freiburg in 1825 and studied at Greifswald (his father had been professor of anatomy in both places), and he also attended the lectures of
J. Müller in Berlin for a short time. He was at one time a lecturer at Halle
and from there was appointed to a professorship at Bonn. He worked with
success there, but died in 1874.

Schultze's field of activities was very extensive; he devoted himself to
microscopical subjects in a number of animal classes. Specially famous are
his writings on the single-celled animals, to which further reference will be
made later on; they form one of the foundations of his cell theory. Furthermore he carried out important investigations into the microscopical anatomy
of worms and molluscs; in the Vertebrata he studied the terminal ramifications of the nervous system, and made weighty contributions to the

knowledge of the structure of the electrical organs. All these works, valuable as they are, are nevertheless put in the shade by a short essay in the *Archiv für Anatomie und Physiologie* of the year 1861, entitled *"Über Muskelkörperchen und was man eine Zelle zu nennen habe."* Schultze has hereby laid the foundations of the modern idea of the cell. "What is the most essential thing in a cell?" he asks at the beginning of the essay. The old theory, which, as we have seen, Virchow still embraced, would answer: "A vesicle surrounded by a membrane, with a nucleus and fluid contents." Schultze refers to the embryonic cells and points out that these consist of a mass of protoplasm with nucleus, but without any surrounding walls; the membrane which had previously been supposed to surround these cells, and which certain investigators had brought out by chemical means, he proves to be an artificial product. He further points out that only cells without any membrane can multiply by division; those cells possessing a membrane which are found in the animal kingdom thus lead a restricted and limited existence — "They may be likened to an incapsulated infusorian or an imprisoned animal." Again, the substance that surrounds the nuclei in the different tissues — muscular fibrillæ, connective substance — is not, as has been declared, a substance foreign to the cell, but a transformation of the protoplasm itself. Accordingly, the protoplasm, in conjunction with the nucleus, is the basis of all the life-manifestations of the cell, and the very name "protoplasm," which had hitherto been used only by the botanists, is introduced as the universal term for the fundamental substance in the cell. And in connexion therewith this substance is characterized with reference to the conditions obtaining in plants, in the lower and higher animals; it is maintained that the cell-mass is by no means a fluid, but an element having a definite form, a consistency which is different in different animal forms and different kinds of cell; it is indissoluble in water and possesses, when it is free to do so, an independent power of motion, which is characteristic for different cases. It is sometimes possible also for a number of nuclei to be surrounded by a common protoplasm, which again can afterwards form cell-boundaries and thus produce isolated cells. The actual word "cell" Schultze reserves for the vital element represented by the nucleus and the protoplasm, and this meaning has also been retained since, illogical though it is, seeing that the word "cell" means a space *with* walls, whereas the living cell is characterized by the fact that it lacks walls.

Improvement of histological technics

THIS contribution made by Schultze laid the foundations on which cell research has since been built, and this marks a new era in the science of cytology. The aids to research that the students of the period just described had at their disposal had been comparatively limited — microscopes of primitive construction, with which cells and tissues were studied in their natural state or at the most after dissection. Just at this transition period, however, there appears a new and far more perfected method; it was not merely that microscopes were rapidly improved, but mechanical and chemical means of a kind hitherto unknown were now beginning to be discovered and to become widely used. Of means of preservation there were already known spirits and certain saline solutions, which, in conjunction with boiling, were used for the purpose of giving the objects of investigation greater durability. Now the method was introduced of "fixing," by various means worked out for each particular purpose, the structures that it was intended to examine. Of these methods, chromic acid was introduced as early as in the thirties by JACOBSON, potassium bichromate at the same period by HEINRICH MÜLLER, and osmium acid in the sixties by Schultze, not to mention a number of similar means that have been discovered since. By the use of different colouring-matter it is possible for the structures thus fixed to be brought out clearly even in the thinnest and most transparent sections; in 1849 carmine colouring was introduced by HARTING, in 1863 WALDEYER's hematoxylin was produced from Campeachy wood, and in the same year BENECKE's colouring with analine associations, which, as is well known, have since then been produced in immense numbers. In 1870 HIS introduced the microtome, an instrument that can make extremely thin sections through the tissues, the construction of which has been varied in many ways. The new discoveries that were made possible by this methodology belonged to the next period.

CHAPTER VIII

THE CONTINUED DEVELOPMENT OF BIOLOGY UNTIL
THE ADVENT OF DARWINISM

1. Experimental Rearch Work

Development of organic chemistry

WHILE BIOLOGICAL RESEARCH was yielding the abundant results which have been described above, it was subject to very important influences from other natural sciences in two special spheres. We have described how, thanks to Berzelius, chemistry had extended its inquiries to the sphere of living beings, and how an immense number of substances of quite a peculiar kind were analysed and described. These substances, which nowhere exist in inanimate nature and might consequently appear exclusively to have "life" to thank for their origin, were called organic associations; their existence was considered to be one of the most palpable proofs that life itself was in its essence utterly distinct from the phenomena that take place in inanimate nature, and even independent of the chemical and physical laws that govern lifeless matter. Organic chemistry thus became, the more it developed, the strongest support for the theory of a special life-force as the essential precondition for all that takes place in animate nature. The theories maintaining this force therefore gained ground amongst an ever-increasing number of biologists; as we have seen, Johannes Müller embraced a theory of this nature, as also did many of his school, and even a scientist like Magendie, opposed to speculation though he was, could not help acknowledging the invalidity of the ordinary chemical laws when applied to living nature. It was in these circumstances that Wöhler made his great contribution to natural science.

FRIEDRICH WÖHLER was born in 1800 near Frankfurt am Main; he became a doctor, but after taking his degree he devoted himself entirely to chemistry. In order to obtain the best training available at the time he went to Berzelius and worked in his laboratory for a year under the strict control of the master. Having returned home, he became a teacher at a *Gewerbeschule* in Berlin and eventually professor at Göttingen, where he died in 1882. He was a very distinguished student of chemistry, but his other activities are overshadowed by his synthesis of urea out of cyanammonium. By this

406

discovery an element of pronounced "vital" character had been produced out of components that in their turn may be entirely produced out of simple, inorganic elements. This first synthesis of an organic element out of inorganic components was naturally succeeded by countless others; organic chemistry, which at one time had been thought to embrace elements produced by life and impossible to arrive at in any other way, thus became a chemistry of the carbon compounds, the unique character of which is due to the nature of the elements with which it operates, but which otherwise has recourse entirely to the methods and theories of general chemistry. The science of chemistry has thus become a knowledge of phenomena that are governed throughout nature by the same laws, with the result that a new possibility arose of combining separate phenomena under one common point of view.

Indestructibility of energy

OF still more radical importance, however, was another discovery, which was made at a somewhat later date than that just mentioned and which led to the well-known law of the indestructibility of energy. In earlier times heat was regarded as an element, a kind of "fluid," like electricity, and although Lavoisier proved its imponderability, both he and his pupils retained the ancient idea as to its essence. Gradually, however, attention was once more attracted to the fact, known ever since ancient times, that heat arises through friction, whence conclusions were drawn as to the connexion between heat and mechanical action. The law-bound condition that arises therein was elucidated in the forties by several investigators working on the subject simultaneously and independently of one another — which only proves how ripe for solution the problem really was. Although the phenomenon falls entirely within the sphere of physics, there were, strangely enough two scientists with an essentially biological training who played a decisive part in its solution, and this fact, as well as the importance of the subject in itself, justifies our going into it in somewhat greater detail.

JULIUS ROBERT MAYER was born in 1814 at Heilbronn, the son of a well-to-do apothecary. He studied medicine and, having passed his examinations, became a practitioner in his native town. He was seized, however, with a desire to see something of the world and he succeeded in obtaining an appointment as a doctor on a Dutch vessel sailing to Java. Having returned home, he once more settled down as a doctor in Heilbronn and died there in the year 1878.

During his stay in Java Mayer had noticed that in venesection the blood in the veins was of a far lighter colour than that in Europe. He started out to discover the reason for this and finally came to the conclusion that the metabolism in the body is dependent on the temperature of the atmosphere; the warmer the temperature, the weaker the conversion of substance is required to be in order that the body may perform its normal functions retaining its normal heat. But then there must also be a definite ratio between the heat produced by combustion in the body and the work that the body performs during a given period, or, in still more general terms, a certain amount of heat must correspond to a certain amount of work. Upon his return home Mayer published in Liebig's *Annalen der Chemie* in 1842 an essay in which he expounded his theory, and in connexion therewith the method, which is still in use, of calculating the dynamical equivalent of heat when the unit of heat represents the amount it takes to heat up a given quantity of water one degree, and the unit of work represents the force required to lift a given weight to a given height — in our days one kilogram one metre. For this ratio he gave a number, which, however, was later found to be incorrect. Shortly after the publication of Mayer's report the English physicist J. P. JOULE published a theory based on years of experiment and having the same gist as Mayer's, but giving a more correct number to represent the heat equivalent; moreover, it was founded on more substantial proofs and supported by a greater number of facts. Then in the year 1847 came out Helmholtz's essay *Von der Erhaltung der Kraft*, in which the law of the indestructibility of energy was elucidated from all points of view and was given its mathematical formula. During the succeeding years, however, Mayer had further elaborated his theory; of particular value to biology was his essay *Die organische Bewegung in ihrem Zusammenhange mit dem Stoffwechsel*, which was printed in 1845 as a pamphlet because it was refused by the editors of scientific journals. In this essay he applies the law of the indestructibility of energy to the vital phenomena in the animal and vegetable kingdoms, gives an account of the mutual relation between muscular action and the digestion in the body's exertion of energy, and at the same time shows the process of assimilation in plants to be the foundation of life on the earth, and solar energy to be its ultimate source. In consequence of this he feels it to be superfluous to assume a special life-force as a source of the metabolism in the living body. This caused but little feeling of satisfaction amongst the biologists of his age; as a matter of fact, the whole theory of the nature of force met with opposition even on the part of the older physicists. When this theory eventually won the day, Mayer considered that due attention had not been paid to his right of priority. This wounded his naturally sensitive feelings, which were exposed to still more serious shocks in the year of the revolution, 1848; he was, in fact, from both a political and a religious point of view, strictly conservative, with the result that he joined a different camp from that of the majority of natural scientists, who were for the most part liberals, and besides he fell out with his brothers, who took part in the revolution. As a result of all these vicissitudes his nerves were completely shattered; finally, after an attempt at suicide, he had to be placed under restraint, and in accordance with the custom of the time he was put into a strait waistcoat. By degrees, however, he recovered his health and in his old age had the satisfaction of being universally recognized as the one who had first laid down the principle of the conservation of energy.

Of Mayer's rivals Joule belongs entirely to the history of physics. Helmholtz, on the other hand, worked both as a physicist and as a biologist and therefore deserves further mention in this place. HERMANN LUDWIG FERDINAND HELMHOLTZ was born in 1821 at Potsdam, where his father was a teacher in the gymnasium. He studied medicine in Berlin, where he was one of J. Müller's pupils, and became first of all an army doctor, afterwards being appointed professor of physiology at Königsberg (in 1849), and later holding the same appointment at Bonn and Heidelberg. In 1871, however, he was made professor of physics at the University of Berlin, and somewhat later he became director of a newly-founded physico-technical institute at Charlottenburg. These two posts he held until his death, in 1894. Being universally regarded as one of the foremost scientists of his day, he was the recipient of innumerable honours both at home and abroad. His research activities were also as multifarious as any that natural science has had to record in recent times. As his career testifies, he was an expert in both biology and physics; besides this, he was not only a mathematician and a philosopher of high standing, but also an excellent stylist and an eloquent speaker. As his doctor's dissertation he published a valuable account of the nerve-cells in ganglia and the nervous ramifications emanating therefrom in different animal forms. His measurement of the rapidity of the reproduction of impressions through nerve-fibres was of fundamental importance. Of still greater significance, however, was his work as a sense-physiologist. Modern physiological optics in particular were in all essentials founded by him. He invented the ophthalmoscope, by the aid of which it has become possible for doctors to examine the retina of the eye; he further explained the mechanism of lens-accommodation and also founded the theory of colours and colour-perceptions that has been adopted in modern times. Physiological acoustics were likewise founded by him; he explained the connexion and mechanical action of the bones of the ear, as also the part played by the organ of Corti in the perception of tone quality. Again, from the point of view of purely theoretical science, he worked out a theory of the sense-perceptions, in which he dealt with such abstract and complicated questions as the relation of the geometrical quantities to the conception of sense, and the justification of the geometrical principles based thereon. His purely physical and mathematical work naturally falls outside the scope of this history.

Helmholtz works out the theory of indestructibility

THE above-mentioned paper *Über die Erhaltung der Kraft* (*On the Conservation of Force*), nevertheless, deserves further mention here. As a result of it, the law of the conservation of energy was given the theoretical formula that has been adopted ever since. Helmholtz, who was physicist, mathematician, and biologist, had, in fact, special qualifications for laying down on an empirical basis an exact formula for this generally accepted principle, and during the succeeding epoch it was his name that was most often associated with this radical change in the general conception of nature. It was this very contribution, however, that caused Helmholtz a good deal of unpleasantness. In his paper he had made no mention of Mayer, because he was not aware of his first essay, but in a later lecture he fully acknowledged Mayer's priority. This, however, was not enough to satisfy Mayer's admirers; one of them, E. Dühring, a lecturer in philosophy, made an extremely bitter and personal attack on Helmholtz, as if he had sought to appropriate an honour to which he had no right. Disciplinary action was taken against Dühring, but Helmholtz kept silent until Mayer, broken in health, had passed away; then he took up the challenge and pointed out how Mayer, with all due acknowledgment both to his genius and to his right of priority, had nevertheless based his views on speculation rather than on empirical research. On this point Helmholtz was undoubtedly right; Mayer was no experimental scientist — he never had a laboratory at his disposal — but he was a brilliant thinker who, with the aid of the observations of others and his own ideas, achieved his epoch-making results by theoretical means. The history of natural science proves, however, that theoretical conclusions of this kind are seldom given the same significance as conclusions drawn from the student's own empirical observations — Swedenborg's brilliant scientific speculation, which, though the work of a genius, was nevertheless forgotten by the immediately succeeding generations, is of course the classical example of this — and besides through his ill health Mayer was prevented from following up his idea, which he would undoubtedly have done had circumstances permitted. As it was, he had to divide the honour with Joule, the experimenter, and Helmholtz, the universally trained thinker and observer; all three contributed towards working out the principle of the conservation of energy — certainly the most important theoretical contribution of the past century in the field of natural science.

Through the principle of the conservation of energy the experimental study of living organisms received a powerful stimulus; immediate steps were taken to apply to as many life-phenomena as possible this new conception, which placed all phenomena in existence, both animate and inanimate, in one single simple and clear causal connexion, and which offered the hope of being able to bring all manifestations of life, even the most complex, under the same simple explanatory principles that physics and chemistry had already adopted. The following decades were therefore a period of brilliant achievement in the field of experimental physiology; both its aims and its methods took definite form during that period, so that in the medical faculties at the universities this science likewise has its own representatives and its own laboratories provided with special equipment. One or two of the most important representatives of this line of research will be cited here as examples showing the aims of physiology during this period and its attempts to realize them.

EMIL DU BOIS-REYMOND was born in 1818 in Berlin. His parents were of French extraction and came from Neuchâtel, which then belonged to Prussia. Some time after their son's birth they moved back to their home district, so that the boy grew up in a French environment, but at the same time, thanks to family influence — his father was a Prussian official — he acquired a strong affection for Prussia. After completing his school studies, therefore, he went to the University of Berlin, where, after some wavering as to a career, he applied himself to medical studies and became a pupil of J. Müller. In 1858 he became Müller's successor as professor of physiology and held this post until his death, in 1896. He never had any very large circle of pupils; but the influence on the educated public which he exercised as secretary to the Berlin Academy of Science was all the greater. The lectures that he had to hold annually in this capacity proved to be brilliantly eloquent; he usually took some subject from the theory of history or natural science, sometimes even discussing political questions of the day, for Du Bois-Reymond was, in spite of his French mother-tongue, a warm German, or rather Prussian, patriot, with an almost devout reverence for the reigning family. These lectures displayed deep scientific general knowledge and keenness of thought and they possessed a lasting value in German literature.

Electric currents in the living body

IN 1840 Du Bois-Reymond was commissioned by J. Müller to study the phenomena of electric currents in the nervous and muscular systems and he was thus led to take up a field of research that he never afterwards abandoned. He recorded his results in an important work entitled *Untersuchungen über tierische Electrizität*, the first part of which came out in 1848, but the work was never completed; the last part was published in 1884. As R. Tigerstedt has justly remarked, it is seldom that an investigator has for so long occupied himself exclusively with so limited a sphere of research. That Du Bois-Reymond must nevertheless be counted amongst the pioneering natural scientists of his age is due to the general principles he expressed and consistently applied in his research work. In the foreword to his great book he expounds his ideas on the innermost essence of vital phenomena, and probably the weaker sides of the vitalistic theory have never before and seldom since been subject to such keen and striking criticism as here. The old life-force is reviewed from all sides and the arguments in its favour are refuted one after another; the finality of the living organisms, which so impressed

J. Müller, is rejected in view of the equal finality prevailing in the inanimate universe; the life-force's quality of resisting the chemical disintegration of the organism — the basis of Stahl's and Bichat's systems — is likewise cast aside in consideration of the fact that a force which without a struggle abandons its material foundation is inconceivable, for force is in reality nothing but a quality of matter; both belong to one another and together represent an expression for natural phenomena, such as science imagines them to be. Force and matter are "*von verschiedenen Standpunkten aus aufgenommene Abstractionen der Dinge wie sie sind. Sie ergänzen einander und sie setzen einander voraus.*" From this he draws the bold conclusion that the difference between organic and inorganic nature is of no importance whatever, and he finds additional support for this assertion in the principle of the permanence of force as formulated by Helmholtz. He maintains also that if the organism presents phenomena which do not exist in inorganic nature, this may be due to the fact that the elements intrinsic in them, though they may be provided with the same qualities and none others, nevertheless enter into new connexions with one another and therefore display new qualities. On these grounds physiology should come entirely under organic physics and chemistry. His own contribution to this plan consists in his investigations into electrical phenomena in the animal kingdom. His most important discovery in this field is the very fact that the muscles and nerves of animals during their state of activity produce electric currents that can be observed and measured with the aid of the usual apparatus of electro-physics. As a result of his study of these currents he demonstrated in practice that phenomena which in the most marked degree belonged to the manifestations of life may be dealt with with quite as much exactitude as the ordinary physical phenomena, thereby providing the most patent proofs of his theory of the physico-chemical qualities of vital phenomena. He based his explanation of these electrical phenomena, however, upon a theory that is untenable, a theory according to which the muscles and nerves are composed of a kind of electrical molecules. Generally speaking, it was a weak point in the physiologists of that era that they overlooked the complex structural conditions of cells and tissues and were thus tempted to deal with the phenomena of life more schematically than accords with the reality.

Besides this limited but nevertheless important specialized research, Du Bois-Reymond, as already mentioned, contributed a number of ideas on questions of general science; specially remarkable is his lecture "*Über die Grenzen des Naturerkennens*" (1872), wherein he seeks to establish the limits of natural research and comes to the conclusion that, though biology might eventually master the laws governing vital phenomena as completely as the astronomers when they calculate the motions of the heavenly bodies, yet science would never be able to determine what matter is or what consciousness is; in regard to both these fundamental hypotheses science must pronounce not only an "*ignoramus*," but also an "*ignorabimus*." While this statement was received with unreserved approbation in many quarters — *inter alia*, by such a keen-minded thinker as Albert Lange — on the other hand, it excited feelings of extraordinary bitterness on the part of radical students of nature, headed by Haeckel. However, this subject, as also Du Bois-Reymond's opinions regarding the theory of origin, belongs to the next period and will be dealt with when the time comes to describe it.

KARL FRIEDRICH WILHELM LUDWIG appears by the side of Helmholtz and Du Bois-Reymond as a pioneer in the field of exact physiology. He was born at Hessen in 1816, studied at Marburg and Erlangen, and, having taken his doctor's degree, spent some time in Berlin, where he joined J. Müller's circle, without, however, being included among his direct pupils. In 1846 he became professor of physiology at Zurich, and was called thence to Vienna and later to Leipzig, where a new institute was founded for his benefit. There he laboured for thirty years as a professor, gathering around him pupils from all countries. He possessed rare powers of organization, which were best displayed in the manner in which he arranged and guided his pupils' work. In his earlier years he developed, with their assistance, a considerable literary production; later on he preferred, modest as he was, to let his pupils publish the ideas he suggested to them. Universally respected, he worked with undiminished powers until the end; he died in 1895.

Ludwig's original investigations primarily concern the functions of the vegetative organs. Thus, he has explained the connexion between the secretion of the salivary glands and the nerves that affect those organs; he investigated the function of the heart in detail and analysed its various phases; he experimented with the circulation of the blood and took valuable measurements of its rapidity. Furthermore, he invented the graphic method, which has since then played an important part in physiology. In a *Lehrbuch der Physiologie* of the year 1852 he summarized his views on the phenomena of life. Characteristically enough, this work starts with a chapter on *Physiologie der Atome*," which is really a survey of animal chemistry, while the following chapter, "*Physiologie der Aggregatzustände*," deals with the phenomena of dissolution, diffusion, and currents. Consequently the functions of the different organs are presented from a purely physical and chemical point of view, the author's special subjects, the phenomena of circulation and secretion, naturally receiving specially radical and expert treatment. Ludwig's teaching was, of course, conducted on similar lines; through his pupils the conception of vital phenomena here described spread to all civilized countries.

Of radical importance for the development of biology was the fact that the vital phenomena were thus explained by means of the same experimental method that had been worked out earlier in physics and chemistry; it was established once and for all that the same processes that take place in inor-ganic nature exist also in the living organism — in other words, every vital process has its purely physico-chemical progress. As usual, however, the great advance thus made led to an overestimation of the possibilities thereby opened up for science; the pioneers of experimental biology, as R. Tigerstedt justly remarks, entirely overlooked the part played by the cell and its various structural forms in the vital processes. They saw in the living body merely the basis for simple physical and chemical processes and they overlooked the extremely complex structures which represent the fundamental condition for the operation of these forces and on account of which the phenomena of life actually become infinitely more complex than the physiologists were pre-pared to admit. As, later on, knowledge of the cell-structure increased, there arose in connexion therewith a mistrust of the over-simplified idea of the phenomena of life, which in certain quarters caused a return to that vitalis-tic biology that exact physiology imagined it had disposed of for all time. We shall shortly make the acquaintance of this so-called neo-vitalism.

2. Morphology and Classification

As far as comparative anatomy and morphology are concerned, the period we are now describing is one of transition, during which the remains of the old natural-philosophical manner of viewing life appears side by side with ideas produced by the great discoveries that have been recorded in the foregoing. Besides this, we find in the research work of this era much that fore-shadows the advent of the origin-of-species theory. A review of the most important events in the morphological research work of this period will therefore form a suitable transition to the great epoch-making discovery of the sixties.

Contemporary with J. Müller in Germany there appeared in England a comparative anatomist of more than ordinarily wide significance. RICHARD OWEN was born in 1804 at Lancaster, the son of a merchant. At school he gave no special indication of genius and was therefore apprenticed to an apothecary; then, after some years, he went to Edinburgh to study medi-cine and there obtained his doctor's degree. Having established himself in practice in London, he devoted his leisure hours to the study of anatomy and as a result of his work on the subject, he became an amanuensis at the Hunter Museum, and later, after the resignation of Charles Bell, he was appointed its director. In 1860 he was made head of the natural-science department of the British Museum and carried out its removal to South Kensington. He held his position for an extraordinary length of time; in his eightieth year he resigned and after that lived for some years in retirement; after failing in health for some time he died in 1892.

KARL ERNST VON BAER

JOHANNES PETER MÜLLER

Owen is generally counted as England's greatest comparative anatomist. His activities were extraordinarily many-sided, and, thanks to his position as head of one of the world's largest museums, he had particularly favourable opportunities for investigating a quantity of rare animal forms, both still existing and extinct, which he described in essays illustrated with very fine and carefully drawn pictures. Among these special investigations may be mentioned his description of the nautilus — the first specimen that had ever been seen of the animal itself, the shell of which, however, had been known since antiquity, and, further, the anatomy of the Brachiopoda and the lung-fishes. Furthermore, he made a thorough study of the gorilla and certain other rare forms of the ape family and the curious finger-animals from Madagascar; and amongst fossil forms, the Saurian bird, Archæopteryx, and the extinct giant birds of New Zealand. His monumental work on dental forms in the Vertebrata should also be recorded.

Homology and analogy

ON the basis of this unique wealth of material he built up a number of theoretical speculations upon the organization of the entire animal kingdom, which had a great influence on the biology of the succeeding age. In one of the courses of lectures which he as curator of the Hunter Museum had to give each year on the subject of comparative anatomy, he takes as his starting-point his predecessor's plan of comparing the same organ through all the animal groups and combines it with Cuvier's principle of examining the mutual relationship of the different organs in one and the same animal form in order to be able thus to ascertain the causes of the changes that the organs have undergone in the different animal types. As a matter of fact, he adheres to Cuvier's type theory throughout, condemning Bonnet's simple evolutional series covering the entire animal kingdom. In making this comparison, he proves that the same function can be exercised in different animal forms partly by similar and partly by entirely dissimilar organs; the dragon-lizard flies with its outstretched ribs, the flying fish and the bird with their extremities, the insects, again, with folds in the skin, which were originally gills. This last idea, according to his own statement, he got from Oken; the gills of the fishes and the lungs of the higher animals possess the same function, but are not the same organs; rather the swimming-bladder and the lungs correspond to one another, as is proved by the lung-fishes. This contrasting relation between the function and the character of the organs he expresses by the terms "analogy" and "homology"; "analogue" denotes "a part or organ in one animal which has the same function as another part or organ in a different animal," and "homologue" denotes "the same organ in different animals under every variety of form and function." The homologies are of course the object of his special interest, chiefly those in the Vertebrata, the bone-structure of which he made the subject of a special work.

He there mentions three distinct types of homologies: *special* homology, or agreement between a part or an organ and a part or an organ in another animal; *general* homology, or the relation between an organ or a series of organs and the general type in conformity with which the animal in question is constructed; and, finally, the series homology, or what would nowadays be called *metameric* homology — that is, the repetition of certain organs in the same individual, segment for segment. He dilates upon these different forms of homology, citing numerous examples. "Special" homology in particular is discussed in detail, a uniform nomenclature, on Linnæan lines, being given for homologous bones throughout the whole vertebrate series; many bones that had hitherto been denoted by a prolix character here acquire for the first time names of their own, besides which a number of other names are rejected as unsuitable.

Anyone with any knowledge of modern comparative anatomy must at once realize how important these terms and ideas created by Owen have been to present-day biology. Indeed, the very idea of homology has proved one of the most fertile grounds for comparative anatomy, although its real meaning has become somewhat altered in the course of time. And the special applications referred to — the comparison between swimming-bladders and lungs, the derivation of insects' wings from respiratory organs — we find amongst the most frequently quoted arguments on behalf of the modern theory of evolution.

Owen's romanticism

YET Owen himself was by no means a modern biologist in his general conception of nature; rather, he stood considerably nearer romantic natural philosophy. He was a great admirer of Oken, whom he extols as being a particularly deep thinker and whose theory of the cranium's being composed of vertebræ he adopted and endeavoured to apply further; also he highly admired Geoffroy Saint-Hilaire, with whom he really has more spiritual affinity than he has with Vicq d'Azyr and Cuvier. Like Geoffroy, he speculates upon a common "archetype" for all vertebrates; he reconstructs one and illustrates it in one of his works, and to this archetype are referred the "general homologies" mentioned above. And he preferred not to recognize the origin of the higher life-forms from the lower and the parallel derivation of the more highly developed organs from the more primitive. He was irreconcilably hostile to Darwin's theory in particular; upon its appearance he challenged it anonymously and in quite a heated controversial spirit, which certainly resulted in the exposure of a few weak points, but showed a great lack of understanding of the true value of the theory. Later, however, while maintaining the purely hypothetical character of the origin-of-species theory, he acknowledged the correctness of Lamarck's assertion that only individuals exist and that the term "species" is relative. On account of this attitude, however, Owen became more and more isolated towards the end of his life, and after his death there were erected in the museum he had founded the statues of Darwin and Huxley, but not his own. Nevertheless, there are to be found deep traces of his influence even in the champions of the origin-of-species theory, chiefly perhaps in Haeckel, whose principal work even in its very title — *Generelle Morphologie* — recalls one of Owen's expressions referred to in the foregoing, and whose method of morphological comparison shows obvious traces of the influence of the English anti-Darwinist.

In the same year as Owen there was born another of the foremost biologists of that period — namely, KARL THEODOR ERNST VON SIEBOLD. At the time of his birth (1804) his father was a professor at Würzburg, but was afterwards called to Berlin. The son studied there under Rudolphi and at Göttingen under Blumenbach, but after attaining his doctor's degree had to take up a practice, first in the provinces and afterwards in Danzig, whither he removed in order to have an opportunity of studying marine animals. On account of his writings he was appointed professor, first of anatomy and physiology at Erlangen, later, in succession to Purkinje, at Breslau, and afterwards of zoology and comparative anatomy at Munich. There he died in 1885, having for some years previously been incapable of fulfilling his duties owing to ill health.

Besides Siebold we should mention his friend FRIEDRICH HERMANN STANNIUS (1808–83). Born in Hamburg, he studied at Breslau and Berlin under J. Müller; he became a lecturer under him and later professor at Rostock. Thanks to his activities there, that small and ill-conducted provincial university acquired a wide reputation. Unfortunately his career was prematurely cut short; after some years of failing health he became the victim of an incurable mental disease and spent the last two decades of his life in an asylum.

The results of the collaboration of these two scientists were recorded in a *Lehrbuch der vergleichenden Anatomie*, in which Siebold took the invertebrate section and Stannius the vertebrates. This work was published in 1846 — that is to say, about the same time as Owen's work just referred to, and manifestly quite independently of it. A comparison between the two works is therefore not without its interest. In both the manner of presentation is largely the same; the organs are dealt with no longer throughout the entire animal kingdom, but each larger main group is considered by itself, and the organs are compared within that group; the natural-philosophical method of comparison of the early years of the century is abandoned. The two German biologists generally avoid all speculation; their presentation is based solely on fact; thus, they do not possess the wealth of ideas of Owen, but their presentation is founded on a particularly many-sided knowledge of detail; they also pay special attention to microscopical anatomy, which

Owen does only in a minor degree. Through their comparative anatomy Siebold and Stannius made an extremely valuable contribution to the development of morphological science. Each also laboured separately and produced important results.

Stannius was especially many-sided, both as a teacher and as an investigator; he taught pathology, physiology, and anatomy and has made valuable additions to our knowledge in all these spheres. As a physiologist he became famous for his attempts at underbinding the various parts of the heart of the frog; further, he investigated the nerves of the tongue with reference to their functions of taste and movement, and also the dependence of muscular contraction upon nervous irritation. As a comparative anatomist he was especially remarkable for the splendid work he wrote on the peripheral nervous system of the fishes, which is still authoritative, in spite of all the more recent work accomplished in that field.

Siebold, as is evidenced by the share he took in the *Comparative Anatomy*, devoted himself mostly to invertebrate research. In this sphere he was without doubt one of the foremost of his age. With a view to furthering the development of research in his own special sphere he founded, in association with Kölliker, the *Zeitschrift für wissenschaftliche Zoologie* — the title being chosen of set purpose in opposition to the soulless method of classification — which was started in 1852 and has since been one of the chief organs for the furtherance of biological science. In many specialized fields of invertebrate research Siebold's contributions have been considerable. Foremost in this regard come his investigations into parasites. It may be recalled that even Rudolphi believed that intestinal worms arose as the result of a diseased process in the host. At an early stage Siebold was quite convinced that this kind of spontaneous generation could not be accepted as rational. He adduced in proof of this argument the existence of large quantities of eggs in the intestinal worms, which obviously indicated that these animals reproduced themselves in the same way as other animals. But the question that still required an answer was how the offspring of the parasites come to harbour in a fresh host. As yet this question was insoluble. To answer it required a knowledge of a phenomenon of evolution — the alternation of generations — which was still unknown at the time and was elucidated by another scientist, who must therefore be described in this connexion.

JOHANNES JAPETUS SMITH STEENSTRUP (1813–97) was the son of a priest from Jutland, in Denmark; he studied in Copenhagen, becoming professor of zoology there after having been a schoolmaster for some years. He was an extraordinarily gifted and many-sided investigator, working in many widely differing fields of research; particularly celebrated are his investigations of peat-mosses, which he explored not only for zoological, but also for botanical, geological, and archæological purposes. He also discovered the ancient shell-mounds from the Stone Age that are called *kjökkenmöddinger* (refuse-heaps), and studied them with valuable zoological and ethnographical results. He was, moreover, interested in marine research and he made a discovery in this field that more than anything else ensured to him a place in the history of biology — namely, the alternation of generations. A. VON CHAMISSO, famous as a poet and circumnavigator, had already proved that in the Salpa free individuals and individuals bound together in chains alternate with one another by generations, but this discovery had been almost entirely neglected. Other observations of a similar kind had also been made before, as, for instance, by MICHAEL SARS in Norway and SVEN LOVÉN in Sweden, and also by Siebold himself in Danzig, but it was reserved for Steenstrup to complete the material for observation and to place it under one common point of view. In 1842 he published his work on the alternation of generations, in which he gives an account of the evolution of the medusæ, the Campanulariæ, the Salpa, and the Trematoda, and finds the phenomenon common to them all — namely, that there exists in them an alternation between the adolescent stages, which he terms "nurses," because without sexual reproduction they develop a new generation, the sexually mature, and this sexual stage, which, in its turn, gives rise by means of ordinary sexual reproduction to new "nurses." Of such asexual generations he sometimes finds many, one after another, particularly in the Trematoda, in which they are often described as one independent genus, the cercariæ. In its details this work certainly required both correction and completion, but its service lies in the fact that it laid down a common principle that proved to be indispensable if a conception was to be formed of the evolution of a great many of the lower animals. The fact that Steenstrup regarded these phenomena from a strictly romantic-philosophical point of view was only in accordance with the practice of the time; he saw in the alternation of generations a striving on the part of nature after freedom and perfection, and on these grounds he accounted for even the insect communities by the alternation of generations: the sexless workers he considered to be "nurses," which devote to the offspring a care of a more ideal character than that which the Salpa series and the cercariæ give to their own progeny.

Evolution of intestinal worms ascertained

IT was Siebold who now succeeded in making practical use of the alternation-of-generations idea, for he realized its fundamental importance for ascertaining the evolution of intestinal worms. He at once began to try to find out by experimental means the connexion between a number of parasite formations that had hitherto been regarded as independent of one another. Thus he proved that the parasite which by insinuating itself in the brain of the sheep causes the disease called "gid" or "sturdy," and which had hitherto been described as *Cœnurus cerebralis*, is actually an adolescent stage of a tapeworm that lives in the intestinal canal of the dog — if dogs were fed on cœnurus-affected sheep's brains they would infallibly be infected with this tapeworm, which is now called *Tænia cœnurus*. Through the excrement of the dog the eggs of the tapeworm are scattered about the pastures and thus infect the sheep. In the same way another tapeworm in the dog was traced to bladder-worms in the liver of hares, and a tapeworm in the cat to similar formations in the rat. Again, the large and often fatal liver-parasite echinococcus, which occurs in man, was found to give rise to an almost microscopically small, and therefore hitherto unknown, tapeworm occurring in the dog, the *Tænia echinococcus*, the eggs of which are conveyed, through too intimate contact with the dog, to the human mouth and thence to the intestinal canal. As a result of these researches the knowledge of the intestinal parasites, or helminthology, as it had already been termed previously, was placed on entirely rational footing, and it only remained for later observers, by working along the lines laid down by Siebold, to collect fresh facts in order to fill the gaps in the knowledge of the subject.

Likewise, as regards insect communities, Siebold produced the idea which has since been pursued up to the present time. At first he held with Steenstrup's attempts to explain the reproduction of the bees as a form of alternation of generations, but he realized his mistake, chiefly through collaboration with a pioneer in the field of practical apiculture, the Roman Catholic priest JOHANN DZIERZON (1811–1906), of Silesia. He is of course famous as the founder of modern rational bee-keeping, and it was he, too, who, notwithstanding his lack of anatomical training, realized before anyone else the true relationship between the sexes in the community life of the bees. He discovered that the queen-bee is fertilized only once in her life and this while in flight through the air, not inside the beehive, and that the drones are evolved out of unfertilized eggs, while workers' and queen-bees' larvæ are developed from fertilized eggs, both having female characters, although in the workers they become stunted, owing to lack of nourishment. These important discoveries, upon which Dzierzon based his reform of apiculture, were received with strong opposition on the part of most bee-keepers and would certainly have failed to win general acceptance had not Siebold given them the support of his authority. Already a long while previously (in 1837) he had carried out a careful investigation into the bees' sexual apparatus, with the result that he had discovered the queen's *receptaculum seminis;* now — at the beginning of the fifties — he placed himself definitely on Dzierzon's side and by means of a series of experiments and treatises on the subject obtained victory for his views. In connexion therewith Siebold elucidated for the first time the conditions obtaining in parthenogenesis in insects in general, thereby introducing into biological science a field of research that, especially in recent times, has aroused keen interest.

By the side of Siebold, Leuckart deserves mention as one of those who contributed towards the progress of biology in the middle of last century. KARL GEORG FRIEDRICH RUDOLF LEUCKART was born in 1822 at Helmstädt, where his father was a business man, and studied at Göttingen, especially under the physiologist RUDOLF WAGNER; he became a lecturer there, then professor of zoology at Giessen in 1855, and was called thence to the same chair in Leipzig in 1869. There he worked until his death, in 1898. Being still keenly interested in the advancement of science even in his old age, he gathered around him large numbers of pupils up to the end; helpful and warm-hearted, original and good-humoured, he won their affection and was universally praised after his death.

Leuckart's scientific activities were many-sided and of deep significance. While still a young lecturer he published an epoch-making work, *Über die Morphologie der wirbellosen Tiere*. "Descriptive zoology," he says, "must permit of the same comparative, morphological treatment as anatomy." From this point of view he discusses the zoological system prevalent at the time, taking as his starting-point Cuvier, whose type theory he unreservedly defends against the earlier belief in one single evolutional series, at the same time upholding the idea of idealistic morphology of a fundamental form after which "nature has constructed" the separate life-forms. Leuckart is definitely opposed to such systematical categories as are based upon negative characters, as, for instance, the Lamarckian group of Invertebrata. Nor indeed do Cuvier's four types meet the requirements of comparative morphology: apart from the Protozoa, whose then still undiscovered structure did not permit of definitive morphological treatment, and the Vertebrata, whose place was already established, Leuckart sets up five fundamental types — namely, Cœlenterata, Echinodermata, Vermes, Arthropoda, and Mollusca. These have, of course, been generally accepted since then, although with sundry modifications: Vermes have been further divided and Tunicata have been separated from the Mollusca. Through this reform, however, Leuckart brought the system a good step nearer the point that it has reached today. In particular his treatment of the Cœlenterata was epoch-making; besides establishing the difference in anatomical structure between them and the similarly radially symmetrical Echinodermata, he explains the curious division of labour that takes place between the individuals in certain colony-building forms within this class, principally in the group Siphonophora, wherein the individuals in a colony are converted for the purpose of performing a number of special functions necessary for the welfare of the whole community. As a result of this elucidation of the structure of these so-called "polymorphous" animal stocks, to a certain extent fresh light was thrown on the term "individual" in the animal kingdom. The fact that Leuckart believed he could compare the plants in general with these colony-formations does not detract very much from the value of the service he rendered in throwing light on an important field in the biology of the lower animals.

Leuckart's discovery of micropyle

LEUCKART made another valuable contribution to the development of biology in his discovery that the thick-shelled egg of insects is invariably provided with a canal through which fertilization takes place. This canal, which Leuckart, associating it with a corresponding formation in the vegetable kingdom, called "micropyle," was studied by him with great thoroughness in a large number of eggs of various insects. His investigation led him to the discovery that the spermatozoa actually do penetrate into the yolk of the egg through the canal — a discovery that essentially deepened our knowledge of fertilization. That the spermatozoa thus play a vital part in fertilization was a point which many investigators have had difficulty in realizing; after all, it was not so long ago that Spallanzani's theory of the decisive importance of the spermatic fluid had eminent supporters. J. Müller, it is true, had already discovered a similar canal in the eggs of the sea-urchin, but it was Leuckart who proved its widespread existence and thereby also its significance. Henceforth there was no doubt that the spermatozoon produced fertilization by penetrating the egg; it was then reserved for the future to ascertain the cytological course of development.

Of Leuckart's other works should be mentioned his investigations of the Spongida, which he referred to the Cœlenterata as a result of detailed study of their structure. Further, his experiments on the intestinal worms, conducted in competition with Siebold; it was Leuckart who found out the evolutional process in the two well-known human parasites *Tænia solium* and *saginata*, as also in the liver-fluke, which is so fatal to domestic animals. It was also thanks to him that the Trichinæ first became thoroughly known. His important text-book on the human parasites is the work on which all subsequent research in this field has been based.

A peculiar form of parasitism is presented by the crustaceans that make fishes their hosts; in these the parasitic degeneration assumes forms such as scarcely exist anywhere in the rest of the animal kingdom. The first real knowledge of these animal forms was established by ALEXANDER VON NORD-MANN. He was born in 1805 of a Germanized Finnish family in the county of Wiborg, Finland, studied at Åbo and from there went to Berlin, where he became a pupil of Rudolphi. There he wrote his work *Mikrographische Beiträge*, in which he deals with certain parasitic Trematoda, but chiefly with the parasitic Crustacea, which were hereby brought to the knowledge of science. The work attracted universal notice; J. Müller in a letter speaks of its "*herrliche Beobachtungen,*" and on account of it the author was called to a professorial chair in Odessa. There he applied himself to exploring the animal world of South Russia, both recent and extinct, which he described in a number of important works. In 1849 he was appointed professor in Helsingfors, where he laboured until his death, in 1866. In his old age he became an original character, and nothing of his later production achieved the fame of his early work.

The epoch now under discussion was on the whole prolific in biological students of high distinction; in the foregoing it has been possible to name only a few of those who took a prominent part in furthering the development of the science in particular branches; we shall now mention a further group of important scientists who made a special study of marine research and, above all, investigated the hitherto practically untouched lower animal world of the ocean. Scandinavia was at the time one of the main quarters of Europe in which interest was awakened in marine biology. As one of its original promoters the Norwegian MICHAEL SARS is worthy of mention. Born in 1805, he studied theology and became a priest on the west coast of Norway. There he began to interest himself in marine animal life and published his observations in a couple of treatises, which attracted widespread attention. He continued to follow the course he had thus entered upon, and in 1854 he became professor of zoology in Christiania, where he worked until his death, in 1869. Among his valuable contributions may be mentioned the discovery of metamorphosis in the marine Mollusca, his observations of the Crinoidea that are found at great ocean-depths and his establishing their likeness to large groups of similar animals from earlier epochs. He was a pioneer in introducing marine research into Scandinavia.

In Sweden the study of marine animal life was taken up by SVEN LUDVIG LOVÉN. Born in Stockholm in 1809, he studied in Lund, where he took his degree, and in Berlin under Rudolphi and Ehrenberg, made a journey of exploration to Spitsbergen in a fishing-sloop — the first of the many voyages of polar exploration that have started from Sweden — and finally became curator of the zoological department of the State museum in Stockholm, where he laboured for over fifty years. He died in 1895. He very greatly enriched the museum's collections and founded Sweden's zoological marine laboratory at Kristineberg.

Of Lovén's observations, most frequently published in the proceedings of the Swedish Academy of Science, may be mentioned his investigation of the metamorphosis of the Annelida — the name "Lovén's larva" still recalls the fact — his study of the evolution of the genus Campanularia, which provided Steenstrup with one of his ideas for the alternation-of-generations theory, and, above all, his investigations into the evolution of the marine Mollusca, in the course of which he established the formation of the so-called polar bodies and their expulsion from the egg — a phenomenon the evolutional universality of which was not determined until a long time afterwards. He was especially interested in the embryology of the lower animals and he carried out many valuable investigations in that field. He retained his youthful interest in polar research and keenly promoted Swedish voyages of polar exploration in his later years. When the theory of the glacial period was first advanced, he embraced it with enthusiasm and produced valuable zoological proofs of it, establishing the existence in some of the deeper inland seas of peculiar animal forms that otherwise belong to the fauna of the polar seas and which have manifestly survived in the lakes since some earlier period when the sea covered the land. These surviving forms Lovén named "relicts," and since then Scandinavian research has been occupied in their study.

The foremost Swedish biologist during this period, however, was undoubtedly ANDERS ADOLF RETZIUS (1796–1860). Born at Lund, where his father was a distinguished natural scientist, he studied under him and the anatomist Florman, and later in Copenhagen under LUDVIG JACOBSON. When still a young man he was appointed professor at the Veterinary Institute in Stockholm and at the same time held a post at the Carolinian Institute, where he at once became the greatest force the Institute had with the exception of Berzelius. This twofold work as a teacher, however, did not prevent him from following up his biological researches both at home and on expeditions, in the course of which he came into close contact with many eminent scientists, including J. Müller, who became a loyal friend, and Purkinje, from whom he learned microscopical technique. He is the pioneer of comparative anatomy in Sweden; he introduced it into the country not merely as a subject for research, but also — after overcoming strong opposition on the part of older authorities — as a subject of medical training. His own works are extraordinarily many-sided. A biographer has said of them that they are seldom consistently worked out and that the whole of his research work was marked by a certain restlessness. As a matter of fact, most of his literary production consists of short articles for journals, written simply in the form of notes without any theoretical reasoning or even observations on the earlier history of the problem under discussion. Nevertheless, many of these articles have had a deep influence on the development of biology, owing to the great value of the facts set forth in them, as, for instance, his account of the anatomy of the Myxinoidei, written in 1822, in which these animals' vascular and nervous systems, head-cartilage, and various other organs are described — a work which formed the basis of J. Müller's important monograph on the Myxinoidei — and, further, his study of the connexion between the spinal and the sympathetic nervous system in the horse — one of the first of its kind, and a beautifully illustrated work. Like Purkinje, Retzius investigated the microscopical structure of dental bone, extending this investigation to a number of animal forms. In 1841 he went with J. Müller to the west coast of Sweden and there applied himself, *inter alia*, to the study of the Amphioxus. In his later years, an ophthalmic disease having put an end to his microscopical work, he entered upon a new field of research, in which he undoubtedly carried out the most important work of his life — namely, comparative anthropology. Till then this science had followed the lines of Blumenbach; mankind had been divided into races for the most part according to the colour of the skin. Retzius began to take an interest in the discovery of human remains in the prehistoric graves of Scandinavia, and in the course of his investigations found considerable variation in the shape of the brain-cap. He extended these researches to include other human races and thereupon found that the skulls may be divided, according to the proportion between length and breadth, into long and short skulls — dolichocephalic and brachycephalic — and, further, after the shape of the facial bones, into orthognathic and prognathic. Several peoples externally akin were found to differ in these respects — thus, for instance, the Germanic are long-skulled, the Slav short-skulled — and as a result of this idea there was created a field of human research based on true comparative anatomy, which was afterwards followed up with splendid results by other investigators: Virchow, Broca, the younger Retzius, and many others.

In France during this period marine biology made rapid progress and contributed much of importance to the general development of our knowledge in this sphere; of the students who distinguished themselves here, two are especially worthy of mention. HENRI MILNE-EDWARDS (1800–85) was a native of Belgium, but of English descent. He came early to Paris and eventually worked as a professor there. Having been a pupil of Cuvier's, he carried on with distinction the latter's work in the sphere of invertebrate research. When quite a young man he published an extremely useful work, a comparative study of the vascular and nervous systems of the Crustacea; after that followed an extensive work on the fauna of the French coast, in which the Annelida in particular were described with minute care. A masterpiece of its kind is his *Histoire naturelle des crustacées*, wherein this order is treated with thoroughness and perspicacity, and a system, based on comparative anatomy, is worked out which is largely applied even today. Further we may mention his work on the coral animals and his study of the evolution of the ascidians, as well as a large number of deep-sea investigations.

Another who worked along the same lines was FÉLIX JOSEPH HENRI LACAZE-DUTHIERS (1821–1901). Born in the south of France of a distinguished family, he devoted himself to the study of medicine and became a professor, first of all at Lille and then in Paris. His numerous works deal chiefly with the Mollusca, the anatomy and evolution of which he worked out in detail; amongst other things he worked out the anatomy of the purpura and its colour-secretion, which was known to the ancients, but had since been forgotten. He is best known, however, as the founder of France's zoological marine stations at Roscoff and Banyuls. On these institutions he spent large sums out of his own purse and laid down the lines and methods on which they were to work. Being a man of essentially conservative views, it was only after a long time and after much hesitation that he accepted the theory of the origin of species. He viewed with scepticism many of the movements of his time; he had written up over the door of his laboratory the words: "Science has neither religion nor politics" — a sentiment that certainly deserves greater attention than has been given to it in modern times.

3. Microbiology

THAT there exists a world of organisms which owing to its small size eludes observation with the naked eye has been known since the invention of the microscope, but our knowledge of these creatures may be said nevertheless to begin from the age with which we are now dealing, when an improved microscopical technique first made possible a more thorough exploration of this extensive field. Leeuwenhoek, the foremost microscopist of the seventeenth century, discovered, as previously mentioned, a number of minute animals, partly in water taken from rivers and lakes, partly in putrefying matter of various kinds. He studied them as carefully as he could, was convinced of their character of living creatures, and declared that they multiplied only by reproduction. During the succeeding century these investigations went on with fresh observations in isolated cases, but without yielding any really novel results. It was found that these minute animals exist especially in water which has been allowed to stand over parts of plants or other similar growths, and from the fact of their existence in such "infusions" they were called Infusoria, or infusion-animals. Buffon, in accordance with his general theory of life, believed them to be products of the life-units existing everywhere, while Spallanzani firmly rejected the idea of their spontaneous generation. The scientist who first made a special study of the Infusoria, however, was the Dane O. F. Müller, who is therefore worthy of further mention in this connexion.

OTTO FREDERIK MÜLLER was born in 1730 in Copenhagen, where his father was a musician. He grew up in poverty, was given an opportunity of studying theology at the university, and then went in for jurisprudence, but the whole time he had to earn his living as a tutor in aristocratic families, on whom his amiable social qualities made a particularly favourable impression. During his visits to their estates he began to interest himself in nature, particularly in insects, which he collected and described in a series of small treatises. As private tutor to a young count he had an opportunity of making a journey through Europe, thereby increasing his knowledge and widening his connexions. Having returned home, he received an official appointment, but resigned from the Government service upon contracting a wealthy marriage, and afterwards lived as a private scholar until his death, in 1784. His most important work was published posthumously. During his lifetime he was generally regarded as an amiable and kind-hearted man, though somewhat vain. Of his immense literary production one or two zoological works have preserved his name for posterity.

As will be realized from the above, O. F. Müller as a zoologist was essentially autodidactic; he had educated himself by studying the writings of Linnæus, but he devoted his research work to spheres that Linnæus and his pupils had overlooked. In two works, *Entomostraca Daniæ* and *Hydrachnæ*, he describes in detail, and extremely well considering the period, two hitherto entirely neglected groups of Articulata. Still more remarkable are his two works on the Infusoria, the last of which, mentioned above, was published by Fabricius in 1786. In these works he makes an attempt for the first time to present a systematic description and classification of the Infusoria, supplemented with detailed diagnoses of genus and species and illustrated with accurate and finely drawn pictures. Quite a number of them, especially the larger Ciliata, he has described so well that they are still recognizable and their names are still in use today. As was usual in his age, he paid but little attention to the internal structure of the creatures; in regard to the origin of the Infusoria, he believes in the spontaneous generation of the lesser forms, while assuming that the larger and more highly developed forms multiply by reproduction.

O. F. Müller's contemporaries and immediate successors made a number of fresh discoveries in the sphere of the Infusoria, as well as various attempts to systematize the forms already known. The same period that gave rise to cell research — the eighteen-thirties — provided also a fresh impetus to microbiology. In this the pioneer was CHRISTIAN GOTTFRIED EHRENBERG. Born in 1795 in the neighbourhood of Leipzig, he studied medicine there and was afterwards given an opportunity of making a six years' voyage of exploration to the East, whence he brought home important collections. This expedition having brought him fame, he was invited to accompany Humboldt on his Asiatic expedition, after which he became professor of medical history at Berlin and secretary to the Academy of Science there. He died in 1876, having long given up active participation in scientific developments.

Infusoria as "complete organisms"

EHRENBERG's great contribution to biology was his work on the Infusoria, the results of which were published originally in a number of brief essays and afterwards in the important and splendidly got up work entitled *Die Infusionstierchen als vollkommene Organismen*, printed in 1838. The result of this and other works of his was that the number of known Infusoria was considerably increased, and their classification essentially advanced. Anguillulidæ and cercaria, which had hitherto been counted amongst them, were excluded, the Rotatoria were separated, and the genus- and species-diagnosis precisely defined, many of them still holding good. The whole of this careful and praiseworthy work, however, Ehrenberg used in support of an utterly unprofitable theory; starting from the then prevalent belief in one single primal type for all animals, he tried to discover in the Infusoria the same organs as in the higher animals; the vacuoles that are visible in the Ciliata, and which partly alter shape, were seen in his imagination to possess canal-shaped outlets and were thus made the basis of an artificially ramified digestive system; he believed he had found sexual organs and eggs in the objects he investigated — in a word, they were to his mind, as the title of his work indicated, "complete organisms." That he entirely rejected the belief in the spontaneous generation of such creatures is self-evident; indeed, this disbelief in the spontaneous-generation hypothesis may have been firmly rooted in him before he began to study the Infusoria and perhaps contributed in some degree towards inducing in him these efforts at finding in them as complete a form as possible. His theory won many adherents among his contemporaries — it was embraced, *inter alia*, by Owen in his earlier works; when eventually it was exploded, Ehrenberg, after spending some years in vainly defending his cause, withdrew entirely from all research work.

The scientist who from the outset came forward as a decided opponent of Ehrenberg and who rapidly won a victory for his views was FÉLIX DUJARDIN (1801–62), professor first at Toulouse, then at Rennes. In certain of his works, the last and most comprehensive of which is dated 1841, he laid the foundations of a new conception of the Infusoria. He achieved this first of all by incorporating with them a category of still lower organisms, which he made the subject of special investigation, namely the Rhizopoda. These, which include types without any external organs, and indeed without any definite external bodily form, offered the best possible proof against the lowest animals' acceptance as "complete organisms." Dujardin found that both these and the higher Infusoria consist of a homogenous mass, which possesses the power of absorbing nourishment, contracting and moving, and reacting to external impressions. This mass he called "sarcode," a name that was at one time used, especially in France, to denote that fundamental substance of which living creatures in general are built up, until it was supplanted by the word "protoplasm." In the sarcode Dujardin found vacuoles and granules, but no permanent organs, and the ciliæ that cover the body of the higher Infusoria possess in his view no affinity with the hairy formations of the higher animals. In all this Dujardin stood undeniably on surer ground than Ehrenberg; on the other hand, he failed to elucidate the

Infusoria's character of simple cells; their nuclei, which Ehrenberg believed to be sexual organs, Dujardin was unable to explain more exactly, but considered them to be simply concretions in the sarcode.

Siebold on Infusoria

IT was reserved for Siebold to put the Infusoria in their right place. As early as his *Comparative Anatomy* of 1845, which has been referred to above, he combines Infusoria and Rhizopoda under the common term Protozoa and characterizes them as "*Tiere, in welchen die verschiedenen Systeme der Organe nicht scharf ausgeschieden sind, und deren unregelmässige Form und einfache Organisation sich auf eine Zelle reduzieren lassen.*" This definition he bases on a careful demarcation of the forms included in the group; the Rotatoria are definitely separated as being more highly organized, and a number of multicellular, but primitive life-forms, which produce chlorophyll — Closterines, Volvocines — are transferred to the vegetable kingdom. It is pointed out in connexion therewith that cilia- and flagella-movements can also exist in the vegetable kingdom, while on the other hand a special free mobility of a higher type is attributable to the Protozoa. The various organic systems that Ehrenberg ascribed to the Infusoria are examined and rejected; there thus remains the simple cell, provided with nucleus and vacuoles, which is hereby proved to be capable of sustaining a free and independent existence, being reproduced by division without any special sexual organs.

In a paper published some years later Siebold further examines the existence of and the relation between single-celled animals and plants, being supported in his views particularly by a work by Nägeli, which had then been recently published, on unicellular algæ, in which these organisms had been thoroughly characterized and described. As a result of these works microbiology was directed on the right way and during the next-epoch came to exercise a great influence on the development of biology in general. In the Protozoa, "the primary animals," had been found a category of living creatures from which, as the name implies, the other higher organisms could be derived, besides which the cell idea was hereby made to cover an entirely new area; it was possible to see in the cell not only the basic element in the structure of the organisms, but also a true elementary organism capable of leading an independent life or, as a transition to the higher cell-structures, of forming colonies of similar elements, such as the Volvox referred to above. Space forbids our going further into the maze of works which were now devoted to the single-celled animals and plants. During the succeeding period in particular, research work in this field went on apace without interruption. Of the works on the Protozoa that appeared during this era may be mentioned those of FRIEDRICH STEIN (1818–85), professor at Prague, on the Infusoria, which form the basis for all later research on the subject.

There still remains to deal with in this connexion one more group of single-celled organisms — namely, the bacteria or Schizomycetes, which, as is well known, have since been found to play a most vital part in human life and which have accordingly been investigated ever since the middle of the last century as a special field of research. In connexion with this branch of research there have existed a number of theoretical problems of the greatest significance; the problems of spontaneous generation, fermenting processes, and the origin of various diseases. The problems of the causes of disease can of course be dealt with only cursorily here; they have of old formed a science of their own — pathology. The questions of spontaneous generation and the fermenting process, on the other hand, have possessed immense theoretical interest and on the decisive occasion mentioned below their treatment has happened to coincide. We may therefore suitably begin our review of the history of bacteriology with a glance at these two questions.

Bacteriology and spontaneous generation

THE belief in spontaneous generation has been mentioned on various occasions in the foregoing: how the earlier naturalists generally believed that the lower animals, especially such as appeared suddenly and possessed more or less the characters of parasites or vermin, could arise through some kind of transmutation process in lifeless matter; Aristotle believed that fleas and mosquitoes originated in putrefying matter — a belief with which even Harvey at least partially associated himself, while van Helmont had seen rats arise out of bran and old rags. In the seventeenth century FRANCESCO REDI (1626–98), court physician and academician in Florence, proved that worms in rotting meat arise, not in consequence of the putrefaction, but out of eggs laid by flies on the meat; if the latter is protected with thin cloth, no worms arise in it, in spite of the putrefaction. On the other hand, Redi believed in the spontaneous generation of intestinal worms and gall-flies. For theoretical reasons Swammerdam denied spontaneous generation; the doctrine of preformation that he founded actually precluded any belief in this kind of propagation and during the greater part of the eighteenth century held it in discredit. Nevertheless Buffon, as we have seen in a previous section, believed in a spontaneous generation through minute life-units scattered throughout the universe and was supported in his belief by his friend the English microscopist NEEDHAM; and Lamarck associated himself with this view. On the other hand, Spallanzani, the preformationist, strongly opposed the doctrine of spontaneous generation and sought to prove that by boiling organic elements in air-tight vessels it was possible to prevent living creatures from arising in them. This theory was put to practical use some decades later by a French chef, APPERT, who invented the still commonly practised hermetical inspissation of food. A French physicist, however, found out that the air in the preserving jars lacks oxygen — this element is really consumed by oxidation processes in the contents — and concluded therefrom that the sterility was due to lack of oxygen. At the beginning of the nineteenth century the belief in spontaneous generation received fresh impetus, not least as a result of the victory of Wolff's epigenesis theory; Rudolphi believed in the spontaneous generation of tapeworms, and the entomologists held the same belief in regard to parasites from the insect world. It was chiefly, however, the increased knowledge of the Infusoria that strengthened the belief in spontaneous generation; Ehrenberg's protests died away unheard when the exaggerations in his description of the organization of these animals had been made manifest.

Chemists on fermenting

THE idea of spontaneous generation received fresh support in the increased knowledge of the fermenting process. Both Lavoisier and Berzelius had studied the fermentation of alcohol and had sought to ascertain the process of sugar-disintegration in alcohol and carbonic acid; the yeast that floats up when it is brewed was believed to consist of albuminous elements, which were separated upon the decomposition of the malt. This conception of fermentation as a purely chemical process found support in the discovery of a substance existing in malt that, when added to a solution of starch, converts the starch into sugar. This substance was called "diastase," and similar substances, "ferments" as they were called, were soon discovered in other quarters: in saliva and intestinal fluids in man and animals, as also in many places in the vegetable kingdom. Chemists believed that they now had in their hands the substances that produce fermentation and similar processes, and these chemical changes, in the course of which albuminous compounds were formed as a by-product, also gave a clear indication as to the direction in which the spontaneous generation of minute creatures might be looked for; fermentation was in fact a part of the process of spontaneous generation.

Then there appeared, in 1836, the Frenchman CHARLES CAGNIARD DE LATOUR (1777–1859), an engineer by profession, with the assertion that the yeast really consists of minute organisms and that it is their activity that causes the fermentation. Shortly after this the question was taken up by Schwann, who tried to demonstrate by experiment that putrefaction and fermentation are processes which are not due to the oxygen in the air, but rather to a special element that exists in the air and is destroyed by heating; he boiled special easily decomposable organic substances and then brought them into contact with air that had first passed through a red-hot pipe, whereupon no chemical change took place, while there was a change if ordinary air was allowed access to them. The leading chemists — Berzelius, Wöhler, Liebig — regarded these theories as a chimera, and they won the day all the more easily because Schwann's experiments were, from the technical point of view, rather poor; other investigators repeated them and obtained results quite different from those published by Schwann. Thus matters stood when a scientist arose who as the result of unchallengeable experiments clinched the matter once and for all and thereby entirely changed the course of microbiology.

LOUIS PASTEUR was born in 1822 at Dôle, a town in the ancient province of Franche-Comté. His parents belonged to the industrial class; his father had served as a non-commissioned officer under Napoleon and after his demobilization had entered the tanning trade, moving his business from one place to another. Young Louis attended the country school and afterwards studied, under circumstances of privation and with numerous interruptions, in Paris. He was mostly interested in the natural sciences and the teaching profession was his aim in life. He became a teacher at the gymnasium at Strassburg and married a daughter of the rector of the school. There he carried out his first chemical work, which procured his removal to the then newly-founded University of Lille, where he became professor of chemistry in 1855. Four years later he was called to Paris, to the École Normale. At the same time he was carrying on his investigations into the fermentation process, which at once brought him world-wide fame, and after that, success and honours came to him rapidly. It was given to him in a specially high degree to make the results of his investigations of practical use to mankind, whether it was a question of inventing a method of preserving food, combating the diseases of domestic animals, or treating rabies, which had hitherto been considered incurable. This last-mentioned discovery made him particularly popular: through an international fund there was founded in 1889 an institute for the purpose of investigating those fields of research to which he had devoted himself, which bore his name and to the management of which he afterwards devoted all his energies. Previously pupils had already flocked to his laboratory from all parts of the world; many of them have themselves won great fame. Nevertheless, this brilliant career had certainly not been entirely free from shadows. In his political views Pasteur was a conservative and a warm partisan of the French Empire, and moreover a strictly faithful Catholic. This caused him a good deal of unpleasantness, owing to radical opposition, producing an atmosphere that is even reflected in the scientific polemics waged against him. Ultimately, however, these hostile voices were silenced, and the more easily as he never meddled in political questions. But instead he suffered in his old age from increasing ill-health; as early as in 1868 he had had a paralytic stroke which impaired the use of one of his arms, but which did not succeed in preventing him from continuing his activities with as great success as before; gradually, however, his powers declined and he passed peacefully away in 1895. His name lives in his work and in the "Pasteur Institutes" which have been established in all civilized countries; he is without doubt one of the greatest scientists of his century.

Pasteur denies spontaneous generation

PASTEUR began his researches in the purely chemical sphere; he investigated organic acids, chiefly the isomeric, and he obtained valuable results in connexion therewith. Thence he was led to study the question of the molecular structure of sugar and the manner in which this substance is converted into different isomeric alcohols and acids — in other words, into the process of fermentation. His first experiments in this field were concerned with the formation of lactic acid; he found on the surface of sour milk minute greyish spots, which he examined microscopically and experimentally. Under the microscope they appeared as a mass of minute globular formations, smaller in size than those of which ordinary yeast is composed; when placed in a saccharine solution they at once disintegrated it into lactic acid. Without at the moment drawing any conclusion as to their origin, he maintains that all fermentation is caused by similar minute organisms. If we plant out such organisms of a definite type in a saccharine solution, we get a definite form of fermentation: alcoholic fermentation, butyric-acid formation; if, on the other hand, a suitable saccharine solution is allowed to stand by itself, there are set up a number of simultaneously disintegrating processes induced by different organisms operating at the same time.

It goes without saying that the chemists of the old school did not feel that they had much to gain from these new discoveries. They at once found a keen supporter in FÉLIX ARCHIMÈDE POUCHET (1800-72), professor at Rouen, who was reputed both as a botanist and as a zoologist. In a series of investigations he tried to prove that the micro-organisms arising upon fermentation and putrefaction are spontaneously generated, and this owing to the very fact of those chemical changes; the fermentation forms the initial stage of the process whereby living creatures arise from the decomposition of existing organic substance. In the view of such a theory Pasteur's fermentation experiments were, of course, pure irrational nonsense, and thus began a lengthy controversy between these two experimental scientists, in which the scientific world and eventually the enlightened public became keenly interested.

Controversy between Pasteur and Pouchet

To Pasteur the specific character of the different fermentative organisms was in itself a proof that they are not the product of chemical change, but actual species of living beings, which come into existence through the multiplication of existing individuals. But whence, then, come all those different organisms which immediately populate saccharine solutions that are allowed to stand, and food that is kept too long? Schwann had derived putrefaction from the air, and Pasteur endeavoured to prove this by experiment; he filtered air by sucking it in through a tube filled with cotton-wool, thereby obtaining a collection of dust particles, which were transferred with great care to a retort filled beforehand with boiled and cooled saccharine solution; the neck of the retort was melted together and after some days there was found in the fluid an abundant vegetation of micro-organisms. Their origin was thus proved; these creatures had existed in atmospheric dust in a dried state. On the other hand, a saccharine solution boiled in a retort the neck of which was melted together in the course of boiling could be preserved for any length of time without changing. Pouchet tried by various means to discredit these experiments; he tried to prove that the organisms cannot stand being dried, that they do not exist scattered in the air as Pasteur declares, that milk becomes sour in spite of being boiled. Space forbids our following this dispute in all its phases — how both parties collected air on high mountains with a view to proving their arguments on that evidence; how Pasteur found that certain organisms can endure heating up to the boiling point of water without perishing (which explains how it is that boiled milk turns sour); how he thought out a whole series of ingenious apparatus to prove his statement that the fermenting organisms always originate in the outer air and that the boiling of the experimental fluids and the heating of the air which comes into contact with them, infallibly exclude the existence of organic life in them. The two antagonists were allowed to carry out their experiments before the French Academy of Science, and Pasteur at once succeeded in convincing some of its foremost members — Milne-Edwards and Claude Bernard, and the chemist Chevreul. Pouchet likewise had his supporters, and especially among the scientifically educated and half-educated public he gained many adherents who regarded spontaneous generation as a "philosophical necessity," indispensable for a natural-scientific explanation of the origin of life, which Pasteur, faithful Catholic as he was, naturally felt himself compelled to explain dogmatically. Thus argument opposed argument, and party faced party. In these circumstances the solution of the problem would never have become possible had not Pasteur been able to put his ideas into practice on a large scale. During the succeeding years he invented his well-known methods of preserving milk by "Pasteurizing" — that is, by heating — of improving the manufacture of wine and beer by controlling the conditions of fermentation, of securing immunity from the silk-worm disease and chicken cholera by eliminating the micro-organisms that produce them. These discoveries, however, belong to the next period, as also the development and perfecting of bacteriology and fermentation research achieved by other investigators — Koch, Hansen, and many others. Finally Pasteur's views on the origin of the micro-organisms received splendid practical confirmation as a result of the development of modern medicine: antiseptics and aseptics in surgery, disinfection, and the treatment of infectious disease. Owing to these facts, which found fresh confirmation daily, spontaneous generation has entirely ceased to exist as a possibility to be reckoned with in modern biology, nor does it come into serious question when we have to explain actual phenomena. That its theoretical possibility nevertheless still continues to be keenly discussed is due to modern natural-philosophical speculation — a subject that will be dealt with in a later chapter.

4. Botany

Plant classification after Linnæus

A SURVEY of the most important data in the history of botany, particularly of plant classification, up to the period covered by our chapter heading, is necessary if we are to obtain a universal view of the development of biology during the period now being described. In order to obtain this view we must first of all return to the days of Linnæus. It will be remembered that Linnæus set up, in the first place, an artificial system based exclusively on the structure of the various parts of flowers and especially intended to be used for practical examination purposes, and, in the second place, a natural system, based on the common forms of plants — a system which he worked at throughout his life, without, however, being able to find a satisfactory conclusion to it. His immediate successors paid but little attention to this latter legacy from him, although it really offers immense possibilities for development; they contented themselves rather with examining as many new plants as possible according to the sexual system. During this time, however, there were published a considerable number of sound systematic works; the study of cryptogams in particular made rapid progress, but nothing was contributed to the development of the system itself until twenty years after Linnæus's days, when Jussieu's systematic work, *Genera plantarum*, was published.

ANTOINE LAURENT DE JUSSIEU came of a family that had already given France two eminent botanists, principally in the spirit of Tournefort; especially Bernard de Jussieu, uncle of the above-named, who had made considerable additions to a natural system, without, however, succeeding in completing it. A. de Jussieu was born at Lyons in 1748, studied medicine in Paris, and became professor at the Jardin des Plantes, after which he held some other botanical and medical posts and finally became professor of pharmacy in the faculty of medicine. He was active for many years and attained a great age; he died in 1836. In his principal work, mentioned above, Jussieu has set up a complete natural vegetable system, the first of its kind. Like Ray, he makes the cotyledons the chief basis of classification for the vegetable kingdom; in his view this is justified, because the plant arises out of the seed, and the latter's most vital part is the cotyledon — it is compared with the heart of animals. Consequently plants are divided into: acotyledons, monocotyledons, and dicotyledons, a system of classification that, thanks to him, has become permanent. These three main divisions are then divided into orders, the names and demarcations of which are taken partly from Linnæus's plan for a natural system and partly from the preliminary work of Antoine's uncle Bernard. Some of these orders are, in fact, quite natural, while others are extremely ill arranged, especially in the acotyledon group, in which are included not only Linnæus's cryptogams, but also the naiads; thus, he includes the ferns among the Cycas, which is placed between the Polypodia and the Equiseta. Jussieu has formed his genera mainly after Tournefort, while his definition of species is reminiscent of Ray rather than of Linnæus. Jussieu, therefore, was not a very original observer, but his service to science lies in the fact that he really worked out a natural system, which he set up in determined opposition to Linnæus's sexual system, and which has, in fact, been the starting-point for all subsequent systematic improvement.

A far more important observer was ROBERT BROWN (1773–1858), who has been mentioned before as the discoverer of the cell-nucleus. The son of a Scottish clergyman, he studied medicine in Edinburgh and became an army surgeon, but at the same time he applied himself to botany and was appointed botanist to an expedition to Australia, which was led by a Captain Flinders. Brown stayed four years on that continent and brought home large collections; on his return he was made librarian to the Linnean Society and curator at the British Museum. In this position he enjoyed the reputation of being one of the finest botanists of his time; he never published any very important work, however, and his papers were, curiously enough, collected and published in a German translation, done by Nees von Esenbeck, with whom he had but little in common from a scientific point of view. Nor did he work out any system of his own; in his works he uses sometimes Linnæus's, sometimes, and more often, Jussieu's natural system. His service to science lies in the care and keen-sightedness with which he works out and analyses the various orders, or families, as he more frequently terms them:[1] his studies of the Compositæ, Asclepiadaceæ, and many other families have been mentioned as models for the research work of the succeeding age and have contributed much towards finding a place for the natural system in the scientific mind. Brown was, moreover, an eminent plant-geographer and made a special study of the distribution of the different families in different climates; in this respect his earlier work on the flora of Australia was unrivalled and attracted the attention of Humboldt, who highly commended it.

As one of the foremost pioneers of botany should also be mentioned AUGUSTIN PYRAME DE CANDOLLE. He was born in 1778 at Geneva, where his family had for generations enjoyed a great reputation. At an early age he began to study the natural sciences, which at that time — the age of Bonnet

---

[1] The term "family" as an expression for the natural groups in the vegetable kingdom apparently comes from the French botanist MICHEL ADANSON (1727–1806), whose attempt, influenced by Buffon, to form a natural system for the vegetable kingdom was somewhat of a failure.

and Saussure — stood in high favour in his native town. After preliminary studies there, he betook himself to Paris in order to continue his education as a botanist. In the company of Lamarck, Cuvier, and Geoffroy he spent ten years there, during which his reputation increased year by year and public commissions were entrusted to him; amongst other things he was sent, with the financial assistance of the State, on scientific expeditions in different parts of France; Lamarck handed over to him the editing of his French flora and he was finally elected professor at Montpellier. In 1816, however, he returned to Geneva, which during the Revolution had become incorporated with France, but after the fall of Napoleon was again united to Switzerland. He then lived in his native town as professor of botany and member of the high council, honoured and respected until his death, in 1841.

De Candolle on plant morphology and physiology

DE CANDOLLE mastered the whole field of botany better than anyone else in his time; he was at once systematist, morphologist, and physiologist. He started a gigantic work, *Prodromus systematis naturalis regni vegetabilis*, which was to describe all known plants, but which for obvious reasons was never completed in his lifetime; his son and many others worked at it after his death. The principles on which he classified the vegetable kingdom he laid down in a work published in 1813 entitled *Théorie élémentaire de la botanique*, which he revised several times and which is without doubt his finest work, worthy 'to be associated with, and at the same time representing a great advance on Linnæus's *Philosophica botanica*, which doubtless gave him the idea. It starts with a general scientific theory, according to which nature is controlled by four great forces: *attraction*, and *affinity*, which are the basis of physical and chemical phenomena, *the life-force*, which is common to all living creatures, and *sensibility*, which is the characteristic of animal life as opposed to vegetable life. Each of these four forces has its own science: physics, chemistry, physiology, and psychology. It is thus a markedly vitalistic conception, which is still more emphasized in such an assertion as that "the life-force annuls or modifies, as necessity dictates, the ordinary laws of matter." De Candolle is by no means a fantastic natural philosopher, however; on the contrary, he has the same sober and critical conception of natural phenomena as Cuvier, whose correlation theory he applies to the vegetable kingdom. He maintains that the two most vital organic systems in the plants, the vegetative and the sexual, are dependent upon one another; a plant with highly developed fertilizing organs cannot possess primitive vegetative organs, and vice versa; therefore a natural system, set up with comprehensive regard to the entire reproductive organization, should at once conform to such a system set up with a view to the vegetative organs, and this, indeed, he proves by means of examples. In connexion herewith he maintains, under acknowledgment to Linnæus, that there is throughout the vegetable kingdom a universal symmetry, a standard of organization, which is modified in the individual by the same organ's being capable of serving different purposes and the other organs' undergoing corresponding changes. Among these changes he includes especially stunted growth, degeneration, and accretion; in his view a flower with free petals is higher than one with accrete petals — a principle which he applied in his system, though it failed to gain the acceptance of posterity. For the rest, he introduced into his system reforms of lasting value; the difference between vascular and cellular plants was established by him, as also the contrast between the bole-plants and the higher plants. Further, his classification of the dicotyledons has been largely accepted by subsequent botanists. Otherwise, de Candolle strongly repudiates Lamarck's theory of one single evolutional chain in the organisms, instead associating himself with Linnæus's idea of the natural system's likeness to a map; in fact, his idea of species is not unlike the Linnæan: according to de Candolle, a species is "the sum total of all the individuals which mutually resemble one another more than they resemble others, which are capable by mutual fertilization of producing fertile individuals, and which are multiplied by generation, so that it is possible by analogy to assume that they have originally sprung from a single individual." Varieties arise, he considers, partly through the influence of local conditions of life and partly through hybridization; moreover, there are in certain quarters varieties which must be regarded as constant, like the species, and which should be distinguished from the accidental local varieties. The genus is defined as a collection of species with a striking mutual resemblance in regard to all organs; families and higher categories are given similar definitions.

Among de Candolle's other works may be mentioned his *Organographie*, an account of the organic systems of plants, and his *Physiologie végétale*, a work of great merit for its time, based on a thorough knowledge of the vital conditions of plants and of chemical and physical processes belonging thereto. We must, however, pass over these works here; as a matter of fact, it is mostly as a reformer and theoretician in the sphere of classification that de Candolle has made his best contribution to the development of biology.

Endlicher's system

THE development of plant classification received further impetus through STEPHAN LADISLAUS ENDLICHER (1805–49). Born of wealthy parents at Pressburg, he studied first of all theology, but afterwards devoted himself both to botany and oriental languages. He became professor of botany and head of the botanical gardens in Vienna, acquiring fame for the splendid initiative he took in furthering the development of natural science in Austria, generously contributing towards that end out of his own private purse. He presented his herbarium to the State and published a botanical journal at his own expense. At the same time he made a name for himself as an expert in the Chinese language. As a teacher he won the affection of his pupils. In the year of revolution, 1848, he took advantage of his popularity to plead the Government's cause before the rebellious students, but they became embittered against him and drove him out of Vienna. This he felt very deeply and he died shortly afterwards, some say by his own hand. His great work on plant classification is his *Genera plantarum*, which comprises all the then known vegetable genera arranged in a natural system; he has given a brief summary of the subject in his *Enchiridion*. His service lies not so much in the new ideas that he produced in the sphere of classification as in the particularly clear, concise, and complete characterization and demarcation of families and genera which he created and which made his work the basis for all later plant classification.

Hedwig on mosses

By the side of this generally systematic and morphological research there developed a keen interest in the specialized study of particular botanical subjects, the hitherto neglected lower plants offering, of course, an especially attractive field of study. As a pioneer in this sphere JOHANN HEDWIG (1730–99) is worthy of mention. He was born in Hungary, but he worked mostly in Leipzig, first as a physician and then as professor of botany. He applied himself to the study of the multiplication of the cryptogams, making careful observations of the propagation and germination of the spores. It was chiefly the mosses, however, that occupied his attention, and in this field he was a pioneer; he divided the large and unwieldy genera that Linnæus had created into a number of well-characterized genera, which are in part still retained, and he found for them a good basis of classification in the shape and marginal formation of the capsules. Many other naturalists have since followed in his footsteps, so that muscology is now a thoroughly elaborated specialized section of botany. ·

The Algæ and the Fungi became subjects of special treatment much later than the mosses. Of the pioneers of Algæ-research we have already described one of the foremost, CARL ADOLF AGARDH (Part II, p. 291). His son, JACOB GEORG (1813–1901), followed in his father's footsteps with credit.

Fries on fungi

In the Fungi as a field of research Sweden has also produced one of the most eminent names, that of ELIAS FRIES, who has likewise been one of the most distinguished Swedish botanists since Linnæus. Born in 1794, the son of a priest in the province of Småland, Fries devoted himself, even as a boy, to the study of botany, and of Fungi in particular; when still a youth he became a lecturer at Lund and in the year 1834 professor at Upsala, where he was active until 1859, when he became professor emeritus; he died there in 1878. When Fries first went to Upsala the University was a centre of romantic reverie and metaphysical speculation; he took up the cudgels with success and honour on behalf of the cause of exact research, by no means allowing his own special sphere to be put in the shade; in speeches and writing he championed the cause of biology, and his plea was heard far and wide. Being a brilliant stylist and an eloquent speaker, he was elected a member of the Swedish Academy, and in his old age he held the position of a recognized patriarch in the sphere of natural science in his own country. Although no friend of fantastic speculations, he shared his age's idealistic conception of nature and thus found it easier to gain a hearing for his high aims; in one of his writings he expressly calls biology a "supernatural" science, for life is something higher, given from above, and its influences must not be explained according to the laws of inorganic nature. To his mind, biology belongs not to the exact sciences, but to the historical; it is more closely akin to theology than to physics and chemistry.[2] In his special research work, however, Fries is quite exact. His chief productions are his great works on the Fungi, which have since formed the basis of classification in this class; he has described quantities of species and given characters to genera and families that still hold good. Next to these works should be mentioned his treatises on the lichens; here he had a precursor in his fellow-countryman ERIK ACHARIUS — born in 1757 and mentioned as Linnæus's last pupil, afterwards provincial physician at Vadstena, died in 1819 — but it was Fries who established the lichen system, which was generally accepted until the eighteen-sixties. Fries also performed a considerable service in producing his classification of the phanerogams; among other things he maintained, in opposition to de Candolle, that the Compositæ are the highest-developed of the phanerogams, and in this posterity has shown him to have been right. His natural system has, with certain modifications, been generally utilized in Scandinavia.

At this point our description of the different spheres of biology has brought us up to the period that is characterized by the launching of the theory of the origin of species. It now remains for us to give a glance at certain phenomena in the sphere of theoretical speculation that have given direction to our modern biological views.

---

[2] *Botanical Excursions* (*Botaniska utflykter*) I, pp. 11–13. Curiously enough, Haeckel from his standpoint arrived at similar conclusions, of which more anon.

CHAPTER IX

Romanticism and positivism

IN HIS *History of the Philosophy of Later Times* Höffding declares that there are two intellectual currents characteristic of the nineteenth century: romanticism and positivism, the former starting from the ideal of thought, the latter from that which is based on fact. This division is undoubtedly in accordance with the actual course of events dominating the whole world of culture; the contrast indicated is discerned no less clearly in the development of biology. The romantic conception of nature that prevailed at the beginning of the century saw the true reality in an idea, of which the actual life-forms were merely modifications; they sought therefore for a primary form or archetype, with which the living forms were compared, as was done, each in his own way, by Goethe, Geoffroy Saint-Hilaire, and R. Owen, the last-mentioned still as late as towards the middle of the century. By that time an entirely different conception of natural phenomena had already appeared, which fought its way year by year into the general consciousness; although champions of the old ideal still survived far into the latter half of the century, nevertheless it may be claimed that the victory of the new conception was already fully confirmed by the beginning of the sixties. Opposition to the old concept first came from the social and political spheres, after which it took in its stride the scientific and literary world. Its original home, therefore, was in the two countries in which public life manifested the greatest mobility — France and England. It was not until later, and then under different forms, that it appeared in Germany and Scandinavia.

In France there set in during the time of Napoleon, and still more immediately after that era, a violent reaction against those radical ideas that the enlightenment of the eighteenth century had created and the Revolution had sought to realize — a reaction especially in the social and political spheres, less in the scientific, although it certainly had its learned theorists, as, for instance, the brilliant and fanatical Joseph de Maistre. But nevertheless the theories of the period of enlightenment could never be wholly suppressed; they survived, as did the longing for the political freedom of the Revolution, and they found support in the natural sciences, which at that time were passing through a brilliant phase in France, being sustained by men who worked for the most part undisturbed by any theoretical speculations.

441

In a previous chapter we have described the most important representatives of biology in France during that period; physics, chemistry, and astronomy at the same time could boast of no less brilliant representatives. And the results that these sciences achieved were very clearly brought home to the world in general, owing to their splendid application to practical life, which was then just beginning and which afterwards represented perhaps the most striking characteristic feature of the century; it was then that the influence of steam-power on industry and communications first began to be realized; it was then that the significance of chemistry in numberless fields of activity began to make itself felt in the daily life of humanity, not to speak of the somewhat later application of electrical phenomena in practical everyday life. As a result of all this the natural sciences began to influence the public mind more than they had ever done before; mankind expected them to lead to new and happier times, while theology and philosophy, which had served the oppressors of the people, reaped nothing but hatred and contempt. Would it not be possible for all forms of human life, for the whole of human culture, to be placed under the ægis of the natural sciences, to be explained through them and developed in their spirit? This question was answered by many with an unreserved yes; foremost of these was Comte, one of the most gifted thinkers that France has ever produced and certainly the most influential during the past century.

Isidore Auguste Marie François Xavier Comte was born in 1798 at Montpellier. He belonged to an ultra-Catholic and strongly conservative family and received a strict upbringing in the same spirit. Even when he was fourteen years old, however, he began to doubt the correctness of the dogmas on which he had been brought up, and when, later on, he went to study at the École Polytechnique in Paris, his oppositional attitude became clearly defined and was all the more strengthened when the Government disestablished that institution, which was feared as a centre of opposition, before he had had time to complete his studies there. Thus he passed no examination, a fact that had a disastrous effect upon his future; nevertheless, in spite of his parents' opposition, he continued his studies in Paris, steadily improving the substantial knowledge he had already acquired of mathematics, physics, and chemistry. He soon won a reputation for his genius and erudition — among his patrons and friends he counted men such as Humboldt and Blainville — yet he was never able to obtain a post in the Government service, but all his life he had to earn his living by private tutoring, except for some years when he was employed as an assistant teacher. This may to a certain extent be explained as due to the peculiar theoretical point of view he adopted; even in his youth he resolved to devote his life to creating a general system, which was to deal, along natural-scientific lines, with the whole of existence, both of nature and of human life, and thus to arrive at a universal knowledge, whereby all the problems of life would be solved, not only the theoretically scientific, but also, and above all, the social and political. The first concept of this system was drawn up in the form of private lectures, to which he succeeded in attracting a large number of listeners; then during the period 1830–42 he worked out his famous *Cours de philosophie positive* in six large volumes. His method of working was peculiar to himself; trusting to his phenomenal memory, he had recourse to no literature of any kind when at work, and he used to write down the contents of a whole volume at a time without corrections, after having worked out in his head the gist of what he was going to write. It is natural that in such circumstances his work should be full of inaccuracies in matters of fact as well as of stylistic redundancies. In several cultural circles the influence of the work has been deep; in view of the important part that biology plays in it, it is worth giving a summary of the book here, all the more so as it had its effect on the biological theories of the succeeding era.

Comte's positive philosophy

IT is a "positive philosophy" that Comte desires to create; by "philosophy" he means, as did Aristotle, whom he greatly admired, a knowledge of the whole of existence; by "positive" he means "*la même chose que réel et utile.*" This "real and useful" knowledge he will substitute for the theological and metaphysical, which he considers to have predominated during previous epochs. He finds the essence of existence to be in the development that has always taken, and is still taking, place; in this instance he paved the way for the explanation of life that has governed human culture since his time. In contrast to so many later positivist thinkers, however, he does not look for this development in nature — it is characteristic that geology does not interest him at all — but in human life. In the history of human thought three successive phases have followed one another: the theological, in which it was believed that personal divine powers were the cause of all that happened; the metaphysical, when for these were substituted impersonal forces; and the positivist, in which men no longer ruminate over the causes of all that takes place, but are content to establish facts and determine their course. The theological stage culminated in the Catholicism of the Middle Ages, for which Comte, in spite of all, expresses great sympathy. As the founders of positivism he cites Bacon and Galileo, whose explanation of nature should, in his opinion, be applied to all phenomena. The middle phase, the metaphysical, is, in his view, the worst of all; it is the belief of the idealistic philosophy in spiritual reality, beginning with Descartes and ending with the romantic philosophy, that is sharply, and for the most part justly, criticized. Instead, his explanation of nature is the same as Galileo's — that it is not possible to find out what the forces of nature are, but only how they operate. In this he is undeniably right; his weakness, on the other hand, lies in his mania for formally reducing to simple formulas all phenomena, even those of the most complex character. He thus forgets Galileo's second great exhortation: to measure what is measurable and to make measurable what is not. He believes, for instance, that every being, and especially every living being, can be studied from two sides, the static and the dynamic — that is to say, as potentially active and as actually active. Thus, biology has a static side, anatomy, and a dynamic side, physiology, and other sciences in like manner. Comte himself declares that he borrowed this division into static and dynamic from Blainville; it again occurs in Haeckel's *Generelle Morphologie*. According to Comte, all science should be classified after the method employed by botanists and zoologists; by this method we get six separate branches of science: mathematics, astronomy, physics, chemistry, biology, and social physics, or, in a single word, invented by Comte and now generally accepted, sociology. Each of these sciences is based on all the previous ones in the series and cannot be mastered without a knowledge of them. The biological section is, of course, the one that affords chief interest to the present history.

His biological ideas

As he repeatedly asserts, Comte's biological speculations are most closely associated with those of Blainville, but are, of course, entirely outside the scope of the control which the theories of that distinguished zoologist exercised in his special research-work. Blainville's view that life consists of *"composition et décomposition"* is thus embraced by Comte, who with its support rejects both Stahl's vitalism and Boerhaave's mechanism. On the other hand, he accepts Bichat's tissue theory, strongly supporting the idea of structure's being the essential factor in the living organism; Bichat's "organic and animal life" is also adopted by Comte. Life itself he defines as "the relation between organism and environment." It can be studied, as to both its static and its dynamic side, after three different methods: observation, experiment, and comparison. Observation is the fundamental method and should be carried out with all available technical appliances. Experiment, on the other hand, is condemned, especially vivisection, which disturbs the relation between the organism and its natural environment and thus merely creates abnormal states and, moreover, leads to cruelty — Comte does not mention Magendie's name, but obviously refers to him. — The finest biological method is the comparative, which is applicable, on the one hand, to different parts and stages of development in the same individual, and, on the other hand, to different life-forms. The latter type of comparison should be concerned with both organs and tissues; Comte assumes a primal tissue from which all other forms of tissue and organ can be derived, but he rejects the cell theories that were just then making their appearance. This derivation, however, turns out to be as idealistic as Cuvier's comparative anatomy;

Comte, indeed, maintains with the latter that the species are invariable, "for the idea of species would inevitably cease to represent an exact scientific definition if we were to allow an unlimited modification of different species, the one in the other." This opposition of Comte's to the theory of the origin of species was undoubtedly the cause of Haeckel's refusing to acknowledge him as a precursor in respect of monism, which he nevertheless was far more than any of those whom Haeckel enumerates.

The details of Comte's biological speculations are, of course, of interest only from the point of view of curiosity. He associates himself with Blainville's animal system, with its exclusive reference to external characteristics; he himself lays down three main types for the entire animal kingdom: Osteozoa, Entomozoa, and Malacozoa — that is, Vertebrata, Articulata, and Mollusca — a classification the clumsiness of which scarcely needs pointing out; indeed, the Mollusca group in particular, a reversion to Linnæus's Vermes, was at the time utterly absurd. Still worse, however, is Comte's attempt to analyse "the intellectual and moral cerebral functions," for here he becomes infatuated with Gall's phrenology. It is only natural from his point of view that he should reject Descartes's theory of the parallel existence of the soul and the body, and the other "metaphysicians" likewise offer many points of attack. In his criticism of the earlier psychology, then, Comte has shown very keen observation, but the psychology that he himself created is all the more lacking in criticism. He denies the possibility of psychical self-observation, for one cannot divide oneself into two parts for the one part to observe the other, and besides one cannot in this way find out the mental life of the animals, which is the vital preliminary stage to that of man. True psychology should, according to Comte, be based on Gall's theory of intellectual and moral areas in the brain, which is the beginning of an entirely new psychology. Here modern psycho-physical research has proceeded along a line which Comte never dreamt of and which led to the complete acceptance of that idea of "self-observation" which he despised.

His sociology

THE last three sections of Comte's work deal with sociology, the doctrine of social statics (that is, organization), and dynamics (that is, progress). Eventually these problems entirely usurped the place of natural science in his life's work. It is true that he produced ideas of value in this sphere too: the actual principle of studying social life from, so to speak, a biological point of view has indeed won adherents in modern times, and a number of items in his program, as, for instance, mixed schools, have actually been adopted. But on the whole his social theory is only a curiosity. This is due mostly to the strange development that he himself underwent. While still a young man he had for a year been a lunatic, but he afterwards recovered his mental balance. When he had concluded his great book, he added to it a general introduction which led to his being accused of megalomania and persecutory paranoia, and after that period he became engrossed in ever stranger social Utopias; he founded a new religion, "a Catholicism without Christianity," as Huxley called it, with catechism, a calendar of saints, comprising great men to whom prayers should be addressed — beginning with Moscs and ending with Bichat and Gall — and a ritual of divine service. His scientifically educated friends deserted him, and only a small group of a less intelligent type gathered round him at his death, in 1857.

The influences that Comte exercised upon the conception of life held by subsequent generations is not easy to estimate. All that in modern times has gone under the name of positivism, monism, utilitarianism, and various other isms has either directly or, at any rate, intermediately been influenced by his doctrines. In conscious opposition to the ideal unity, in which romanticism saw the connexion of existence, he took evolution to be the connecting force in life. He certainly did not view biological evolution in the same light as modern biology — if he had, he would not have rejected Lamarck — but he observed with all the keener vision the evolution that is taking place in human culture. He was thus able to do justice to the various stages of history — a thing which the period of enlightenment of the eighteenth century was unable to do — while, on the other hand, he could point to a goal in the future towards which to strive. And this belief in the evolution of mankind was, as we shall find later on, a precondition before the theory of evolution in nature could gain a hearing; Comte therefore paved the way for the doctrine of the origin of species more than most others did. And though his own biological concept was deficient, it has nevertheless had its influence; we have already pointed out traces of it in Haeckel, and these could probably be supplemented; even in later times there has been a corresponding tendency reminiscent of such characteristics as a preference for comparative investigation and a dislike for experiment. Comte's name has, in fact, a definite place in the history of biology.

English positivism

THE other representatives of positivism in France — Comte's pupils — devoted themselves principally to social and general cultural problems and may be passed over here, however deep their influence may have been on the general conception of life, both inside and outside their own country. The same is to a certain extent true of the precursors of the same realistic trend of thought in England. That country was indeed the cradle of eighteenth-century enlightenment, and the ideas of the era of enlightenment never quite died out there, even in the days of romanticism. These ideas took rather the form of strivings after practical social reforms, as in JEREMY BENTHAM and JAMES MILL, who are named as the founders of utilitarianism, a general philosophy of life with a social aim, based upon the highest possible happiness for the greatest possible number of people, a happiness that would infallibly be attained through the activities of the individual being as little as possible restricted by the various organs of the community. Here, then, we find the same belief in evolution as in Comte, although in a more practical form. The foremost supporter of these ideas, however, is JOHN STUART MILL (1806-73), son of the above-mentioned James Mill. He was taught by his father and never studied at a university, but as a young man entered the Civil Service. For a time he was a friend of Comte and was influenced by him. Among his works should be first mentioned his *System of Logic*, an analysis of the laws of thought, which had a great influence on the generation that felt the first effects of Darwin's theory; Haeckel especially cites its doctrines frequently. Mill derives all knowledge from experience, and this in its turn from sense-impressions; of special interest is his analysis of the different ideas of the natural-scientific systems, particularly the idea of species; he considers a well-defined species to be a reality, not merely a conventional term, but, on the other hand, he maintains that the species should be based on characters and not on any imaginary ideal type. The closer study of these extremely detailed analyses of ideas is, however, more a concern of philosophy than of the history of biology. For the rest, Mill was active both in theory and in practice as a liberal social politician and as such possessed a wide influence.

Downfall of romanticism in Germany

THE advent of the realistic conception of life took an entirely different turn in Germany. It will have been seen from the foregoing how education in that country had for half a century been entirely dominated by the romantic philosophy, with the result that even natural science came to a great extent under the influence of its modes of thought. The Schellingian polarity-theory certainly had very soon to give ground, but in the world of speculation the Hegelian philosophy, with its dialectical method and its contempt for all empirical research, prevailed all the longer. But after the death of the master the school was divided against itself, and many of its members developed their doctrines along distinctly radical lines, as, for instance, KARL MARX, the famous founder of socialism, and LUDWIG FEUERBACH (1804-72), whose views closely approached the positivism of Comte and who was otherwise a thinker mostly engaged in problems of religious philosophy, and also D. F. STRAUSS, the well-known Bible-commentator. Other philosophers remained on the old ground, while still a number, including some of the most keen-sighted, returned to Kant's critical studies. Whereas, then, the romantic philosophy was being internally disrupted, the natural sciences made the splendid advance that has been described in the foregoing. It is no wonder, therefore, that natural-science students took courage; the results of philosophy had resolved themselves into vain squabbles; why not, then, let scientific research be self-sufficient and solve the riddle of existence on its own account?

Good progress had already been made when science was backed by such sound victories as Mayer's law of energy and Wöhler's organic syntheses.

German materialism

IT was in these circumstances that the new realistic natural philosophy arose, whose different ideas have occupied the attention of so many thinkers and writers up to the present day, and which has gone under so many different names, such as positivism, materialism, monism, agnosticism, and other isms. Its main characteristic has been the endeavour to build up, on an exact natural-scientific basis, an explanation of *the whole of* existence — that is, to base on the limited results of research an explanation of the illimitable, by means of weights and measures to explain the immeasurable and imponderable. These natural explanations might have been fully justified as expressions of a personal view of life, if their originators had clearly realized the difference between facts and hypotheses, between manifestations that are actually capable of being observed and turned to practical use, and theoretical constructions of such as are inaccessible to any observation whatsoever. This clear thinking, however, has unfortunately been somewhat rare; far more common has been the tendency to work up explanations of nature and then insist upon having them regarded as the results of natural-scientific research — a weakness that has often been apparent in men who in their own special sphere have been keen and conscientious observers. From the outset this temptation was no doubt due to the influence of romantic philosophy, which had confidently proclaimed the infallibility of its absolute natural explanations. Another factor, especially as regards the more popular scientific literature, was the rivalry with the ecclesiastical tenets, which maintained the absolute truth of the words of Scripture, even in questions of natural science. And, finally, there were in Germany the political points of view to be reckoned with; the ruling powers did their utmost to preserve the old belief in authority, which was considered to conduce to obedience to government; the opponents of this belief were consequently on the side of natural science. The contrast was still further sharpened by the revolutionary outbreak of 1848 and was by no means softened by the stern measures which the Governments adopted after their victory, in order to maintain their authority. Considerable light is thrown upon these conditions by the so-called materialist dispute in the beginning of the eighteen-fifties, a controversy which not only caused great excitement at the time, but also produced after effects that have been felt ever since. It may therefore be worth glancing at, all the more so as the parties to the dispute were exclusively scientific investigators, some of whom were very distinguished, while philosophical and theological opinions do not come into consideration at all.

Among those who became involved in this dispute JUSTUS LIEBIG (1803-73) ranks first; on the whole, he may be considered one of the greatest scientists of the century. The son of a colour-man at Darmstadt, he acquired in his father's shop even as a child an interest in chemistry and its practical application. At one time he endeavoured to follow his bent as an apprentice to an apothecary, but did not get on well there, and he then studied for a couple of years at German universities, during which time he associated himself with Schellingianism; he soon wearied of this also and went to Paris, where he eventually found the training he sought for in the laboratory of the famous Gay-Lussac. On the recommendation of Humboldt he was called to the chair of chemistry at Giessen, and after struggling for years against jealousy and hostility he succeeded in bringing into being the first chemical university-laboratory in Germany. As a teacher he resembled J. Müller in his capacity for gathering around him and educating numbers of pupils; indeed, the revival of chemistry in Germany is attributable to him. Towards the close of his life he became a professor at Munich. He was a pioneer in his purely chemical discoveries, especially in the field of organic chemistry; he gave to organic elemental analysis the form that it has retained ever since, while his investigations into organic acids were of epoch-making importance, as were also his discoveries in the sphere of zymurgy. These latter discoveries made him the foremost supporter of the chemical fermentation theory, and Pasteur's stubborn opponent. He is of greatest importance, however, as the creator of practical agricultural chemistry; hitherto it had been thought generally that plants absorbed their principal nourishment out of the surface-soil, but he proved that the surface-mould was rather augmented by cultivation, that carbonic acid was the plants' sole source of carbon, and ammonia its source of nitrogen, and to prove his theory he instituted experiments with manure on an expensive scale. As a result of these experiments he placed agricultural economy on a natural-scientific basis, but he certainly shot far beyond the mark — partly owing to his ignorance of vegetable anatomy — and he gained many enemies on account of his overbearing polemic, especially against the plant-physiologists. These in their turn exposed a number of Leibig's inaccuracies; he denied the value of nitrogenous manures, he wanted to supply the earth with insoluble instead of soluble phosphoric acid and potassic salts, and he entirely ignored the respiration of plants. On many points he received sharp criticism at the hands of Schleiden and Mohl. As an animal-physiologist Liebig also acquired fame for his pioneer studies of the preparation and utilization of foodstuffs; he ascertained the chemical compounds that are conveyed to the body through the food, but here, too, he often went wrong, as when he divided food-substances into "plastic" and "respiratory," including albuminous substances among the former, and fats and carbohydrates among the latter.

In this sphere Liebig was opposed by a young Dutch physiologist, JACOB MOLESCHOTT (1822–93). The son of a physician, he studied physiology

— and at the same time Hegel's philosophy — at Heidelberg and became a lecturer there, but was dismissed on account of his "materialistic" views. He then became professor at Turin and afterwards at Rome. He introduced research in experimental physiology into Italy and carried out valuable investigations, especially in the sphere of the phenomena of respiration. These brought him into conflict with Liebig, whose theory of the influence of food-substances upon breathing he rejected. But at the same time he made a violent attack upon Liebig's entire conception of the cosmos. In a series of popular papers, *Chemische Briefe*, the latter gave an account of the progress of chemistry, in the course of which, confirmed Schellingian that he was, he extolled in fervent eulogy the wisdom and might of the Creator. In opposition to these letters Moleschott wrote a book, *Kreislauf des Lebens*, in which he attacked Liebig in vigorous though courteous terms, and in connexion therewith produced a purely materialistic conception of the world. This he bases on the theory of the permanence of energy and on the syntheses of Wöhler; on the other hand, unlike Comte, he propounds no original ideas on evolution. To him life is a magnificent process of metabolism; thought is a product of the activities of the brain. As a confirmed Hegelian he delights in abstract speculations; through combining these with physiological theories he often becomes involved in a helpless confusion of thought. Albert Lange in his *Geschichte des Materialismus* quotes some amusing instances of Moleschott's muddled attempts to get away from the contrasts between subjective mental impressions and objective reality, and of his still more confused ideas of matter and energy; after quoting a more than usually vague page of Moleschott's book, he asks: "What part of the philosophical backwoods are we in now?" In fact, Moleschott has no idea of the limits of scientific research; in accordance with the idealistic philosophy that he once embraced he imagines that he can explain the whole of existence by a few artificial ideas. On the other hand, Liebig certainly had no thought of letting natural science hold its own and leave it to religion to satisfy the ideal requirements of life — showing that he too was a victim of the vagueness of thought that romantic philosophy left in men's minds.

Another important naturalist who was involved in a similar controversy was RUDOLPH WAGNER (1805–64). He had studied medicine and taken his degree at Würzburg and afterwards worked under Cuvier in Paris, eventually being appointed Blumenbach's successor at Göttingen. He was a creditable investigator and teacher; among his pupils were such men as Leuckart and the philosopher Lotze, and among his works his investigations into spermato- and ovogenesis and into the tactive corpuscles are especially worthy of mention; he was also reputed as an anthropologist, in the spirit of Blumenbach. At a scientific meeting at Göttingen in 1854 he gave a lecture on *Menschenschöpfung und Seelensubstanz*, in which he discussed the question of the origin of man from one single pair in accordance with the Church's doctrines of creation — a question which he certainly believed anthropology to be incapable of proving or disproving, but which gave him an opportunity of making a violent attack upon the materialistic soul-theories of the time, which he inveighed against from the point of view of both science and morality. He himself worked out a theory of the soul as a kind of ethereal substance, which leaves the body at death and imparts itself to the children that are born — an idea somewhat reminiscent of Swedenborg's spirit theory. His antagonist on this subject was KARL VOGT (1817–95), who had been professor at Giessen between the years 1847–9, but had been removed on account of his having participated in the revolutionary movements of that period; he afterwards became professor at Geneva, gaining a reputation especially as an author of sound text-books and popular scientific works. Between him and Wagner there ensued a controversy on the question of the creation and the soul of man, which rapidly degenerated on both sides into sheer lampoonery, involving personal insults of the basest kind. In this Vogt maintained that the different human races cannot have a common origin and in support of his argument adduced a number of proofs of the constancy of species and varieties, which were not quite in the spirit of the theory of the origin of species. Further, there was considerable discussion as to the fertility of hybrids, which Vogt upheld and Wagner denied, and finally Vogt found an easy butt for his witticism in Wagner's divisible soul-substance, and at the same time maintained the assertion that the soul was a product of the brain, which "produces ideas as the liver produces bile and the kidneys urine." On the whole, Vogt seems to have been entirely unmoved by the earlier natural philosophy; this frees him from having to solve a number of problems that his philosophically trained contemporaries felt themselves bound to take up for discussion, but, on the other hand, it involved him in gross self-contradictions. The most painful feature of this polemic, however, was its markedly political character; on the one hand, a Christian conservative professor, holding a good position and boasting of his friendship with statesmen and ministers, and, on the other hand, an exiled revolutionary, embittered by the shipwreck of his ideals and by his own misfortunes. It would almost appear as if the whole of this scientific controversy was merely an excuse for giving two individuals from opposite political camps an opportunity of coming to grips. In fact, the antagonism of the two ideas, materialism and idealism, retained this character in Germany not only during the decade with which we are dealing, but also up to a far later period; the points of view as to the soul's "to be or not to be" coincide with the attitude: supporter of the Government or supporter of the opposition. During the eighteen-fifties, as we have seen, the representatives of radical ideas at the universities found themselves in quite a difficult position as far as regards educational freedom; this state of things was certainly improved at a later date, but for some considerable time to come Christian conservatism was an officially approved standpoint.

A radical thinker who never succeeded in acquiring any permanent right to give instruction was Büchner, one of the most widely read of the authors who wrote on the materialism controversy. FRIEDRICH KARL CHRISTIAN LUDWIG BÜCHNER (1824–99) belonged to a very gifted family, especially in regard to literature. He studied medicine and concurrently also philosophy, was a lecturer for a time, but having been dismissed, he earned his living by taking up a medical practice. He was of noble character and a keen upholder of liberty and justice, and from his early youth he enthusiastically adopted materialistic ideas, in which he saw a means of bringing humanity out of darkness and superstition. His famous work *Kraft und Stoff*, one of the most widely read popular scientific works of his age, is really a collection of talks on various theoretical questions in connexion with natural science, written in an attractive form, but without any very great originality. The old theme — the indestructibility of energy, the permanency of matter, the soul as a combination of cerebral functions — is played upon with constant variations and in a spirit of incessant controversy against theologians and philosophers. Büchner certainly has a better idea of the limitations of natural science than Vogt; he admits that existence is full of riddles that cannot be solved; but like Moleschott and Vogt he never attained to that clearly formulated self-limitation that Comte in his great work imposed upon positivism. Nor did any of them realize the importance of evolution as Comte did. All of them hailed the advent of Darwin with enthusiasm; his doctrine gave to their conception of nature an impetus that it never had before. The fact is, the idea of the origin of species gave to the realistic natural philosophy the connexion that the idealistic conception of nature had in its theory of ideas. Energy and matter were far too abstract and difficult ideas to support a popular theory of life, all the more so as the above-mentioned champions of their omnipotence lacked that thought-training which would have made them capable of mastering a subject so hard to elucidate. Their service to natural science and their labours for its propagation among a larger public are at any rate deserving of recognition.

FROM DARWIN TO OUR OWN DAY

CHAPTER X

THE PRECONDITIONS OF DARWINISM

1. Modern Geology

URING THE ZENITH of the power of Darwinism it was considered in certain quarters that one of the chief missions of cultural history was to seek after "pre-Darwinists." It was obvious that in such circumstances aspirants to this honour should come forward in large numbers; to begin with, the old Greek natural philosophers Anaximandros and Empedocles were named, and the number increased the nearer one came to modern times. There came another period when the list of personalities thus accumulated could be used to depreciate Darwin, as Kohlbrugge used it.[1] If, however, we damp our enthusiasm somewhat and have regard to actual facts, we shall find that the precursors of Darwin were far fewer. He himself has acknowledged the influence that he derived from Lyell's geological theories and Malthus's studies of population, and it seems only fair when reviewing a scientist's development to take into consideration his own remarks on the subject. If we do this, we get two preconditions for the origin of the Darwinian theory—a natural scientific, or, more exactly, a geological, and a socio-political. We shall now proceed to consider the former of these two.

Compared with biology, modern geology is a young science. Some of its pioneers have been mentioned in the foregoing: da Vinci, Steno, Buffon. The creator of geological study as a special branch of science is without doubt ABRAHAM GOTTLOB WERNER (1750–1817), professor at the mining academy at Freiberg, a teacher of Humboldt and many other geologists and mineralogists. He systematically explored the geology of his own district, determined the sequence of the rock-beds, examined their composition, and on the results thereof based a rational mining-industry. He never actually printed his theories; it is only through the medium of his pupils that the world has become acquainted with them. He is best known as the advocate

---

[1] Kohlbrugge, "*War Darwin ein originales Genie?*" *Biologisches Zentralblatt*, Vol. XXXV, p. 93.

of "Neptunism"; he believed that all mineral species, even basalt, are precipitated in water. The narrowness of his conclusions was largely due to
the fact that he never made any journeys; he presumed that the geological
conditions all over the world were like those in his own country. The energy
with which he defended his views was, however, impressive, and his pupils,
who came from all parts of the world, endeavoured faithfully to apply the
master's doctrines, however difficult they might prove to be in practice. The
whole of the earliest generation of geologists, as a matter of fact, shared
this failing of Werner's — even the scientist who is named with Werner as
the creator of geology, Hutton, had never been outside his own country.

JAMES HUTTON (1726–97) was the son of a Scottish landowner, and
studied medicine in his youth, but, having inherited a fortune, he afterwards devoted himself entirely to scientific research, especially geology. It
was not until late in life that he published the work *Theory of the Earth*, in
which he expounds his original ideas, though in a not very clear form. He
considers that geology has nothing to do with the history of creation; its
function is to describe the rock and earth strata now existing and to account
for their origin. He believes that the present rock-beds have arisen through
the destruction of older strata, similar to that which takes place daily
through the influence of water. This principle of explaining the past out
of the present represents his most valuable contribution to the development
of geology, though his own applications of that principle were often not
very successful.

It was not possible to ascertain the reciprocal age of the different rock
strata, and thereby also to create a history of the evolution of the earth's
surface, until attention had been paid to the remains of living creatures that
are found in the various geological beds. This, indeed, Buffon had already
done, but the one who really systematized palæontology was Cuvier. His
work in this sphere has already been described and its deep significance
pointed out; his catastrophe theory, the gist of which has likewise been
explained above, had disastrous consequences. Its influence was felt least in
England, where geology was developed independently in this field also. The
scientist who introduced into that country the knowledge of fossils as a
guide to geological research was WILLIAM SMITH (1769–1839). Born of poor
parents in the country, he received a deficient school-education and afterwards became apprentice to a surveyor, who taught him sound professional
knowledge, with the result that he was sought after as a surveyor and level-
ler, making a fortune in that profession and at the same time having opportunities for studying very different geological strata and rock formations.
He quickly came to realize that these possessed a settled order of succession
and that different animals and vegetable remains characterize the different
stratifications. The fossils he himself was unable to determine, this being

done by some of his friends, but he had a keen eye to the place into which each form should fall in the strata system. Eventually he published the results of his life-work in a great geological atlas of England, which cost him his whole fortune. For a time he suffered want, but was eventually granted a government pension, which ensured him a peaceful old age. It was through him that the use of guiding fossils for identifying the age of geological formations was introduced into science.

An investigator who surveyed, and in a high degree developed, the geological knowledge of his age was CHRISTIAN LEOPOLD VON BUCH (1774–1853). He belonged to a distinguished and wealthy Prussian family, and studied under Werner at Freiberg together with Humboldt with a view to entering the mining service, but he soon applied himself entirely to geology, which, thanks to his inherited wealth, he was able to study without having to earn his living. He made extensive expeditions, in the course of which he made a particularly fine collection of comparative material from various countries. The result of this research work soon led him from the Neptunism of Werner to the opposite extreme; he ascribed to volcanic activity an important, and indeed far too important, part in the history of the earth's surface. His investigations, carried out in different regions, are nevertheless of lasting value; he was, moreover, an eminent palæontologist, making valuable investigations of special subjects, particularly of fossil invertebrates: Cephalopoda, Brachiopoda, and others.

CHARLES LYELL is, however, the scientist that is first worthy of mention as the founder of modern geology and thereby as a pioneer of the descent theory. He was born in 1797, the son of a Scottish landowner, who was also interested in botany and who inspired in his son a passion for nature study. The latter took his degree at Oxford and afterwards adopted the profession of a lawyer. But he did not go far in that career, for eye-trouble compelled him to give up public work of any kind. Long before this, however, geology had attracted him, W. Smith's investigations especially interesting him, and for the rest of his life he devoted himself to that study, bearing with unparalleled courage the severe deprivation that defective vision always means to a scientist, especially a natural scientist. One source of comfort in these circumstances was the fact that his wife with devoted self-sacrifice dedicated her life to helping him in his work. He thus became one of the many brilliant private scholars in which the cultural history of England abounds, and he was the recipient of not a few honours. He undertook a number of long voyages of exploration; he considered them to be indispensable for a geologist, for it is only thus that he can gain that living idea of the various forms of the earth's surface which may serve as a basis for a theory of its history. The rest of his time he spent in London, where he was a member of many learned societies and was also otherwise held in high repute. He died in 1875, after having some years previously lost both his sight and his wife, who had been the mainstay of his life and his work.

Lyell's actualistic geology

LIKE Hutton, Lyell takes as his starting-point the present form of the earth's surface, studies its changes as the result of various natural influences, and finally draws the conclusion that the same forces have always, and approximately in the same degree as in our own time, been operating on the earth's surface; he who declares otherwise must substantiate his argument with proofs; it is the upholders of the catastrophe theory whose duty it is to prove the correctness of their views, and not vice versa. This conception of the evolution of the earth — it has been named the "actualistic" — forces Lyell to follow it through to its extreme consequences and far beyond what science in modern time is prepared to admit. Thus, in his *Principles of Geology* he absolutely denies the possibility of the earth's having originally existed in an incandescent state; he likewise definitely rejects Lamarck's theory that the animal world in earlier ages consisted of entirely different species from those in modern times and declares that mammals and birds have existed from the very earliest times. But apart from these extravagant statements, which as a matter of fact he afterwards partly corrected, his strict adherence to the principle that the phenomena of past ages should be explained from what is known from the phenomena of the present time has formed the basis on which it has been possible to construct a truly scientific geology. The earlier geological theories, both ingenious and foolish, had all been mere products of the imagination; Lyell introduced the principle, which must inevitably be adopted by every empirical science, of starting from what is known and has been investigated and thence proceeding gradually towards the more remote and the unknown. If past natural phenomena in general are to be calculated or at least reconstructed with fair probability, it is necessary to start from the present, whose course of events it is possible to survey. This astronomy has long done with its calculations of the position and motions of the heavenly bodies in past ages; and modern geology has in certain spheres, as, for instance, in the determination of annual stratifications out of water during preceding periods, reached a degree of accuracy that should not be far inferior to astronomical calculation. And this principle essentially represents Lyell's service to science.

His criticism of Lamarck

MOREOVER, Lyell has made important contributions in his above-mentioned work to problems of the development of life upon the earth. His criticism of Lamarck's theory undeniably touches the latter's weakest spot, when he maintains that Lamarck never even attempted to find out the origin of a single vital organ, but merely occupies himself with modifications in those already existing. He can hardly be blamed for the fact that he does not consider that he had found any actual proof of the transition of one species to another, since it has indeed scarcely been possible to discover one even in our own time. He does not believe in the possibility of the various species' being able to vary beyond a certain limited extent, and this limit is soon reached; if we try to force a form beyond this, it perishes; as an instance he quotes the adaptability of species to different climates. Man's domestic animals have from the beginning been especially suitable for taming, while other equally or more intelligent animals, the apes, for instance, have to be left at liberty. It is primarily, however, the rare existence of and sterility in hybrids that to Lyell's mind gives proof of the constancy of the species. The similarity between embryos of various kinds merely testifies to a common plan in their structure, but no common origin. He believes that every species has been created in a locality suitable to it and has spread from there under the constant influence of the climate, means of subsistence, and competition with other life-forms. In disproof of Lamarck's theory of species-modification he maintains that an alteration in the climatic changes or other alterations in the conditions of life would give certain species advantage over others, so that the adaptability assumed by Lamarck would never be realized in the latter. If a lake were to be converted into a swamp, already existing marsh-plants would be ready to overrun its area, while the aquatic plants would die out before they had time to adapt themselves to swamp conditions. How the species came to be created is a question that Lyell refuses to discuss; he speaks of "creative force," though he attributes no personality to it, regarding the whole problem as insoluble. Instead he discusses in detail the conditions governing the distribution of species, their development and extinction during different geological epochs. The whole of this exposition exercised a very great influence on Darwin, both positive and negative, by calling forth a contradiction from him — a point on which more light will be thrown later.

But the main point is that Lyell's theory of geological evolution offered at the time particularly valuable support to the idea of evolution, which was one of the watchwords of the age; here indeed there was confirmation in nature herself of the idea of an uninterrupted development as the fundamental force in existence. The result was that Lyell's name became one of the most popular at the time, and he himself enhanced his reputation by his ability to keep pace with scientific developments; he, the opponent of Lamarck, associated himself directly and without reservation with Darwin. His activities as the promoter of Darwinism will be dealt with in the next section.

2. The Ideal Preconditions of Darwinism

Failure of Lamarck's theory

THE question has not infrequently been discussed: Why did not Lamarck's theory of evolution succeed? The reason has been put down to the opposition of the Church, but certainly without justification; as far as is known, Lamarck was never interfered with by the Church, and the latter's opposition to the theories of origin and species-modification is, as we shall find, of a far later date. We are, then, far more justified in blaming the romantic natural philosophy, which, seeking, as it did, after one common idea for every life-form, lacked all feeling for material development. For herein lies the real gist of the problem: if a theory of evolution is to attract general attention, there must naturally be evinced an interest in evolution. Our next duty, therefore, will be to try to explain how this interest arose and how it expressed itself at the time of Darwin's appearance.

It is common knowledge that mankind is always ready to fix its ideals in antiquity — "the good old times." One is most inclined to deplore the present and to view the future with feelings of anxiety. And just like individuals, the public opinion of the different epochs has done the same; if man has carried out reforms, it has mostly been done under the form of reviving the ideal conditions of ages long past; so it was during the Reformation, when people wished to revert to the conditions of early Christianity, and so too during the French Revolution, when people raved over the republics of antiquity, and imaginative popular leaders called themselves Anacharsis or Gracchus. If one has dared to cast a glance at the future, one has most probably expected to find happiness in some vast catastrophe resulting in the total annihilation of the present; thus all apocalyptical enthusiasts of antiquity and ever since, and thus too the political extreme tendencies of modern times. Belief in a gradually progressive, law-bound development has always been limited to a few, and these perhaps are to be found among the men of action rather than men of thoughts and words. The most pronounced faith in progress that has ever existed has been the liberalism of the nineteenth century, a current of ideas which had just reached its zenith by the middle of the century, when the theory of origin came to the fore. The coincidence is of course not accidental; on the contrary, the one idea is dependent on the other, and therefore the victory of Darwinism is inexplicable without some insight into the general intellectual conditions at the time of its birth.

Liberalism of the nineteenth century

THE optimistic belief of liberalism in the progress of the human race had its true origin in England, where throughout the entire eighteenth century prosperity and enlightenment increased slowly but surely, where humane

CHARLES DARWIN

CHARLES LYELL

legislation and democratic social development were demanded and even gradually achieved without violent upheavals. This belief was strongly influenced by Rousseau's doctrines of the natural goodness of man, which has only been perverted by social life and by the oppression of evil kings and priests; it found expression in the democratic reforms of the French Revolution, but it acquired its true character through the great technical and material progress made during the nineteenth century, to which reference has already been made above. The new big-scale industrial development and world trade, rendered possible by steam-power, created an intelligent middle class, which felt well satisfied with the present and hoped for still greater benefits from the future; the labouring classes were not yet organized and their discontent was thus perceptible only in isolated instances. The vast production of material values set its mark upon the age and was met by the belief, adopted from Rousseau, in the natural goodness of the human race and in Bentham's doctrine of happiness for as many as possible as the chief aim in life; happiness was made the synonym for material welfare, and this could best be attained by letting mankind, endowed by nature with goodness and intelligence, look after themselves, undisturbed by oppression and superfluous regulations. Human life thus came to be regarded as a dominion of impersonal forces guiding humanity with the necessity of a natural law towards better times, if only they were allowed to operate freely. The people — the impersonal summary of the individuals living in a country — were better advised than any single person; if only they were allowed to look after themselves, their activities would conduce to a successful development, to which there seemed to be no limits. Free competition both in the material and in the spiritual world and no interference with the individual's liberty of action were the watchwords of the age; how the free will of the individual was eventually to be reconciled with the popular will was a question that did not bother the minds of many; for the time being, the individuals looked up to the popular will as to a higher power, the only fault of which was that it had not yet had sufficient time in which to operate.

This conception of life, which naturally appeared under quite different forms in different quarters — in historians like Buckle, in thinkers like Mill and Spencer, not to speak of their pupils and imitators on the Continent — was without doubt the most favourable soil possible in which to cultivate a general theory of evolution. Evolution, Progress, were in fact the slogan of the age. It had been employed in Comte's system, described above, but only as far as regards human culture; through Darwin evolution was elevated to a natural law governing all life. It is no wonder, then, that his theory was hailed with enthusiasm by all those who cherished the ideals of the new age. It was indeed the ideal itself that was hereby sanctioned to embrace the whole of nature; on the other hand, it affords an explanation for the violent opposition on the part of all adherents to the old order of society, which was not yet won over and which towards the close of the century was to muster increasingly stronger forces. But Darwin himself was influenced by the new conception of community life; it was from one of its theorists that he obtained the actual idea for his theory of selection — namely, Malthus, wherefore he is worthy of a place in the history of biology.

THOMAS ROBERT MALTHUS (1766–1834) was the son of a landowner, took his degree at Cambridge, was ordained priest, and obtained a curacy, but at the same time devoted himself to the study of national economics. On account of his writings he was given a professorship in London, where he afterwards worked with great success. His father had been a personal pupil of Rousseau and entertained rather radical views on the improvement of the human race by a fair distribution of wealth — an idea which had then, as it had later, many supporters. Against them there appeared the younger Malthus with his chief work, *The Principle of Population*, which came out in many editions and has been very widely discussed. Although himself a liberal, he is in no wise a revolutionary optimist; he sees the cause of human misery not in an unfair distribution of property, but in man's own habit of living thoughtlessly and frivolously. The cure for this he sees in bringing mankind up to exercise self-control, every man being taught not to raise a family without definitely guaranteed means of subsistence. It is, he believes, a fact throughout nature, in plants, animals, and human beings, that natural procreation is stronger than the possibilities of maintaining life; from this there arises in nature a violent competition for the maintenance of life, and in human life there is, further, helpless and ever-increasing misery among the poverty-stricken classes, which no philosophical measures can remedy. He then tries by means of historical and geographical-statistical investigations to find out how it is that the increase in the population never has followed, and does not follow now, its natural course, but is restricted, more or less owing to the fact that the supply of maintenance is limited, with the result that want, with its concomitant vice and crime, thins out a great number of the poorest in each community. Upon its first appearance this doctrine was violently opposed from both conservative and radical quarters; it is not, however, within the scope of this work to enter into a detailed discussion of the subject; it is sufficient to point out the above-mentioned theory of competition, which gave Darwin the idea for his theory of selection. To this latter we shall now proceed.

CHAPTER XI

DARWIN

CHARLES ROBERT DARWIN was born in 1809 at Shrewsbury in the west of England. His father, Robert Waring Darwin, was the son of the physician and natural philosopher Erasmus Darwin and was himself a physician. He was married to Susannah Wedgwood, daughter of the famous procelain manufacturer Josiah Wedgwood, who, from being a poor and ignorant apprentice to a potter had made a successful career, acquiring a splendid fortune and a famous name in the history of ceramics. Charles was the sixth out of eight children. He went through a school of the usual English type, his education consisting almost exclusively of the classical languages, and was afterwards sent to Edinburgh in order to study medicine in the family tradition. The Latin he learnt at school did not interest him very much and he was utterly bored by the anatomy lectures. Darwin broke off his medical studies after a couple of years, so that he never became an anatomist, to his own great loss. He now decided to try his hand at theology at Cambridge, where he spent three years and took his degree of bachelor of arts, but he spent most of his time pursuing the usual occupation of the well-to-do English undergraduate — sport, especially shooting. He also collected insects and plants for his own amusement, but he chiefly interested himself in geology, receiving a sound elementary training in that subject under the guidance of the eminent professor ADAM SEDGWICK (1785–1873), whom he accompanied on several expeditions. On the recommendation of a friend he was offered in 1831 the unsalaried post of naturalist on board the cruiser *Beagle*, which was to circumnavigate the world for mainly cartographical purposes. This voyage, which lasted five years, gave him, as he himself says, his real training as a naturalist, as it also determined the direction that his future work was to take. He worked with zeal and sent home from the various stopping-places on the way both notes and collections. Of these the geological possessed the greatest value; the zoological and botanical were regarded by contemporary judges as nothing extraordinary. This persevering activity was so much the more praiseworthy as Darwin suffered throughout the journey from incurable seasickness, which gradually irremediably impaired his health. On his return home he devoted himself for years to the working up of the natural objects and the material for ideas that he had gathered in the course of the voyage. During that period there slowly developed in his mind the theory which bears his name. In 1839 he married his cousin Hannah Wedgwood. Her wealth added to his own made it possible for him during his remaining years to lead the quiet life of a private scholar, which in fact became in time an absolute necessity, owing to his increasing ill health. Three years after his marriage he left London and settled in Down, a small town in Kent, where he spent the rest of his life in his own comfortable house, with a delightful garden. Even in these circumstances, however, his health did not improve; he suffered from a nervous stomachic trouble, which occasioned constant vomitings and frequent insomnia. It was only through living a painfully regular life under the self-sacrificing care of his wife that he was able to hold out as long as he did. His days passed with brief but intensively concentrated periods of work, alternating with medical attention, walks, and literary diversion; journeys and social life were restricted to a minimum. During this period there was given to the world that unique production — considerable even in its extent — which made his name immortal. His bodily existence, so full of suffering, was compensated for throughout his life by a rare spiritual poise; complete freedom from passion, from hate, envy, and ambition, and an almost tender amiability, which certainly found it difficult to refuse a petition, however unreasonable, but which also made it easy for him to enjoy and find child-like pleasure in the narrow life to which his ill health restricted him. His was no critical character; towards the statements of others he used to show, as Johannsen says, "an amiable credulity," and his own experiments were often consciously childish. His sensitiveness, however, was in no way associated with weakness of character; on the contrary, few students of nature have striven with such unbending determination for years and years towards a given goal, and adhered to a point of view when once adopted with such firm conviction. His ideas were, as is well known, both unreservedly praised and violently vituperated; attacks were met by him with unfailing steadfastness and a noble calm, so that he never allowed himself to be involved in personal polemics, but he always took note of and parried material objections. Thanks to these qualities, Darwin came in the course of years to enjoy personal esteem such as seldom falls to the lot of scientists. Occupied in constant work, his life moved quietly towards its close. He died in 1882 and was buried in Westminster Abbey, not far from Newton, followed to the grave by the most distinguished men in the country both in the social and in the scientific world. Shortly before his death he had written down in some notes on his own life the oft-quoted words: "As for myself, I believe that I have acted rightly in steadily following and devoting my life to science. I feel no remorse from having committed any great sin, but have often and often regretted that I have not done more direct good to my fellow creatures."

In his youth Darwin was a confirmed lover of the open-air life; a good shot, an enthusiastic huntsman, and a keen observer of life in nature. This love of animate life in the open air he retained even in his old age; long after ill health had compelled him to give up shooting and voyages of exploration, he applied himself with indefatigable devotion to the care and observation of life in his park and garden. Dogs and cats, birds, insects, and earthworms, no less than plants of the most varied kinds, were to him a never-wearying source of joy and observation; all their manifestations of life in the minutest detail were the object of his most careful study; animals' actions, instincts, and manifestations of intelligence were observed, analysed, and summarized by him day by day and year by year with never-failing interest. His theoretical training, on the other hand, was deficient — most thorough in the sphere of geology, whereas in biology it was, on the whole, limited to the systematic side. His observations made during the circumnavigation of the world also bear witness to this restricted basis on which his education was founded. He was, moreover, in his youth a firm believer in the Christian faith — he intended, in fact, to become a clergyman — and he accepted without criticism the traditional dogmas, including, of course, the doctrine of the origin of living species as the result of a divine act of creation. During his voyage, however, he found that this belief conflicted with the results of his observations. His diary contains many proofs of this; in particular, the existence of many species with a small area of distribution, of forms closely allied to one another, but not alike, and taking the place of one another in different localities, yet not existing together, seemed to him difficult to reconcile with "nature's great plan." Why had it been necessary to create all these slightly differentiated and narrowly distributed species? He spent one month on the desolate Galápagos Islands, situated a long way off the coast of South America and composed of volcanic lava comparatively recently cast up out of the ocean; here he felt himself "placed in proximity to the very act of creation itself." But here he found a fauna of markedly South American genera, though possessing peculiar species; of many birds each separate island had its own species. That one species should have been created for each small island seemed to him irrational; but how, then, had the different species arisen and why did they belong to the South American genera? This problem, having once penetrated his mind, gave him no rest. Upon his return home he at once started to record in a separate book his experiences in connexion with the question of the formation of species, and he sought long and restlessly for proofs of the correctness of his ideas. In 1844 he writes in a letter to his friend the botanist Hooker: "I have read heaps of agricultural and horticultural books and have never ceased collecting facts. At last gleams of light have come, and I am almost convinced (quite contrary to the opinion I started with) that species are not (it is like confessing a murder) immutable." Lamarck's theory of the modification of species, however, Darwin was unable to accept; it appeared to him to be "rubbish" — "Heaven forfend me from Lamarck's nonsense of 'a tendency to progression.'" Nor indeed in any other biological literature accessible to him could he find any way out of the difficulty involved in the origin of species.

Darwin's experiments to prove the mutability of species

DURING this period he was closely associated with Lyell, the scientist who most influenced him — he too, as we have seen, no friend of Lamarck — and resolved to deal with the species as Lyell had dealt with the geological strata of the earth — namely, to collect as many facts as possible regarding the transition from one form to another. In this respect domestic animals seemed to him to give the best suggestions: that each separate domestic animal was a true species no systematist had ever denied and it was likewise acknowledged that man had produced a mass of different forms of every species of that kind. Darwin placed himself in communication with a great many animal-breeders, and himself for years bred different races of pigeons, all for the purpose of discovering how the different races arose. Expert breeders believed that by a selection of suitable parents it was possible gradually to modify the progeny at will. Darwin also came to accept this view; all the young in a litter of domestic animals are indeed somewhat unlike one another and their parents — they "vary" as he says — and by selecting the suitable variations it is possible to guide the breed in the required direction. But if man was able by selection to produce out of the uniform canine type that still exists among wild tribes such a large quantity of different forms, should it not then be possible for species to be modified by nature in the same way? The difference between a greyhound and a bulldog is far greater than that between many wild life-forms which without doubt pass for good species. But is there in nature a force operating in the same direction as the breeder when he selects new forms of domestic animals? Here lay the worst stumbling-block. Then Darwin happened to read Malthus's above-mentioned work on population: how both in nature and in human life there are produced individuals in far greater numbers than there are means for maintaining, and how the weakest perish in the competition for food. This gave him his idea; in the struggle for existence those life-forms are destroyed that are least capable of adapting themselves to prevailing conditions, while the strongest individuals survive and reproduce those qualities that have a greater chance of survival. Thus the external conditions themselves come to multiply the differences brought about by the variability of the offspring in relation to the parents, until new varieties and new species arise. Consequently, the struggle for existence induces a natural selection that operates similarly to the choice of races among domestic animals exercised by man, only with this difference — that vast expanses of time are available for natural selection, which justifies the assumption that all the manifold forms of life on the earth, both those which have existed and those which still exist, have been developed through its influence. These facts, then — the dissimilarity between the offspring and the parents (that is, variability) and the struggle for existence, with the resultant natural selection — explain, according to Darwin, the origin of species.

His zoological works

FOR two decades Darwin kept this theory to himself in an unceasing search for fresh proofs of its universal application. Finally, in 1859, he published it in a work entitled *On the Origin of Species by Means of Natural Selection*, one of the most famous works of natural history that have ever been written. Even before that, however, he had won a high reputation as the result of a number of monographs on various subjects. Among these may be mentioned two geological treatises: *On Volcanic Islands* and *On Coral Reefs*, the latter being specially famous for its universally accepted theory of the arising of atolls or circular reefs through the sinking of the land area around which the coral reefs had originally grown up. Among his zoological works may be especially mentioned an extensive work on the Cirripedia, in which he gives a detailed and exhaustive description of the system and evolutional history of these animal forms — their peculiar dwarf males discovered by him — besides which he has also dealt with the fossil forms of that animal group. Moreover, as editor of the scientific results of the *Beagle* expedition he contributed much of great value. It was thus a naturalist with a good reputation who came forward with the work on the origin of species. The violent controversy that it occasioned brought immediate world-wide fame to its author. A somewhat detailed account of the main ideas of the work is therefore called for, all the more so as, in spite of its immense popularity, it would seem to have been less widely read in recent times than one might suppose, and the exposition of the theory of origin to be found in the usual text-books has been strongly influenced by that comparative morphology with which Darwin himself was more or less unfamiliar.

The theory of origin that Darwin created is decidedly characterized by the personality of its founder. Darwin brought to his work, as we have observed above, a deficient theoretical training, particularly in the sphere of anatomy, an intense geographical and systematical interest, and, as a standpoint beyond which he had already advanced some way, a somewhat ingenuous orthodox-Christian belief in the creation. Being a systematist, he saw in the problem of species the central point of biology, and to him the centre of this problem was, in its turn, the problem of creation. This must be borne in mind if we are to understand Darwin's relation to the earlier morphologically inclined generation of scientists of the Cuvier school.

Whatever their view of life, the species idea was to them essentially a practical basis for comparative morphology, whereas the problem of creation was a question that was entirely put aside as not concerning natural science. It must at once be admitted that it was certainly due to Darwin's dilettante conception of nature that he thus adopted just the problem of creation as a starting-point for many years of research work and cogitation; on the other hand, it was his treatment of the problem that caused such a public sensation over his work.

Immutability and creation

LINNÆUS in his youth defined the species as the progeny of those animals that had been created in the beginning; he afterwards altered his view, in that he assumed a few species to have been created, out of which the others were evolved at a later period. To the systematists who succeeded him it was the immutability of the species that was the essential point, the actual basis of the system, while the problem of creation was seldom discussed. De Candolle, it will be remembered, has a definition of species based on mutual similarity and fertility between the individuals, but without any mention's being made of the creation. To Darwin, however, "immutable" and "created," in regard to species, are inseparable terms; doubt of the immutability of the species is induced by doubt of the creation, which in its turn has been caused by the species' conditions of distribution and not by any doubts as to the assumption of a supernatural act of creation being in itself an explanation of nature.[1] Then he gets the idea of the variations which by means of natural selection are adapted to prevailing external conditions, thus giving rise to, first of all, new varieties, and then new species. Even earlier systematists had taken it for granted that varieties are produced by external conditions, flourish, and disappear; what is novel in Darwin's theory is that the species are nothing but more fully developed varieties, which selection, resulting from the struggle for existence, has determined, while the intermediary varieties, as being less capable of defying competition, have perished. He adduces a great many arguments to prove that those species which are most widely scattered and, where they exist, are richest in individuals are also those which produce the most varieties, which in his view is the same as initial species. And he points out how vague the boundaries between species and varieties really are in the minds of different systematists and how difficult it is to define what is meant by species. He considers that this name is given arbitrarily and for the sake of convenience to a number of individuals which highly resemble one another, and that it

---

[1] In his diary of the voyage Darwin in one place explains the absence of certain fossils in a geological deposit by assuming that animals of that kind had not been created at the time when the deposit came into being (*Life and Letters*, II, p. 1). Again, in the *Origin of Species* a Creator is mentioned as the ultimate cause of life.

is not essentially different from the term "variety," which is used for less distinct and more fluctuating forms.

Variations in progeny

THE causes of these variations, which by means of selection — natural in wild life, human in domestic animals — are developed into varieties and species, involve a problem that occupied the mind of Darwin a great deal. He at once points out that only *hereditary* variations have any significance and also that the essential causes of them have never been really ascertained. On this question he adopts a somewhat hesitant attitude; it is true, he asserts, and collects ample material to prove, that external conditions produce variations in the progeny, which is a view strongly reminiscent of Lamarck, but, on the other hand, he definitely rejects all Lamarckian ideas. As a matter of fact, this theory of the heredity of variations is the basis of the Darwinian theory, but it is also one of its weakest points; in this connexion modern research into the problem of heredity has passed severe judgment on him — often indeed unfairly severe, it being forgotten that he had not that accumulation of facts to build upon which is available in modern times, but here he certainly does touch upon extremely vague conceptions, which make the chapter on the law of variation difficult to comprehend. Among the circumstances that influence the individual's reproductive organs and thus affect the offspring, he mentions climatic conditions of various kinds and alimental conditions, as well as the correlation between different parts of the body. Nevertheless, he always insists upon the importance of natural selection as being greater than the direct influence of environment. For instance, a number of insect forms on islands in mid-ocean have restricted powers of flight as compared with their relations on the mainland; this has arisen through the fact that those specimens that are best at flying have been blown out to sea and perished, while the weaker fliers have continued to propagate, rather than through the animals' not having dared to use their wings, with the result that their growth has become stunted. Correlation, again, compels other organs to follow suit when one organ has been modified as a result of selection. Further, he holds that parts of the body that have become especially developed in one species, as compared with corresponding parts in closely related species, are liable to peculiar variation; thus, the length of the arms of the orang-utan varies, just as, in general, every strongly developed characteristic indicates strong variation in the previous generation. On the other hand, the wings of the bat do not vary, abnormal though they are in comparison with the extremities of other mammals, for the entire group has wings of the same kind; the law would hold good only if one species had longer pairs of wings than other species of the same genus. On these grounds he believes also that species-characters vary more than genus-characters, but the variations of species of the same genus are analogous. In connexion with this point he accounts for what he calls the "tendency to reversion" — the frequent and unexpected tendency, especially in domestic animals, for forms to arise having the characters of the wild species: tame pigeons resembling the wild rock-pigeon, horses with zebra-like streaks, and other similar instances.

Difficulties of the evolution theory

DARWIN having thus sought to determine the laws of variation, he takes up for study the difficulties offered by the theory of evolution by means of natural selection. The chapters devoted to this task comprise more than half the book and represent a strange miscellany. As a matter of fact, Darwin acknowledges no limitations to his duty to answer all objections to his theory, and he always finds some way out of a difficulty, however desperate it may at first appear. He himself considers that the most difficult phenomenon to explain according to the theory of selection is how the ants' workers have acquired their intelligence; they cannot reproduce themselves and thus transfer their favourable variations by heredity to any offspring. The difficulty is solved by the assumption that here it is actually the community as a whole which derives the advantage from variations; the sex-individuals that have produced workers with the best and most advantageous qualities have been victorious in the struggle for existence, and thus have arisen both the highly cultivated worker types and the strongly developed instincts to make slaves, tend aphides, etc. Darwin undertakes another particularly different task in seeking to explain how such a complicated organ as the eye came to be formed. This explanation, which was ill received by contemporary critics, is certainly rather far-fetched; there is no direct transition between the vertebrate animals' type of eye and that of the Arthropoda —it is not stated why association is not sought with the molluscs instead, in which order the highly developed visual organs of the ink-fish might have served as a transition. — And so the whole work concludes with some general assurances as to the metamorphosing power of selection. It is much easier, of course, to explain the origin of the lungs from the swimming-bladder; on this subject earlier comparative-anatomical observations have been available as a basis of study. Darwin even undertakes to defend the old objection of the sterility of hybrids, which has so often been brought forward in favour of the constancy of species. He differentiates between infertility as the result of crossing species on the one hand, and the sterility of hybrids on the other; as far as the sterility between the species is concerned, he finds that it varies greatly in different organisms — Koelreuter's and Gärtner's experiments are especially cited as instances — and the final conclusion is that "accidental and unknown circumstances" are the cause of it in the different cases. The sterility of hybrids, again, is compared with the infertility of wild animals in captivity; each is attributed to the direct influence of external conditions upon the sexual organs. Here, too, reference is made to the varying results to which different experiments have led. The fertility of variety-crosses, on the other hand, is attributed to favourable conditions of variations in closely related characters. The result of this is a proof that transition forms exist between species and varieties.

Darwin and Mendel

IF we compare these discussions of Darwin's on heredity and hybridization with the experiments that Mendel concurrently carried out for the same purpose, the English scientist naturally gets left hopelessly behind — on his part, widely vacillating speculations; on the part of Mendel, clearly conceived and exact experiments. The very starting-point brings this out clearly; Mendel starts from a few simple and easily determined characters and establishes their appearance in different generations in various combinations; Darwin, on the other hand, starts from the ideas of species and variety — that is, from the most abstract terms in biology and the most difficult to define. In fact, in this starting-point lies the whole weakness of Darwin's research work and speculation. His successors, indeed, almost immediately abandoned this standpoint and instead sought for proofs of their theory by recourse to the material and methods of comparative anatomy; Cuvier and his successors had already studied the changes undergone by one and the same organ in a series of different animal forms. It was through this comparative method's being placed at the service of the theory of origin that Darwinism, especially through Gegenbaur and his school, came to use for purposes of investigation objects of a definite and concrete nature. But Darwin himself had but little mind for comparative anatomy; he certainly cites for the purposes of his theory a number of proofs derived from morphology, but in quite a brief and summary fashion. He was more interested in embryology. Although he himself had never worked practically as an embryologist, he nevertheless realized the value of comparative investigations into different stages of development and he works out the basis for a "biogenetic principle," which Fritz Müller and Haeckel only had to supplement.

Darwin on questions of geography

DARWIN is, however, far more at home in the sphere of geology and geography and he firmly rejects any attempts at employing the results of these sciences to disprove his theory. The incompleteness of palæontological remains he considers to be sufficient argument against those who inquire after "missing links" between now existing genera and species, while the conditions of distribution of living creatures seemed to him from the very outset to be the surest guarantee of the truth of his doctrine. Climatic changes have in the course of the ages given the most powerful impulses both to new variations and to the struggle for existence under new conditions, while newly-formed natural barriers, mountain ranges and encroachments of the sea, have split up uniform groups of life-forms and created isolated areas of development with accompanying new forms of genera and species. And always variation and natural selection are sufficient to explain all phenomena; since Aristotle produced his explanation of nature, no biologist has ever conceived it his duty to the extent that Darwin did to explain anything whatsoever. In this respect he takes the most extraordinary trouble to achieve his purpose; in a letter to Lyell he expresses surprise that the bats of New Zealand — of old the sole representatives of the mammals on the island — had not made their home on the ground and developed into land-animals, seeing that they had no competitors. And he ascribes to selection the most remarkable powers; a traveller had seen a bear swimming in a North American river and snapping at insects in the water; Darwin thinks it not impossible that, if food of this kind were abundant and there were no competitors, a number of bears would become aquatic animals and would gradually acquire larger and larger mouths, eventually becoming as monstrous as whales. This strange conclusion, which is given in the first edition of the work, but was modified in succeeding editions, gives striking evidence of another weakness in Darwin's speculation: his lack of sense of a law-bound necessity in existence. "I believe in no law of necessary development," he expressly declares. The variations are certainly guided by laws, as mentioned above, not, however, in any given direction, but in all possible directions, and they are influenced, depending upon every chance, quite incalculably by natural selection. But if, then, natural selection were guided by chance, it would exclude the possibility of any law-bound phenomenon in existence. Herein really lies the greatest weakness of the Darwinian doctrine of selection. It has, in fact, been sharply criticized — in modern times especially by Oscar Hertwig in his work *Das Werden der Organismen*, the subject of which is indicated by the subtitle: *Eine Widerlegung von Darwins Zufallstheorie*. A similar judgment was passed by Rádl, who, moreover, points out that Darwin really applied the social conception of contemporary liberalism to life in nature; which, as a matter of fact, is at once realized from the acknowledged part played by Malthus's social doctrine in the working-out of Darwin's theory. This human-social conception of nature stands out clearly in the above-mentioned statement regarding the bear, which, if the chance offers, can take to swimming and develop into a whale. More applicable to human-social life than to nature is also the form that his utterances often take of fancies thrown out at random, which reminds one of a social reformer's improvement schemes rather than of the binding conclusions of a scientific investigator; "it would be easy," "it would offer no difficulty to suppose," and other similar expressions frequently occur. This, of course, is also due to his oft-recurring tendency to allow his thoughts to dally with all kinds of possibilities — a tendency which, when combined with a belief in the ability of his theory, once advanced, to explain practically any biological phenomena whatsoever, is bound to lead to far-reaching conclusions. Another result of the selection theory is the constant reference to the greatest possible adaptability to prevailing conditions, with the consequent insistence upon the finality existing in nature. It has already been pointed out how unsatisfactory this explanation of nature is and further reference will be made to it later on; here it need only be observed that this belief in a purposeful adaptability to prevailing conditions has in no small degree contributed towards retarding the development of biology into an exact science.

The influence that Darwin's *Origin of Species* exercised will be described in the following. His fame in no wise induced the author to rest on his laurels; on the contrary, he laboured indefatigably throughout his life still further to develop the theory that he had created and to apply it to different life-phenomena. The greatest and most important of his subsequent works was published in 1868 in two volumes and bore the title *Animals and Plants under Domestication*. In the first volume he gives a detailed account of his intensive racial-biological studies of domestic animals and cultivated plants. The systematic biology of preceding ages had, as a general rule, depreciated these beings: they were not true species, only a medley of varieties that no one could make anything of. Darwin then showed how great is the interest that this racial research possesses and what important results can be produced from it. All later racial research is, in fact, based on his initiative. In point of exactness these investigations of Darwin's certainly cannot be compared with those carried out concurrently by Mendel, but they are far more many-sided, as regards both material and conclusions, and they also caused an immense sensation, especially amongst those who led a practical life. Darwin himself largely had recourse to data provided by animal-breeders and gardening experts, and he was certainly not very particular about weeding out their alleged results. In the second part of this work he makes fresh contributions to his descent theory. Here he dilates at length upon his conception of heredity, which played such a radical part in the cultural history of the nineteenth century, although it is now entirely abandoned. As has often been pointed out, heredity is to him equivalent to the direct transmission of qualities from the parents to the offspring, a transmission that is influenced by a vast number of external circumstances. Further, he characterizes atavism — the recurrence of qualities similar to those of earlier generations — as due to the contrast between the transmission of qualities and evolution, and, moreover, he points out a number of other hereditary phenomena — the transmission of qualities confined to only one sex, and the inheritance of qualities that come out at some special period in life. He also sought to explain that phenomenon which is now termed the "dominance" of certain qualities; he calls it the "prepotency of transmission" and finds

its existence extremely hard to explain, but puts it most closely in connexion
with the age of the qualities in question. Also "latent," or, as they are called
nowadays, "recessive" qualities, he made the subjects of observation and
speculation. Particular care was devoted by him to the problem of hybridization; he is all the time procuring from his experiments proofs of the overlapping of varieties and species. Further, he investigated the cross and
self-fertilization of plants, which he was to deal with more fully in a subsequent special work. His speculations on fertilization and hybridization should
not be judged by modern standards; he knew as little as his contemporaries of
the true course of fertilization and so easily became deeply involved in speculations as to the consequences of the effect upon the egg of adequate or inadequate quantities of sperm. He then discusses his favourite theory of the laws
of variation, which he now considerably expands, with an increasing tendency towards Lamarckism, external circumstances — climate, food, and
even the use and non-use of organs — being definitely stated as influencing
the forms of variation. Even hybridization and atavism are cited as causes of
variation, besides which the phenomena of correlation are more closely
analysed in connexion with variability.

Pangenesis

THE anxiety to find a universally applicable explanation of the phenomena
of heredity and variation led Darwin to think out what he called a "provisional hypothesis of pangenesis." In this theory he gives to the cytology
of the time, with which he otherwise had had nothing to do, a new and
curious interpretation. He believes that every cell, every tissue- or organ-unit
in the body, produces and gives off minute "atoms," which he calls gemmules, and that these latter, scattered throughout the body by the currents
of blood and other juices, conjoin as required, and then re-create those
"units" from which they are derived. The sexual products thus contain
"gemmules" from all parts of the body, and these are combined in the embryo,
and it is for that reason that all the latter's parts resemble those of the father
or mother, according to whose gemmules have constructed the part of the
body in question. Unused gemmules may be transmitted to the next generation, with the result that some individuals resemble their father's or mother's
parents. In the same way the bud of a plant is formed by the gemmules of
those parts that are evolved out of it, and the regeneration of the severed
foot of a salamander takes place through the extremity gemmules accumulating at the mutilated end; if, as sometimes occurs, a malformation takes
place, then the wrong gemmules have come into operation. This theory has
been shattered by modern research in the sphere of heredity and need not
therefore be discussed any further in this place; Darwin himself, it is true,
considered it to be only provisional, but he holds that it explains the problems at issue better than any other theory and should therefore be allowed

to stand. This is characteristic of him; the more a theory takes it upon itself to explain, the more convincing does he consider it to be.[2] But exact and critical research has not dealt thus with the theories; it has set up theories according as special research has required them, but it has never expanded them beyond the bounds of absolute necessity. Darwin is here, as so often elsewhere, a speculative natural philosopher, not a natural scientist.

Darwin on the descent of man

THIS speculative characteristic is still more conspicuous in his next work, *The Descent of Man, and Selection in Relation to Sex*, which was published three years after the former book, but which was likewise written after many years of preparation. In *The Origin of Species* he had already in passing expressed the opinion that natural selection would without doubt eventually throw light also on the origin of man — an assertion that was enough to excite very great attention. The subject had already been taken up by others: by Huxley and, above all, by Haeckel, and it was thus no longer a matter of real urgency. Darwin's presentation of it, however, possesses an interest of its own. His arguments that man has through natural selection by means of the struggle for existence been evolved from a series of animal forms are, of course, the same as those he had previously developed in regard to the animals; the anatomical and embryological argumentation he was able to borrow from his above-mentioned predecessors. It may be pointed out, however, that he does not insist upon man's relationship with the anthropoid apes, as Haeckel has done; he observes, it is true, physical and psychical agreements, but otherwise maintains for the most part man's character of a mammal. Of greater interest, however, is his derivation of the human psychical qualities; he analyses a number of such qualities of different kinds — curiosity, the tendency to imitate, memory, imagination, reflection — and he finds them existing also in the animals. He even notices an equivalent to religious feelings in the dog's awe of his master. On the whole, he falls into the same error as innumerable animal psychologists since then, of letting qualities that man has through training inculcated into his domestic animals be regarded as spontaneous manifestations of the intellect. As to the existence of moral qualities, he refers to the characteristics of self-sacrifice and social sense to be found in many animal forms — in regard to the ants he holds in this respect the same exaggerated ideas as many of his contemporaries —

---

[2] As an instance of how boldly Darwin takes up the most difficult problems for discussion, and how casually he afterwards solves them, the following may be cited (*Variations*, I, p. 8). He maintains that, in spite of natural selection, very simple life-forms have nevertheless been preserved through the ages by adapting themselves to very simple conditions of life: "for what would it profit an Infusorial animalcule for instance or an intestinal worm to become highly organized?" It must be admitted that, if the problem is difficult to solve, the answer certainly makes it none the easier.

and maintains that even amongst wild tribes only social virtues are respected. He has no interest in individual soul-life — a lack of interest which he likewise shared with many scientists of his age and which involves him in anthropomorphitic interpretations of purely instinctive phenomena, not to mention the credulity that he shows towards the statements of other owners of domestic animals regarding the purely human intelligence manifested by their four-legged friends. As to the time and place of the first appearance of the human race he expresses himself with a certain amount of caution, as he does also in regard to the racial problem.

Sexual selection

By far the greater portion of the work under discussion deals, however, with another question — namely, the origin of the secondary sexual characters. To these Darwin considers that the theory of natural selection in the ordinary sense cannot be applied; he does not believe he can use it to explain the origin of such features as the horns of the stag-beetle and the males of other coleopters, the brilliant coloration of male butterflies, the cock's-comb, the horns of the stag, and other similar characteristics. He considers rather that these features have arisen as a result of special sexual selection; the males have competed for the favour of the females, and the most attractive or the strongest have gone off victorious and been allowed to propagate and to transmit their characteristics by inheritance to their offspring. He finds proofs of this in the playing and the fighting that takes place between the males in the mating-season; the butterflies' sport in the air, the combats of cocks and stags, the song of the nightingale and the lark, the play of the wood-grouse, and the stately mating-dance of the cock of the rock. But it is not only the male qualities, but also certain common characteristics that he attributes to this kind of selection, as for instance the coloration of the butterflies, which he believes to have arisen owing to the females' also having acquired their share of the inheritance of sexual selection. This doctrine of sexual selection was rejected even earlier than the general theory of selection and is nowadays embraced by hardly any true scientists, although popular literature shows traces of it. What really brought about its rejection is the increased knowledge of internal secretion and the connexion of the secondary sexual characters with it; both sexual coloration and mating-play have their explanation in this. That Darwin knew nothing of this cannot, of course, be laid at his door, but even apart from this fact, the sexual-selection theory certainly gives strong evidence of his tendency to attribute without criticism purely human ideas to the animal kingdom, to believe in "beauty competitions" among butterflies and beetles, fishes and newts, or that grasshoppers and crickets have a musical ear. It has also been pointed out that it is purely physical strength and not beauty at all that makes cocks and stags successful with the females, besides which it may often happen that the strongest males spur or butt one another to death, with the result that afterwards comparatively weak specimens win a place among the females. In support of his theory Darwin placed the male intelligence at a radically higher value than the female. He overlooked the fact that the females also exercise an important function, which likewise demands intelligence, in the care and protection of their offspring. His theories on this subject nevertheless won strong support in certain literary quarters; it is well known that, among others, Strindberg has referred to them with enthusiasm.

In connexion with the last-mentioned work Darwin published another book, entitled *Expression of the Emotions in Man and Animals*, in which he records a large number of facts regarding emotions in man and the animals, which he had amassed and compiled and to which he, of course, applies his theory of selection and descent. Further, in his later years he published a number of works on special subjects that are in part extremely valuable. Among these may be mentioned his work on insectivorous plants — it was he who first pointed out that these plants really digest and resorb the imprisoned animals — another on the climbing organs of plants, in which these organs are described with exhaustive thoroughness and from numerous fresh points of view, and finally a work on cross- and self-fertilization in plants, as also a book of fundamental importance wherein he continues Sprengel's work, which he had rescued from oblivion, and paves the way for modern heredity-research. In the year before his death he also published a brief but ingenious work on the formation of vegetable mould through the action of worms, in which he establishes, on the strength of a large number of observations and experiments, the important role played by these animals as re-formers of the earth's surface, in that a considerable portion of the earth's outer layers passes through their intestinal canal and is thereby influenced physically and chemically — facts which research had previously failed to observe, but which have latterly been fully confirmed.

Darwin's general opinion of life

DURING the greater part of his life Darwin devoted himself to his own particular field of research more thoroughly than most other scientists. He never went in for teaching nor took up any other public appointment, while owing to his ill health he had to give up social life, with the result that his activities became more and more confined to biological speculations and experiments. This may explain why he embraced with such intensity, but also with such limitations, the theories he set up. He was but little influenced by other natural-scientific tendencies, eventually losing interest even in general cultural problems. In his youth he had been interested in art, poetry, and music, but in his old age even these lost their attraction for him. True, by way of diversion he used to have novels read to him, requiring only that they should have a happy ending; he paid but little attention to literary faults. And his religious interests went the same way as the literary; the Christian faith of his youth had undoubtedly been traditional from the very beginning, without any feelings of personal experience; his faith died gradually and without any crisis, leaving behind a peaceful and untroubled resignation in face of the ultimate problems of existence, a resignation which was never disturbed by anything except the innumerable senseless and irrational inquiries he received on the subject and to which he invariably replied conscientiously. It is worth quoting the following out of one of these replies as a final touch to the description of his character: "The safest conclusion seems to me that the whole subject is beyond the scope of man's intellect; but man can do his duty."

Judgments on Darwin

VERY different judgments have been passed on Darwin. Even on his first appearance he was either extolled as one of the greatest geniuses in the world or abused as an ignorant and unreliable dilettante, according to the different points of view. Nor have subsequent generations been any more unanimous; especially since the theory of selection has been condemned — at least in its original form — hard words have not been spared against its creator — as a matter of fact, a natural reaction against the adoration meted out to him towards the close of his life, which received confirmation in his being buried beside the grave of Newton. Was this apotheosis justified or not? This question has been answered and can still be answered either way. To raise the theory of selection, as has often been done, to the rank of a "natural law" comparable in value with the law of gravity established by Newton is, of course, quite irrational, as time has already shown; Darwin's theory of the origin of species was long ago abandoned. Other facts established by Darwin are all of second-rate value. But if we measure him by his influence on the general cultural development of humanity, then the proximity of his grave to Newton's is fully justified. It is certain that since the days of the latter no scientist has so deeply influenced man's general conception of life as Darwin has done; it is his theory of evolution that has taken the place of the idealistic theory of romanticism and made the common descent the connecting link in existence instead of ideas and archetypes. In all spheres of knowledge the development from earlier to later stages has been the one clue for research; history, which had previously sought for "guiding ideas," is now an evolutionary science, just as is philology, and even philosophy has at least one school that has followed the same principle. Everyone knows the important role played by the idea of evolution in naturalistic literature. The influence of Darwinism on biology will be described in the next chapter. Of its weaknesses a certain number have already been referred to above; it shared with all new ideas the illusion that it could do too much; this was so with Darwin himself, modest though he personally was, and still more so with his admiring successors. We shall now proceed to describe the differences of opinion caused by the new doctrine.

CHAPTER XII

Why Darwin's theory prevailed

MODERN CRITICS have often asked themselves how it is that a hypothesis like Darwin's, based on such weak foundations, could all at once win over to its side the greater part of contemporary scientific opinion. If the defenders of the theory refer with this end in view to its intrinsic value, it may be answered that the theory has long ago been rejected in its most vital points by subsequent research. It has also been pointed out, for instance by Rádl, that the objections made against the theory on its first appearance very largely agree with those which far later brought about its fall. The factors governing the victory of Darwinism thus represent a problem of the greatest importance, not only in the history of biology, but also in that of culture in general — a problem that would require far more exhaustive treatment than can be given to it here. In this work we can only endeavour to throw light on some of the circumstances that appear to be specially remarkable surrounding this important episode, the history of which it will largely be the duty of future generations to write.

Darwinism and liberalism

DARWIN's origin of species contains many points that were likely both to win the applause of and to give offence to his contemporaries. A factor that without doubt very largely contributed to both the one and the other was the book's relation to the political movement of the time, to which reference has already been made. From the beginning Darwin's theory was an obvious ally to liberalism; it was at once a means of elevating the doctrine of free competition, which had been one of the most vital corner-stones of the movement of progress, to the rank of a natural law, and similarly the leading principle of liberalism, progress, was confirmed by the new theory — the deeper down the origin of human culture was placed, the higher were the hopes that could be entertained for its future possibilities. It was no wonder, then, that the liberal-minded were enthusiastic; Darwinism must be true, nothing else was possible. But beside this there was a good deal more in it that could attract radical cultured views, chiefly its strongly worded polemic against the doctrine of creation, which could be employed to counteract theological obscurantism, and also the very idea of a material

477

connexion in existence, a principle that could be set up in opposition to the theories of ideas held by reactionary romanticism. The deficiencies in Darwin's work were therefore readily overlooked — his vague starting-point, his uncritical material, his weak arguments based on loose assumptions, his belief in the power of chance and of finality as an explanation of nature. As a matter of fact, the natural explanations of the preceding ages failed still more in that respect; they were generally based on the wisdom of the Creator and the benefit of man as the cause of all that exists and takes place — that is to say, an explanation without the slightest trace of scientific treatment. Darwin's theory, then, was at any rate an immense advance; its weaknesses could be overcome by continued research, its vagueness and casualness removed by fresh discoveries and replaced by firmly established facts, while the finality in nature could thus be made synonymous with natural law. Briefly, no one was prepared to doubt the possibilities of the theory's future development, and for the moment it entailed a freedom from the pressure of prejudice which there had previously seemed to be no means of avoiding.

Defiance of the conservatives

WHILE, then, liberal tendencies felt themselves closely bound up in Darwinism, the new movement was for that very reason all the more repugnant to the conservative social elements. Those who looked for their ideal in the past and in tradition must have been appalled to see the good old times depicted as a kind of half-way station along the road from the ape stage; and that free competition which to their mind only led to all manner of licence, was *that* to be the true creator of the life that is lived today, instead of the divine reason which has governed the world and preserved law and justice? And, again, this vague, indeterminate idea of evolution, was it to be substituted for those firmly established and eternal ideas that governed the creation of nature and its forms? Thus reasoned many, and Darwin's theory was therefore challenged from pulpit and professorial chair, at scientific gatherings, in journals and newspapers. This first polemic against Darwinism has its own peculiar interest; it is dazed and not particularly keen-sighted, it clings despairingly to the old ideas and as yet lacks orientation as to the exact position adopted by its opponents. In the present history it is only possible to give attention to some few of the more representative scientific contributions, whereas the miscellaneous mass of protests from other quarters can have no place here.

Owen's opposition

THE highest scientific reputation among the opponents of Darwin was undoubtedly that of Richard Owen. It was, of course, impossible for the latter's idealistic morphology to be reconciled with the Darwinian doctrines of descent, and if anyone was to discover and demonstrate the weaknesses underlying the new theory, it was he. The fact, however, that his influence was not so great as his scientific reputation might have warranted was mostly due to the way in which he conducted himself; instead of openly defending his views he wrote anonymously, repeatedly referring to "Professor Owen" as his authority in opposition to Darwin. This gave his contribution a tinge of lampoonery that detracted from the effect it might otherwise have had. The article (it appeared in the *Edinburgh Review* of 1860), which much embittered Darwin, is chiefly interesting as being an expression for the sharp contrast between the romantic natural philosophy and the realistic evolutional theory. Owen points out with strong emphasis how few are the facts and how weak the proofs that form the basis for the new theory, how the problem of species-formation must still be considered unsolved in spite of the theory of selection, how it was possible to assume other factors governing species-formation besides variation and natural selection. As such factors he suggests parthenogenesis and alternation of generations; he believes it possible to suppose that the various stages in such a cycle — polypus and medusa, or sporocyst, redia, cercaria — might, so to speak, liberate themselves from the series and begin to give rise to forms similar to themselves, with the result that the whole cycle would disintegrate into a number of widely differing life-forms. He even adopts Pouchet's spontaneous-generation experiments in his support against Darwin: if the Infusoria spontaneously generate daily, how can it be assumed that all higher beings could have been evolved in one single series originating in primitive forms? Owen's suggestions in regard to species-formation are certainly not very happily conceived from a modern point of view, and indeed they are only presented as experiments with ideas in order to prove how complicated and difficult of solution the problem of species-formation really is, but the worst of it is that Owen brings into the field the whole of the thought-systems of the old idealistic natural philosophy; as a factor that actively operates in the creation of the symmetrical forms of the higher animals he adduces a "polarizing force," the true essence of which need not be analysed here, as the name itself explains it. Even the old doctrine of "the ideal type" is brought forward for the same purpose. But one who has recourse to such empty phrases to explain the origin of life-forms has no right to accuse Darwinism of weak argumentation and of making false hypotheses. A controversy such as this best shows what an immense advance Darwinism nevertheless involved at the time, and at the same time explains why it is that even the authorized objections of the old school must die away unheard.

One gets the same impression from the criticism of Darwinism offered by Agassiz, another important representative of the old biological school. JEAN LOUIS RODOLPHE AGASSIZ was born in 1807 at Motier in Switzerland, of French parents, and even during his school-time devoted himself to natural science. He studied at several German universities, his teachers including both Schelling and Oken, but principally Döllinger, mentioned above as von Baer's master; he became doctor of medicine and afterwards spent a couple of years in Paris in lively discussion with both Cuvier and Humboldt. His chief object of study was the fishes, both recent and fossil; a large work that he had commenced on the fishes of Europe was never finished, while another on fossil fishes proved a pioneering work in its own sphere. But besides this, glacial research proved of special interest to this many-sided scientist, and in this field too he was a pioneer. He proved that the glaciers had in earlier times been far more extensive in his native country than they are now, and during a journey to Scotland he found that large glaciers had existed there too in past ages. From this he drew conclusions regarding the general glacialization of Europe, which afterwards led to that highly developed research-work on the glacial period which has been especially noteworthy in Scandinavia. During the years 1832–46 Agassiz was a professor at Neuchâtel; he then moved to America and became professor at Harvard University. There he did splendid work as both a zoologist and a geologist, making extensive journeys and producing works on the animal world and the zoology of America, as well as on theoretical problems. He died in 1873.

In his theoretical writings Agassiz shows himself a true romantic natural philosopher, as might be expected from the education he received. The problem of species engaged him a great deal and is solved by him in a markedly idealistic direction. In such circumstances it was obvious that he could not hail the advent of Darwinism with any great enthusiasm. In his polemics against it he makes a great point of its weaknesses; lack of observation of the real transition from one species to another, lack of obedience to law in its theory of natural selection, the weak conclusions drawn from similarity in the embryonic stage to similarity of origin. But the most serious mistake to his mind is that the new theory fails to realize the creative idea running through all animate nature. The individuals perish, but hand over to their posterity, generation by generation, all that is typical, with the exclusion of what is merely individual; therefore, while the individuals have only a material existence, species, genera, families, and so on upwards exist as the thought-categories of the Supreme Intelligence, and as such possess a truly independent and immutable existence.[1] Here, it will be seen, speaks the pure romantic idealism, whose supporters, thanks to their intensive professional insight, have no difficulty in discovering the weaknesses underlying the new biological theory, though only to maintain in its stead their own

---

[1] In his "Essay on Classification" Agassiz, speaking of rudimentary organs, maintains that these exist not for any purpose of function, but to complete the design, just as in a building certain details are introduced for the sake of symmetry, without any idea of their serving a practical purpose.

speculation, equally unworkable in form as in contents and therefore inevitably doomed to failure.

It would hardly be worth while to carry this account of the attacks against Darwin any further. We might still mention the contribution of S. WILBERFORCE, Bishop of Oxford, owing to the sensation it created at the time. Having himself studied natural science, and with the indefatigable Owen as prompter, he reviews the weaknesses of Darwinism in an easy and fluent style, though somewhat superficially, but at the end of his treatise he spoils his case completely by sermonizing on the subject of the origin of man, bringing forward all the persons of the Trinity as arguments to prove a special creation in the image of God. From such opponents Darwin clearly had nothing to fear. But even scientists with a truly modern conception adopted from the outset an attitude of criticism towards this theory — KÖLLIKER, for instance. In a brief examination, substantiated by numerous facts, Kölliker submits in a concise and determined style his objections to the theory of selection, at the same time acknowledging the great service of Darwin in having sought to base the knowledge of the origin of organisms upon experiments and in having made descent the foundation thereof, so that the life-forms might be regarded as a series of evolutionary phenomena. He expressly declares that the earlier attempts of natural philosophy to construct a history of evolution are weak in comparison with Darwin's, and, moreover, he appreciates the far-reaching insight and the splendid conscientiousness on which his theory is founded. As its weak points he mentions first of all its teleological conception; the principle of finality as applied to life-forms, which has already been pointed out above; further, the absence of transition forms between the species, both extant and fossil, the lack of proof that characterizes the entire hypothesis of selection, and finally the circumstance that nothing is known of unfertile variety-hybrids, which would nevertheless be bound to appear somewhere if the varieties were transitions to species. Moreover, Kölliker holds that it is possible to imagine other ways of evolution than Darwin's. He considers that the idea that all species have been created as they are is not worth discussing, but it is conceivable either that all organisms have arisen each out of its own primary form, or that the species have come into existence through one primary form or through a few. The latter alternative he considers to be more probable, but then there must be a common law governing formation, according to which forms of one kind may in certain circumstances give rise to entirely different forms, either by a larval form's adopting an independent course of development, or by an egg or embryo of a lower form's giving rise to a higher type of life. This creation theory of Kölliker's is merely a concept and is, moreover, based on hypotheses that have never been confirmed. Of real value, on the other hand, is his criticism of Darwin's theory which is founded on a truly exact, and not on a natural-philosophical, basis.

Huxley versus Kölliker

THIS criticism of Kölliker's was opposed by Huxley, who vehemently denies that there is any teleological explanation at all in Darwin, whose entire theory is based rather on the absence of any creative purpose in nature. And in proof of his view Huxley cites exactly the same quotation out of the *Origin of Species* as Kölliker does for his own argument. From this it is obvious that the two antagonists must be standing in some essential respect on different ground, and the question is of such great general interest that it deserves closer examination. Strictly speaking, Huxley is right, in so far as no creative design in the romantic natural-philosophical sense is ever referred to by Darwin; but this does not prevent his constant assertion as to the adaptability of life-forms and organs to certain given conditions from implying a teleological explanation of phenomena; not only the entire theory of sexual selection, but also most of the doctrine of natural selection actually rests on this assumption. The contrast between the romantic and the Darwinian teleology is best explained by an example. It is asked: Why has a cat claws? For the sake of the creative design, say the romanticists, and in order to serve the purposes of the cosmic order. For its own sake, says Darwin, and in order to enable it to survive in the struggle for existence. But it is really the question itself that is absurd — as absurd as the question: Why does a stone fall? or Why does the earth revolve round the sun? Biology can only endeavour to find out the conditions under which cat's claws are developed and used, but never anything more; those who question beyond that fail to fulfil Bacon's requirement that we should "ask nature fair questions." But Darwin and his contemporaries are constantly putting such wrong questions to nature. This is, of course, due to the fact that they were unable to free themselves entirely from the influence of romantic philosophy, which, indeed, they desired to abandon and the weaknesses of which they fully realized, but its grasp of the problem of life was really too firm for them to loosen. Natural philosophy had, indeed, found in its plan of creation an explanation for everything, and to resign in face of the causes of the phenomena of life would have meant, to the new direction in which biology was moving, almost the same thing as a declaration of bankruptcy in face of its opponents. And in contrast to the idealistic plan of creation Darwin's teleology involves possibilities of development, in so far as a number of the so-called purposeful adaptations have since, mainly through modern researches into the problem of heredity, found its law-bound explanation, while other phenomena have had to accept that resignation in face of the inexplicable, which is the hall-mark of exact and critical science. Darwin's theory of adaptation, which is now so often condemned for its credulity, has thus in reality formed a necessary transitional stage, which has freed biology from the illusions of the past and made a more exact research possible in the future.

Among Darwin's other opponents in Germany may be mentioned AL-BERT WIGAND (1812–86), professor of botany at Marburg, a pupil of Schleiden, and well known as a capable plant-anatomist and plant-physiologist, as well as a leading expert on cryptogams. He was, moreover, deeply religious and on that account was unable to accept the theory of spontaneous generation. It was therefore inevitable that Darwinism should have been odious to him from the start, and he wrote many treatises against it. Eventually, after ten years of preliminary work, he summarized his views in a work comprising nearly thirteen hundred pages, entitled *Beiträge zur Methodik der Naturforschung*. He here shows himself to be a keen-sighted student of nature and a keen critic of the old exact school. Cuvier is his ideal as a scientist and he definitely associates himself with him in his opposition to Geoffroy's efforts to attain natural-philosophical unity. He has a keen eye for the weaknesses of Darwinism and analyses them objectively and in detail; he especially brings out the weaknesses underlying the theory of selection, and in contrast to the lack of design in the phenomena of variation and selection, as presented by Darwin, he maintains the existence of a definite course and plan in evolution — a plan that excludes both chance and explanations of finality. This criticism is, indeed, on the whole justified, and even Wigand's assertion that Darwinism is natural philosophy rather than exact research is quite a fair judgment; but when it comes to trying to justify the idea of conformity to law urged in opposition to the doctrine of chance, the former is ascribed to a personal deity, for natural science cannot get away from an ultimate cause of existence. This, of course, should not be used as grounds for a natural-scientific explanation, but the doctrine of the creation and the theory of the immutability of the species, which Wigand would urge in opposition to Darwinism, are nevertheless based upon it. In doing so, however, he has vitiated the effects of his criticism; his ideas were capable of satisfying neither the natural philosophers of the old school nor the exact scientists, and he himself lived just long enough to see Darwinism reach the height of its influence upon human culture.

Even the aged von Baer entered the lists against Darwinism, complaining of its lack of conformity to natural law; he sees in evolution a striving after a definite goal — "*Zielstrebigkeit*," as he calls it — and this, indeed, explains the finality in existence, but presupposes in its turn a common scheme for all natural phenomena, which is only conceivable with a personal creator as the primary cause. The high respect in which this octogenarian student of evolution was held, exempted him from harsh criticism at the hands of the younger generation; his contribution was added to the records in silence.

Darwinism was least appreciated in France, where Cuvier's pupils held sway in the realm of zoology and where even representatives of experimental research — Bernard and others — had little sympathy for the speculative and hypothetical elements in the new theory. It is striking that Darwin was not elected to the French Academy of Science until after he had published his works on plant-physiology, and then under reference only to them and not to the descent theory. And when this theory — or *"transformisme,"* as it was called in French — eventually found acceptance in the country of Lamarck, it was with him rather than with Darwin that the followers of the new tendency associated themselves. Of the earlier critics of Darwinism in France the first name is that of JEAN LOUIS ARMAND DE QUATREFAGES DE BRÉAU (1810-92), first a physician and finally professor of anthropology at Paris, and famous as a leading specialist on marine fauna, particularly the Annelida, but foremost as an anthropologist. As this last he carried out valuable investigations into special subjects, all, however, governed by a firm conviction as to the unity of the human race and its independence of other life-forms. He wrote a number of treatises against Darwinism, the chief of which was one entitled *Charles Darwin et ses précurseurs français*, in which he begins by describing several transformistic authors of French nationality: de Maillet, Buffon, Lamarck, and others. In regard to Darwin, Quatrefages admits that there is a struggle for existence, but does not believe in its power to create new life-forms. He sharply criticizes Darwin's habit of adducing the probable and the possible — purely personal conviction instead of facts proved on conclusive evidence — and he particularly points out that, when it comes to the question of the life-phenomena of past ages, Darwin constantly appeals to "the unknown." And Quatrefages concludes his critical examination with the words: "Let us not dream of what *may be;* let us instead assume and seek what *is!*" Among the earlier critics of Darwinism Quatrefages is worthy of respect on account of the considerate and objective manner in which he passed judgment on the theory. Eventually, however, the descent theory gained ground even in France, chiefly, as mentioned above, in the Lamarckian form, which at the same time became known in other countries also, and which will be described later on.

Vast quantites of polemical writings against Darwin and his theory appeared during the period immediately after he first attracted public attention; most of these were of practically no scientific value, since they were based on religious arguments, which were the most usual, or else on quasi-scientific or other grounds. Of the really objective contributions to the subject it would be possible to name many others besides those referred to above, but space forbids a more detailed review of them. At the same time there came forward in defence of Darwinism many distinguished scientists, who made weighty contributions to the discussion and assisted in the rapid advance along the new lines laid down by Darwin. Even of these it is possible only to name a few; in the present chapter reference will be made to the English contributions in favour of Darwin, while one or two separate chapters will be devoted to the development of Darwinism in Germany, where it acquired an essentially novel character.

Darwin's supporters

AMONG the first to associate themselves with Darwin was the aged Lyell. In a work entitled *Geological Evidence on the Antiquity of Man*, published in 1863, he takes up the question of the origin of species by means of variation. He refers briefly to Darwin's theory and in support thereof cites a number of facts, especially geological and palæontological; of these he bases his argument mainly on the extinct proboscideans of the Tertiary period, while he further adduces a number of fossil insects, as well as the saurian bird Archæopteryx. In regard to man, whose primitive history had been the real subject of the book, sympathetic reference is made to the statements of, *inter alia*, Huxley, as to man's anatomical agreement with the higher apes; similarly, mention is made of Darwin's theory of the origin of the intelligence by means of natural selection, and the work concludes with a refutation of the accusation that Darwinism would lead to materialism. Darwin himself highly appreciated the support thus given him by Lyell, and the influence of the aged geologist certainly contributed much towards bringing the new doctrine to victory.

Among those who, besides Darwin, should be named as supporters of the theory of selection, the first place is due to ALFRED RUSSELL WALLACE. Born in 1823, he was originally an engineer and afterwards a schoolmaster, and he was besides interested in collecting plants and insects. In 1848, in company with his friend HENRY WALTER BATES (1825–92), he made a voyage to Brazil for the purpose of exploration and the collection of scientific material. After a year the two friends parted; Bates remained in Brazil, while Wallace returned home and shortly afterwards made a journey to the East Indian archipelago, where he remained for a number of years, continuing his comparative biological studies on the various islands. Upon his return home he found himself already a famous man and continued his biological research-work, partly on voyages and partly in his own country. He never received any permanent appointment, but had to earn a living as a writer and lecturer. In his old age he was assured a means of subsistence through a government pension. He died in 1913, over ninety years old.

Wallace's discoveries in animal geography

THE result of Wallace's Indian journey proved to be of the greatest importance; he thereby became one of the pioneers of modern animal-geography.

He established the fact that the western half of the archipelago possesses an essentially Indian animal world, whereas the eastern half has an equally marked Australian fauna; the border-line he found to lie in the narrow but deep sound between the islands Bali and Lombok, and northwards from there in the Macassar Strait between Borneo and the Celebes. He afterwards compiled, with the aid of the results gained during this and subsequent voyages, an animal geographical system, in which the globe was divided into separate regions based on the distribution of animal forms both in recent times and in preceding periods. This animal geographical system, which is universally known from the text-books on zoology, is a contribution of lasting value to the development of biology.

But in the course of his studies of the distribution of animal life in the East Indian islands Wallace found himself faced with the same problems as Darwin in the Galápagos Islands; the various islands and island-groups possess their peculiar animal species. The distribution of species on the earth is thus governed by geological conditions, and if we consider the animal life of earlier periods we find that, instead of the new extant forms, there were other similar forms — in fact, that, as he says, every species has been preceded in time and space by a similar species. These reflections he recorded in a treatise which he sent home and which was printed in 1855. The explanation of the phenomenon he found — like Darwin — when meditating upon Malthus's theory of competition; it is the struggle for existence that has compelled living creatures to develop themselves in order not to perish in the struggle against other species; if a variety has been equipped with more powerful qualities than the main species, it drives out the latter and usurps its place. This theory Wallace expounded in a report, which he sent to Darwin for perusal; the latter was struck by the agreement with the ideas that he himself just happened to be working out and found the situation highly embarrassing. At the suggestion of some friends he published Wallace's treatise together with a report of his own results, which he submitted to the Linnean Society in 1858, thus giving science an opportunity of seeing the same theory presented by two investigators working independently. Much surprise has been expressed at the incident, which has often been put forward as a proof of the undeniable truth of the theory. It is possible to find an explanation of the phenomenon by making a comparison between the two originators of the theory; they were both self-taught men with essentially systematic interests, but without any anatomical training; they had both explored an island region and received their impressions therefrom; both had consequently been confronted with the problem of the distribution of species, and, finally, both had been influenced by Malthus; and though the fundamental view-point is the same in both scientists, yet Wallace has a conception of the problem that is in many respects peculiar to himself. He manifestly never felt so deeply moved by the actual doctrine of the creation as Darwin had been, and, further, he has by no means the same interest as Darwin in domestic-animal varieties, with which he himself had never experimented; rather, having studied in the richest tropical regions, he had gained a far stronger impression of the wealth of life-forms and their adaptation to environment. This especially comes out in the mimicry theory that he and his friend Bates created.

Theory of protective resemblance

AFTER a ten years' sojourn in the tropics of South America Bates returned home with rich collections and eventually became secretary to the British Geographical Society. He wrote an essay in which he propounded the idea of protective resemblance in the animal kingdom — an idea that was afterwards taken up and further developed by Wallace. It is known that a large number of animals possess external characteristics that correspond to conditions in the natural surroundings in which they live; the white fur of polar animals, the sandy yellow of desert beasts, the likeness of many insects to the bark of the trees on which they live, are all examples of this. In the more abundant plant-life of the tropics there appear still more remarkable instances of this similarity, especially among the insects; well-known examples are the "wandering leaves" and "wandering sticks," which, owing to their likeness to the undergrowth, often elude the observation of even the most experienced collectors. Wallace believes that all these forms have arisen through the circumstance that natural selection in the struggle for existence has favoured those individuals that, owing to variations in the direction of greatest likeness to their surroundings, have been better protected than others and have thereby had a better chance to propagate. But Wallace considers that even the obvious exceptions from the rule which quite often occur — animals with strikingly brilliant colours — only still further confirm the law, seeing that they really possess some other characteristic which acts as a powerful protection against their enemies and which thus converts their splendid colours into a kind of warning signal to the latter; as, for instance, an offensive odour, as in the skunk of America, the natterjack, and the salamander, as well as a large number of insects, including our common lady-bird, with its magnificent red-and-black spotted wings; or, again, a hard shell, as in many brilliantly coloured tortoises; or poison, as in many of the vividly marked snakes in the tropics. The most remarkable application of this law Wallace sees, however, in the mimicry or disguise whereby certain animals protect themselves against their enemies by resembling other more dangerous animals in their outward appearance; there are flies that in form and manner of flying resemble bumble-bees, butterflies that resemble wasps; and this disguise is demonstrated still more in the tropics, where non-poisonous snakes are often misleadingly like the poisonous ones and many insects exhibit similar |congruities. Even this protective resemblance Wallace, of course, derives from natural selection. Furthermore, he points out in this connexion the females' need for protection during the period when they are ministering to their young as a cause of their less conspicuous coloration, as in the birds, whereas the cock birds, which do not require this protection, have developed greater splendour of colouring. Wallace thus explains the external dissimilarity of the sexes without having recourse to Darwin's theory of sexual selection, which he rejects.

The whole of this theory of protective resemblance, which was once cited as one of the strongest arguments in favour of Darwinism, has naturally been discredited concurrently with the theory of selection itself; the mimicry theory in particular had already been vehemently attacked by scientists who did not find it accord with their observations and experiments; the enemies that through their pursuit of prey were supposed to have called for a protective resemblance have in many places been found to be nonexistent, and remarkable instances have been discovered of resemblances of this kind in animals in different parts of the world, which could not therefore have influenced one another's appearance. Further, in order to maintain the theory it has been necessary to ascribe to a great many animals powers of observation and distinction as weak as man himself possesses. In the latter respect Wallace was extremely credulous; he states, *inter alia:* "The attitudes of some insects may also protect them, as the habit of turning up the tail by the harmless rove-beetles (Staphylinidæ) no doubt leads other animals besides children to the belief that they can sting." This comparison between the animal's power of observation in nature and that of a child is certainly very naïve. But Wallace was on the whole more of an imaginative than critical nature; very soon he had astonished the world by becoming a convinced spiritualist, although he was a free-thinker in religious questions, and in later years he became entirely engrossed in spiritual seances and a number of similar fantastic ideas of spiritual life in nature, while at the same time he expressed his utter contempt for the results of modern heredity research. He thereby placed himself definitely on the side of natural-scientific development.

A personality of an entirely different character was Darwin's other champion in England — Huxley, one of the most famous biologists of his time. THOMAS HENRY HUXLEY was born in 1825 in a London suburb, the son of a poor schoolmaster. After two years at school, which he himself described as "a pandemonium," he had from the age of ten to pursue his studies by himself and he did so with such success that seven years later he gained an entry into the medical faculty in London. Having passed his examinations, he became a surgeon in the English fleet and served in that capacity in a vessel that was exploring the channels north of Australia. In these tropical

waters there existed abundant animal life, which induced the young doctor
to investigate it. Among other works he brought out a book on the medusæ,
which brought him wide recognition. At the age of thirty he was appointed
professor at a School of Mines and this led him to take up palæontological
research, but he further had an opportunity of teaching physiology and comparative anatomy, which he had already thoroughly studied during his student days. Full of energy and initiative as he was, he was able to make
practical use of his science to an extent that few have equalled. Not only
by means of popular lectures and text-books, but also as a member of school
committees and an expert on fishery questions, he laboured to expand the
knowledge of biology and to increase respect for its methods and mode of
thought. His authority ultimately became very great and honours of all
kinds were showered upon him — more, in fact, than he cared for. After
a long period of suffering he died in 1895. His marble statue stands by the
side of Darwin's in the South Kensington Museum in London.

Huxley's work on the medusæ

HUXLEY was a highly gifted scientist, though critical rather than creative.
His first work on the anatomy and affinities of the medusæ was that of
a pioneer; he therein demonstrated the connexion between hydroid polypi
and hydromedusæ and combined them into one order, the Hydrozoa. Of
still greater value was his idea of comparing the dermal and intestinal layers
of these animals with the germinal layers in the embryonic stages of the
higher animals; out of this comparison eventually arose the general theory
of germinal layers. His sea voyage likewise produced a series of valuable
studies on the Tunicata. He also made important contributions in the sphere
of vertebrate anatomy; especially well known are his comparative studies
of the structure of the cranium, whereby he proved, on the support of the
preparatory embryological works of Rathke and others, the absurdity of
the Oken-Goethe theory of the cranium's being composed of vertebræ, while
at the same time he admitted its original metameric structure. This proved
a severe blow to Owen's archetype theory; from that moment the aged, cap-
tious anatomist became Huxley's enemy, all the more so as the latter had
already rejected the idealistic morphology. When, later on, old Owen declared that the human brain has certain parts which no other animal can be
shown to possess and sought on these grounds to claim for the human race
a special position as towards the rest of the animals, Huxley proved in a
sharply critical way that the anatomical details of Owen's account were entirely inaccurate, and this nullified all the latter's efforts to isolate man from
the animals.

Huxley embraces Darwinism

NEVERTHELESS, in his youth Huxley was an upholder of the immutability
of the species, and an opponent of Lamarck's theories of evolution. Upon

the advent of Darwin, however, he was one of the first to be convinced and from that time onwards became one of the most zealous champions of Darwinism — its general agent, as he himself jokingly remarked. He took part in the earliest controversy on *The Origin of Species*, contributing a paper that proves how understanding and at the same time how independent was his attitude towards Darwin's theory from the very outset. In the first place, he has not Darwin's blind faith in the absolute dominance of the *small* variations in nature. He cites as an example of sudden changes the oft-quoted ancon or otter sheep of America, whose sudden appearance is a well-known fact, and further, borrowing from Réaumur, the story of a family that had a child with an excessive number of fingers and toes, which phenomenon was afterwards inherited by its descendants. The arising of the otter sheep is an obvious mutation; from the appearance of the supernumerary fingers, again, conclusions might have been drawn in the spirit of Mendel. This, however, was not done; even in the moderate form that Huxley gave to his divergences from the true selection-theory, they attracted no attention; small indeed would the variations have to be for the struggle for existence and selection to have any material effect on them, and what interest could be awakened by the story of the inheritance of six fingers? On the basis of such exact observations of detail one came no closer to the theory of creation, which, indeed, was the main idea at that time. Regret has often been expressed that Mendel's observations were published in such an out-of-the-way place that no one noticed them; it is more than likely that the result would have been the same wherever they had appeared; the fact is, the time was not yet ripe for them. — However, Huxley's objections to the master's theories were not numerous, nor were they bitter; he took far greater pains to defend what good he found in them, which, indeed, was a very great deal. And in contrast to Darwin himself, Huxley was a born controversialist, with an ever-wakeful pugnacity, a never-failing promptness in reply, an extensive knowledge of books, and a rare gift of putting the most involved questions in a fluent and easily understood style. For the rest, his polemics are always courteous; sceptic as he is, he confronts his opponent with a supercilious, but not always a friendly, smile, and he never allows his composure to be ruffled, nor himself to be reduced to silence. He particularly enjoyed crossing swords with men of the Church, and on that battlefield there were certainly to be found opponents *en masse* so long as he lived. Among them was Gladstone, the great Liberal statesman, who was also an extremely learned and highly conservative theologian. At one period during the eighties Huxley entered into a controversy with him — the foremost biologist against the leading statesman in England at that time — concerning the gospel story of the Gadarene swine, which were drowned after the Devil had entered into them. Rather more urgent problems were dealt with in his dispute with another famous politician, the Duke of Argyle, who in a work entitled *The Reign of Law* had opposed Darwinism's lack of conformity to law in the sense given to it by idealistic natural philosophy. But Huxley sought to influence the contemporary world of ideas also in a positive way; he enunciated the same social ethics that Darwin had taught and that their age so largely embraced; he laboured to make the results of modern natural science the basis of school education instead of the traditional classical languages, and he endeavoured by means of popular writings and lectures to bring them to the knowledge of the general public. As a popular scientific writer he is unrivalled for the clearness, warmth, and honesty of his style; he never expresses a view that he cannot defend and never tries to disguise the fact that the capacity of science for explaining phenomena is limited. The same honesty he displayed also as a specialist. He once described a gelatinous substance taken from the bottom of the sea, which he thought was a kind of undifferentiated, but living plasm, and which in honour of Haeckel he named *Bathybius haeckelii;* when it was later discovered that the substance was an inanimate precipitation, he frankly and boldly acknowledged his mistake, which Haeckel found it rather difficult to admit. But Huxley was also interested in purely philosophical problems; he was a great admirer of David Hume, the famous sceptic of the eighteenth century, Kant's predecessor as a critic of knowledge, and Huxley described his life and teaching in a monograph. He himself represented his theoretical standpoint as agnosticism, as a strict insistence upon the impossibility of knowing anything beyond the actual observations of the senses. And it must be admitted that he succeeded in an unusually high degree in keeping free from the materialistic dogmatism to which the opponents of the traditional ideals of thought and religion are so easily addicted.

Gray on Darwinism

AMONG the professional botanists, also, Darwin at once found valuable supporters. One of these was ASA GRAY (1810–88), professor at Harvard University and well known as a leading writer on systematic botany and American flora. Immediately after Darwin's appearance he came forward on his behalf and in opposition to his own colleague Agassiz; the contributions he wrote on behalf of the new theory in the course of a number of years he collected in a special book, entitled *Darwiniana*. In his first review of *The Origin of Species* he first of all attacks Agassiz's idea of species and then goes on to point out that, even though Darwin was unable to prove his theory of descent, he at any rate made an origin of species far less incredible than before. Gray was at great pains to prove that Darwinism could be reconciled to the belief in a personal God, and he likewise sets great store by the Darwinian theory as being better adapted to finding an explanation of the finality in nature than earlier theories; he expressly points out that Darwin rehabilitated the teleological explanation of nature — that is, the same as Kölliker maintained and Huxley denied. To such divergent interpretations could Darwin's theory give rise.

Darwin had also found a convinced supporter and a lifelong friend in JOSEPH DALTON HOOKER (1817–1911), a keen explorer in various exotic countries, a distinguished plant-systematist, and finally director of Kew Gardens in London. Darwin was engaged in an almost constant interchange of ideas with him during the many years he worked at his theory; when eventually *The Origin of Species* was published, Hooker zealously defended the new theory of descent, both in articles in the press and in his own more important works. In particular, the introduction to his book on the flora of Tasmania, which was published in the year after *The Origin of Species*, represents a defence of these doctrines, based partly on geographical arguments and partly on evidence derived from classification, with special reference to the many phanerogam genera which are so rich in species and, owing to their numerous middle-forms, are so difficult to classify — Rubus, Rosa, Salix, and others. Darwin's later researches into plant physiology were also supported to the utmost by Hooker; in fact, of all the champions of Darwinism he was, on the whole, in closest personal contact with its founder.

Development of Darwinism in England and Germany

IT was from many different quarters, then, that Darwin won support for his theory of descent; in fact, as has often been emphasized before, it became the most widely cherished scientific idea of the age. In these circumstances it was obvious that many forces were destined to make for its further development. Towards this end, however, it was possible to follow different methods; of these there were really two that immediately came into use — one in each of the two countries that came to be the principal centres of Darwinism; in Germany recourse was had to the method which was only hinted at by Darwin himself, of seeking fresh proofs of the descent theory in comparative morphology and embryology, while in England his supporters followed the experimental and statistical method, in which Darwin himself had expressed the greatest confidence. These two courses will be described in the following. In this connexion, however, we must first mention one thinker who, although not a professional biologist, yet sought more consistently than most people to apply the descent theory, in the form that Darwin had given it, to all the phenomena of life, and who was regarded by his contemporaries as the philosopher of evolution above all others.

HERBERT SPENCER was born at Derby in the midlands, in the year 1820, the son of a schoolmaster. His parents were both Free Church people, but belonged to different sects, and this lack of harmony induced feelings of doubt in the son at an early age; in political radicalism, on the other hand, he was fully in accord with his home throughout his life. He received a good school education and especially distinguished himself in the exact sciences, the classical languages having no attraction for him. He chose engineering as his profession, distinguishing himself by a number of minor inventions. His restless and insatiable desire for knowledge, however, soon induced him to abandon that career, and he resolved to devote himself to working out a general scientific system. In order to carry out his purpose he studied many different sciences, chiefly those of an exact character, and during that period he earned a livelihood by writing for newspapers and journals. He never received any public appointment and he consistently declined the honours that were offered him, especially towards the close of his life, from many quarters. In a constant struggle with poverty he lived in solitude, being also during the latter part of his life a sufferer from a severe nervous afflic-tion. He was ruthlessly radical, not only in his political views, but even in his personal behaviour; he always gave his opinion straight out, and if a conversation bored him, he put stoppers into his ears. In spite of his ill health he lived to a good old age. When he died, in 1903, his body was cremated without any funeral ceremony.

Spencer's idea of evolution

HERBERT SPENCER was not a specialist in biology, and his speculations on biological problems have not advanced that science to any very great ex-tent. He nevertheless deserves a place in the history of biology as a rare example of a consummate and typical representative of that evolutional mode of thought which was awakened to life by the general tendency of the times in the middle of last century and which was promoted by Darwinism. He is commonly called the most consistent philosopher of evolution which that period produced — evolution forms the very groundwork of his system. In its essential features this system was already pretty definite before the ad-vent of Darwin; it was promulgated in a number of small articles in periodi-cals, often characterized by masterly penetration and lucidity, afterwards brought together to form an imposing work entitled *A System of Synthetic Philosophy*, which was the fruits of thirty years' work and which gives "a broad, often too broad, development of what is recorded in the short trea-tises" (Höffding). When Darwin produced his theory, Spencer associated himself with it, although he interprets it after his own mind, and he became one of the most influential promoters of the new doctrine of evolution. Otherwise he is said not to have been in favour of extensive studies; he preferred to think for himself and was very jealous of his independence. Nevertheless, there is no doubt that Comte and his contemporary English positivists exerted some influence upon him, and he himself admits that he discussed biological problems with both Huxley and Hooker.

Law of differentiation

Of Spencer's shorter articles there is one dated 1852, "The Development Hypothesis," in which he clearly and definitely dissociates himself from a belief in the immutability of species; a hypothesis of creation is unscientific because it is incomprehensible, and the probability is that the various forms of life on the earth have been modified in the course of the ages by the influence of different external conditions of life. In a couple of other similarly pro-Darwin essays, "Progress, its Law and Cause" and "Genesis of Science," he gives a more general presentation of his evolutional theory, which was afterwards further developed, in view of the selection theory, into his great philosophical work. According to him, the function of philosophy is to combine under one common standpoint the results achieved by all other sciences: physics, chemistry, and biology, as also psychology and sociology. This unity common to all sciences exists in evolution. All existence is evolution; the heavenly bodies are undergoing change, the earth was once incandescent and has since then gone through a series of evolutional forms, and all things existing on it, both animate and inanimate, are doing the same; the separate plant and animal individual is being evolved, just as species and genera and humanity are being evolved, individual for individual and generation after generation. The question of what "evolution" is, Spencer has in such circumstances to try to get answered as exhaustively as possible. In the above-mentioned treatise on the law of progress he endeavours to formulate the answer from a biological standpoint; starting from the evolution theories of C. F. Wolff, Goethe, and von Baer, he finds in agreement with them that the development of the individual proceeds from the homogeneous to the heterogeneous; out of the egg, which is uniform throughout, both in structure and composition, is evolved an individual possessing various parts and organs, which are the more differentiated the further the development proceeds. This law Spencer believes holds good for everything; the earth was once uniformly incandescent, but after having cooled off, it acquired an increasingly different and varying surface; all living creatures were originally primitive and homogeneous, but out of these primal forms there has since been developed an ever greater multiplicity of life-forms; the life of the human society offers the same picture, and differences in language and other manifestations of intellectual life have similarly developed. But whence is this differentiation produced? Spencer answers this question with the contention that every cause invariably has more than one effect; if a candle is lighted, it is one simple chemical process, but it produces a number of different effects — heat, light, chemical products. Thus there are created on the earth an ever-increasing number of phenomena. The whole of this discussion on causality is, of course, a purely metaphysical problem; against the theory of evolution on which it is based it may be remarked from a biological point of view that Spencer deliberately threw himself into the arms of the Wolffian epigenesis theory. If the standpoint of the preformation theory is adopted, then the whole foundation of this doctrine of evolution is destroyed. Now, in modern times, the egg is certainly not regarded as non-differentiated; rather, with its numerous hereditary factors and the orientation given it from the very beginning, it is a tremendously complex structure.

Process of consolidation

AT a later period Spencer tried also to expand his evolution theory. He sees in it a process of consolidation; the egg-cell absorbs nutriment from surrounding tissues, the embryo from the yolk of the egg, both under a process of increasing consolidation. In the same way the celestial bodies have been consolidated out of nebulous masses, and the human communities out of scattered groups. Further, evolution may be regarded as a transition from the indefinite to the definite, as indeed is demonstrated in the life of individuals, species, and communities. But, above all, in his later years Spencer began to realize that evolution does not always advance; it can also show the exact opposite phenomenon, that progression and retrogression succeed one another in evolution. This speculation suffers on the whole from the attempt to bring all phenomena on the earth without exception under one common definition, which in the circumstances becomes far too abstract: it says too little because it is meant to embrace too much. The same fault underlies the definition of life that is given in the biological section of Spencer's system. Various characteristics of life are examined, and finally the definitive characteristic is formulated thus: "Life is a continuous adjustment of internal conditions to external conditions." The higher the life, the stronger is the connexion between the internal and the external; the intellectual life represents the highest degree of relationship between internal and external changes. His detailed application of this theory of life offers little in the way of interest; although controlled by Huxley and Hooker, it corresponds but little to modern ideas. As an instance may be quoted the assertion that life precedes organization in the matter in which it develops, whereas in reality life and organization are indissolubly bound up in one another.

Limitation of the capacity for knowledge

A LIKING for abstract conclusions has often been held to constitute Spencer's chief weakness; it is in accord with the above-mentioned tendency to bring together the most dissimilar phenomena in existence under one viewpoint. He himself has defined knowledge as the bringing of every separate phenomenon within the compass of a more general and previously known one — the operation of muscle, for instance, is explained if one has a chance of comparing it with the already known lever-mechanism — and he contends that in consequence hereof the ultimate and most general phenomena must remain incomprehensible because there is nothing more general with which to compare them. He repeatedly and with almost passionate emphasis affirms that our capacity for knowledge is limited: what matter, force, space, and time really are we shall never know, for our mind cannot grasp them; we can only investigate the phenomena that our personal experience of them educes. But for that reason Spencer also gives religion the right to hold its own views on this "unknowable." Religious problems, however, have little interest for him. He is all the more occupied with social questions, and it is in this sphere that his evolution theory finds its most curious expression. His belief in the progress of humanity is boundless and he is prepared to apply to it unreservedly Darwin's theory of natural selection — that is, as he himself says, that the fittest shall survive. The freedom of the individual he places above all else: "Every man is free to do that which he wills, provided he infringes not the equal freedom of any other man." The State is a survival from the primitive conditions of earlier ages, and its interference with the life of the individual is purely wrong and merely hinders the operation of free selection. All measures adopted by the Government are worse than if they were carried out by individuals; public poor-relief is expensive and badly administered compared with private charity; State schools are always inferior to private schools; in a word, the State should gradually be done away with, but for the present it is necessary to maintain a police force to ensure domestic security, and a military force to protect the country from invasion, though on no account should there be compulsory military service. So much the higher, then, must be the claims laid on private morality, and, in fact, Spencer claims much from it. He holds, in conformity with his belief in the heredity of acquired qualities, that the intellectual capacity of the individual becomes the common property of the race; the quality of the intellect corresponds to certain structural conditions in the brain; if the former is perfected, then the latter develop, are inherited by the descendants, and thus benefit humanity. The aim of morality is to create as much happiness as possible; happiness, however, must not be sought in material prosperity — the more so as the latter leads to dishonesty. To be allowed to contribute, in however small a way, towards the advancement of general evolution should be the highest happiness to which the individual can attain. Morality thus benefits the community more than the individual, according to Spencer, as indeed according to the positivism of the age as a whole. Both his and his contemporaries' limitation in this sphere lay in an insufficient sense of the purely personal; he had but little sympathy for the individual's longing for personal release from his confined and trying environment or from his inner qualms of conscience; he thought that one and all should take things calmly in the hope for better times to come — which, indeed, seemed a far more likely prospect for the people of those days than for those of our own.

As a matter of fact, Herbert Spencer himself lived to see the future of the world darkened. The march of militarism, which he hated, went on apace towards the close of the century; the colonization of tropical countries, of which he also disapproved, was carried still further afield; while socialism, with its State production, must necessarily have been equally distasteful to him. And even philosophy began in his lifetime to strike along paths other than those he had marked out. But though his ideas are now for the most part out of date, he will always be remembered as one of the most persistent, disinterested, and courageous champions of the theory of evolution.

CHAPTER XIII

THE DOCTRINE OF DESCENT BASED ON MORPHOLOGICAL GROUNDS. GEGENBAUR AND HIS SCHOOL

Leading position of Germany in biological research

IN HIS *Geschichte der biologischen Theorien* Rádl declares that Darwinism was born in England, but found a home in Germany. The statement is certainly justified in so far as, during the decades immediately succeeding the first appearance of the descent theory, Germany came to take a leading position in the sphere of biological research; here England and America rapidly came under German influence, as also did Italy, while France, which kept itself isolated, nevertheless could not entirely avoid being influenced. There were undoubtedly many reasons for this: on the one hand, the great economic and technical development that resulted from the founding of the German Empire, which in many ways proved beneficial to research, and, on the other hand, the splendid manner in which the work at the German universities was organized, which became a pattern for other countries, especially as a result of the careful and methodical guidance given by the teachers to their pupils' theoretical studies, practical work, and scientific production. And especially as far as biology in Germany is concerned, this organization had reached a very high standard — chiefly in the sphere of comparative anatomy — even before the appearance of Darwin. Originally, of course, comparative anatomy had been based on idealistic morphology, on the assumption that ideas formed the existing basis for the various forms of life, but we have already seen how this form of romantic natural philosophy was gradually supplanted by a more realistic manner of viewing life. What Darwinism gave to this realistic morphology was, as we know, a hitherto lacking connexion in existence; common descent took the place of the common ideal types. The fact that it was the representatives of comparative morphology in Germany who hailed the new doctrine with such deep enthusiasm is explained by the insistent demand that they had of old felt for a uniform conception of nature, a heritage from the, at one time all-prevailing, romantic philosophy. But it is just this never entirely eradicated romantic element that gives to German Darwinism, with its application of the descent theory to comparative anatomy, a character of its own. Again, the general cultural situation in Germany at the time of the launching of the new doctrine must of course have had a considerable influence on the form

CARL GEGENBAUR

RUDOLPH ALBERT VON KÖLLIKER

it was to take. This is essentially bound up in the two names Gegenbaur and Haeckel, each of whom in his own way represents a different side of the influence of Darwinism upon contemporary culture.

CARL GEGENBAUR was born in 1826 at Würzburg, of an ancient and well-to-do family closely connected with the Civil Service. After being at school in his native town he graduated at its university and applied himself, at variance with his family traditions, to the study of medicine, with a view to fulfilling his ambition to pursue scientific studies, in which he had early shown a keen interest. With his natural bent for science, he at first derived but little pleasure from his country's educational system; the gymnasium in Würzburg was run by Jesuits and was conducted in the Jesuitical spirit; nor were things much better at the University, until Kölliker arrived there with Leydig as his assistant. From that time onwards biology received a powerful stimulus, which was still further increased when, a couple of years later, Virchow began his activities as a teacher there. Within a short time these men made of Würzburg a nursery for biological research, and amongst their pupils Gegenbaur at once took one of the foremost places. In 1851 he wrote a dissertation for Kölliker and shortly afterwards accompanied his master on his research expedition to the Mediterranean coast, a trip that resulted in the young explorer's being ever afterwards attracted to comparative anatomy. The immediate result of the voyage was a number of valuable investigations into marine animals of various kinds, and the consequence of these was a summons to a professorial chair at Jena in 1855. At this little Protestant university, maintained by a liberal-minded Government, Gegenbaur at once succeeded very well, although himself a Catholic; he had had enough of the conditions prevailing at home, where the hospitals were under ecclesiastical administration and the doctors were subject to clerical control. At Jena he gathered around him a host of like-minded friends and pupils, chief among them being Haeckel. Here he worked out a scientific system, which he afterwards applied throughout his life, and here too he produced his finest works. In 1872, however, he accepted an invitation to Heidelberg, where larger resources were placed at his disposal and where he afterwards laboured until the close of the century, when, owing to increasing ill health, he resigned. He died in 1903, having been paralysed by repeated strokes.

Gegenbaur was a forceful personality, a friend to his friends, and an enemy to his enemies. As the founder of a school he is worthy of mention with J. Müller, but while the latter taught his pupils the method he had invented and in theoretical questions allowed them to go their own way, Gegenbaur permitted no divergence from the general principles he had once and for all made his own. Moreover, he succeeded in inspiring his disciples with such "boundless admiration" (Fürbringer) that most of them were prepared for the rest of their lives to swear by the master's word. This influence he won not through his lectures, for they were not very perfect in form, but through the keen interest he showed in his pupils' work, provided that it followed the right direction; in the laboratory he was a friend and comrade to his pupils and followed their careers in after life with never-flagging interest. But he could never endure contradiction; as a controversialist he was bitter and irreconcilable, although he invariably controlled his language. At Jena he collaborated loyally with Haeckel, and the exchange of ideas that took place between them was mutual. After he removed to Heidelberg, however, this co-operation ceased and even their friendship cooled off, as Haeckel devoted himself to popular agitation, of which his friend never approved. In his old age it was the Hegelian philosopher Kuno Fischer who was in closest contact with Gegenbaur; he described the latter as a deep thinker, which in its way characterizes him correctly.

Gegenbaur's first works came into being during his visit to the Mediterranean and comprise studies of the anatomy and evolution of various marine animals; in particular, medusæ and other Cœlenterata, Ascidia, and worms were investigated by him during this period, many of them with important results. Soon, however, he went over entirely to the study of the Vertebrata. One of his earliest works in this field is an essay on the evolution of the egg, published in 1861; in this he shows that all eggs of the Vertebrata are simple cells; hitherto it had been supposed that the egg of the bird, for instance, was a multicellular organ, whereas the granules in the yolk were held to be independent cells. In this connexion he strongly supports Max Schultze's view that the cell need not necessarily possess a membrane, but that the plasm and the nucleus are its principal components. This investigation, which of all Gegenbaur's writings is perhaps of the greatest value in the field of discovery, was followed by a long series of other works, wherein he applies to different organic systems in the Vertebrata the comparative method which he worked out in order to confirm Darwin's theory, and the main principle of which is to discover by means of anatomical comparisons the affinity between the animal forms due to descent. His finest productions on this subject, taken as models for the whole generation of research students, were his comparative studies of the skeleton, which were brought together in the work *Untersuchungen zur vergleichenden Anatomie der Wirbeltiere*. Among these studies the most notable is the essay entitled "*Carpus und Tarsus*," in which he compares piece by piece the bones of the hand and foot in different vertebrate animals, establishes their identity, and endeavours to reconstruct the form of extremity at one time possessed by the ancestors of the Vertebrata. This primal extremity he terms in a subsequent treatise "archipterygium"; he holds that it has been developed out of the gill apparatus and reconstructs the modifications by which have been evolved therefrom the fins of the fishes on the one hand, and the motive organs of the land-animals on the other. He supplemented this investigation with another on the scapular apparatus and the pelvis, in which the bones of these parts are similarly compared. The last and biggest section of this work is called *"Das Kopfskelett der Selachier"* and is described by Gegenbaur's disciples as the climax of his production. In it he examines and condemns after the style of Huxley the old theory of the skull's being composed of vertebræ; instead he makes the cranium of the sharks the archetype from which the same part in all higher vertebrates must have been derived; it comprises in the sharks throughout their lives a cartilaginous capsule and is formed of the same elements also in the higher vertebrates, while the latter's definitive cranium is constructed with the co-operation of a number of covering bones, originating in the skin. On the other hand, the visceral skeleton of the head — gill-arches and jaw-bone — is compared with the ribs, so that a part of the head, at any rate, possesses a segmented character. In conjunction with these skeletal investigations Gegenbaur also carried out a number of comparative studies in the sphere of the anatomy of the nerves and musculature and the organs of digestion. The entire results of his research work he collected in that great work which was published towards the close of his life, *Vergleichende Anatomie der Wirbeltiere*, in which he gives the most complete expression to his ideas and aims. As parts of this work he added his previously published text-books *Grundzüge* and *Grundriss der vergleichenden Anatomie*, which present his views, in concise form, and the method of presentation and the contents of which have been imitated by many later authors. In these, as also in his monographs, Gegenbaur's style is always heavy and sometimes hard to understand, which his admirers held to indicate depth of mind; nevertheless, consistency and set purpose are the most conspicuous features in his scientific writings — which indeed explain their success with his contemporaries.

Gegenbaur's general principles

IN some essays written late in life Gegenbaur sets forth the principles on which he considers that biological research should be carried out. To him, comparative morphology is the essential science, not to say the only road to the knowledge of life; and the final goal of this knowledge is the determining of the mutual relationship of the different life-forms by discovering their common origin. "The ultimate aim is phylogeny," he says in an account of the relation of anatomy to ontogeny. After the fashion of Darwin, he ascribes the actual formation of species to natural selection, though, practically speaking, he does not discuss this problem, but confines himself to tracing the individual organs back to common archetypes, which he seeks in the lower organisms, in the vertebrates particularly in the sharks. Investigations in homology with a phylogenetical purpose are thus the aim of his researches; the form and modification in the form of the organs are all that interests him; physiological problems are thrust aside, experimental investigations are unnecessary; histology is to him simply microscopical anatomy and he fails to understand its efforts to discover the phenomena of metabolism in the elementary parts of the body. Even embryology, which has nevertheless made such weighty contributions to the theories of descent, is given no independent position, but is recommended to adjust itself carefully to the results of the comparisons between the outgrown organs. But, owing to the fact that these investigations into the problem of origin can, of course, never be verified, Gegenbaur's research work proves in reality to be a theoretical speculation, which differs from that of idealistic natural philosophy only in appearance, but not in reality. Gegenbaur's archipterygium and Owen's archetype are practically alike fictitious, only that the former is believed to have existed some time in the beginning of the ages, whereas the latter had its existence located in the ideal world. But Gegenbaur and his school are the last people to attribute unreality to their primal types; provided one could once get the evolutionary series in order and the gaps filled up with suitably reconstructed forms, it could be urged that the primal type had as real an existence as if it had actually been dug up out of one of the earliest fossiliferous deposits. Here is undoubtedly demonstrated an intellectual contact with the romantic natural philosophy, and Gegenbaur himself was without doubt influenced from that quarter; that he as a thinker should have been approved by the surviving representatives of Hegelianism was in this respect striking enough, and, as a matter of fact, he himself has clearly expressed his sympathies for the romantic tendency — Goethe's morphological schemes found in him a warm admirer; the former's and Oken's theory of the skull's being formed of vertebræ is referred to with unreserved acceptance, in spite of its not being tenable any longer. The worst of such evolutional constructions, however, is that they are never allowed to live long undisturbed, owing to the discovery of fresh facts, and Gegenbaur's life's work has to a great extent had to suffer that fate. His archipterygium theory was soon supplanted by another, which derived the extremities from a lateral fin instead of from the gill-bones, and which perhaps nowadays has most supporters. Besides, palæontological finds in recent years have proved that the earliest amphibious types had seven digital bones instead of the five that Gegenbaur assumed. Further, it has been shown in our own day that the earliest fossil fishes possessed a cranium of bone, wherefore the theory of the shark's cartilaginous cranium as a primal type is no longer tenable.[1] In certain other cases, too, he has been alternately right

---

[1] For further reference to these questions see: BRAUS, "*Die Entwicklung der Form der Extremitäten*" in O. Hertwig's *Handbuch der Engwicklungslehre der Wirbeltiere;* HANS STEINER, "*Die Entwicklung des Vogelflügelskelettes,*" in *Acta Zoologica* (Stockholm, 1922); E. STENSIÖ, *Triassic Fishes from Spitzbergen* (Vienna, 1921); and the literature referred to in those works.

and wrong, according as new facts have come to life. Thanks, however, to his firm convictions and will-power, Gegenbaur succeeded in compelling a whole generation to follow his line of thought. Research on the subject of origin was regarded as the most important function of science, and thus, to quote his foremost and most independent disciple, Oscar Hertwig, hypothesis was made the main point of evolution in science. And it must be admitted that these theoretical speculations on the problem of descent have had a highly stimulating effect upon morphological research; a number of practical discoveries of the greatest value and of the highest significance for the development of biology have been made by the Gegenbaur school. Even to this very day comparative anatomy contains problems still unsolved and still attracts investigators of worth. And though the purely speculative problems of descent do not, it is true, predominate to such an extent as formerly, the presupposed common origin nevertheless still forms the basis on which rest present-day homological investigations. But comparative anatomy has certainly had to abandon its monopoly of biology and to recognize other biological tendencies and methods as being equally justified in their existence.

Fürbringer on the system of birds

OF Gegenbaur's disciples the majority naturally came from Germany, but students also flocked to his institute from Scandinavia, England, and Russia. Chief of these was MAX FÜRBRINGER. Born in 1846, he eventually joined Gegenbaur as a pupil at Jena and accompanied him to Heidelberg as prosector. Having for a time been professor at Amsterdam and Jena, he succeeded his master at Heidelberg and faithfully preserved the latter's traditions. He carried out comparative investigations in many fields; the excretal organs of the vertebrates as compared with those of the Annelida, and the evolution of the scapular regions are two of his best-known contributions to comparative anatomy. He is chiefly to be remembered, however, for his *Untersuchungen zur Morphologie und Systematik der Vögel*, a monumental work in both size and content. The first half of it consists of a comparative study on the Gegenbaur model of the region of the breast, shoulder, and wing throughout the whole order of birds. To this is added a general systematic section setting forth the natural bird-system based on a comprehensive comparative investigation of representatives of all the bird families. This system, which is now universally accepted, has entirely re-formed the bird class; the old orders are for the most part exploded — the owls, for instance, are transferred to the nightjars; the falcons and the vultures are placed next to the petrels, herons, and storks — an example of how intensive anatomical investigations may give to the family relationships an entirely different value from that of ancient tradition. One of Fürbringer's advantages is that he avoids the fanciful elements in the descent theories in which his age otherwise abounds; he investigates the extant birds, but produces no reconstructed middle-forms and archetypes, letting the scientific material that actually exists speak for itself. His method thus acquires a soundness that has given to his results a lasting value. He died in 1920.

Hubrecht on phylogeny

GEGENBAUR had a far more imaginative disciple in the person of the Dutchman A. A. W. HUBRECHT (1853–1915), a professor at Utrecht. In his younger days he was occupied mostly with the invertebrates, especially the worms, but he afterwards devoted himself entirely to the evolution of the Mammalia. In this field he made valuable contributions by collecting material in the course of expeditions in tropical countries and by investigating, with special reference to their embryonic development, a large number of rare and little-known animal forms. On the basis of this material he speculates deeply upon the origin of the Vertebrata from lower animal forms, producing a number of theories that diverge considerably from what the earlier evolutionists regarded as indisputable truth. The sharks, for instance, he places for palæontological reasons in an isolated position in the system — that is to say, in direct opposition to Gegenbaur's view — one of many proofs of how fresh facts in this sphere have produced fresh difficulties, which it has not been possible to solve with uniform results.

A complete account of the works produced by the pupils of Gegenbaur or in his spirit would fill a volume; even in modern times comparative anatomy is largely under the influence of his method. Pupils had flocked to his laboratory from all parts of the world. In the following chapter will also be mentioned some pupils of Gegenbaur, who in their younger days followed in his footsteps, but in later years struck out new paths of their own, sometimes with brilliant success.

Even the man who was in closest touch with Gegenbaur in his best days and who exercised most influence on him, just as he was most influenced by him, went his own way towards the end; this was Haeckel, a man who gave to Darwinism a peculiar stamp, extremely characteristic of the age, and who contributed much to its success, though perhaps still more to bringing it into discredit.

CHAPTER XIV

HAECKEL AND MONISM

Ernst Heinrich Haeckel was born at Potsdam in 1834. His father had taken part as a volunteer in the Prussian War of Independence against Napoleon, and afterwards, having adopted a public career, he advanced to the rank of "*Regierungsrat*." His mother was the daughter of a Civil Servant, who had been dismissed from his post and arrested for having opposed the French conqueror. His parents' home, in spite of its bureaucratic character, had nevertheless preserved the liberal-minded traditions of earlier times and the literary interests acquired in the days of greatness of the German world of letters. Young Ernst received his school education in a provincial gymnasium, where, as he himself says, mathematics were neglected for philosophy and the classical languages. Even when grown up, he still enjoyed reading Homer in the original and throughout his life delighted in interlarding his writings with Greek terms. His greatest pleasure, however, he found in nature, both in reality and in poetry; he was a keen botanist and at the same time read Goethe's works, Humboldt's travels, and Schleiden's popular writings. He was especially interested in Schleiden and was very anxious to go to Jena in order to be trained as a botanist under him. After he had matriculated, in 1852, however, his father insisted upon his going to Würzburg to study medicine. He spent two years there and, in spite of his dislike for professional studies, devoted himself with interest to anatomy and histology under Kölliker and to pathology under Virchow. He then spent one year in Berlin studying under J. Müller, whom he regarded as his true master and who inspired him with a love for marine research, particularly in regard to the lower animals. Having been for some time assistant to Virchow, he went on an expedition to the Mediterranean at the suggestion of Kölliker and Gegenbaur and collected at Messina material for his first important work, *Die Radiolarien*, which resulted in his being called, upon the recommendation of Gegenbaur, to the chair of zoology at Jena in 1862. There he worked until his resignation, in 1909, after which he lived for another ten years, continuing his literary work until he died, in 1919.

The life which falls within these dates is without doubt one of the most remarkable during the epoch just closed; there are not many personalities who have so powerfully influenced the development of human culture —

and that, too, in many different spheres — as Haeckel. He has been much disputed — now praised to the skies, now vilely abused. As a matter of fact, it is not at all easy to grasp the true value of his life's work. No important scientific discovery attaches to his name, and the ideas he promulgated are largely borrowed from others. The works that once brought him fame are now hopelessly out of date, but it must be admitted that much in them has now been incorporated in our general knowledge. The idea of evolution, in the form given to it by Darwin, found in Haeckel its most devoted champion; his personality and his trend of thought have set their mark on the elaboration of this theory, especially on the continent of Europe, and they are therefore worthy of closer examination.

Light is thrown on Haeckel's early development by two collections of letters, which have since been published, the one addressed to his parents in his student days at Würzburg, the other to his betrothed during his Italian journey. This development is highly characteristic of the generation to which he belonged and therefore explains in some degree how it was that he acquired such an influence over his age. Young Haeckel at Würzburg is by no means a German "corps" student of the ordinary type; on the contrary, he was a very nice youth, abhorring duels and drinking-bouts, diligently attending lectures and exercises, writing tender and affectionate letters to his parents, regularly attending church, and comforting his lonely hours with pious thoughts. True, he could cause his parents anxiety on account of his dislike for medicine and his propensity for unpractical dreaming, but, on the other hand, he was always ready, with a somewhat rhetorical and precocious eloquence, to confess his weaknesses to his old parents and to promise to make them happy in the future. The most striking feature of these letters is their Christian piety, which contrasts strongly with the hatred that Haeckel felt for Christianity in later years; the youth expresses his indignation against Karl Vogt and other "materialists" of the time in terms that were afterwards used almost word for word against himself. It is, of course, the opinions held in his parents' home that here recur — the old-fashioned, serious, moral-religious atmosphere pervading the home of a Prussian Civil Servant, with its literary and patriotic traditions. At Würzburg young Haeckel was enraged at the Catholic propaganda, which was carried on at that time, during the period of reaction after 1848, with extreme ruthlessness, and at the same time as his father was deploring the unhappy political situation. In the letters from Italy the whole aspect is altered; that was in 1859, the year of the liberation of Italy. Haeckel is full of enthusiasm over Germany's unification and raves against her opponents, vassal princes and Prussian junkers, who were serving the reactionary politics of Austria. His religious attitude is now something quite different; Christianity has been superseded by a worship of humanity in general, combined with enthusiasm for the enlightened minds of classical antiquity and hatred against the ecclesiastical reaction — a very common trend of thought at that time — which found expression in many quarters in literature, as, for instance, in the works of Haeckel's contemporary fellow-countryman Paul Heyse. His biographers declare that Haeckel's religious change of front took place in the course of spiritual struggles, but there is little trace of them in his letters; it would appear more likely that with him, as with countless others, religious free-thinking was induced by political independence of thought; it was difficult in those days to reconcile Christian belief and political liberal-mindedness, owing to the Church's intimate connexion with the reactionary forces in society and her obstinate resistance to all movements of reform. Through his free-thinking, however, Haeckel lost that conviction which had kept him going before, and he felt himself beginning to doubt the possibility of penetrating any deeper into the essence of natural phenomena.

Haeckel embraces Darwinism

THAT guiding line for his thoughts which he thus lost Haeckel rediscovered when he made the acquaintance of Darwin's theory. In Germany as in England this theory had been received with mixed feelings; instances of this have been given above. Haeckel at once became an ardent supporter of the new doctrine; in it he found not only the means to understand existence, but also the confirmation of the progress he desired to find in it. It was mainly through his promulgation of it that Darwinism became a watchword for all supporters of the idea of a liberal-minded development in the sphere of social and cultural life, and obviously an abomination to its opponents, the clerical and conservative elements in the community. The course of social development in Germany took an unexpected turn, however; the unification of the country, the long-cherished dream of the free-minded, was brought to reality through Bismarck, but in such a manner that the power of the princes and the junkers was preserved. It was not thus that the liberals had imagined things would turn out; their opinion now became divided; the majority of them sided with the new work of unification and its leaders, while a smaller group still insisted upon their demand for liberal-minded social reforms. To this latter group belonged some of the leading scientists in Germany, and among them Haeckel, although, living as he did in the small town of Jena, he never took an active part in politics. It was with all the greater enthusiasm, then, that he devoted himself to promoting this radical development in the sphere of general culture, and he rapidly gained a following of people with similar ideas to his own, who took up the struggle against dogmatic conservatism in both the social and the religious sphere, employing the evolution theory of Darwinism as their principal weapon. Of course the authorities viewed with anything but friendly eyes this natural-scientific opposition, with its social tinge; the employment of these hostile elements in the government service was opposed with all their power, and many of the radical party lived for the rest of their days as independent writers. Haeckel himself, however, was protected by the liberal-minded Weimar Government from all unpleasant consequences and was thus enabled, though a professor, to take a leading part in the struggle. It goes without saying that the natural-scientific contents of this doctrine were influenced by the political and social views of the antagonists, but, on the other hand, this circumstance contributed towards making Darwinism popular and creating a widespread interest in its problems and arguments. Before proceeding to describe the part Haeckel played in this struggle, however, we must take a glance at the subject of research that he made his own and determine how far his general scientific conclusions were based thereon.

His work on Radiolaria

HAECKEL began as a microscopist; when he was at Würzburg his father gave him a microscope and he could not find words to express his delight at all that he saw in it. In fact, both the papers he wrote for his degree were on microscopical subjects — his dissertation on the tissues of the river crayfish, and an essay on the pathological changes of the venous system; two school essays, the former worked out under the guidance of J. Müller, the latter under that of Virchow and noteworthy as being Haeckel's only specialized investigation in the sphere of the vertebrates — both papers creditable in their form and contents, but not very original. His appearance as an independent investigator is marked by his monumental work on the Radiolaria, which is without doubt his best. It is dedicated to the memory of J. Müller and is written in his spirit; he was, in fact, his foremost predecessor in that field. It contains about one hundred and fifty new and carefully described and illustrated species, as well as abundant material derived from observations of their structure and mode of life. It makes what were at the time valuable contributions to the problem of the biology of single-celled animals, and moreover, the identity established by Max Schultze of the protoplasm in the higher animals and the sarcode, which had already been described by Dujardin, is hereby confirmed with fresh proofs. In connexion therewith several cytological observations are quoted that are of considerable general interest — on the phenomena of currents and the manifestations of assimilation in pseudopods and protoplasm, and also on the power of cells to absorb solid bodies. Haeckel has observed how the blood corpuscles in a mollusc absorb indigo-particles injected into the blood, but he did not follow up this important fact any further, it being left to Metschnikoff a couple of decades later to take up the subject and make it the basis of his theory of phagocytes. In regard to classification, Haeckel tries to found a natural system based on affinity; it is in connexion with this that he announces for the first time, though tentatively, his association with Darwin's theory and endeavours to find a primal form from which the rest of the Radiolaria could be derived. On the whole, however, the system was set up in the traditional way.

Later on Haeckel continued the work on the Radiolaria, adding two new parts (1887-8), and in the special section of these he describes several hundred new species, which are splendidly illustrated. The complaint has been made against his descriptions that they keep too much to the skeleton, and against the illustrations that they are over-simplified, but on the whole both text and illustrations compare creditably with the former volumes of the work. The general section of the new work, on the other hand, is strongly characterized by the natural-philosophical speculations which Haeckel had produced in the mean while, and to which we shall revert later on.

His work on sponges

ANOTHER field that Haeckel made the subject of systematic research was the sponges, of which he dealt especially with the Calcarea in his monograph *Die Kalkschwämme*, of 1872. In this work he has recorded his most consistent attempt to create a Darwinistic classificational system — a true "natural" system based on descent, instead of the old "artificial" system. The group had been very little investigated and the facts contributed by Haeckel are of some importance, considering the age when they were published, although they have, of course, undergone a good deal of modification as a result of subsequent research. On the other hand, this natural system has its curious features. The order of the Calcarea is divided into families according to the shape of the canals in the walls of the sponge, and this mode of classification has been retained by subsequent naturalists. The division into genera, on the other hand, is based on the calcareous spicules of the skeleton. These two features, the canals and the calcareous skeleton, are, according to Haeckel, the only systematically employable elements, for their form is inherited, whereas the artificial system has taken account of mouth formation and colony or solitary life, which are dependent upon "adaptation." No evidence, however, is offered in proof of these statements, and it certainly does seem decidedly artificial to base the division into genera upon one single character, without the slightest attempt to test the theory by means of comparative morphology. Nevertheless, the system has its peculiar interest as an attempt to separate entirely from the Linnæan system. The terms hitherto employed have been entirely abolished; instead of "genera" he uses "generic varieties," besides which there are differentiated and nominated "specific, connective, and transitory varieties" or "initial, binding, and transitional species." One must admit the logical consistency of this attempt to get away from Linnæanism; the latter rests entirely upon the immutability of the species, and if it is once denied, it is necessary really to set up a new system with a different idea of species. Haeckel's attempt, however, has proved unsuccessful and has failed to gain the acceptance of more recent systematists; in his own later systematic works he himself uses the old traditional terms of genus and species, in spite of all the assurances of Darwinism.

On the medusæ

A THIRD subject in which Haeckel worked as a systematist is the medusæ; here, too, he summarized his results in a monograph of huge dimensions, entitled *Das System der Medusen* (1879), containing a large number of newly-described forms and a system of classification that in part is of some value. In particular, the two main groups that he classifies in it, the Craspedota and the Acraspeda, have been retained by later systematists. On the other hand, it has been found upon examination that some of the diagnoses of species are full of serious mistakes, which is explained by the fact that Haeckel has in general a far keener eye for the demarcation of the large groups in the system than for genera and species; careful detailed examination was never his strong point.

There is still another group of life-forms which engaged Haeckel's interest and which perhaps appealed more to him than any other — namely, the order Monera. To this order he refers single-celled organisms without nucleus — that is, those formed of only a homogeneous mass. He has described a great number of these — generally amœboid organisms — many of them with systematic validity. Nevertheless, the improved microscopy of modern times has actually discovered in the majority of these a nuclear substance, either in the form of a single nucleus or divided into minute particles, and modern biology, which has learnt by experience to count the nuclear substance among the essential components in a cell capable of life, has in general presupposed the existence of the nucleus even in cells in which, owing to its minimal dimensions or indistinct cell-content, it has not been possible to confirm its existence. Haeckel, however, stubbornly held to his non-nuclear Monera, the existence of which he regarded as an essential qualification of that spontaneous generation by which he believed life to have arisen, and which he looked upon as ''a logical postulate for philosophical natural science.'' This brings us to Haeckel's natural-philosophical speculations — that part of his activities which, far more than his specialized research-work, brought him both fame and ill fame.

The essentials of his opinion

As has already been mentioned, Haeckel declared his adherence to Darwinism in his work on the Radiolaria. At a scientific congress in 1863 he expounded Darwin's theory in a manner that considerably enhanced its success in Germany. The lecture really comprised a brief summary of the *Origin of Species* — of the doctrine of selection and the struggle for existence. In its essentials the argumentation is Darwin's own, taken from the theory of domestic animals, from animal geography and palæontology; but striking indeed are the radical conclusions that Haeckel draws in regard to the origin of man; they represent what eventually became one of his chief interests and immediately caused a great sensation. There are two more peculiarities of Haeckelian thought that come out clearly in this lecture: his political radicalism, which induces him to call progress "a natural law which no human power, neither the weapons of tyrants nor the curses of priests, can ever succeed in suppressing" — the words were uttered just at the moment when the struggle between Bismarck and his liberal opponents waxed hottest — and his predilection for the romantic natural philosophy, which makes him praise Goethe, Geoffroy Saint-Hilaire, and Oken as "deep-thinking men possessing prophetic inspiration," and as supporters of "philosophical theories of evolution" foreshadowing Darwin. These elements — Darwin's theory of evolution, political radicalism, and romantic natural philosophy — really impress the whole of Haeckel's subsequent pronouncements with their character, whether they concern "general morphology," "cosmic riddles," or "artificial forms in nature." The doctrine of natural selection forms the groundwork, which he never takes steps to reconstruct or add to, however great the progress made by research. His political radicalism mostly finds expression in a violent hatred of priests and Christianity, but also, though not so apparent, in opposition to the undue interference of government authorities. The influence of romantic natural philosophy comes out most clearly in his utter incapacity to grasp the relativity and limitations of human knowledge, which Herbert Spencer among others so forcefully and repeatedly emphasized; Haeckel's way of constantly trying to solve the "riddles of the universe" is far more reminiscent of Schelling than of the contemporary positivist trend of thought, just as his overbearing self-confidence and his abusive polemics are more representative of romanticism than of exact research. Thus through Haeckel's influence romantic natural philosophy experienced a revival in the century of exact science.

His Generelle Morphologie

HAECKEL struck his great blow for "philosophical scientific research" with his *Generelle Morphologie der Organismen*, with its subtitle *Kritische Grundzüge der mechanischen Wissenschaft von den entwickelten Formen der Organismen, begründet durch die Deszendenztheorie*, which was published in 1866. The first part of the work was dedicated to Gegenbaur, the friend with whom he had constantly exchanged ideas and who had inspired much of its contents. The latter part of the book is dedicated to Darwin, Goethe, and Lamarck, those "scientific thinkers who founded the theory of descent." As this trio was afterwards constantly referred to by Haeckel, it may be worth while examining the combination more closely. Lamarck and Darwin may both be regarded as founders of the theory of descent, although the latter, it is true, positively rejected the former's explanation of nature and was but little concerned with its materialistic speculation. But the idea of seeing in Goethe a precursor of a "mechanical science of the organisms" certainly needs some explanation. The great poet was universally looked upon by his age as an idealistic natural philosopher; the biologists who acclaimed him did so under that assumption and he himself had adduced "*geistige Kräfte*" as a cause of the origin of and modifications in the life-forms, and otherwise also given utterance to markedly spiritualistic views. Whence, then, Haeckel's assertion to the contrary? The reason is no doubt to be found partly in Haeckel's own natural-philosophical turn of mind, which could never be induced to take the idea of "mechanism" in existence really seriously, and partly in the position Goethe enjoyed in the cultural life of the period — his influence as a poet and a cultural personality, which was highly admired even in Haeckel's home and circle, and which was opposed by no one beyond the extreme-orthodox ecclesiastical authorities, who found free-thinking and libertinism in his poetry, something which in its turn increased the sympathy of the liberals for the really somewhat conservative poet-minister. And the liberal opposition became once and for all one of the leading motives in Haeckel's system of thought.

The very choice of subject and the consequent title of the work — *General Morphology* — is also obviously borrowed direct from Goethe, who, in fact, invented the word in question and from whom Haeckel also derived the philosophical conception of morphology that he develops in the book. For, strictly speaking, Haeckel was no professional morphologist in the modern sense. He had till then worked almost exclusively on the classification of single-celled animals; and in the comparative anatomy of the higher animals, especially the Vertebrata, he practically never carried out any special investigations, at least none of which the results have been published. That he nevertheless based his theoretical speculation not on classification, as Darwin himself did, but on morphology, was no doubt due, as hinted above, to his admiration for Goethe, but also, of course, to the influence exerted on him by his friend Gegenbaur, who was no doubt responsible for the best contributions of facts in the work. But a speculatively inclined student who concerns himself with second-hand knowledge will, of course, easily succumb to the temptation to let his imagination get the better of his critical sense — a fact that finds strong confirmation in Haeckel.

His ternary division of nature

THE *General Morphology* begins with a chapter on the relation between morphology and other sciences. First comes an assertion that every natural object possesses three qualities: *matter, form,* and *energy* or function. In connexion with this idea natural science is divided into three disciplines: *chemistry,* or "*Stofflehre,*" *morphology,* and *physics,* or "*Kraftlehre.*" Then the knowledge of inorganic nature is divided into *mineralogy*, *hydrology*, and *meteorology;* and biology is divided into *zoology*, *protistology*, and *botany*. Thus we have here four threefold groups, all extremely ill-grounded. One is tempted to assume that it is really Schelling's romantic ternary mysticism haunting him here — though, of course, indirectly and unconsciously. The division into plants, animals, and protista is, of course, entirely useless, nor did it ever succeed; instead of one vague line of demarcation such as that between plants and animals, we here get two. Then the aims and means of morphology are described; the aim is a mechanically causal explanation of the forms and phenomena of life, whereby a "monistic" explanation of the universe will be made possible, which indeed, it is declared, is already so in the other natural sciences, but in biology is for the time being replaced by a "vitalistic" and "dualistic" view, the incorrigibility of which is depicted in vivid colours. The means of attaining this monistic explanation of nature is declared to be by "philosophical thinking," by the aid of which, facts should be capable of interpretation, whereas the mere observation of natural phenomena is deeply despised. As a matter of fact, this philosophizing constitutes Haeckel's great weakness, which gradually induces him to abandon exact research. The insistence upon interpreting the phenomena of life according to purely mechanical laws is in itself fully justified; physiology had already pursued that method before Haeckel's time, and his claim that the other branches of biology should follow its example was quite reasonable. But Haeckel's great mistake lay in his refusal to realize and acknowledge the limited possibilities of the mechanical explanation of nature. He certainly admits in one passage (p. 105) that the human capacity for knowledge has its limits: that we cannot reach the ultimate grounds for a single phenomenon, and that the origin of a crystal down to its ultimate causes is just as inexplicable as the origin of an organism. But he does not stop for a moment to think that in such circumstances natural philosophy should endeavour to determine these limits and see that they are not exceeded. Shortly after the above admission he confidently asserts that no essential difference between animate and inanimate exists; after making a close comparison he comes to the conclusion that the crystal and the living cell are in all respects comparable, as to their physical and chemical composition, their growth and individuality. The restriction that should follow from the limitation of the human capacity for knowledge is entirely forgotten. The memory of it certainly reawakens now and then, but, generally speaking, he entertains a blind faith in the power of "mechanical causality" to explain anything whatever.

Mechanical interpretation of nature

WHAT Haeckel chiefly bases his conviction upon as to the unlimited possibilities of the mechanical explanation of nature is Darwin's theory. His enthusiasm for it knows absolutely no bounds; once he assures us outright that, thanks to this theory, there is now not a single fact in organic life that cannot be explained, although many are still unexplained. This enthusiasm, indeed, he shared with the whole of his generation; in a previous chapter light has been thrown upon the hopes that the selection theory aroused on its first appearance; the fact that in Haeckel they reached such dizzy heights was due, of course, to his personal temperament, in which enthusiasm, a naïve self-satisfaction, and a blind confidence in the correctness of his own ideas had been the predominant features since his youth. Otherwise, he desired to a certain extent to modify the selection theory itself, in so far as he would define more precisely the actual term "struggle for existence." He urges the exclusion of all conditions belonging to surrounding nature; the competition with other living creatures is all that should be considered in this connexion. He further maintains the existence of a competition *within* the individual, between its various parts — that is, an adaptation of the theory of correlation to Darwinism, which was later developed in certain respects by others. And, finally, Haeckel insists, far more emphatically than Darwin, upon the transformation of the individual through the influence of environment and the inheritance of the modifications thus brought about; he defines evolution as a co-operation between an "*innerer Bildungstrieb*" — heredity — and an "*äusserer Bildungstrieb*" — the influence of environment. These expressions, which have a very natural-philosophical and not a very mechanical sound, he borrowed, as he himself admits, from Goethe's *Pflanzenmetamorphose*, which he considers represents Darwinism *in nuce*, and which to his mind still forms the basis of plant morphology — a view which at that time was shared by only a few supporters of natural philosophy, but which has been repeated on Haeckel's authority up to modern times by literary historians and other non-professionals. For the rest, he attributes to Darwinism an infinite mass of new determinations, with their attendant terminology. Haeckel almost surpasses Linnæus in his mania for classifying and naming, but he is entirely lacking in the incomparable gift for form that the great systematist possessed; most of his categories and nomenclature have not survived their originator, although a number of them have been universally adopted, as for instance, the terms "ontogeny" and "phylogeny," the former denoting the individual's, the latter the race's development, and "œcology," as an expression denoting the relation of living beings to their environment. Utterly absurd, on the other hand is his "promorphological" classification of the life-forms according to a symmetrical plan intended still further to confirm the alleged similarity between the structure of crystals and organisms; the details of this system, which, as a matter of fact, give evidence of a very superficial knowledge of the foundations of crystallography, may be compared with

Oken's wildest flights of imagination; infusorians, pollen granules, corals, flower-spikes, are cited, amongst other things, as examples of the various supposed crystal-symmetrical forms. Undoubtedly more successful is the natural system that he afterwards sets up for the organisms, wherein is employed for the first time the method, so frequently used since, of representing by means of a graphic chart in the form of a genealogical tree the mutual agreement of the different life-forms, as if derived from an assumed common origin. Haeckel has certainly had to endure a good deal of chaff for his genealogical trees and they will not, of course, bear too close examination, but it cannot be denied that the method itself has proved of good service to scientific works aiming at a natural system; we need only mention how Fürbringer employed it in his great work on the birds. Here, as in many other respects, Haeckel has had a rousing and stimulating influence on subsequent research.

Haeckel identifies spirit and matter

THE genealogical tree that now, as henceforth, interested Haeckel most is, however, that of man; already at this stage he sets forth the ideas concerning it that he was later to develop still further. From man he proceeds to the universe and God, and now makes the entirely unexpected assertion that "no matter can be conceived without spirit, and no spirit without matter." It is hard to make out how this idea is to be reconciled with his earlier assurance that every natural phenomenon, both animate and inanimate, can and is to be explained mechanically; ever since the days of Galileo, indeed, all spirits have been outlawed from mechanics. Haeckel, nevertheless, makes use of his spirit-matter to decree unity between God and nature — a unity which denotes true monism and which admits of a true divine worship. It is again from Goethe, of course, that these pantheistic reveries are borrowed, so that in this first philosophical work of Haeckel's, romantic idealism has the last word.

Generelle Morphologie, which in Haeckel's own views is his principal speculative work, had but little success; only one edition was published. Darwin, it is true, was delighted, although he complained mildly of the vehement style in which the book was written, but the German biologists were enraged at the natural-philosophical daring, the dilettante treatment of detail, and the scurrilous language. After some years of silence, however, Haeckel resumed his natural-philosophical activities, this time in a more popular form, with the result that he was extremely successful with both his series of lectures *Natürliche Schöpfungsgeschichte* (1868) and his *Anthropogenie oder Entwicklungsgeschichte des Menschen* (1874); the former work especially became extraordinarily popular, being translated into many languages, and it really represents perhaps the chief source of the world's knowledge of Darwinism. It reproduces the ideas and the arguments from *Generelle Morphologie*, but in an easier style and excluding his extensive speculations on symmetry. He gives instead a special account of the descent of man, which Haeckel regarded all along as the centre point of the theory of evolution and of all science in general. This problem likewise represents, as the title indicates, the subject of his *Anthropogenie*, a work of far greater significance than the *History of the Creation* and certainly the one in which Haeckel has set forth his most brilliant and most important ideas — those of his that most deeply affected the development of biology. The intention of the work is to give a comprehensive idea of the origin of man, based on the evidence of morphology, embryology, and palæontology. Haeckel definitely takes as his starting-point his well-known "biogenetical principle": that the ontogeny not only of man, but also of every living creature is a recapitulation of its phylogeny; "the development of the embryo is an abstract of the history of the genus." This idea in itself is not new; as we have seen, it had already been propounded by Meckel, and Darwin gave it an important place, although it was formulated in summary fashion, in his *Origin of Species*. It was then taken up and further elaborated by FRITZ MÜLLER (1821–97), one of the more peculiar representatives of biology during last century.

Fritz Müller on the development of crayfish

BORN in Germany, Fritz Müller had studied medicine — among other things, biology under J. Müller — but afterwards went out to Brazil, where he remained for the rest of his life in various occupations and with varying fortunes. Having from the very beginning been entirely won over to Darwin's theory, he resolved to prove it by applying it in detail to a suitable animal group, for which purpose he chose Crustacea, which in his adopted country exist in a multitude of forms. He paid special attention to the different types of development to be found in closer related forms within this class: the river crayfish creeps out of the egg like its parents; the crabs have one or two larval forms, while the prawn has many — a *nauplius* stage similar to the larvæ of the lowest Crustacea described under that name, a *zoea* stage, like that of the crabs, and a *mysis* stage, like the perfected form of the schizopod crayfish. Various other Crustacea likewise possess peculiar metamorphoses, especially the strangely formed parasite crayfish, whose early stages resemble those of the independently living Crustacea. All these facts, especially the fact that the larvæ of certain higher Crustacea resemble the fully grown individuals of lower Crustacea, convinced Fritz Müller that the evolution of the individual is a "historical document," which is sometimes effaced, owing to the development's striking into a more and more direct path from the egg to the fully grown creature, and which is sometimes "counterfeited, owing to the struggle for existence that the independently existing larvæ have to maintain." A case such as that of the prawn he regards as typical; the prawn's ancestors in past ages possessed the form that its larvæ now possess, and that, too, in the same sequence as that in which the larvæ now succeed one another, whereas the history of the river crayfish has been obliterated and that of a number of other Crustacea has received fresh contributions in respect of form.

This theory, which Fritz Müller expounded in 1864 in a paper entitled *Für Darwin*, aroused Haeckel's ardent enthusiasm. To him it became a "principle for the origin of life," the main support of the theory of descent and a particularly weighty argument in the controversy over the struggle for man's "natural creation." It was then chiefly to human evolution that he sought to apply the theory and in his *Anthropogenie*, as also previously in his *Natürliche Schöpfungsgeschichte*, he works out the embryonic development of man from the egg to birth with a view to collecting proofs of the conditions governing man's descent and affinity. Haeckel was never a specialist in embryology and its points of detail were of no interest to him in themselves, but only in so far as they could serve as evidence to prove the descent of man. His ideas of embryology could in such circumstances only be one-sided and deficient; the professional embryologists offered serious objections to them, which he either affected to overlook or else answered with personal abuse. Complaints were made especially against his illustrations, which, contrary to usual practice, he hardly ever borrowed from monographs on the subject, but drew himself. Being designed exclusively to prove one single assertion, his illustrations were naturally extremely schematic and without a trace of scientific value, sometimes indeed so far divergent from the actual facts as to cause him to be accused of deliberate falsification — an accusation that a knowledge of his character would have at once refuted.[1]

Haeckel's theory of germinal layers and gastræa

Two specially remarkable details in Haeckel's doctrine of the biogenetical principle are the theory of the germinal layers and the gastræa theory. We have previously described the investigations into the embryonic germinal layers carried out by Pander, von Baer, Remak, and others, and also how Huxley compared dermal and intestinal layers in the medusæ with the germinal layers in higher animals. Besides these facts Haeckel had for material on which to work his own researches into the Calcarea, the embryonal development of which he had studied. On all this he now bases the theory of the origin of the animals, and especially that of man; since man originates from a single cell, the egg, then in the beginning of time the original form out of which the human race has evolved must also have been a unicellular animal. Out of the egg-cell there is developed by segmentation a cell-group;

---

[1] It is nevertheless difficult to understand such an action as this: allowing in his *Natürliche Schöpfungsgeschichte* (ed. i, p. 242) the same cliché, reproduced three times, to represent an egg of a man, an ape, and a dog. This absurdity was removed from subsequent editions, albeit only after Haeckel had rewarded with abuse those who pointed out the fact; and the incident was for ever afterwards a theme on which his enemies constantly harped.

this stage the primal forms of the higher animals and man have also passed through, and during that period they have resembled such cell-colonies, as, for instance, the Volvox. Out of the simple cell-mass there evolves in the sponges, by means of invagination, a stage of development with double walls, a gastrula, which corresponds to the simplest form of an animal possessing an intestinal canal; the original form of the higher animals must likewise have passed through this stage. This original form common to all higher animals is called gastræa. From each of the walls of the gastrula there splits off through segmentation a fresh layer; these two secondary layers combine and form the mesoderm, which gives rise to the musculature and various other organs in the higher animals. This process has also taken place at some time or other in the primal form of the higher animals, and therefore all these three layers and their derivatives are homologous throughout the entire animal kingdom.

Importance of Haeckel's biogenetical principle

THIS evolutional theory is undeniably Haeckel's most brilliant and most important contribution to the history of biology. O. Hertwig was right in saying that for fifty years biological literature was under the influence of this idea; the abundant facts that were amassed on the subject of embryology during this period were mostly intended to confirm the biogenetical principle or the "recapitulation" theory, as it has also been called, and biologists strained every effort to apply it to every detail in the development of the embryo. And the application was "strained" in the fullest sense of the word. Haeckel knew from the outset that the gastrula stage of the mammals is not formed through invagination, as the theory claimed, but through delamination, or splitting off; he consoled himself, however, with the thought that in the lancet-fish invagination generally takes place, and from this primitive animal he derives the Mammalia, with the assertion that their gastrula form is due to later adaptation — to the "falsification" of documents, of which Fritz Müller had spoken. He also explains a number of other facts of a similar kind according to the same method. Matters became still worse when the embryologist HIS came forward with an attempt to explain the entire cause of embryonic development on purely mechanical grounds. Haeckel was furious and replied with a shower of abuse, quite forgetting all his own utterances, in which he insisted upon a mechanical explanation of nature. In reality this mechanical, or, in other words, physiological, side of embryonic development is of very great importance, though Haeckel quite overlooked the fact in his anxiety to explain natural creation; later on, however, it received all the greater attention. But, even apart from this, time has dealt hardly with Haeckel's ontogenetical theories. The gastrula formation by means of invagination has proved far less general than Haeckel believed — *inter alia*, it is lacking in most of the Cœlenterata — and the far-fetched homologization of the germinal layers has been considerably restricted, the same organs in a number of different animal forms having been found to possess an entirely different origin. In particular, the mesodermal formation has now been resolved into a number of different processes. In fact, the entire "biogenetical principle" is nowadays severely challenged, even as a hypothesis; in the vegetable kingdom it has received no confirmation, which is indeed strange for a theory proposed to hold good as a general explanation of life, but even those zoologists who in general give any support at all to the recapitulation theory do so with considerable reservations, called for by the results of modern hereditary research and experimental biology. Nowadays one does not compare without question, as they did in Haeckel's time, the ideas of similarity and affinity; similarity can demonstrably arise through the influence of very different factors, and it is preferred to follow His in seeking for the mechanical conditions governing the development of form instead of seeing therein resemblances to the animal life of past ages. But this should not involve our depreciating Haeckel's influence on the development of embryology; it was his theory which evoked that interest in those phenomena that brought about the immense revival of this form of research, lasting up to the present day. In this connexion we may remember von Baer's words that "inaccurate but definitely pronounced general results have, through the corrections which they call for and the keener observation of all the circumstances which they induce, almost invariably proved more profitable than cautious reserve." It is just herein that Haeckel has benefited his science most; here he has made his most important and historically most valuable contribution. But with it he gave all that he had to give; the years that he lived afterwards produced nothing to increase his reputation, but detracted much from it.

His work on "perigenesis"

FOR as early as in his *Anthropogeny* Haeckel displays his increasing weakness for vague and profitless speculations. Talk of a mechanical explanation of nature is certainly kept up, but it becomes more and more empty words, while the spiritual qualities of matter appear increasingly in the foreground; energy and soul are now consistently identified, and are generally denoted by the term "energy," in a manner which testifies to his absolute contempt for the simplest grounds of physics. And this fault is still more intensified in a treatise published in 1875 entitled *Die Perigenesis der Plastidule*, the natural-philosophical confusion pervading which it is truly difficult to reproduce in a summary. The title itself is supposed to mean "the wave-production of life-particles" and this is intended to explain the same phenomena as those upon which Darwin tries to throw light by means of his pangenesis theory — that is, heredity and adaptation. The pangenesis theory fails to satisfy Haeckel, who instead endeavours to explain heredity by an analysis of the molecules, or plastidules, as he calls them, of living matter. Life is due to their atomic structure, and "*jedes Atom besitzt eine inhärente Summe von Kraft und ist in diesem Sinne beseelt.*" Energy and soul are thus identified anew, and this having been done, all difficulties disappear. Haeckel now explains reproduction, as always, with the old definition: growth over and above the individual; heredity is a transmission of the motion of the plastidules, and adaptation a change in this motion. This certainly sounds somewhat mechanical, but some pages further on we suddenly find a new definition: "*Die Erblichkeit ist das Gedächtnis der Plastidule, die Variabilität ist die Fassungskraft der Plastidule.*" And yet still later we are told that "*das Gedächtnis*" is a transmitted motion. It would, of course, be superfluous to judge these fancies according to scientific standards; Haeckel himself admits that he got the idea of "the memory of the atoms" from Goethe's famous romance *Die Wahlverwandtschaften*, and indeed the whole plastidule theory sounds like a romance; in producing it Haeckel had abandoned himself entirely to romantic natural philosophy and there he remained for the rest of his life. The phenomenon might seem to have only a psychological interest and might be passed over with a reference to Haeckel's æsthetic turn of mind — he was, in fact, something of an artist, a gifted dilettante in water-colour painting and an admirer of beauty both in art and in nature — but it might also be pointed out that a pioneer in science may be considered justified in entertaining some strange thoughts on general problems — this has been acknowledged throughout the ages. Yet this does not explain how it was that this speculative side of Haeckel's activities should have proved capable of creating such an extraordinary sensation among his contemporaries — that people should have been so loud in their praises and in their abuse. This point demands an explanation by itself, wherefore we must cast a glance at the political and social conditions of the time.

Political radicalism of the Haeckelians

In Germany the seventies were a somewhat restless decade; the recent victories had certainly confirmed Bismarck in his power, but he nevertheless had opponents in two directions: the Catholics, whose ultramontane politics were regarded as a menace to the unity of the Empire, and the international labour movement, which had recently found expression in the communal riot in Paris, that had so scared the world, and not least Germany, where some attempts against the lives of distinguished people were placed to its account. In such circumstances the liberal-minded apprehended a further reign of terror, and the friends of domestic peace still further social upheavals. And Darwinism in particular, which indeed had from the beginning been strikingly characterized as a theory of progress, through Haeckel's boisterous attacks on the authorities of State and Church and through his dogmatic description of the contrast between the doctrine of creation and the theory of evolution, had been suspected by the conservatives and looked upon as a socially dangerous hypothesis, the truth of which, moreover, could be disputed on the grounds of the plastidule theory and similar ideas contained in it. The exchange of ideas on Darwinism became in these circumstances more and more lively; round Haeckel there gathered a crowd of young naturalists who preached the new doctrine with enthusiasm. Among them may be named A. BREHM, the author of the universally known work *Animal Life*, F. VON HELLWALD, known as a geographical writer, G. JÄGER, famous for his curious hygienic theory, and others. Since the universities were mostly closed to them, they carried on their agitation by means of popular lectures and polemical writings, in which they expounded their views, willingly associating their natural-scientific radicalism with a political radicalism, and with this party the old radicals Vogt and Büchner associated themselves. But the new theory claimed also politically conservative adherents, as, for instance, Du Bois-Reymond; he had, it is true, embraced Darwinism with enthusiasm and had declared that its appearance had freed biology from all explanations of the vexed problems of finality, but at the same time he had expressed disapproval of "Haeckelism." In his above-mentioned lecture on the limitations of our knowledge of nature he had uttered a warning against belief in a possibility of definitely solving the riddles of nature and life. Haeckel, who, it will be remembered, had nevertheless himself admitted the limitation of man's capacity for knowledge, became enraged at the word "*ignorabimus,*" in which he scented political reaction. The foreword to his *Anthropogenie* is directed against it and treats the expression entirely politically. The situation became still more tense some years later, when the Prussian Government was engaged in drafting a new educational law, the provisions of which were bound to affect the future of science in Germany. Then Haeckel came forward at a scientific meeting at Munich in 1877 with an address on the relation of the evolution theory to science in general. In it he presented his old theories, including the plastidule hypothesis, and expressed the assurance in connexion with them that biology, as conceived evolutionally, is not an exact, but a historical and philosophical, science, and as such aimed at uniting natural-scientific research with the psychical sciences and thus forming the basis for a uniform view of life, which would gradually reconstruct the whole of human existence on general humanitarian lines, and which should therefore constitute the foundations of all education.

Virchow opposes Haeckel

THIS proposal was opposed by Virchow in a speech in which he points out all that is hypothetical and unproved in Darwinism, and on these grounds he uttered a warning against incorporating it in a scheme of school education, for such a program should only concern itself with indisputable proofs.

Virchow's speech was greeted with cheers by the conservatives; as a matter of fact, its criticism of Haeckel's fantastic ideas was justified, but its pedagogical program was of doubtful value; it might reasonably be asked what would be left if all hypothesis were banned in the schools — every explanation of nature is fundamentally hypothetical, and much of the results of historical research rests, of course, upon disputed facts. And even more unacceptable sounds his passing reference to the spiritual affinity of Darwinism to socialism — a denunciation which at that time was equivalent to an accusation of high treason. Shortly afterwards the Prussian Minister of Education sent round a circular strictly forbidding the schoolmasters in the country to have anything to do with Darwinism, and in the new educational law biology was entirely excluded from the curriculum for the highest classes in the schools, with a view to protecting schoolchildren from the dangers of the new doctrines. Haeckel replied to Virchow's speech in a pamphlet, *Freie Wissenschaft und freie Lehre*, in which he again formulates the antithesis "Creation — Evolution," brings forward "certain proofs" of the correctness of the theory of descent, declares that cell-psychology can be traced to Virchow's own ideas, and finally urges the freedom of education and Darwinism's independence of the political questions of the day. His reply was hailed with enthusiasm by the free-thinkers and it is easy to realize the eagerness with which the friends of the freedom of thought and word must have gathered around him in spite of his many delusions, when such measures as the school regulations mentioned above were adopted by the opposite party. All the more so as the outcome proved Haeckel's justification; Darwinism might be prohibited in the schools, but the idea of evolution and its method penetrated everywhere, in historical research and linguistic studies, and even in the scientific treatment of religious documents and religious history. And to this result Haeckel has undeniably contributed more than most; everything of value in his utterances has become permanent, while his blunders have been forgotten, as they deserve.

Victory of Darwinism

DURING the eighties the dispute as to the justification of Darwinism died down; Haeckel himself spent most of this period in studying the Radiolaria, and his partisans likewise began to pursue other activities. Instead, that decade was to witness the undisputed domination of comparative morphology in biological research and training; it was at a time when Gegenbaur's and Haeckel's ideas universally prevailed without opposition and were applied to various groups of the animal kingdom. But the results were in no wise what Haeckel had anticipated. Instead of simple and easily comprehended proofs of the indisputable validity of Darwinism, the younger generation of scientific students found masses of involved facts, which only contributed to confuse the biogenetical principle, the gastræa theory, and the other "natural laws." This was not at all what Haeckel had expected. Self-confident by nature and spoiled by the successes of his earlier years, he was lost amongst all these developments; intensive study of detail had never been his strong point, and the minute methods and detailed observations of the young morphologists aroused his keen opposition. In a letter dating from this period he expresses the opinion that modern morphologists in general, and "*Querschnittler und Anilinfärber*" in particular, possess far less "logical schooling" than the systematists of the old school. And, as always, special research was followed by a waning interest in theoretical speculations; instead of paying attention to Haeckel's watchword — either creation or evolution — students preferred to leave the theories to their fate and to go over to practice. When, then, even Haeckel's favourite idea of man's origin from the higher apes and his affinity to the gorilla and the chimpanzee began to be doubted by scientific students, who found man to be in anatomical respects highly isolated and traced him back direct to lower mammal forms, it is not to be wondered at that the old master lost patience. He was no longer capable of controlling developments, or of obeying them; to withdraw from the struggle, which would have been the wisest thing for him to do, was more than his unbounded energies could endure — perhaps also he was too vain to do so — and so he continued the struggle on behalf of his natural philosophy, becoming, as the years went on, more and more isolated from his old friends and disciples in the world of science. In compensation he gained from another quarter a new and grateful public. The old political radicalism had died out towards the close of the century; most of the liberal party ceased altogether from offering opposition; instead the struggle was taken up with increasing success against the government authority by the socialistic labour movement, which, violently persecuted by Bismarck, sometimes counteracted, sometimes favoured by his successors, waxed stronger and stronger, until in the revolution of November 1918 it destroyed the old social order. With youthful idealism its members embraced the modern natural science; they too were enemies of the conservative State Church, which was friendly to the Government and which condemned them to show humble obedience to superiority. There was all the more reason, then, for their being drawn together by a natural-scientific explanation of the world which made progress the aim of life. To them Haeckel's monism was a welcome ally; that its cosmic view was over-simplified and falsely depicted it was not in their power to control, owing to their lack of special studies, but its founder's ardent belief in natural science and intense hatred of the State Church, combined with his oppositional attitude in politics, sounded irresistibly attractive. It is against this background that Haeckel's later scientific activity must be viewed in order that its influence may be understood aright.

Haeckel's Welträtsel

IN the nineties Haeckel returned to natural philosophy; he published one or two papers on monism and an important work on "systematic phylogeny," comprising a genealogical tree for all living beings — that is, a detailed application of his earlier, and even then somewhat out-of-date, theories. In 1899 he published his famous work *Die Welträtsel*, which was intended to be a summary of his ideas and at the same time a farewell to his activities; being a child of the nineteenth century, he wished to conclude his work with its exit — a promise that unfortunately he failed to keep. *The Riddle of the Universe* had extraordinary success; in Germany the book was sold by the hundred thousand and in England by tens of thousands; special emphasis has been laid on the fact of its widespread distribution among the working-classes, and in Japan it is said to have been used as a school text-book. Nevertheless, from a scientific point of view it must be regarded as utterly valueless. Its biological section is a rehash of the history of the creation, anthropogeny, and the monograph on the plastidule, as little attention as possible being paid to the immense progress made by scientific research since then. As a matter of fact, biology takes up only one-quarter of the volume; the rest is devoted to psychology, cosmology, and theology. The cosmological section gives evidence of the author's hopelessly confused ideas on the simplest facts of physics and chemistry; final judgment has been passed on it in a widely distributed polemical paper, which has never been challenged by trustworthy authorities, written by the Russian physicist CHWOLSON, to whom we refer those who desire to gain an insight into Haeckel's standing in regard to the exact sciences. The philosophical section of the book has been no less severely criticized by specialists on the subject; philosophers of different schools have pointed out its utter lack of clarity in point of theoretical knowledge and logic, its incapacity to define even the simplest ideas. In passing, it may be mentioned that this time "monism" is based mostly on Spinoza, the great dogmatist and repudiator of evolution, whose purely metaphysical idea of substance is at once placed on a par with the "matter" of physics. True, the real character of substance is said to be inexplicable, but, notwithstanding this, everything between heaven and earth is explained with its aid. If we add to this Haeckel's total lack of historical sense and critical judgment — his views on events and persons are derived from the simplest vocabulary of contemporary political and cultural radicalism — the final impression of *The Riddle of the Universe* will be an utterly depressing one. The cause of the book's popularity is obviously to be found in the political and social sphere. Its very introduction points in that direction, the progress in the scientific world being there contrasted with a dark picture of the political situation of the time: government, administration, courts of justice, and education are depicted as appallingly behind the times, and, above all, the Church is, of course, represented as the centre of all kinds of obscurity, superstition, and tyranny. From all quarters, both radical and conservative, the signal to open hostilities was eagerly awaited. Some years after the appearance of *The Riddle of the Universe* there was founded the Monist League, a widely ramified association formed for the purpose of working for the ideas to which Haeckel gave expression in this book and in a sequel to it, *Die Lebenswunder*. Since then it has laboured, by means of meetings, lectures, and papers, and in some circles by devotional exercises, with a view to taking the place of the ecclesiastical cult. Haeckel's colleagues, however, for the most part kept aloof from the league; only a few scientists of importance have joined it. From the side of the conservatives violent attacks were made on the league; in the Prussian Diet REINCKE, the professor of botany, made a strong stand against it, characterizing it as a menace to society and subversive of morals. This started the battle in earnest. To counteract the Monist League there was founded the Keplerbund, so-called after the great astronomer. The very name, however, proved fatal; Kepler, it is true, was at the same time a great naturalist and a devout Christian, but all the same he was so saturated with the grossest superstitions of his time that he cannot by any stretch of the imagination be held up as the ideal seeker after truth in modern times. And the Keplerbund failed no less than the Monist League to attract scientists of any weight; the latter kept more strictly than ever outside the struggle and showed on the whole — the biologists, at any rate — their sympathy for Haeckel, whose work, in spite of all his mistakes, nevertheless seemed to them to represent a struggle for enlightenment and liberty of doctrine against the constant menace of the powers of reaction.

And Haeckel certainly did maintain the radically liberal-minded standpoint of his youth undisturbed through all these changes — which it was all the more easy for him to do as he had never taken part in practical politics and therefore had not to solve any political or social problems of detail. But the shock caused him by the Great War proved all the greater on that account; the idea that the fellow-countrymen of Darwin should have sided with the enemies of Germany drove him to despair. A few more works came from his pen, among them one entitled *Fünfzig Jahre Stammesgeschichte*, with which he celebrated the fifty years' jubilee of *Generelle Morphologie*, and which testifies to his having learnt nothing and forgotten nothing. His final work, *Kristall-Seelen*, is sufficiently characterized by its title. It came out in 1917; two years later he died, his death being hastened by an accident, which delivered him from the infirmities of old age and the misery of those unhappy years.

Hartmann on Haeckel

THE well-known philosopher EDUARD VON HARTMANN, in an otherwise sympathetic character-sketch of Haeckel, describes the latter's "monism" thus:

"He is an ontological pluralist in that he conceives nature to be a multiplicity of separate substances (atoms), a metaphysical dualist in so far as he assumes in every substance two combined metaphysical principles (energy and matter); a phenomenal dualist in that he assumes two distinct spheres of phenomena (external mechanical happening, and internal sensation and will), a hylozoist because he ascribes to all matter the possession of life and soul; further, he is a philosopher of identity, a cosmonomistic monist, and a materialist." Thus the Haeckelian monism, if closely looked into, will be found to contain a little of everything. It may therefore be worth pointing out in this connexion that even more deeply elaborated monistic systems have appeared in our own day. As a matter of fact, monism as a philosophical view of life is a comparatively ancient doctrine; the neo-Platonists, who ascribed true existence only to ideas, were undeniably monists, as was also Spinoza, and so, too, Schelling and his successors, including both Goethe and Hegel. Monism based on natural-scientific grounds, however, has undoubtedly become an especially widespread conception in modern times. As one of its leading representatives may be mentioned ERNST MACH (1838–1916), professor originally of physics and then of philosophy at Vienna. As a physicist he applied himself, *inter alia*, to mental-physiological studies after the pattern of Helmholtz, but he also studied Kant's writings and was led through them into the sphere of the theory of knowledge. He thereupon felt himself called upon to create a method of scientific thinking, not as a philosopher, for he was unwilling to call himself that, but as a student of science. He will have nothing to do with transcendent spheres of thought. Through the analysis of different sense-impressions he came to the conclusion that everything is phenomenal; nothing exists in itself; the outer world consists of a series of phenomena, and the ego, the personality, likewise of a series of phenomena, which we call perceptions; the phenomena stand in a relation to one another, which is expressed by the functional terms of mathematics: one change brings about another; the phenomena inside and outside the personality are mutually interdependent. Mach denies the principiant contrast between appearance and reality; the most fantastic dream is just as much a phenomenon as a real event; he likewise denies the contrast between ego and non-ego, for both are a series of mutually interdependent phenomena. The manner in which Mach explains on these postulates such phenomena as will and thought has been much discussed by philosophers who have made a special study of the subject, and has often been characterized as lacking in seriousness: by Höffding, for instance, who points out that the elements common to physics and physiology are in Mach indefinite and mystical, like a shapeless nebula. Now, Mach, as already mentioned, claims to be only a natural scientist and to try to solve only natural-scientific thought-problems. But even as such he exposes himself to the same criticism; his biological reasoning must thus be regarded as out of date even for his age, and partly also somewhat ingenuous; he argues about evolution and heredity without taking into account the results of contemporary research, he believes in much of the old, childish animal-psychology in the spirit of the earlier Darwinism, and he speculates, like Haeckel, upon the possibility of explaining the origin of the sense-organs by means of the theory of selection. Strangely enough, he also defends the teleological explanation of nature, as far as biology is concerned — though as a provisional explanation only, until a true causal explanation is forthcoming. His references to all that teleology has achieved in arousing interest in problems and collecting facts with which to solve them may not be devoid of truth, but he certainly overlooks the confusion it has caused by inducing vitalistic explanations of nature; such, in fact, were revived under the influence of Mach, as we shall see later. Finally, with regard to his monism, it possesses, in spite of his own assurances, more philosophical than scientific interest; the practical scientist should at any rate be allowed to treat things as really existing and the changes that take place in them as having been causally effected.

Another monistic theory was set up by RICHARD AVENARIUS (1843-96), professor of philosophy at Zurich. He, too, elaborated a kind of theory of function, but in contrast to Mach he gives to its elements a material nature. His theory, which suffers from having been presented in very difficult language, has had less influence than Mach's.

Natural-philosophical theories of this kind may offer some interest as thought-experiments and besides may have their ideal value, if they give expression to the conception of life of a consummate personality. Exact scientific research, on the other hand, carves out paths of its own, its progress sometimes hindered, sometimes furthered by the different conceptions of the world, according to how they deal with existing facts. Pasteur, for instance, in the controversy over spontaneous generation, undoubtedly derived advantage from his Catholic dogmatism as against those who saw in spontaneous generation a "philosophical necessity." And his very example shows, too, how in the long run the practical utility of observations is the most conclusive criterion of their value. Those facts will last which contribute, however indirectly, towards extending man's dominion over nature, whereas the "theories of life," after surviving for a time, find a haven in the archives of cultural history, provided they are found worthy to be preserved there.

CHAPTER XV

MORPHOLOGICAL SPECIALIZED RESEARCH UNDER
THE INFLUENCE OF DARWINISM

1. Anatomy and Embryology

Development of anatomy

THERE IS NO DOUBT that the power of Darwinism reached its zenith in the seventies and eighties. By then the opponents of the earlier school had for the most part said their last word, and the younger generation of scientists who had embraced the new doctrine as yet found no difficulties in its application. Rather, efforts were made, by means of exhaustive investigations in every possible field, to collect fresh proof for it. These endeavours resulted in an extraordinarily abundant and many-sided production, chiefly in the sphere of morphology, with its various special subjects, though also in those of geography and œcology, as well as in the purely systematic sphere. In this chapter we shall give a comprehensive review of this specialized morphological research-work, which was as many-sided as it was rich in results.

Anatomy developed as the outcome of a number of investigations, the results of which were recorded in numerous memoirs. To give an account of all the valuable facts that were brought to light in the course of this ceaseless work would be impracticable within a reasonable compass; a mere list of the anatomical works that were published during that period would run into hundreds of pages. In the field of the invertebrates especially, innumerable new and important anatomical discoveries were made; hitherto unknown, or at least neglected, animal forms were now studied and often produced undreamt-of ideas for the furtherance of comparative research. Chætognatha and Enteropneusta, Tunicata and Brachiopoda may be mentioned as examples of such forms, which, though insignificant in their appearance and scope, are nevertheless interesting for their structure and development. But the Vertebrata also continued to provide valuable contributions to comparative anatomy, which, for the very reason of its morphogenetical aims, found every animal form, however insignificant, worth while investigating and examining for the circumstances of its origin and evolution. But, on the other hand, by reason of the aims they had in view, these investigations ultimately became somewhat monotonous, with the result that this line of research finally became quite unmodern and the interest began to turn in other directions. Among the investigators who compiled the results of this work may be mentioned ROBERT WIEDERSHEIM (1848–1923), a disciple of Leydig and professor at Freiburg, well known for his comprehensive work on the anatomy of the Vertebrata as well as his studies of special subjects, particularly of the bone-structure of the Batrachia, and the Swiss, ARNOLD LANG (1855–1916), a disciple of Haeckel and professor at Zurich, who wrote a widely referred-to work on the anatomy of the invertebrates and a number of monographs on various groups among the worms.

Embryology

THAT branch of morphology, however, that was specially developed under the influence of the descent theory was embryology. The biogenetical principle and its related subjects, the theories of germinal layers and the gastræa, were applied to different spheres and gave rise to ideas in many directions, besides which the new microtechnics offered a means for detailed discoveries of hitherto undreamt-of results. Embryology, therefore, proves to have been the most productive of the morphogenetical specialized spheres and attracted to it the most eminent biologists of the time.

Among these representatives of phylogenetical embryology only a few of the more important can be mentioned here. ALEXANDER KOWALEWSKY (1844–1901), an academician of St. Petersburg, worked in the spirit of Haeckel, encouraged by his commendation; his detailed investigations into the development of ascidians and salpæ covered an immense amount of detail and the same is true of his work on the development of the lancet-fish, with the result that even the ontogeny of this much-discussed animal became known. Kowalewsky was a firm supporter of the theory of the germinal layers and developed it by making contributions of his own in the theoretical sphere.

The same line of research was also followed by the two brothers Hertwig, and it led them both to make discoveries of fundamental importance and to produce theoretical ideas of a very different nature from those from which they had started. OSCAR HERTWIG was born in 1849 and RICHARD HERTWIG in 1850, the sons of a merchant at Friedberg in Hesse. They both studied at Jena under Haeckel and became lecturers there and finally professors, Oscar of anatomy at Berlin, Richard of zoology at Munich. Both carried on, each in his own subject, extensive and important activities as teachers and investigators. At Jena they worked together in the sphere of evolution in the manner of Haeckel and published a series of papers entitled *Studien zur Blättertheorie*, which dealt especially with the problem of the middle germinal layers. Here they expounded their famous "cœlom" theory, which was intended to be a universal answer to the question: "How does the two-layered embryo develop into a higher organization?" The theory takes as its starting-point the two primary germinal layers, the ectoderm and the entoderm, between which there arises at an early stage an originally structureless layer formed by immigrating cells, which is here termed "mesenchyme." The animals are now divided, in respect of their development, into two groups, dependent upon whether the mesenchyme participates in the formation of tissue or not. The former takes place chiefly in the coral animals, the flat-worms and the molluscs, in which the muscular and nervous systems are formed out of the mesenchyme, whereas in most other animal types, chiefly the Articulata and the Vertebrata, the said tissues are of purely epithelial origin and are formed out of a dual evagination of the entoderm, the inner cavity of which gives rise to the body cavity, or the cœlom. The theory was afterwards applied, after a series of special investigations, to the organic formation of different animal forms and won general acceptance at the time. It is true, His declined to accept it, but did not succeed in substituting any better explanation. Later research, however, has found this theory to be far too schematical; students have given up referring the various organs to the three germinal layers and now instead seek their origin, each separately, in so-called primitive rudiments. Furthermore, the formation of the cœlom through simple invagination has been found upon closer investigation to be far less frequent than the two brothers imagined. Their theory has nevertheless played its important part and has called forth abundant special research-work of value for all time. In the following pages we shall repeatedly find their names mentioned in connexion with valuable contributions to the advancement of biology. Among their pupils may be cited the scientific collaborators EUGEN KORSCHELT (born in 1858, professor at Marburg) and KARL HEIDER (born in 1856, latterly professor at Berlin), who together published an exhaustive summary of the knowledge of their time regarding the evolution of the invertebrates. Moreover, both have distinguished themselves as specialists, particularly in the sphere of experimental research.

During this period England was also the scene of valuable embryological research-work. Among her representatives may be mentioned EDWIN RAY LANKESTER (born 1847), a professor at the British Museum and author of a number of papers on evolution, dealing especially with the fishes and the Articulata. He especially took up for study and further elaborated the cœlom theory and has brought it to the highest point it has yet reached, having sought to base on it the classification of the animal kingdom. A very distinguished name in the sphere of evolution has been won by FRANCIS MAITLAND BALFOUR, who was born in 1851 and died, as the result of an accident, in 1882, the younger brother of the famous statesman Lord Balfour.

During his short life he found time to carry out a number of extremely important works on evolution, including a study of the evolution of the sharks and a *Treatise of Comparative Embryology*, giving an account of the evolution of the egg and the embryo throughout the animal kingdom, a work that was of unrivalled importance at the time; an application of modern genetical embryology to the whole animal kingdom and at the same time a powerful defence of the Darwinian morphology in its classical form. Balfour, in fact, definitely maintains that phylogeny is the goal of evolution, while at the same time in certain details, as, for instance, in the theory of extremity-formation previously mentioned, he adopts a dissentient attitude towards the contemporary Gegenbaur school.

Even by then the morphogenetical embryology had met with decided opposition on the part of the naturalists who desired to substitute for phylogenetical conclusions the study of function in those organs whose evolution was under investigation, and thus to give evolution a more or less physiological direction. To this group belongs the afore-mentioned WILHELM HIS (1831–1904), who was born at Basel, became professor of anatomy there, but was afterwards summoned to Leipzig, where he worked until his death. Famous both as an anatomist and as an embryologist, he paved the way for a new line of research, particularly in the field of embryology. First of all he expected to see in the evolution of the embryo a physiological process, the course of which should be so studied that each later stage of development must necessarily proceed from the immediately preceding one. The changes whereby the simple egg-cells are formed into complex organisms are, to his mind, purely mechanical; as the result of a series of flexions, fold-formations, and accretions the embryo arises out of the originally lamellate germinal layers, and its folds are in their turn produced entirely from variformed growth. Every organ possesses its given rudiments in the germinal layers and these layers thus consist of a quantity of "*organbildende Keimbezirke*"; they are therefore not indifferent, as C. F. Wolff and, after him, Haeckel declared. In connexion herewith His sharply criticizes the biogenetical principle; embryos of different animal forms are as easily distinguishable from one another as the fully developed animals; Haeckel's proofs to the contrary, both verbal and pictorial, are examined and found to be untenable, and finally the question is put: "If we possessed a complete genealogical tree, would our own or any other extant organic form be fully explained thereby?" In reply His declares that if in a case of near-sightedness it is possible to establish the fact that the individual in question has inherited the defect, little will have been gained therefrom as regards our knowledge of the character of that defect; rather, the eye's capacity for accommodation and other concomitant circumstances must be investigated for this purpose; in the same way, the physiological side of embryonic development is more important than any phylogenetical speculation. The whole of this way of thinking won but little acceptance in his own time, when research was being directed along phylogenetical lines; in the eyes of posterity, on the other hand, His stands out as precursor of the mechanical method of evolution, which has since won so many adherents and which will be dealt with in the following pages.

His, however, was by no means alone, even in his own age, in his conception of evolution. There were others who also opposed the one-sided phylogenetical line of research; among them may be mentioned ALEXANDER WILHELM GOETTE (1840–1922). He was born at St. Petersburg of Baltic origin, studied at Dorpat and at Göttingen, and finally became professor at the German university in Strassburg, where he worked for the greater part of his life. As an embryologist he was influenced from the beginning by his fellow-countryman von Baer. In his principal work, *Die Entwickelungsgeschichte der Unke*, he seeks to make the evolution of *Bombinator igneus* the basis of a purely mechanical theory of evolution, freed from both Haeckel's "*formbildende Kräfte*" and Gegenbaur's phylogenetical constructions. Starting from the old, but at the time commonly accepted, delusion that the nucleus of the egg is dissolved before fertilization, he declares that the egg is an "unorganized, inanimate mass," wherein are formed by purely mechanical forces — osmotic currents and resultant pressure-changes — the first divisional furrows, and together with them fresh nuclei as centres for the development of the new cells. Thus is explained the origin of life out of lifeless substance. Similar mechanical explanations are then invented for the formation of the germinal courses and organs. In a later work Goette deals in the same method with the stages of development in certain worm-forms. The interest attaching to these investigations lies in the mechanical conception of the embryonic development, which is not only maintained theoretically, but is also in many respects successfully applied. Unfortunately, Goette was so delighted with his theory that he let all criticism go by the board; his false conception of the nature of the egg he still maintained long after it had been proved untenable; his detailed research was extremely arbitrary and was severely criticized by Gegenbaur. His theory has nevertheless not been without its effect; his disciple Roux especially, doubtless under his influence, formulated a mechanistic conception of embryonic and organic development that received widespread support.

Another opponent of the universally current embryological conception was NICOLAUS KLEINENBERG, born in 1842 at Mittau, a disciple of Haeckel's, and eventually professor at Palermo, where he died in 1897. Of his literary production, which was small in extent but original in character, may be mentioned his monographs on the evolution of the fresh-water polypus, in which the ontogeny of this primitive animal is elucidated for the first time.

In a subsequent work on the course of development of an annelid he has propounded a curious theory of its embryonic evolution in strong opposition to Haeckel's gastræa theory and the cœlom doctrine of the Hertwigs. He starts with the sentence: "*Es gibt kein mittleres Keimblatt,*" and adduces a number of examples of how various organs that had been supposed to be mesodermal, originate directly or indirectly from the ectoderm or entoderm. At the same time he maintains that the form of one organ depends upon its function and not upon its origin; "A comprehensive tissue-system is possible only on a physiological basis." These views were at the time at which they were expressed (1886) so utterly opposed to those of his age that they scarcely caused any sensation; their time came later, in connexion with the altered view of evolution that has become prevalent in our day, which will be described in a subsequent chapter.

2. Cytology

Development of microscopical cell-research

MICROSCOPICAL cell-research is undoubtedly the branch of biology that received the greatest stimulus during the last decade of the past century, and that has seen the most important results and in many ways set its mark upon the whole of biology in general. Its highly perfected methodics, with its minute technical preparation of material for investigation, carefully adapted to suit each particular case, and its careful microscopical study of the smallest details, employing the highest possible magnifications, became a characteristic feature of the research work of that period. The purely technical side of biology thereby received an entirely new character; whereas formerly skill in dissection was the most essential qualification of the biologist, this ability now became to a certain extent superfluous, thanks to the development of the technique of microtomy. On the other hand, the student of cells, if he desires to create something new and to work independently, must acquire a chemical knowledge of the means of fixing the tissues, as well as a colour technique for the purpose of their further treatment. It was, of course, possible for the whole of this method of research to degenerate into a mere unintelligent dexterity, as biologists of the old school in particular called it, but it has also made possible more than any other method the obtaining of results that have entirely transformed our conception of the phenomena of life.

We left cell research at the point to which Max Schultze had brought it — the knowledge of the cell as a limited quantity of protoplasm with concomitant nucleus. Schultze is also remarkable inasmuch as in his cell studies he was still working without a microtome; he brought cytology to the farthest point possible with the old methods. During the period immediately succeeding his death, cell research made rapid strides in regard both to the value of the discoveries made and to the number of workers engaged in it. Limitations of space make it impossible to do justice to all the truly distinguished minds that during this period laboured for an increased knowledge of the life and structure of the cell. Some of the most prominent cytologists will be mentioned here, after which a description will be given of the most important discoveries in their field of research; the fact that their activities and the rivalry to achieve the most important results were contemporaneous would indeed render it extremely difficult to retain here the biographical method of presentation of the subject that we have followed hitherto.

Its representatives

IN the sphere of plant cytology EDUARD STRASBURGER takes the first place. He was born in 1844 of German parents at Warsaw, received his school education there, and studied partly in Paris and partly at German academies, first at Bonn and then at Jena, where Haeckel won him over to Darwinism and even procured him a professorship. He afterwards became professor at Bonn, where he worked until his death, in 1912. Equally distinguished as a research-worker and a teacher, he attracted to his institute a large number of pupils from all countries; he was a leading writer of text-books, and his scientific production included, besides his epoch-making cell-studies, a number of branches of vegetable anatomy.

Among students of animal cytology the above-mentioned brothers Hertwig take high rank; besides them there is WALTHER FLEMMING (1843–1905), professor first at Prague, then at Kiel, distinguished not only as an observer, but also as a technician and teacher. Further, HERMANN FOL (1845–92); a native of Geneva and the son of wealthy parents, he studied in Berlin and became professor in his native town and a scientist of high repute. Being specially interested in marine research, he equipped at his own expense a vessel for the purpose; in the course of a voyage, he, together with the vessel and the crew, disappeared and were never heard of again. OTTO BÜTSCHLI (1848–1920), after having studied chemistry and mineralogy, devoted himself to zoology and became professor at Heidelberg. It is possible that his earlier occupying himself with inorganic elements and processes induced in him that liking for comparison between organic and inorganic structures which characterized his later research work. Besides these names should be mentioned that of the Belgian EDOUARD VAN BENEDEN (1845–1910); the son of a highly reputed zoologist, who was especially known as an expert on parasitology, he applied himself to the study of medicine and eventually became a professor at Liége, famous as a many-sided investigator and publisher of the well-known journal *Archives de biologie*. Finally, reference should be made to THEODOR BOVERI (1862–1915), a disciple of the brothers Hertwig and professor at Würzburg, as well as the two HEIDENHAINS — RUDOLF (1834–97), a pupil of Ludwig and professor of physiology at Breslau, but active also as a cytologist, and his son MARTIN, born in 1862 and professor at Tübingen, who devoted himself exclusively to cell research. All the above have advanced their science by making valuable discoveries and important technical improvements.

Strasburger on the formation and division of cells

IN 1875 was published the first edition of Strasburger's pioneer work *Zellbildung und Zellteilung;* a third completely revised edition came out in 1880. The main problem that occupied cytological research during this period was that of the origin of the cellular nucleus. As we have seen, Nägeli had already observed the division of the nucleus, but neither his own nor other similarly extensive observations were able to possess general application. Even in the first edition of his said work Strasburger makes the nucleus of the egg-cell in the plants he investigated dissolve upon fertilization and its mass disperse into the plasm of the cell; in the latter are then formed a number of concretions, which give rise to fresh nuclei. In the third edition, on the other hand, it is asserted that examples of independent cell-formation can no longer be cited from the vegetable kingdom; fresh nuclei invariably arise through the division of older ones. This established one more of the principles of modern cytology. Even before this students had begun to observe the curious phenomena attending nuclear division in the majority of cells, but apart from these scattered observations, it was Strasburger who, as far as the vegetable kingdom is concerned, elucidated this process, which, though complicated, is now widely known and is set forth in all text-books. This process — indirect nuclear division, also called "mitosis" or "karyokinesis" — is as follows: the nucleus, having lost its membrane, concentrates its colourable contents around its middle plane, after which the latter divides itself and the two halves go each its own way and thereupon again concentrate into two daughter-nuclei. The main principles of this process were already elucidated in the above-mentioned first edition of Strasburger's book, and in the third a number of further details are given. In the field of zoology Bütschli, O. Hertwig, and Flemming during the same decade made their decisive contributions to our knowledge of nuclear division, and, besides, certain isolated details were discovered by Fol, van Beneden, and others. As a result of this research work the elements composing the nucleus were also investigated; filament substance and nucleolus, nuclear juice, and nuclear membrane were the constituents that were distinguished to begin with. Of these the first-mentioned was, owing to the part it plays in the nuclear divisions, the object of greatest attention, and especially on this subject Flemming's studies of the cells of amphibious larvæ were conclusive.

It was he who ascertained after detailed study the process of nuclear division and actually gave to its various phenomena the names that have been in use since then; owing to its strong colourability he called the filament substance "chromatin" and non-colourable substance, which also appears upon division, "achromatin"; the names of the different phases of division — spirem, aster, metakinesis, dyaster, and dispirem — were also invented by him. He also proved the conversion of the chromatin from a network into a convoluted filament and further into a number of bent staves, and proved that the actual division consists in these latter's splitting along their length. These chromatin staves were afterwards called by WALDEYER "chromosomes" and have, as is well known, come to play a decisive part in modern heredity-research. The processes of the fusiform achromatin filaments during division were also studied by Flemming; it was not until later that the minute central body, the centrosome, which is of such vital importance for their transformation, was investigated, primarily by Flemming and Boveri, who together with van Beneden discovered its division in cell-reproduction. Boveri's studies of the centrosomes especially were very intensive and have proved to be of fundamental importance. This formation has been characterized by him as the cell's dynamic centre, which facilitates the nuclear and cellular division. He also discovered that the centre of the spermatozoon is formed thereby.

While the nucleus has been found in the different forms of life to represent the conservative element — the chromosomes are, as is well known, equally numerous and similarly formed in all cells in the same individual — the protoplasm of the cell and its many and various derivatives have offered fresh problems, owing to their wealth of form, which has proved all the greater, the more these formations have been investigated. The actual basic substance, which still has to bear the clumsy and illogical name of protoplasm, has been investigated by a vast number of students and has called forth many attempts at an interpretation of its essence. These attempts have for the most part concentrated upon three different theories based on observation, which have been named after their founders: Bütschli's froth theory, Flemming's filament theory, and Altmann's granule theory; to say nothing of the purely speculative attempts to discover the fundamental substance of life. The chief difficulty that revealed itself in these explanations and that brought out their mutual contradictions is actually caused by the inconstancy which the living protoplasm always displays and which is a necessary consequence of its role as bearer of all the metabolism in the cells and the organisms composed of them. Even the nucleus displays phenomena of substance-renewal and it has been found that the vital manifestations of the cell ultimately receive their impulses from that quarter, but it is in any case in the plasma substance that these manifestations of life are essentially expressed. And they are as much of a chemical as of a physical nature; the physical phenomena — movement of various kinds — are invariably induced and brought about by chemical reactions and in their turn produce new reactions.

Plasma theories

BÜTSCHLI's froth theory is an essentially physical attempt to explain the structure of protoplasm. He certainly repeatedly points out the chemical reactionary phenomena of the cell, but he pays little attention to them. Taking as his basis the strongly vacuolized substance of the lowest protozoa, especially of the amœbæ, with the current-phenomena visible therein, he conceives the living protoplasm as a fluid mass identical in its structure with the emulsion that is obtained when oil and soda-solution are shaken together. This purely mechanical emulsion-theory he afterwards elaborated after making a series of experiments of a very ingenious character. Through the mixture of variously composed liquids both he and a whole school of investigators after him succeeded in imitating in a surprisingly natural way a great many of the most complicated movements and structures of the living cell-substance. It cannot, of course, be denied that the mechanical phenomena which were found in these experiments to cause the movements in the given substratum may also be capable of asserting their influence upon the plasma movements, but as a reproduction of the phenomena of life these experiments possess the fundamental fault of entirely disregarding the chemical reaction that is incessantly going on in living substance; the mobile oil-emulsion remains chemically what it was, whereas a creeping amœba is continually changing its chemical composition, so that movement and chemical reaction are indissolubly dependent upon each other. In connexion herewith we find also the belief, which has proved unsatisfactory from the very beginning, that the fundamental substance of life is fluid — a theory that has been considerably revised by modern colloid chemistry, of which we shall have more to say presently.

Flemming's plasma theory undeniably takes more account of chemical conditions. According to this theory, protoplasm consists of a network of fibres embedded in a homogeneous substance. These structures he found particularly in the cellular mass in various tissue-elements: in egg-cells and in cartilaginous and glandular cells in higher animals. He believes the phenomena of metabolism in the cell to be accompanied by changes in the filament mass and in the basic substance, which should be examined in detail in different subjects. The threads may sometimes be dissolved into canals and vacuoles and thereby convey the assimilation products not only to different places within the cell, but also between various cells, for these latter are in most cases demonstrably connected with one another by bridges of filaments. Thus the cells become the structural elements in the body, though also elements incorporated in one and the same vital unit; their independence need not be over-stressed.

In opposition to these two theories, which belong to the eighties, there appeared somewhat later the granule theory of Altmann. RICHARD ALTMANN (1852–1901), professor at Leipzig, devoted his attention chiefly to the fundamental substance in which the above-described network of plasm lies embedded, and with the aid of suitable colouring-matter he found in it a mass of grainlike formations — the Latin *granula* — of different kinds in different cells. In these he sees the true substance of the cell; he even finds that the threadlike structures which can be produced by Flemming's method are composed of similar granular formations. Indeed, many of his observations have been confirmed; in the glandular cells especially, the forthcoming secretion first appears in the form of homogeneous granules, which gradually increase in size and assume the form of drops. In most other cells, too, such granular structures appear as expressions of the cell's change of substance; as in nerve- and muscle-cells, of which we shall have more to say later. Altmann, however, sees far more than this in these granule formations; he calls them bioblasts and considers them to be the true elementary organisms of which cells and tissues are composed, just as bacterial colonies are composed of various bacteria. He even believes these protoplasmic granules to be of equal value to micro-organisms and would make this his contribution towards the solution of the riddle of life, a contribution that he further supplements by finding a resemblance between bioblast and crystal; these two are in fact compared, though hypothetically. These fantastic ideas have naturally been given but little support; Altmann is on firmer ground, however, when he emphatically states that the living substance must be solid and not liquid — an assertion that he bases upon his granule theory in opposition to Bütschli's above-mentioned experiments and speculations.

These granular structures of the cell-substance have, as a matter of fact, been studied by numerous later investigators, who have given them innumerable names: "mitochondria," "chondriosomes," etc. They are brought to light by the use of special colouring-methods, but in favourable circumstances they may also be visible in the living subject, which justifies the assumption that they are not pure artificial products. The same, indeed, is true of the other two plasmic structures: the fibre and the froth structures. M. Heidenhain rightly points out the possibility of all three structural forms' existing in one and the same cell. But this would also go to show that none of the structural theories is capable of forming the basis for a uniform conception of what living matter is really composed of. Heidenhain therefore holds that the common structure of the living plasm must be sought beyond what is microscopically visible — that it consists in a system of minute particles that possess the qualities of life, principally that of multiplication by fission, and that build up those structures of which the cell is composed. These ultra-microscopical particles, which therefore cannot be observed, but the assumption of which he considers to be an incontrovertible necessity, he calls "plasomes." It is not worth while going further into his speculations; it will at once be realized that we here have a name for Haeckel's plastidules, Darwin's gemmules, and innumerable similar ideas — unknown quantities that can be used neither for the purposes of observation nor for theoretical calculation and that are therefore automatically eliminated from the problem of life. True, ultra-microscopical technics have since given us some insight into the composition of the living substance over and above what the microscope has been able to provide, but no one has succeeded in isolating any vital unit in this way, and up till now the cell, with all its complications, remains the smallest form under which the living substance has been found to exist by itself and independently of other living entities. Of undoubtedly greater value have been those facts in regard to the composition of the cell that have been contributed by modern chemical research, which will be discussed later.

While, then, the fundamental substance of the cell has remained in its innermost essence undiscovered, careful and extensive studies have been devoted to the mass of cell products of which the bodily tissues are built up. Of the pioneer research-work in this field may be mentioned the investigations of the elder Heidenhain into the glandular secretion in man and the higher animals, as a result of which light has been thrown for the first time especially upon the microscopical structure of the salivary glands and the relation between the composition of their cells and the nature of their secretion. In his footsteps followed Flemming, Altmann, the younger Heidenhain, and many other cytologists, who observed and compared the different phenomena in the epithelial cells, both the secreting and the resorbing, in the various organs of the body and the cells covering the surface of the body. In this sphere the study of the origin and development of the granular formations has been most intensive.

Nerve investigations

ONE field of inquiry that has especially occupied the attention of modern cytology is the nervous system. Its highly complicated structure long resisted all attempts at an explanation, until methods were discovered whereby it is possible to colour only certain special elements, which can thus be examined in their entire length. These "elective" methods include impregnation with metallic salts, which has been applied in various forms by the Italian CAMILLO GOLGI (1844–1926), professor at Pavia, the Spaniard SANTIAGO RAMÓN Y CAJAL, born in 1852, professor at Madrid and an unusually thorough expert in the elements of the nervous system throughout the animal kingdom, the author of a number of papers on the subject, as well as the monumental work *Histologie du système nerveux*, and the Hungarian STEFAN APATHY (1863–1923), professor at Koloszvár. Another method that was productive in this respect has been the intravital methylene blue-dyeing method, which was discovered by PAUL EHRLICH (1852–1915), disciple of Koch and principal of the laboratory of hygiene at Frankfurt am Main, and which has been further applied especially by A. BETHE, professor at Strassburg. Others who have studied the nervous system include the aged Kölliker, the Frenchman LOUIS ANTOINE RANVIER (1835–1922), professor at Paris and active worker in many branches of cytology, and also GUSTAF RETZIUS (1842–1919), son of the above-mentioned anthropologist and early in life an assiduous worker in this sphere. Neurological research has to a certain extent sought to ascertain the structure of the actual nerve-cells and their internal modifications during different stages of activity; as expressions for the physiological condition in the protoplasm of these cells have been characterized the granular formations amassed in stages of rest and disappearing upon irritation, which are called tigroid substance, owing to their appearance, or "Nissl's granules" after their discoverer, FRIEDRICH NISSL, hospital doctor at Frankfurt am Main (died 1919). Still greater interest, however, has been devoted to the problem of the connexion between the nerve elements, which indeed is of vast importance also from the physiological point of view. In this field there have been two mutually opposed theories. Even His had observed that there grow out from the embryonic nerve-cells threads, which become longer and longer. Later on, Kölliker, Cajal, and Retzius, among others, held the view that these threads give rise to the nervous fibrillæ and that the nervous system is thus formed of a number of mutually independent elements, consisting of a cell with its concomitant nerve-thread and connected with its neighbours only by contact. In opposition to this view, Apathy in particular has maintained that the nerve-thread is formed of a whole series of cells and that its ramifications extend not only up to, but also into, the plasm in the ganglion-cells. The conflict between these two lines of thought was at one time quite lively, but apparently died down without either party's being able to claim a decisive victory.

Muscle investigations

BESIDES the nervous system, the musculature early attracted the attention of the cytologists, especially the cross-striated musculature, the complex structure of which had long withstood all attempts to interpret it. WILLIAM BOWMAN (1816–92), professor of physiology in London, was the first to make any weighty contribution towards the solution of the problem. In a treatise printed in 1840 he describes how the muscle is composed of fibrillæ, surrounded by a substance that he calls sarcolemma, and how the fibrillæ are divided crosswise into laminæ of various degrees of density. During the time that has elapsed since then, muscular histology has had many students.

The most important progress is coupled with the names of ALEXANDER ROL-
LETT (1834–1903) and THEODOR WILHELM ENGELMANN (1843–1909), both
professors of physiology, the former at Graz and the latter first at Utrecht
and afterwards in Berlin. Both of them have done service in ascertaining
the regular sequence of the cross-stripes in the muscles. Rollett is responsible for these formations' being denoted, as they still are, by letters. Engelmann, a disciple of Gegenbaur and a distinguished investigator in many fields
of research, made a special study of the physical qualities of muscle — the
condition of the various elements in normal and polarized light, upon contraction and relaxation. These results led to a one-sided physical view of
muscular action, which was still further advanced by Helmholtz's and other
physiologists' investigations into the mechanics of muscular action. On the
other hand, EMIL HOLMGREN (1866–1922), professor of histology at Stockholm, Sweden, held a more morphological conception of the muscular
process; by careful experimental and microscopical studies of the granular
formations which, thanks mostly to G. Retzius, were already known, which
are situated between the cross-sections of the various fibrillæ, he discovered
that the granules are the organs which bring about the change of substance
in the muscle during action; his views were accepted and elaborated by
AUGUSTE PRENANT, professor at Paris and well known as an unusually many-
sided cytologist and author of that both extensive and intensive work entitled *Traité de cytologie*. We can deal only briefly with the various categories
of supporting tissue — connective tissue, cartilage, bone. In this field of research a number of investigations, important from the point of view of principle and masterly in their technique, have been carried out by, *inter alia*,
Ranvier, Flemming, STUDNICKA, and the Dane F. C. C. HANSEN; these have
discovered especially the origin of the categories of supporting tissues and
their transitions into one another.[1]

Discovery of fertilization

UNDOUBTEDLY the greatest service to biology that has been performed by
modern cell-research, however, is its having given us our present knowledge
of the course and significance of fertilization — a discovery worthy to be
placed by the side of the explanation of the circulation of the blood in the
seventeenth century. If, however, we compare the course of these two great
achievements in the field of research, we get a striking impression of the contrast that exists between scientific activities nowadays and those of a couple
of centuries ago. On the one hand, Harvey, who spent twenty years or so
quietly and peacefully examining the idea that had been kindled in him in
his youth, and who afterwards submits it to the world in its perfect and

---

[1] Accounts of the development of cell research in more modern times will be found in
Prenant's above-mentioned work, in M. Heidenhain's *Plasma und Zelle*, and in other histological
text-books, to which the reader is referred.

consummate form. On the other hand, a number of scientists of today, some of them admittedly of the highest rank, working all at the same time in mutual rivalry at the problem of fertilization, expound their results practically every year, often in a half-finished state and sadly in need of emendation, and then disputing each other's claim to the honour of having produced the various details of the discovery, each one seeking to interpret according to his own lights statements that are often found to have been originally formulated as mere assumptions and suggestions. The account of how our knowledge of the origin of individual life was finally acquired can therefore hardly be so attractive a task as that of describing Harvey's lifework.

That science had so long to wait before the phenomena of fertilization were fully elucidated is, of course, primarily due to the fact that the knowledge of its basis, the cell, was for so long incomplete. And in particular the idea as to the nature of the nucleus of the cell was, as O. Hertwig so weightily observes, still extremely vague as late as in the seventies; a cystic, homogeneous formation was seen in the middle of the cell, and no really clear idea was obtained as to its meaning. It was supposed to have been observed that on certain occasions this formation disappeared and that this was particularly so within the egg-cell; moreover, it was postulated by Haeckel's biogenetical principle, according to which every living being arises out of an entirely undifferentiated mass of plasm. It was thought to be probable that spermatozoa, one or several, penetrate the egg upon fertilization, but the part they played in the process was utterly vague; on the whole, it was deemed sufficient to assume some kind of chemical or physical influence upon the egg-cell, whereby its stages of segmentation, which had already been studied, were produced. Indeed, the phenomena of nuclear division, referred to above, had been partially investigated in the case of egg-cells; Fol particularly had observed radial phenomena accompanying division, and Bütschli the actual nuclear pole, but no one had as yet gained any clear idea of the process.

In 1875 O. Hertwig spent some months by the Mediterranean Sea and there discovered an object particularly suitable for studies in fertilization and egg-development in the sea-urchin, whose eggs are transparent, occur in large numbers, and are rapidly developed. The results he obtained from his investigations of this material he recorded in a dissertation written for the purpose of obtaining a lectureship at Jena. Among the theses that accompanied the paper, according to the German custom, the first runs as follows: "*Die Befruchtung beruht auf der Verschmelzung von geschlectlich differenzierten Zellkernen.*" This statement, upon which further light is thrown in the paper itself, really contains the essence of our modern theory of fertilization. There is indeed still another of these theses that is of importance: the assertion that the egg does not pass through any "monera" stage — a statement directly conflicting with the master's (Haeckel's) theory. Otherwise this work is in many parts somewhat incomplete; thus, the so-called pronucleus in the egg is supposed to be developed out of the germinal spot in the ovarial egg, while the rest of the germinal vesicle is assumed to have disappeared — this being stated to apply to the entire animal kingdom. No polar bodies have been observed, nor has it been possible to prove the penetration of the spermatozoa into the egg. On the other hand, what is described and illustrated is how two nuclei in the egg gradually approach one another and coalesce, one of which comes from the extreme part of the egg-cell and is therefore characterized as the nucleus of the male sexual product. As a matter of fact, Bütschli and others had previously observed two nuclei unite in the fertilized egg, but had not utilized their discovery for the purpose of a general interpretation; simultaneously with O. Hertwig, van Beneden had published an account of certain fertilization-phenomena in the egg of the Mammalia and had therein expressed the view that, of the two nuclei which he also had observed in the newly-fertilized egg, one is of male and the other of female origin, but he made this statement under reserve as being only a hypothesis, "which may be accepted or rejected." The principle that fertilization consists in the union of the male and the female nuclei was thus without any doubt first set forth by Oscar Hertwig; he was the first to realize the significance of the phenomena and he therefore deserves all the honour for it.

Our knowledge of fertilization thus made slow progress, with the collaboration of different investigators. Fol was the first who actually saw (1879) the spermatozoon penetrate the egg, thereby establishing what O. Hertwig had already concluded, that one single male cell performs the act of fertilization. The latter scientist followed up his studies of fertilization and gradually succeeded in arriving at a clearer view of the subject; in a work on the fertilization of the worms he gives an account particularly of the expulsion of the polar bodies; these bodies, which have already been described by Sven Lovén — and possibly still earlier by the aged Carus — were now found to arise through indirect nuclear division, but were still regarded by Hertwig as one of the more incidental phenomena of fertilization. It was not until later that he discovered them in his first subject of investigation, the egg of the sea-urchin. The next great step towards a solution of the riddle of fertilization was taken by Flemming, who in 1879 established the longitudinal cleavage of the chromosomes in indirect cell-division, which was afterwards confirmed by Retzius and Strasburger. In 1887 van Beneden published the results of investigation into the fertilization of the lumbrical ascarid worm of the horse, *Ascaris megalocephala*, well known on account of its few but large chromosomes. In this animal he found, and afterwards established in other quarters also, the important fact that every animal has an equal number of chromosomes in each cell. In connexion therewith he discovered the reduction in the number of chromosomes in the sexual cells: that upon the latter's maturation-division the number of chromosomes in both the male and the female elements are reduced to half of the normal number, which is again restored upon fertilization, when the male and female chromosomes are united. Somewhat later KARL RABL (1853–1917), professor at Leipzig, detected the individuality of chromosomes: that in a cell every chromosome originates from a given chromosome, like itself in form and size, in the mother cell. And finally, in 1902, the American W. S. SUTTON discovered the so-called accessory chromosome, which at the nuclear division assumes a place for itself. All these facts have played a decisive part in modern heredity-research and will be further developed later on in this work.

As a result of these investigations into fertilization,[2] very briefly referred to above, our knowledge of the phenomena of life was so considerably enhanced that it is difficult to overestimate its value; this not least because the same fertilization-phenomena were established in the vegetable at the same time as in the animal kingdom: the union of male and female nucleus, the reduction of the chromatin, and the individuality of the chromosomes; all these processes take place with a certain number of modifications, but on the same principle in every multicellular organism. Life has thereby been given a uniformity far more demonstrable and real than the hypothetical common descent of Darwinism. Even in the lowest unicellular organisms, whether they belong to the animal or the vegetable kingdom, a similarity in their evolution has been definitely established. We shall now proceed to discuss these forms.

3. Microbiology

THE Darwinists of the earlier school, chiefly Haeckel, largely interested themselves, as we have seen, in the very lowest animal forms; it was expected that they would produce fresh ideas in regard to the origin of life upon the earth, discoveries that would fill the gap between living and lifeless substance and would thus make the great evolutional series in the universe entirely uniform. These expectations, however, whether associated with Huxley's bathybius slime or with Haeckel's Monera, have not been fulfilled; bathybius turned out to be a lifeless calcareous deposit, and in the Monera have been found nuclei and other organic details giving evidence of ordinary

---

[2] In an article entitled *"Dokumente zur Geschichte der Zeugungslehre"* (in *Archiv für microscopische Anatomie*, Bd. 90), O. Hertwig has given a comprehensive account of the history of fertilization-research up to the year 1917.

cell-structure. Indeed, the cellular structure of these lowest organisms has proved to be highly complex, in many of them competing with the fundamental elements in the highest organisms. Thus there remains nothing to be done beyond widening, by strenuous intensive research, our knowledge of these beings, whose vital manifestations have in many respects proved of service in answering questions of the greatest theoretical interest, not to speak of the important practical problems that have been solved through an extended knowledge of the subject.

Protozoa

OF the unicellular organisms, the Protozoa have, as we know, been longest known; the earlier progress made in this field of research has been described in a previous chapter. Of those who have worked at the subject at a later period the most conspicuous and successful investigators and distinguished teachers were Bütschli and Richard Hertwig. The Protozoa have proved to possess a wealth of different forms and structures, both in protoplasm and nuclei, which has provided the science of general cytology with an invaluable material for purposes of comparison. Bütschli's investigations into their plasma and the changes that take place therein in different stages have already been mentioned above. Earlier naturalists were inclined to see in the Protozoa radically undifferentiated plasm, but this assumption has been utterly disproved by experience; on the contrary, the higher unicellular animals possess a great number of plasmic formations of a markedly organic character — cilia and flagella, vacuoles and muscular fibrillæ. And their vital manifestations have after careful investigation been proved to exist in undreamt-of numbers; their irritability and way of reacting to different impressions offer an important field for experimental biology to investigate.

The nuclei of the Protozoa have been of special interest owing to their immense wealth of form, to which the higher animal cells have not attained, and owing to their correspondingly numerous functions. It is in this sphere that R. Hertwig has made his most important contribution to the advancement of biology. To start with, he has demonstrated that the nucleus in the Protozoa contains the same constituents as the cell-nucleus in general — chromatin, linin, and nuclear body. Moreover, the chromatin substance is in many cases found to be divided up in the cellular plasm in the form of granules or a network, and sometimes the nucleus is entirely incorporated in this latter — a phenomenon that is reminiscent of bacterial chromatin's being invariably distributed over the cellular body, and that at the same time explains part of Haeckel's Monera. Upon the presence of the nucleus depends the Protozoa's capacity for assimilation; a fragment of such a cell without the nucleus would perish for lack of metabolism. Of still greater importance is the condition of the nucleus in the propagation of Protozoa; since this generally takes place through division, the process is started by the nucleus, which then displays a number of different phenomena in different forms: on the one hand, direct division through interlacing; on the other hand, a regular mitosis, with a division of the centrosome and spindle-formation; and, between the extremes, a number of transition forms. After the division the nucleus and plasma grow at different speeds until a certain ratio of bulk arises between them; then a fresh division takes place. This "nucleus-plasma relation," as R. Hertwig calls it, is thus a decisive factor in reproduction and not, as formerly supposed, a growth beyond the normal standard, for this can vary considerably even in the same species. Still more remarkable are the phenomena that R. Hertwig discovered upon the conjugation of the Protozoa — in the fusion of two individuals which precedes division in certain circumstances. In many Protozoa there exists, besides the ordinary large nucleus, a small nucleus, called the micronucleus, which, previous to conjugation, divides itself twice; three of the divided nuclei perish, while the fourth unites with the corresponding nucleus in the conjugating neighbouring cell, whereupon out of the unifying product fresh nuclei are formed in the cells, whose large nuclei meanwhile disintegrate. In the three disappearing divided nuclei R. Hertwig has seen the equivalent of the polar bodies in the eggs of higher animals, while the likewise moribund large nucleus has been held to correspond to the body in a higher animal, which dies, whereas the sexual cells, here equivalent to the conjugated small nuclei, reproduce the life-form. Whether or not these comparisons may perhaps have been carried too far, the future must decide; it is certain that through them a number of vital phenomena of general interest have been viewed in an entirely new light, and the uniformity of the fundamental phenomena of life has received further confirmation.

Bacteriology

Of even greater importance, however, has been the progress made in the sphere of bacteriology; it was during this period that light was thrown upon the part played by bacteria as producers of disease and that their biology was discovered. Theories had long been in circulation that minute living "seeds of disease" were the causes particularly of the great plagues; such a hypothesis had been set up during the Renaissance by the Italian physician GIROLAMO FRACASTORO (1483–1533); Linnæus, it will be remembered, had embraced similar ideas; these theories had been encouraged by the discovery of yeast-fungi in the eighteen-thirties; Henle had been specially interested in "parasites" as producers of disease, and as a proof of his assumption of such a cause of disease he had formulated the principle: constant existence, isolation from foreign interference, reproduction of the form of disease by means of the isolated parasite. These conditions, however, were found to be difficult to fulfil; even Pasteur, who was nevertheless the founder of modern bacteriology, had not succeeded in finding a means of safely isolating the micro-organisms that were to be examined and thus obtaining pure cultures of them. He had certainly adhered to the idea of constancy of species in the micro-organisms, but other investigators of the highest reputation had maintained in contrast thereto the "pleomorphism" of these beings — that one form could pass unrestrictedly into others of an entirely different nature; this had been the view of LISTER, the famous inventor of the antiseptic bandage, as also of the well-known botanist Nägeli, for reasons which will be mentioned later on. It was in these circumstances that Koch made his important contribution to the development of bacteriology.

HEINRICH HERMANN ROBERT KOCH was born in 1843, the son of a miner in the Harz mountains; he studied at Göttingen under Henle and became a district doctor in a provincial town in Posen. In that district there was a serious outbreak of anthrax among the cattle, and the young doctor was thus faced with the problem of this disease. At an early date a French physician, CASIMIR JOSEPH DAVAINE (1812–82), a practitioner in Paris, had discovered small stick-shaped formations in the blood of animals affected with anthrax and through experiments had found them to be producers of the disease, but had not succeeded in ascertaining their course of evolution and method of distribution. Koch took up the problem for fresh treatment; after victoriously struggling against the difficulties that a provincial doctor always experiences when he proposes to carry out experimental research-work, he succeeded in elucidating the entire evolutional history of the anthrax microbe; how, when introduced into the blood of an animal, it propagates by repeated division on a vast scale, and then, when the animal has died, these microbes are converted in favourable circumstances into spores possessing great powers of resistance to external influences and the ability, after migrating into a fresh animal host, to start the process of evolution all over again. Koch's genius in these experiments lay in the simple and yet extremely effective technique that he worked out; indeed, it was in this sphere that he afterwards won his greatest successes. The anthrax microbe was at first cultivated in a damp chamber in serum, but Koch soon invented the method of planting bacteria on a gelatin solution; on this substratum, which could be made solid or liquid at will by a slight alteration of temperature, it was easy to isolate the bacteria and produce absolutely pure cultures. The method, which in its simplicity is one of the most brilliant inventions of modern times, has been the foundation on which the whole of present-day microbe-research has since then developed. But, in addition to this, Koch introduced the aniline-dye method into the study of bacteria, a method which since that time has been perfected in many ways and one whereby innumerable, otherwise invisible micro-organisms have been discovered and described; and, furthermore, he invented the microscopical illuminating apparatus, constructed to his order by the physicist ABBE, of Jena — nowadays an indispensable aid to all who work with strong magnifications.

Koch's first discoveries won him immediate fame; he was elected a member of the Kaiserliches Gesundheitsamt in Berlin, whose leading force he at once became, and had munificent sums placed at his disposal. During the period that followed, he was responsible for two great achievements — the discovery of the tubercular bacillus and the cholera microbe. In the case of the former he tried to produce a specific cure in tuberculin, but, as is well known, without success. Extremely valuable, on the other hand, was his reform of the technics of disinfection: the abolition of the earlier ineffective means of disinfection, such as fumigating with sulphur, and spraying with carbolic, and the substitution of new experimentally tested and therefore effective methods.

Koch's pupils

KOCH's activities included also the training of a host of pupils; from all countries there flocked to his laboratory students, who have since diffused his methods everywhere. Among these may be named F. J. S. LÖFFLER (born in 1852), the discoverer of the microbes of diphtheria and swine-fever, and EMIL BEHRING (1854–1917), professor at Marburg, the founder of serum therapy. In his later years Koch himself was the accepted authority on everything concerning problems of infection, and he undertook many voyages, especially to the tropics, with a view to investigate infectious diseases existing there and to try to find a cure for them. Of a despotic disposition and spoiled by his early successes, during his last years he did not always take into account the most recent discoveries, nor who had made them, which sometimes resulted in disputes that proved of little benefit for the advancement of science. He laboured, however, up to the last, in spite of impaired health; he died in 1910 of paralysis of the heart.

While thus the disease-producing micro-organisms were giving rise to an entirely new branch of research, the yeast-fungi, which were allied to them, likewise became the object of close study. The pioneer in this field, next to Pasteur, was the Dane EMIL KRISTIAN HANSEN (1842–1909). Born of a working-class family, he became at first a secondary-school teacher, matriculated when he was near the age of thirty, and afterwards applied himself to the study of chemistry and biology. When the famous Karlsberg laboratory was instituted by the brewer Jacobsen, of Copenhagen, Hansen became its leading force, and, in compliance with the wishes of the founder, devoted himself entirely to the study of the fermenting process. In this sphere he created what has ever since been the accepted technology of the subject; in particular, he perfected the pure cultivation of the yeast-fungi by an ingenious method of isolating a single specimen of these organisms, which occur in masses and are only visible under the strongest magnifications; by allowing this specimen to reproduce itself there came into being a "pure line" of yeast-fungi, possessing fully controllable characters. This technique of yeast-cultivation has entirely reformed the brewing industry and the manufacture of yeast. Moreover, Hansen has added much of value to our knowledge of the enzymes, which play an active part in fermentation phenomena, and a number of other kindred manifestations. A new conception of the process of fermentation has been produced by EDUARD BUCHNER (1860–1917), professor of chemistry at Berlin. He has proved that the alcoholic fermentation of sugar is not, as was hitherto believed, caused directly by vital action on the part of the yeast-fungi, but that these organisms produce a chemical ferment which brings about the yeasting and which can be isolated and made to function even in the absence of the fungi. As a result of this discovery the classical fermentation-theory set up by Pasteur has been considerably modified and has been transferred from the sphere of biology to that of chemistry. A number of other phenomena in the same category will be discussed later on in this work.

Of far later date than the knowledge of bacteria and yeast-fungi is our knowledge of another group of organisms, which are usually referred to the animal kingdom and which have been found to resemble bacteria in being producers of disease — namely, the Sporozoa. Even Meckel was aware that the well-known disease "ague," or malaria, was accompanied by a peculiar darkening of the blood corpuscles and of certain other tissue elements in the infected subjects. But the disease itself was considered to be "miasmatic" — that is, due to poisonous vapours emanating from the humid districts in which it occurs. Then the French army surgeon ALPHONSE LAVERAN (1845–1922), while serving in Algeria in 1880, discovered that the said pigmentation is caused by a parasite which occurs under various forms, but which, owing to its mobility, he thought belonged to the animal kingdom. His accurate description of the newly-discovered producer of disease was worthy of the closest attention, but it threw no light on the causes of its distribution. This point was definitely answered by the Englishman SIR RONALD ROSS, who was born in 1857 in India, where he was serving as an army surgeon, and who was afterwards elected to a professorial chair at an institute of tropical diseases at Liverpool. He discovered the alternation of generation in the malarial plasmodium: how, after developing in the human blood, it is absorbed by blood-sucking mosquitoes, is conveyed from the mosquito's intestinal canal into its salivary glands, and thence passes into the blood of human beings, who thus become likewise infected. An important contribution to the problem of malaria has been made by the Italian GIOVANNI BAPTISTA GRASSI (1854–1925), professor of zoology at Rome, who made his name especially on account of the effective measures of protection he adopted against malaria in his own country, which was so terribly ravaged by that disease; by excluding the mosquitoes from human dwellings and eradicating their larvæ, he has succeeded in making inhabitable districts that were formerly dangerous to live in, and, further, his exhaustive studies of the biology of the malarial mosquito have made it possible for other countries also to take energetic measures for its extermination.

A number of other parasites have latterly been discovered and described, as, for example, the Flagellata, which produce in tropical Africa the fatal sleeping-sickness, which is transmitted from one person to another by the tick; and, further, the producer of the cattle-plague, also an African disease, transmitted by the tsetse-fly, which had made cattle-breeding impossible in extensive districts. These parasites were specially studied by Koch and his pupils.

To the beginning of our century belongs the discovery of *Spirochæte pallida*, the carrier of syphilitic infections, one of the most dangerous enemies of man. It was discovered by FRITZ SCHAUDINN, who has thereby ensured for himself a place in the cultural history of the world. Born in 1871 in East Prussia, he studied at Berlin, and after taking his doctor's degree he was given an appointment at the Gesundheitsamt. Labouring under constant difficulties and in frequent dispute with the old despotic Koch and other bureaucrats in the Civil Service, who neither would nor could appreciate the value of his ideas, he worked his way up to a brilliant reputation as a microbiologist. It was not until shortly before his death that he received the permanent post worthy of him as head of a research institute at Hamburg. He made valuable contributions to our knowledge of the life of the malarial parasite; by means of experiments upon himself he studied the dangerous *Amœba histolytica*, the producer of a serious form of intestinal catarrh — an experiment that cost him his life. He also published the valuable results of his researches into the reproduction of the Foraminifera and Heliozoa. The above-mentioned discovery of *Spirochæte pallida* he made in the year before he died. Moreover, he had a number of distinguished pupils, as, for instance, M. HARTMANN, born in 1876, who took up for further study his theoretical research-work on the Protozoa, and S. PROWAZEK (1875–1916), who continued his work on the disease-producing Sporozoa.

4. Vegetable Morphology

Development from romanticism to exact investigation

IT is necessary to take a brief glance at the method of morphological research as applied in the sphere of botany, especially in view of the part played by plants as a basis for modern evolutional theories. For this purpose we must go back to the period before the appearance of Darwinism, when romantic idealism still prevailed in biological research-work. At that time Goethe's spiral visions and metamorphosis theory were still playing their part as foundations on which various naturalists based their conceptions of the form and growth of plants and the position and development of their leaves. These imaginings produced appalling confusion in ideas and theories. "It is remarkable," says Sachs in his *Geschichte der Botanik*, "that as soon as there was any mention of the metamorphosis of plants, even gifted and clever men gave way to nonsensical gibberish." But even clear-thinking investigators sought to solve these problems from a purely ideal point of view; the position of the leaves of the plants was created by an idea which expressed itself in mathematically formulated relations between the leaves. KARL FRIEDRICH SCHIMPER (1803–67), for the greater part of his life a private scholar, was one of these speculative plant-morphologists; he expressed the "spiral tendency" in the position of the leaves by means of a serial fraction. His ideas were further developed by ALEXANDER BRAUN (1805–77), who studied at Munich, among others under Schelling, and finally became professor of botany at Berlin and a distinguished teacher. Among his disciples was Haeckel, who highly admired him and was largely influenced by him in the romantic direction. Braun, who was otherwise a specialist of some merit, recorded his morphological speculations in a treatise *Über die Verjüngung in der Natur*, a curious blend of exact knowledge and romantic imaginative thought. The work contains a number of, for the time, excellent studies of lower plants, especially unicellular Algæ, the growth and reproduction of which are carefully described. Upon these observations, as well as some studies of the position of the leaves in buds and flowers and on the stem of higher plants, is based a "living view of nature," which tries to find in nature "*nicht bloss die Wirkung toter Kräfte, sondern den Ausdruck lebendiger Tat.*" This conception of nature is based on "rejuvenation" as the driving force in life, whereby the old is constantly being converted into a new: the child's "old" milk-teeth into new ones, the "old" pupa of the butterfly larvæ into a new butterfly, to say nothing of the spring's rejuvenation of leaves and herbs, which, of course, gave rise to the whole of this speculation. "The spirit that develops in man is not outwardly united to nature, for its appearance is already indicated in the lower stages of natural life, especially in the animal kingdom; rather the spiritual life is the purest representation of the same basis of life as that which in previous stages confronted us as natural life." This, of course, is pure natural philosophy; it is no wonder, then, that Goethe's metamorphosis doctrine finds its application here, both in ascending and descending metamorphosis and in the spiral arrangement of the leaves, which on the model of Schimper is expressed in mathematical formulæ. It is strange to note how exact observations are mixed up with this fantastic terminology, especially when it is applied to the then newly-discovered cytological details. "*Alle Verjüngungen im Zellen-leben sind mit einer mehr oder minder tief eingreifenden Entbildung der bereits be-festigten und der Fortbildung widerstrebenden Teile der Zelle verbunden.*" By this "*Entbildung*" is meant simply the dissolution of the cell membranes upon the segmenting of the cells, which, of course, represents the "*Verjüngungs*" phenomenon. These same principles are also applied to vegetable geography, and the system becomes merely a link in this magnificent unity whereby "the whole of nature's course of development from the first manifestations of life through an infinite number of rejuvenations gradually rises to the emerging of man." All this speculation is interesting as being an intermediate link between the old natural philosophy in the spirit of Goethe and the new philosophy of Haeckel, who differs from his master only in his materialistic tone, though but little in fact. Haeckel's symmetry ideas in particular are certainly in imitation of Braun.

Exact research, however, must eventually come into its own even in vegetable morphology. The scientist who has contributed more than anyone else towards producing an exact conception of the forms and development of plant life is Nägeli, though even he was in close contact with the old idealistic philosophy. CARL WILHELM NÄGELI was born in 1817 near Zurich, where his father was a physician, and it was intended that he should be trained for the same profession at the college in his native town. His attendance at the lectures of the aged Oken, however, induced him to take up a more speculative career, which he was finally permitted to do. He studied botany at Geneva under de Candolle and wrote as his dissertation a work on vegetable classification; he then went to Berlin and spent a couple of years studying Hegel's philosophy, which, according to his own statement, did not attract him very much, and he finally spent some time working at Jena under Schleiden. He was a friend of Kölliker and accompanied the latter on a trip to Italy, afterwards becoming professor, first at Freiburg, then at Zurich, and ultimately at Munich, where he spent the rest of his life in work of an unusually many-sided and productive character. Since his childhood his health had been poor, but he worked with indomitable energy up to the last ten years of his life, when sickness compelled him to abandon his activities. He died in 1891. Ill health brought with it an irritable temper, which made it difficult to associate with him, either as a teacher or as a man of science; indeed, his personal pupils were few in number, but the influence exercised by his writings was all the greater. Indeed, he must without doubt be counted among the foremost botanists of the century, and that, too, in many different spheres, being at the same time anatomist and cytologist, morphologist and systematist; moreover, his natural-philosophical speculations have proved of deep significance.

Nägeli's cytological investigations

NÄGELI was certainly greatest as a cytologist; his studies of the dividing of the pollen-grains and of the unicellular Algæ have already been mentioned as epoch-making in their sphere, and through them he became one of the pioneers of modern cytology. We may also include in this category his studies of the sexual reproduction of the cryptogams, a problem that has largely been elucidated by him. Nevertheless, his cell research had its weaknesses; the fact that he long maintains the old belief in independent cell-formation is of less importance in this respect than the influence which he permitted his theoretical speculations to exercise on the observations that had already been made. In the field of vegetable anatomy a series of essays that he wrote on the growth of the stem and root forms the basis of our present-day knowledge of the subject. As a systematist he was especially occupied in studying genera that possess abundant forms, but are difficult to elucidate, chiefly Hieracium, the numerous and mutually overlapping microspecies of which he sought to explain by means of both natural observations and horticultural experiments. His experiences in studying this difficult genus led him to speculate upon the term "species,"[3] which formed the basis of his evolutional theories. As a plant-physiologist he distinguished himself chiefly in his investigations into the growth of starch granules, whereby he laid the foundation of our knowledge of that curiously organized structure in these elements of stored nutrition, which, as far as their chemical composition is concerned, are comparatively simple. Even in this line of research, however, he became involved in theoretical speculations of that abstract kind which had interested him since his youth.

In a treatise *Über die Aufgabe der Naturgeschichte*, dated 1844, Nägeli has given an account of the theoretical standpoint from which he started. He lays down as the aims of natural research, firstly the discovery of fresh facts, and secondly the creation of new laws of thought. His interest in these latter are clearly reminiscent of his studies in the Hegelian school, referred to above. True, he indignantly repudiates the accusations of Hegelianism that were directed against him, but the likeness is nevertheless unmistakable and gives his speculations a character all its own, which is strongly divergent from, for instance, Haeckel's; while the latter speculates upon forms of symmetry, the psychic qualities of matter, and other ideas reminiscent of Schelling, Nägeli is ever seeking to create fixed categories of thought, preferably with reference to mathematical deductions. Above all, he strives to create "absolute ideas," in which the various phenomena are to be defined. All life is movement, and so all biology must be evolution, and from the

---

[3] In contrast to Lamarck, Darwin, and even Haeckel, Nägeli speaks in every way depreciatingly of Linnæus and his work for the advancement of classification. As is well known, these views, which have but little justification, have since recurred in many German botanists.

evolution of the individual it is possible to conclude that of the species. The species is a summary of all similar individuals and, as such, an absolute idea; similarly, there are many circles, ellipses, and other geometrical figures, but their ideas are essentially different — and in the same way the species have no intertransitional forms. "*Die absolute Verschiedenheit der Arten scheint mir durch die Erfahrung hinlänglich bestätigt, und allgemein genug angenommen, um auch ihrerseits die Absolutheit der Begriffe zu bestätigen.*" And like the species the higher systematical categories also possess their absoluteness; the vegetable and the animal kingdoms have no transitions, for "*dieser Annahme widerspricht schon die Absolutheit der Begriffe.*" If this is not Hegelianism, it is at any rate very near it.

His micella theory

THE ideas expounded in this work of his early years proved in many respects to have a decisive influence on Nägeli's future development. In his work on starch granules he presents his once famous micella theory, which again shows his tendency to transfer the deductions of geometry to biology. According to this theory the cells and their derivatives are composed of particles called "micellæ," which are supposed to be composed of a number of molecules, possess a regular crystalline form, and in a dry state keep close to one another, owing to their mutual attraction; in certain circumstances, however, the micellæ attract water, which penetrates in between them and surrounds each one of them, and through this action arises tissue. This theory, which was irreconcilable with the findings of physics and consequently had to be abandoned even during the lifetime of its originator, shows how he strove to compare animate and inanimate matter; in actual fact he denies any principial difference between them. In consequence he believed also in spontaneous generation, stubbornly maintaining that doctrine throughout his life; he certainly recognized Pasteur's experiments, but he considered that they did not demonstrate the *impossibility* of spontaneous generation, and to his mind spontaneous generation is "*nicht eine Frage der Erfahrung und des Experiments, sondern eine aus dem Gesetze der Erhaltung von Kraft und Stoff folgende Tatsache.*" In his youth he believed in the spontaneous generation of unicellular sponges and believed that it could be demonstrated by experiment; when he did not succeed, he assumed the spontaneous generation of very primitive unicellular creatures, and in his old age he continued his retreat, inasmuch as he assumed as products of spontaneous generation a kind of extremely primitive life-units, whereof countless numbers go to form one cell, and which he termed "*probiæ.*" But he was undeniably more consistent than Haeckel in not moving spontaneous generation back to the beginning of the world, holding that it could just as well take place now as then, "for the difficulty of letting a cell arise out of formless chemical substance is not a jot greater for primeval times than for modern times." We must also take in conjunction with this theory his positive assertion, previously referred to, as to the pleomorphism of micro-organisms; it cannot, of course, be denied that the absence of any species-characters in them was a feature of primitive organization, which would make them closely akin to lifeless matter.

His descent theory

NÄGELI's descent theory is also in line with his theory of spontaneous generation; since spontaneous generation goes on incessantly, it is possible to suppose that the most highly developed organisms are really the oldest, while the primitive organisms have been evolved later. His descent theory has thus acquired a decidedly polyphyletic character, and, strictly speaking, it does not presuppose any transition from one species to another. The most interesting feature in Nägeli's phylogenetical speculation — recorded in an essay on the genesis of the natural-historical species (1865) and in a large work, *Mechanisch-physiologische Theorie der Abstammungslehre* (1884) — is his criticism of Darwin's theory of selection. In the course of his experiments on the Hieracium he had discovered that the external conditions of life which cause the struggle for existence do not alter the life-types; species that are placed in new conditions of life do not assume any similarity to kindred species previously brought under these conditions. Natural selection, therefore, cannot possess any form-building power; it does exist, but has only an extenuative effect on middle forms. Instead, evolution takes place out of the inner being of the organisms in virtue of an internal force, which Nägeli most frequently terms "*Vervollkommungskraft*," and once, in imitation of Blumenbach, "*Nisus formativus*," a force by means of which the development is led in a certain direction, not, as Darwin holds, to variations in every possible direction. This force, however, is by no means a special life-force; on the contrary, it is compared with the inertia in inorganic nature; just as a rolling globe goes on until it meets an obstacle, so organic evolution — it, too, being a movement — advances until an obstacle comes in the way. These obstacles can be either the struggle for existence, or else direct material influence due to irritation; according to Nägeli the ruminants have got horns as a result of striking their foreheads against one another — an explanation in the spirit of Lamarck.

His heredity theory

IN connexion with his doctrine of descent Nägeli propounds a heredity theory of his own. At variance with Haeckel's view on the undifferentiated character of the egg-cell he definitely maintains that the egg-cell is as complicated as the creature which is to be evolved therefrom; the qualities of the coming individual are all united in the egg-cell. But because this latter and the sperm cell, in spite of their difference in size, have an equally large share in the qualities of the new individual, these qualities cannot be allocated to the whole protoplasm of the egg; there must be a certain part of them that is particularly responsible for the specific qualities. This constituent of the cell Nägeli calls idioplasma; he believes that through segmentation it is imparted to every fresh cell and gives the latter its character; it is through it that every organism is such as it is and not otherwise. The idioplasma is, according to Nägeli, a solid body, not semi-fluid like the rest of the cellular mass, and it has, of course, its peculiar composition of micellæ, the shape and size of which give rise to the most subtle calculations. All evolution consists in changes in the micellæ of the idioplasma, and these changes go on incessantly, although they are not at once perceptible, for the energy amassed through these changes is released intermittently, and therefore the alterations in species likewise take place, not gradually, but suddenly.

From the structure of the idioplasma Nägeli gradually passes to atomic structure in general, and he here becomes involved in speculations as to the atoms' being composed of still smaller particles, which are called "*amera*"; of these latter the simple chemical basic elements are composed, and Nägeli builds up a kind of phylogeny for these elements, according to which the heavy metals must have originated first, and afterwards the other elements in succession. Further, he speculates upon the form of the atoms, upon ethereal atoms, ethereal heat, upon the impossibility of entropy, and various similar subjects, which contemporary physics and chemistry had naturally passed over in silence.

His influence

NÄGELI's mechanical-physiological theory was his last work, so that he concluded his life's activities, in spite of his expressed intention to deal with natural phenomena on a mathematically exact basis, in a mass of thought-constructions of just as impractical a nature as those of the master of his youth, Hegel. His influence, however, has been of deep significance, not only on account of the immense number of important facts that he established, but also in the purely theoretical sphere. He was the first unreservedly to venture to reject the doctrine of natural selection as the sole cause of the evolution of life and to demand that it be replaced by another theory capable of producing a more convincing confirmation by way of observation and experiment. The "*Vervollkommungskraft*" on which he would base his explanation of the origin of species was really nothing but a word, but behind it there lay at any rate an insight into the fact that evolution is a quality in life itself, not a movement that is thrust upon living creatures from outside. And in connexion therewith Nägeli points out that life need not necessarily evolve as a result of minute imperceptible variations, but the changes might just as well take place suddenly and on a larger scale — an idea which is certainly not very strongly brought out in him, but which nevertheless afterwards survived; de Vries especially inherited it through his theory of mutation, but, above all, Nägeli's idioplasma theory was an idea that was utilized by subsequent investigators with much profit. Here, again, he really only invented a word, but the idea of a special substance's being the bearer of the cell's hereditary qualities received remarkable confirmation in the above-described discovery of the role played by chromatin in cell-division and its importance for the vital processes of the cell in general. What prevented Nägeli himself from drawing from his speculation conclusions of practical value was undoubtedly his belief in "absolute ideas" and the derivation of facts from them — a belief that he never really succeeded in eradicating. Herein, too, we must obviously seek the cause of his attitude towards Mendel, violently criticized at a later date. The latter had reported to him the results of his epoch-making experiments and received in reply an inquiry as to whether the formulæ he had set up were not "empirical rather than rational." In these words is clearly shown the weakness of Nägeli's abstract-speculative method: he could not grasp Mendel's incontrovertible results based on fact, since they did not agree with his own theories, and the correspondence, which went on for some time, though in courteous terms, produced no result. With all his weaknesses Nägeli nevertheless stands out as one of the foremost biologists of his time, and his ideas had an influence long after his death.

Among Nägeli's pupils the first that deserves mention is his fellow-countryman, SIMON SCHWENDENER (1829–1919), for a long time an assistant to his master and finally professor at Berlin. Of his works should be mentioned one entitled *Das mechanische Prinzip im anatomischen Bau der Monokotylen*. In this he describes the mechanical functions of the cells and tissue elements and shows how the structure of the plant closely follows the general laws of mechanics governing its sustaining power and strength. In doing so, however, he sometimes interprets the structure and functions of plants from a too narrowly mechanical point of view. Thus, for instance, he sets up a mechanical theory in regard to the position of the leaves, wherein he examines the above-mentioned idealistic spiral theory and finds that the leaves' spiral position is caused by conditions of mechanical stress and is altered if the stress alters. In spite of its one-sidedness, however, this work contributed in its own sphere towards overcoming the romantic belief in an idea's being the cause of a natural phenomenon and substituting a mechanical explanation. Schwendener's works in the sphere of lichenology, however, caused a still greater sensation than the above investigation. Hitherto the lichens had formed a class in the vegetable kingdom by the side of Algæ and Fungi. Schwendener now declared, as the result of a series of microscopical investigations, that the lichens are really a kind of double organisms, consisting of fungous hyphæ, in which cells of Algæ lie embedded and which jointly contribute to the existence of the whole, the Fungi by forming the substratum, the Algæ by assimilating carbonic acid with their chlorophyll. This discovery, which upon publication aroused the keen opposition of the lichen-systematists, has gradually received confirmation and is now universally accepted as correct.

Among those who, as far as the vegetable kingdom is concerned, paved the way for a uniform conception of its vital manifestations, must also be mentioned WILHELM HOFMEISTER (1824–77). Born in Leipzig, he was educated with a view to taking up a commercial career and became a music-seller in his native town, but he spent his spare time studying botany, and eventually became a professor, first at Heidelberg and then at Tübingen. His great achievement is his comparative investigations into the reproduction of plants, which he carried out while he was still a music-dealer and which resulted in his being appointed professor. He closely studied the phanero-gams, as well as vascular cryptogams and mosses, especially observing their formation, development, and combination of the sexual products, and he established in all these phenomena a bond of agreement that made possible in all essential respects the adoption of a uniform conception of sexual re-production throughout the vegetable kingdom — an achievement that is all the more remarkable, seeing that his knowledge of the cell was not in advance of the stage at which his own period had arrived. Hofmeister's work on the reproduction of plants was followed up by several later natu-ralists. Among these may be mentioned NATHANAEL PRINGSHEIM (1823–94), at one time professor at Jena and then a private scholar in Berlin. He found out the method of reproduction of the Algæ and published several valuable works on plant physiology. Also, HEINRICH ANTON DE BARY (1831–80), professor at Strassburg, who discovered the sexual reproduction of the Fungi and the alternation of generation in the rust fungi, and also solved a large number of important problems in the sphere of mycology and bacteriology.

Considerations of space forbid our continuing the account of the develop-ment of plant morphology up to modern times; in fact, all the details will be found in the text-books on the subject. We shall therefore proceed to another branch of biological research, which has also played an important part in modern times.

5. Geographical Biology

IN the foregoing, Humboldt and Wallace have been named as founders, in the modern sense, of vegetable and animal geography. Like all other branches of biology, these fields of research have in our day become highly specialized, while fresh fields have been opened up for the employment of their methods on an extensive scale. Among these novel spheres we note not only new land-areas — it was not until our own period that the entire globe can be said to have been explored and described — but also to a still greater degree the oceans, the deeper areas of which have only recently been known as regards their physical character and conditions of life. Our knowledge of them has been gained partly through the work carried out at the zoological marine stations, of which the most famous in recent times has been that founded by ANTON DOHRN (1840–1909) at Naples, and partly as the result of oceanographic expeditions specially equipped for the purpose, among which may be recalled in particular the important English *Challenger* expedition (1872–6) and the German Valdivia expedition (1898–9), not to mention the results obtained by polar expeditions equipped by a number of countries, both large and small. It is through these voyages of exploration that the life-forms of the ocean first became known — the life in the vast depths, whose denizens live in constant darkness and under high pressure and often assume amazing forms; the actual inhabitants of the vast ocean-expanses, the so-called plankton fauna and flora, with their often transparent and fragile forms, which are constantly swimming in the water, forms of widely differing systematic groups; and, lastly, the life that moves around the coasts. Innumerable workers have devoted themselves to this branch of study, which has often been carried out under the leadership of committees, national and international, with the consequence that the work and its results have to a certain extent acquired an impersonal character. The pioneer in this field is KARL AUGUST MÖBIUS (1825–1908), professor first at Kiel and then in Berlin. By his great work *Die Fauna der Kieler Bucht* (1865) he has created the modern system and methodics of œcology. By way of introduction he describes the topography of the estuary that he investigated, various sections of it being surveyed and characterized in regard to position, depth, and their plant and animal life. He then presents in systematical order the creatures that inhabit each locality. Others have continued along the path thus beaten by Möbius. Among them may be named VICTOR HENSEN (1835–1924), who was professor of physiology at Kiel and in that capacity studied the structure of the auditory organ, but afterwards he devoted himself entirely to marine research, chiefly with the idea of improving the fishing industry. He made a special study of plankton life, with particular reference to its microscopical forms. In order to advance the study of these creatures, which are of importance as food for fish, he worked out a statistical method of his own. Of others who laboured in this sphere, to some extent practically important, may be cited the Dane C. G. J. PETERSEN (born in 1860), who investigated animal life in the sounds and bays of Denmark on a method of his own, and J. SCHMIDT (born in 1877), who after lengthy and difficult exploration succeeded in elucidating the reproductive process of the eel — a problem that many before him had tried to solve in vain.

Touching the continental life-forms, the classification into large geographical regions already drawn up has on the whole been retained, and both plant and animal geographists have for the most part devoted themselves to the study of conditions within smaller areas belonging to these regions. Among investigators of this category in the sphere of animal life may be mentioned the explorer KARL SEMPER (1832–93), professor at Würzburg, who studied the problem of the life-conditions of animals from various points of view.

Plant geography: its floristic and morphological courses

IN the field of plant geography, research has taken especially two courses, a systematical, which is ultimately based on Linnæus's observations and theories in connexion with the distribution of the plant species, and a morphological, which has its origin in Humboldt's theories on the morphological association of different vegetable types with different countries and forms of landscape. These two tendencies have exerted a mutual influence and have, each in its own way, been influenced by the doctrine of descent and its attempt to explain the origin of species out of conditions of geographical distribution. And at the same time valuable results were gained by the comparison between the distribution of existent plant-forms and that of the corresponding genera and species of earlier geological periods. All representatives of modern plant-geography have been compelled more or less to take these conditions into consideration. It is still possible, however, to trace two main tendencies in this sphere, which nevertheless incessantly touch and cross one another. The first of these, the systematic or floristic, which rests upon the systematic entities, treats of the distribution of the species within larger or smaller areas and their variations in different parts of one area under the influence of certain factors. It endeavours to find out the causes of the changes in the character of species in certain localities and countries and for this purpose studies the migrations of species, such as occur through the distribution of land and sea in recent times and through the changes that have taken place in the distribution in earlier ages, in so far as it has been possible to trace these shiftings of the world's surface. Further, it examines the distribution of extinct and fossilized species, from which those of our own time may possibly have originated. Special interest has been devoted to the immigration of plants in those parts of Europe that were once visited by the glacial period, as well as those vegetable remains in the mountain ranges of the polar regions that give evidence of a previous warmer climate there.

Morphological or œcological plant-geography does not investigate the nature of the flora, but of the vegetation. It works, not with species, but with plant communities, by which is meant plants of very different systematical categories that, on account of a uniformity in the alimental conditions, have adapted themselves to living together within a certain area. The aim of this tendency is to analyse such plant-associations and to ascertain their relations to the climate, the soil, and other environmental conditions. Both these tendencies have made considerable progress up to recent times and can claim a number of distinguished representatives, of whom it is possible only to name a few. The two previously mentioned English botanists Brown and Hooker made valuable observations as to the distribution of plants, especially in extra-European countries. The Swiss OSWALD HEER (1809–83) made a special study of the conditions of the flora of the glacial period and also of earlier geological strata. There followed in his tracks the Swede ALFRED NATHORST (1850–1922), who did very creditable work in investigating the fossil vegetable world of the polar countries. ADOLF ENGLER (born in 1844), professor at Berlin and founder of an important school of plant geography, has endeavoured, by studying the recent and fossil vegetable world, to gain some insight into the evolution and changes of the flora, especially in the temperate countries. AUGUST HEINRICH GRISEBACH (1814–79), professor at Göttingen, sought to carry out an investigation into the influence of climate on the vegetation and a classification, on a climatic basis, of the flora in certain areas. The Dane EUGEN WARMING (1841–1924) performed a considerable service to science by his study of plant-associations, which he classified and analysed in respect of plant forms and conditions of life. By this work he made a contribution to œcological plant-geography of fundamental importance. ANDREAS FRANZ WILHELM SCHIMPER (1856–1901), professor at Basel, made long voyages for the purpose of studying tropical vegetation, and from climatological and modern physiological points of view he worked out the vegetation of the entire globe in his *Pflanzengeographie auf physiologischer Grundlage* (1898). His geographical and œcological classifications have exerted great influence upon subsequent development.

CHAPTER XVI

NEO-DARWINISM AND NEO-LAMARCKISM

Decline of Darwinism

TOWARDS THE CLOSE of the nineteenth century the influence of Darwinism began noticeably to wane. The signs of this are many: partly internal, in that the actual theory, as had so often happened before and indeed always will happen with dominating views, becomes split up into a number of mutually conflicting tendencies in different directions, and partly external, in phenomena manifested in the general cultural situation. The optimistic belief in progress as a law governing nature and human life, which prevailed in the middle of the century and formed the basis of the success of Darwinism, had some decades after been essentially disturbed. The unlimited progress that was to follow upon political and economic freedom had proved to be somewhat relative; democracy, which had been introduced in many countries, had led to disappointments, out of which much capital had been made by its political opponents, while free competition had called forth, not a friendly and stimulating rivalry with a universally acknowledged precedence for the best, but an inimical and severe struggle between rival enterprises, social classes, and nations, wherein people sought rather to do one another the greatest possible injury. It was quite natural that the confidence in liberalism that had but recently been so strong should in such circumstances begin to waver; the belief that progress goes on by itself began to be regarded as a matter of course; instead men of courage were required to remove the increasing difficulties. So there arose a long line of opponents to liberalism, from the strange romanticist CARLYLE, with his demand for hero-worship, to NIETZSCHE, with his paradoxical "superman" ideal; both deserve mention as men who made violent attacks on Darwin and his theory. Their successes in the sphere of literature may thus be registered as defeats for Darwinism, and they were by no means the only ones of their kind; on the contrary, there appeared in th nineties a literary tendency that was wholly intended to be a contrast to the naturalistic literature of the preceding decade based on natural science. And while the popularity of Darwinism among the general public thus began to wane, its champions among the scientists had to defend themselves against the obstacles that the results of fresh research placed in the way of the old theory.

As we know, both Darwin and Haeckel had based their doctrines of

MENDEL

From *Gregor Johann Mendel: Leben, Werk und Wirkung* by Dr. Hugo Iltis,
published by Verlag von Julius Springer, Berlin

RICHARD OWEN

descent partly on the theory of variability and natural selection brought about by the struggle for existence among the variations, and partly on the assumption of the direct influence of environment upon the individual, and the inheritance of the changes thus brought about — that is, a Lamarckian conception. Here at once, in this double explanation, lay the seeds of dissension: one could with prejudice emphasize the idea of selection, or with equal prejudice maintain the influence of environment. And this was exactly what happened; the period immediately preceding and around the turn of the century witnessed the birth of the two evolutional schools of thought called neo-Darwinism and neo-Lamarckism, whose advocates sought to convince the biologists of the absolute validity of their own views. Out of these two main directions there further originated a number of special attempts to explain the causes of evolution, so that the situation in which the doctrine of descent eventually found itself was somewhat chaotic. We shall here describe some features of this internal dissolution of Darwinism.

In Germany the theory of selection found a highly gifted and powerful advocate in the person of AUGUST WEISMANN (1834–1914). He studied medicine, being a pupil of Leuckart, who inspired him with an interest for biology. After working for some years as a practitioner he was invited, on account of a useful treatise on the evolution of flies, which he had written in the mean while, to be professor at Freiburg, where he laboured until his death. His special subject was the evolution of the lower animals; in this field he particularly distinguished himself in his studies of the reproduction of the Daphniidæ, as a result of which he elucidated the peculiar egg-development in these crustaceans and the no less curious "cyclic reproduction" that characterizes them. An ophthalmic disease, however, soon precluded him from using the microscope and compelled him to apply himself partly to experimental and partly to purely speculative activities. One result of this was his strange theory of evolution, which placed him among the very foremost of Darwin's successors.

Weismann's theory of descent and heredity is based, firstly, on his above-mentioned special investigations, and secondly on Nägeli's idioplasma theory, referred to above. Nägeli had sought for a material substructure for the inherited dispositions, out of which are developed in every individual certain given qualities, and he believed he had discovered it in his hypothesis of the idioplasma, which, existing equally in the egg and in the sperm, through their union forms in the new individual the basic material for its special qualities. Weismann, who as a result of his studies and his own research work had acquired a deep insight into contemporary cytological knowledge, came for that very reason, when he was forced to devote himself to purely theoretical speculations, to take up the question of cell-structure as a basis for the evolutional theory that Darwin and his school had promulgated. In a series of lectures and monographs dating from the eighties he endeavours to find out what it is that produces heredity in a biological sense; how, he asks, are we to account for the characteristic in organisms of transmitting to their offspring their own essential being, for the fact that from the eagle's egg is invariably hatched an eagle, and, moreover, one of the same species as its parents? In imitation of Haeckel he starts from the unicellular animals and finds that in these the mutual resemblance of the different generations is due to the individuals' propagating by division; to the fact that every infusorian is a segment of a previous one, that there thus exists in them a *"Continuität des Individuums."* And the same is true of the multicellular animals in virtue of sexual reproduction; for the individual's life the sexual cells are without significance, but they preserve the continuity of the species through the ages; out of them arises in certain given circumstances a new individual of the same kind as the old.

Weismann's germinal-plasm theory

FROM this it may be concluded that there exists a special "germinal plasm" which corresponds to the individual series in unicellular animals and which, like them, preserves the species by repeated dividing, whereas the corporeal plasm of the individual gradually falls into decay. Originally the differentiation of sexual and corporeal cells had been due to a division of labour in the simplest cell-colonies, such as we still see in the primitive colony-forming animals; for the sexual cells that perform the function of reproduction contain both germinal and corporeal plasm, which separate when, in the earliest embryonic stage, the rudimentary cells of the sexual organs separate from the rest of the cells. Out of the germinal plasm, therefore, arises the long series of analogous individuals, and these resemble one another for the very reason that their form is governed by the character of the germinal plasm, which is determined once and for all; if changes appear in the external bodily form, they correspond to and are induced by changes in the germinal plasm. These changes are brought about by fertilization, in which the germinal plasm of two different individuals is united; through this "amphimixis," as Weismann calls it, is formed a new germinal plasm, with both the parents' qualities, which accordingly appear also in the offspring. But if the qualities of the individual are thus due entirely to the germinal plasm, there can be no possibility of influencing the individual series from outside; the organs of the individual that are formed of corporeal plasm can be influenced by practice, in so far as the germinal plasm has created possibilities therefor, but changes of this kind exercise no influence upon the germinal plasm. Consequently, Lamarck's theory that the character of the species is created by the habits of the individual is untenable.

This denial of the heredity of acquired characters became one of the corner-stones of Weismann's biological theory and he sought in many and various ways to procure proofs for his argument. He bred large quantities of rats, whose tails he cut off at birth, but he never succeeded in finding a rat born tailless, nor did other malformations brought about by outward interference ever reproduce themselves. He therefore felt fully justified in maintaining his standpoint that all changes in the outward appearance of the individual compared with other individuals are due to changes in the germinal plasm; that every so-called acquired character is really produced by a change in the germinal plasm, whereby the body becomes capable of adapting itself to the different external conditions of life. But how, then, have the various life-forms arisen in the course of ages? By means of natural selection, answers Weismann, and by that means alone. The variations that are brought about especially through the amphimixis of sexual reproduction, but also through other changes in the germinal plasm, are advanced or retarded by natural selection and thus give rise to new forms, whose germinal plasm is better adapted to the conditions of existence than that of the old forms. Natural selection is thus the cause of the evolution of animate beings, Weismann rejecting Nägeli's assumption of internal causes of evolution inherent in the organisms themselves, for such a theory "cannot explain the finality of the organisms. And yet this is the very riddle that the organic world gives us to solve." Numberless instances are quoted of this adaptability, this connexion between form and function, and every instance is likewise made to serve as evidence of the creative power of natural selection.

"The continuity of the germinal plasm" and "the omnipotence of natural selection" are two phrases in which Weismann's theory of life used to be summed up. As a result of the former of these ideas — that of the germinal plasm as the preserver of heredity — Weismann has reached by way of speculation conclusions to a certain extent foreshadowing those that modern heredity-research has since arrived at by means of exact observation. His subsequent attempts to expand this theory gave him similarly happy inspirations, as when he localizes the germinal plasm — that is, the preserver of heredity — in the chromosomes of the sexual cells. "The idea in itself was sound," says Johannsen in this connexion. Even Weismann, however, succumbed to the danger of basing his conception of a phenomenon on mere speculation; in his continued efforts to extend his germinal-plasm theory downwards he works out a highly complicated plan to show the structural nature of living substance; every one of its minutest entities consists of a mass of chemical molecules; they are termed "biophores," and he assures us that they are not hypothetical: "They must exist, for the phenomena of life must be bound to an entity of matter." Of biophores are composed the determinants: those units in the germinal plasm that govern the various qualities in the smallest parts of the individual; the determinants in their turn build up the ids, which form larger groups of qualities, and these again the chromosomes, which unite in themselves all hereditary qualities. With these we have at last arrived at something that can be observed; in fact, the foregoing has been pure imagination, of the kind that the biologists of the past century had such difficulty in avoiding when they had to explain the phenomena of life. Darwin and Haeckel were content each with one hypothetical unit of life; it can hardly be said that Weismann would have done any special service to biology by burdening it with three of them.

Theory of germinal selection

However, the germinal-plasm theory and its conclusions, both the ingenious and the false, only served Weismann as a means for proving the doctrine that gradually came to mean for him the very corner-stone of biology: the doctrine of the omnipotence of natural selection. The championship of this theory, and the fight against that of the inheritance of acquired characters eventually became his chief aim in life; all that could serve his purpose he took to be good, while all that militated against it was rejected. He went through many a hard struggle on behalf of his favourite theory; in the nineties he was especially attacked by Herbert Spencer, who maintained the doctrine of the transmission of acquired characters, chiefly for social reasons; it was, in fact, the precondition of human progress. But from many other quarters also there arose the cry of "the impotence of natural selection," and this cry was again taken up after the turn of the century. Weismann's defence was often somewhat laboured; against Spencer he defended himself mostly on the old argument about the intelligence of the workers among the bees, which cannot be transmitted by inheritance, since they are sterile, and which therefore cannot be directly "acquired" either. It was more difficult to answer the question as to how that finality arose that shows itself in occasional encroachments upon an organism; how, for instance, a fracture heals in certain definite ways; the fracture certainly cannot be traced to natural selection. Here Weismann found support in a theory that was produced by the afterwards famous experimental biologist Roux, who in his youth published a work entitled *Der Kampf der Teile im Organismus*. Here an attempt is made to explain what Roux calls "functional adaptation" within the organism: that every organ, even every cell, possesses its given structure, which changes if the conditions of the organ's function are changed, so that in normal circumstances the life of the body runs its even course; if this is disturbed by interference from outside, cells and tissues adapt themselves as required to repair the damage. This fact Roux considers to be due to a "struggle for existence" between the cells in the body and even between the molecules in every cell, each of which strives to force its way forward at the expense of its neighbours, an effort that is controlled by the general requirements of the body, the weakest elements being thrust aside and destroyed. This theory, to which we shall revert in another connexion, at once won Weismann's keen approval, but it met with opposition from other quarters; as, for instance, from O. Hertwig, who held that upon the first division of the egg it should be possible to see something of this struggle between the cells, but that, on the contrary, the first easily observable embryonic cells show no inclination whatever for mutual strife, but rather each one has its carefully defined form and place. Weismann, however, adopted the idea of natural selection within the organism and combined it with his germinal-plasm theory. In a work entitled *Über Germinalselektion* he declares that between the various parts of the body and their "determinants" in the germinal plasm there exists reciprocal action; if, now, an organ is not used, its determinants are weakened and annihilated by the struggle within the organism, and the organ disappears in succeeding generations; in this way, for instance, the posterior extremities of the whale have been lost. But, all the same, Weismann comes in this way, although indirectly, to accept the inheritance of acquired characters, which indicates that the theory of selection finds it difficult to do without this auxiliary hypothesis. We shall here leave the omnipotence of natural selection and pass on to its diametrical opposite, neo-Lamarckism.

Lamarck's theory of the direct influence of habits of life upon the bodily structure of the individual and its offspring gained strong support towards the close of the century, especially in France. When the supporters of Cuvier finally left the arena, it was to Lamarck that people turned for a basis for their biological ideas. When the belief in the constancy of species had to give way to the theory of evolution, the form that this was to take was readily sought from a fellow-countryman, and, moreover, an older man than Darwin; thus "transformism," as it was here called, could also claim to be an originally French science. Lamarck's theory found an eloquent supporter in ALFRED GIARD. Born in 1846, he studied at the École Normale in Paris and eventually became professor of zoology at the Sorbonne and head of the marine laboratory at Wimereux, near Boulogne; he held that post with success until his death, in 1908, being especially known for his profound studies of a number of marine animal forms. Under the characteristic title *Controverses transformistes* he collected some years before his death a series of contributions to the problem of descent, in which he examined and further developed Lamarck's doctrines. According to Giard, evolution proceeds under the influence of two categories of factors; namely, the primary, which directly influence the individual and indirectly its offspring, and among which are mentioned light, temperature, food, and relations to other beings — that is to say, the struggle for existence — and the secondary, which include everything that is adapted to remove less suitable forms of life — that is to say, natural selection. Giard now takes upon himself to prove the existence of the primary factors, and he adduces quite a number of proofs.

Among the positive proofs he includes a number of experiments carried out by the physiologist Brown-Séquard, who believed that by interfering with the nervous system in guinea-pigs he had induced epilepsy in their young; these experiments, however, have been found by other investigators either to have miscarried or to have been misinterpreted. Giard has better success when, in a controversy with Weismann, he declares that the secondary factors alone cannot explain the origin of the forms of life; he points out a number of phenomena that cannot be explained by selection alone. On the other hand, he does not deny the existence of selection, as already mentioned above; he only considers its importance to be "secondary."

O. Hertwig against natural selection

A FAR more severe judgment than that of Giard and several other Lamarckists — for example, the famous American palæontologist E. D. COPE (1840–97) — is passed upon the theory of selection by Oscar Hertwig, who devoted the latter part of his life particularly to attacking the common belief in it. In fact, in his great work of 1916, referred to above, *Das Werden der Organismen*, he finally settles his account with that theory and at the same time gives a summary of the natural philosophy which he had produced in the course of a long life that had been unusually rich in experience. Being mainly a cytologist, Hertwig attaches decisive importance to the cell and its structure as the groundwork for all speculation upon evolution. To him the cell is the elementary organism and he vehemently sets his face against all theories of biophores, plastidules, and such lower vital entities. Every form of life has its peculiar cell-structure: there are in nature as many "species cells" as species; it is the character of the species cell that causes every form of life to be what it is and produces descendants of the same kind. Evolution is regulated in each separate case by the character of the species cell, and those phenomena that coincide therewith are described as the "biogenetical law of cause," which precludes the possibility of any such biogenetical principle as that which Haeckel conceived; in its embryonic development a mammal certainly does not pass through a series of stages identical with the lower animals, but rather the egg of every mammal species is just as fully specialized as the animal itself, and similarly with the embryonic stages. The egg contains within itself all the characters of the organism as rudiments; Hertwig, following Nägeli, terms the bearer of the rudiment within the cell "idioplasma," but he is generally content to speak of the rudiments of the species cell. Towards the question of heredity he adopts a decidedly morphological attitude and insists that the material basis of the relative phenomena must be observed and explored, thereby opposing Johannsen's physiological view of the phenomena of heredity. He most emphatically maintains the heredity of acquired characters — that is to say, the metabolistic influence of environment upon the hereditary dispositions. In support of this assertion he cites a number of experiments, which, however, have been interpreted differently by modern heredity-research, such as, for instance, Kammerer's experiments with colour changes in the salamander, and Tower's experiments in connexion with the evolution of the beetle. If Hertwig's views on this point approach those of Lamarck, he refuses all the more definitely to have anything to do with Darwin's theory of selection. He brings out the latter's weaknesses in a strong light; he especially points out how much of the theory is borrowed from human conditions and is, moreover, utterly misinterpreted. The breeder who selects suitable variations creates nothing new thereby, but only chooses what suits him, and this procedure has no counterpart in nature — the struggle for existence does not destroy creatures; the masses that die do so from quite different causes. To declare that selection favours certain variations postulates mere chance as an operative cause, but chance is no natural-scientific explanation. And he views with equal disfavour the above-mentioned theory of the struggle of the parts within the organism; this, too, is found to rest upon utterly false conclusions.

Once again Oscar Hertwig attacks the theory of the struggle for existence and natural selection, in his polemical paper *Zur Abwehr des ethischen, des sozialen, des politischen Darwinismus,* wherein the old student of evolution sharply criticizes the outgrowths that Darwinism had induced in the sphere of social life. The fact is that a number of writers, partly biologists with a deficient social grounding, partly newspaper-men and political authors of various kinds, had made use of the theory of the struggle for existence and of selection to proclaim a new social theory, on the one hand rejecting activities based on Christian charity and strivings after social equality, and on the other hand extolling war, social want, and ruthless competition as phenomena destined to thin out the weaker and less hardy human beings, and thereby to further human progress. Against these assertions Hertwig maintains that natural phenomena cannot be made the standards of human culture; justice and morality have their origin exclusively in human community life; in nature no such principles exist; the beast of prey that tears its victim to pieces acts neither justly nor unjustly, but according to its nature. To make a merciless struggle for existence the basis of social life would therefore be equivalent to destroying all that the cultural efforts of the past have built up. War and economic misery arc of no constructive value; on the contrary they ruthlessly destroy both the capable and the incapable. Thus in the very bitterest days of the Great War Hertwig dares to hope for a peaceful settlement between the nations. That, however, he did not live to see; when he died, in 1922, the unhappy consequences of the war — unhappy for the whole of humanity and most of all for his own fatherland — had been brought out in all their frightful clearness, and his distress at these misfortunes is said to have broken the aged patriot and philanthropist and to have appreciably shortened his life.

Eimer's "orthogenesis" theory

WE shall mention in the following some further tests of the descent theory, though more as proofs of the increasing difficulties with which the successors of Darwin had to contend than for any actual value that the results possessed. THEODOR EIMER (1843–98), a fellow-countryman and disciple of Kölliker's and eventually professor at Tübingen, endeavoured to solve the difficulty of the descent theory chiefly along the lines laid down by Nägeli. He rejects Darwin's theory of variation in all possible directions as the basis of selection; the development of the organic forms must rather, he holds, depend upon a force operating in a definite direction, a force induced and modified by outward influence, such as light, air, heat, nourishment, and thus one that provides selection with the material for changes which it influences. This definitely directed evolution he terms "orthogenesis" and he seeks to prove that it is indispensable for the building up of species; selection alone cannot produce anything new, but rather it is this inner force, lawbound, yet affected by external influences, that is the true origin of lifeforms. In proof of this he cites a large number of observations dealing with the colour changes in butterflies, as well as the development of the skeleton of vertebrates, which attracted great attention at the time and brought their originator a large following, but which are now no longer up to date.

Semon's "mneme" theory

RICHARD SEMON (1859–1919), a pupil of Haeckel and at one time a professor at Jena, pursued another line of thought. His theory of evolution is also based entirely on the belief in the transmission of acquired characters, but he endeavours to give this theory a direction more suited to the time by submitting it in a new form. To his mind, the weakness underlying previous theories of this kind had been due to lack of clearness in the term "character"; instead of transmission of acquired characters it should be called transmission of acquired reactions. For the hereditary transmission depends upon the nature of the germinal plasm, and this reacts under natural laws to the influence of the general condition of the body. The power of living substance to react, its "Reizbarkeit," is the primary cause of evolution. Even if the external influence upon it is transitory, there nevertheless remains an impression, which becomes a decisive factor in its future development. This impression is termed "Engramm." And the altered construction of the new generations' form, upon which natural selection has an extenuating influence, is due to the co-operation between the outwardly induced Engramm of the body and that o the germinal plasm. This influence of the corporeal substance upon the germinal plasm is called "somatic induction," and the assumed power of the living substance to preserve external impressions, upon which the whole theory rests, is called "*mneme.*" Semon adduces a mass of arguments to prove his theory of the transmission of external influence, but all of them have since been rejected or given a fresh interpretation by the representatives of experimental heredity-research. The purely evolutional proofs are borrowed partly from palæontology — for example, the process of stunted growth in the toes of certain animal forms — and partly from embryology — the skin of the human sole is even in the embryonic stage thicker than that on other parts of the body — but as these proofs cannot possibly be tested as regards the evolution that has actually taken place, they cannot be considered binding according to exact methods. The experimental proofs, however, will be more closely dealt with later on; as a matter of fact, they recur quite regularly in all Lamarckists and are adduced by them as being positive, whereas other investigators have pointed out fallacies either in the experiments themselves or in their interpretations.

Pauly's Lamarckism

WE must mention only in passing one more line of thought based on Lamarck, represented, *inter alia*, by AUGUST PAULY (1850–1914), professor at Munich, who seeks the cause of evolution in a conscious psychic striving towards a certain goal on the part of the organism and all its elements. This theory can hardly come within the scope of natural science; it has crossed the border of metaphysics, but is worth mentioning as a further example of the desperate expedients that the speculation on the problem of origin was finally compelled to adopt. In support of his views Pauly further cites the traditional examples of the inheritance of acquired characters.

The struggle between the champions and the opponents of the theory of the inheritance of acquired characters is to some extent reminiscent of that between Pasteur and Pouchet regarding spontaneous generation — people could not agree upon the interpretation of the experiments and a theoretical standpoint based on them. And in this case, too, it would seem as if the practical consequences must eventually determine the value of the theory; neither animal-breeders nor horticulturalists have really succeeded in applying the theory of the transmission of acquired characters, those of them who believed in it having really remained at the same grossly empirical standpoint as their colleagues had adopted long before Darwin's time, while modern racial research, working on the Mendelian method, has attained results of an entirely different practical value and has, in fact, completely revolutionized the methods of racial breeding.

Plate's Darwinism

WHILE both the theory of selection and Lamarckism thus had their supporters, who carried the conclusions of their several lines of thought to extreme limits, the middle course pursued by Darwin himself has also been followed by scientists of repute up to modern times. As one of these may be mentioned LUDWIG PLATE (born in 1862), Haeckel's disciple and successor at Jena. After having applied his theoretical ideas to a number of special investigations, both morphological and experimental, he summarized his main arguments in an extensive work, *Selektionsprinzip und Probleme der Art-bildung*, which may be said to contain all that can be adduced in modern times in defence of the old Darwinism. And as its champion Plate has done a great service, thanks to his wealth of knowledge, his strong convictions, and his honesty. From the imperious disposition of his master, Haeckel, he kept entirely free; he refused to employ the theory of selection to explain the fundamental qualities of the living substance: assimilation, growth, respiration, etc., or indeed to explain variability or heredity; its sole function, to his mind, is "to explain the origin of the teleological organizations, in so far as they are not elementary qualities nor can be placed in the category of Lamarckian factors." Darwin's greatest service, in his opinion, lies in the fact that "he sought to explain organic finality out of natural forces, to the exclusion of any metaphysical principle operating with conscious intelligence." Finality is thus the principal quality of organic life; adaptations are expressly declared to represent a main difference between animate and inanimate bodies. In his anxiety to defend the theory of selection, Plate has, obviously without realizing it, hereby come perilously near Johannes Müller's old doctrine of finality and a far cry from Haeckel, who was at one time prepared to characterize rudimentary organs as being the opposite to profitable [1] and the knowledge of them as "dysteleology." Plate, however, examines all the different kinds of finality — phenomena of correlation, mechanical equiponderant apparatus, embryonic structures, instincts, protective resemblance, and a good deal more — all this in order to find proofs of the operation of natural selection. But if the theory of selection is thus to stand or fall by the question of finality in nature, the result will, of course, be that the function of selection automatically lapses if a different view of natural phenomena is advanced. And, as has already been pointed out, ever since the days of Democritus of old, research has constantly aimed at seeking the existence of law-bound necessity in nature without any explanations of purpose. But then, as Johannsen says, finality in an organism becomes merely an expression for the fact that "organisms must be systems in dynamic equipoise," that finality in general is self-evident in the very fact of organization. From this standpoint one does not, of course, explain by external causes the origin of functional adaptation in the organisms,

---

[1] Haeckel has made a special point of the human appendix as a proof of nature's lack of finality; it serves no purpose, but produces only dangerous inflammations — an extremely ingenuous argument. A healthy appendix, of course, plays its part in the renewal of substance, and the danger of sometimes becoming inflamed is one to which any section of the intestine whatever is exposed.

which differentiates them from inanimate natural objects; on the contrary, this becomes part of the problem of life itself, but instead we get rid of the anthropomorphistically childish speculations upon teleological and non-teleological forms of organizations in nature, which actually imply a confession that we cannot grasp the idea of nature's being law-bound.

In such circumstances it is of no particular interest to follow Plate in his attempts to meet all conceivable objections to the theory of selection. Among other things he honestly admits that it has never been possible actually to observe selection, but he consoles himself with the numerous indirect arguments that could be quoted in its defence. Furthermore, he emphatically maintains that, besides selection, environment co-operates in renewing the organisms and their descendants; he thus rejects Weismann's theory of the omnipotence of selection, as well as Eimer's and other neo-Lamarckists' denial of its metabolistic power. He also tried to adopt an attitude, consistent with his own point of view, towards the results of experimental heredity-research, dealing with them in a monographic *Vererbungslehre*.

With this we can leave the doctrine of descent in the old Darwinistic sense. Modern heredity-research has introduced quite a different and essentially experimental treatment of the problems of evolution, and the old morphological speculation upon the origin of species and genera has proportionately lost ground — as it has always happened in the history of the exact sciences that speculation must give way to facts. The old doctrine of descent actually possessed the weakness of insisting upon *external* grounds of explanation for the phenomena of life; selection as well as direct outward influence were to explain phenomena that must really be an expression for manifestations of life itself. Nevertheless there appears here also an increasing self-deliberation based on a progressive knowledge of facts; there is a vast difference between Haeckel's belief in the power of Darwinism to explain anything whatsoever and Plate's modest limitation of the function of selection to a mere explanation of the teleological arrangements of living creatures, "in so far as they are not elementary qualities." Indeed, why not let them all be elementary qualities? One can trace here the human age-old tendency to look for outward causes for everything; formerly one sought in the phenomena of life manifestations of a divine creator; when this was no longer perceivable, one had to look for a material creative power — it was difficult to realize that evolution is a part of life itself. We shall revert below to the problems of evolution in the form in which modern heredity-research has presented them.

CHAPTER XVII

EXPERIMENTAL BIOLOGY

1. Experimental Morphology

THE HISTORY OF BIOLOGY might really close with the establishing of the dissolution of Darwinism. This theory undeniably constitutes a construction of thought of its own, and the ideas and methods of research that have succeeded it are still in the developing stage, and their possibilities of development can at best be only guessed at. A summary glance at these conquests of the newest biology is nevertheless defensible as a further justification for the defeat of the old ideas and in view of the intrinsic interest that the new discoveries possess. The author, who himself laboured exclusively in the sphere of the old morphology and who accordingly does not feel justified in competing with the many splendid presentations of the development of experimental biology that already exist, proposes to make only very brief reference to the results of modern research and give a short account of the theoretical reflections to which they have given and may give rise.

In a lecture given in 1900 in celebration of the birth of the twentieth century, Oscar Hertwig gave a summary of the history of biology during the past century. In it he sharply criticized physiology, which, in his view, had created a brilliant experimental technique and had discovered a number of facts concerning the chemical and physical processes of the organisms, but, on the other hand, had neglected all other vital phenomena, with the result that a whole category of the most important physiological problems — namely, fertilization and embryonic development — had fallen entirely into the hands of the morphologists and been dealt with by anatomists, zoologists, and botanists. While the professional physiologists had thus reverted to "an empty mechanism" and imagined that the explanation of life was only a chemico-physical problem, the morphologists discovered the structural conditions in the fundamental elements of the body that are characteristic of the phenomena of life and thereby extended the knowledge of life in a sphere in which the methods of chemistry and physics have no part. This accusation against the old classical physiology has certainly been to some extent justified and is indeed confirmed by a statement made by one of its representatives already quoted above; owing to this exclusiveness physiology came to be an antithesis to morphology, which proved disastrous, not least because the experimental method and the physiological point of view no longer served the purposes of morphology, which in consequence reverted to narrow phylogenetical speculations. Botany was the first to free itself from this narrowness of view; indeed, there appeared at quite an early stage students of this subject who were capable of not only realizing but also solving physiological problems.

JULIUS SACHS was born in 1832 at Breslau, of poor parents. His studies at school were embittered by privation and could be continued only thanks to the kindness shown him by the aged Purkinje, whose sons were his school-fellows. In their home he found help and encouragement, especially in his interest in botany, which he displayed at an early age. When Purkinje moved to Prague, young Sachs's prospects looked gloomy, especially as he was now an orphan, but fortunately he was not forgotten by his old benefactor; he was allowed to go to Prague after him and to work in his institute as an assistant and draughtsman, while he completed his studies. Having received his degree, he was given a post as teacher of botany at the Saxon academy of forestry at Tharand, afterwards obtaining a similar situation at Bonn and finally being called in 1868 to Würzburg, where he laboured for nearly thirty years, gaining a brilliant reputation both as an investigator and as a teacher. In his best days Würzburg was an international centre for botanical research. Towards the close of his life his powers waned, and at the same time he lost his ability to follow the development of evolution; his self-conceit had always found it difficult to keep within reasonable bounds, and in his old age he simply could not endure any other opinion than his own. This eventually resulted in isolation, which embittered his existence. He died in 1897.

Sachs was one of those who was early won over to Darwinism, and throughout his life he viewed biology from the angle of the doctrine of descent. This was in fact the reason why in his otherwise praiseworthy *Geschichte der Botanik* he speaks so contemptuously of Linnæus, who to him was conspicuous only as a narrow-minded apostle of the constancy of species. But Nägeli has also exercised a great influence upon Sachs, who, for instance, associates himself with a view that there are internal causes of form-development in living creatures, and he also embraced the theory of protoplasm's being composed of solid particles capable of absorbing water in their intervening spaces. In regard to heredity, Sachs holds views most closely reminiscent of Weismann's germinal-plasm theory.

Sachs creates experimental plant-biology

SACHS, however, is best known as the creator of experimental plant-biology. This science had really made but little progress since the days of Saussure, for both the botanists of the romantic school and their immediate successors had mostly applied themselves to morphology. Sachs, on the other hand, conscientiously devoted himself from the beginning to the problem of plant physiology and the means of exploring it. Among his earliest works may be mentioned his investigations into the physiological part played by the chlorophyll granules, which he elucidated in its most essential features: he established the fact that starch formation is the first product of carbonic-acid assimilation, and, further, that sunlight plays the decisive part in this process. He investigated the different parts of the solar spectrum with the view to discovering their influence upon the alimentation of plants. Again, the continued metabolism and conveyance of nutrient substances within the plant has been worked out by him in all essential features. He studied all the vital processes in the vegetable kingdom with a view to ascertaining their intensity, which he expressed in graphic form. It was mainly, however, the movements of flowers that he systematically investigated; he established the dependence of the direction of growth upon the law of gravity and invented a rotating apparatus by means of which this growth can be studied under abnormal conditions. He is responsible for the method of investigation and thorough exploration of all that goes by the name of tropisms — at least as far as the vegetable kingdom is concerned. As a basis for all these phenomena he mentions irritability, a quality which he believes originates only in the living plasm and which he carefully analyses in respect of its various manifestations — an investigation that has been of immense theoretical importance for subsequent research. Finally, it may be mentioned that Sachs was the finest text-book writer of his time. Through his manuals the facts that he discovered and the ideas that he defended have spread far beyond the circle of his personal pupils.

Among these pupils WILHELM PFEFFER (1845–1920) is worthy of special mention. When still quite young he became a professor, first at Tübingen and afterwards at Leipzig, where he subsequently laboured as a brilliant teacher and investigator. He made a special study of the phenomena of growth in plants and the influence exerted upon them by both external and internal factors. He investigated a number of external influences and made valuable contributions to our knowledge of them, at the same time considerably improving the technique employed for their study. However, he emphatically declares that external influences do not act directly in a developmental way, but only cause a change of activity in the plant itself. He sees in the study of this reciprocal action between external influences and internal manifestations of life the very aim of biology, and "*die formative Determinierung der Zellen und der Organe*" becomes the object of his close analysis. In this study very careful attention is paid to all co-operating factors, so far as they can be calculated, and unwarranted attempts at simplification are rejected; it is thus definitely declared that cell-division is a physiological process and not merely a question of increased superficial distention within the cell, as mechanistically inclined investigators have tried to explain the phenomenon. Generally speaking, clear traces of Pfeffer's influence as regards the presentation of problems and general points of view are also to be found in experimental zoology, a fact, indeed, that some of its exponents have openly acknowledged. His labours have proved of still greater importance to botanical specialized research-work; he is generally acknowledged as one of the leading personalities in botany in modern times.

Another eminent pupil of Sachs was KARL EBERHARD GOEBEL (born in 1855), who was at one time an assistant at Würzburg and has since worked as a professor at several universities, latterly at Munich. He is an investigator of many parts, being a plant-geographist with a wide experience of the tropics, a morphologist, and a physiologist. As a morphologist he has always maintained the dependence of form upon function; morphology should no longer be kept separate from physiology, as formerly. He employed the old word "metamorphosis" to denote organic development, but not in the old idealistic sense; by it he would express the idea that a change in function produces a change in form. "Our idea of metamorphosis is thus in the first instance ontogenetical, and thereby experimentally conceivable and demonstrable," he says. According to him, no "indifferent rudiments" exist, for every rudiment has its peculiar qualities, which determine its development, and this can suffer change only as the result of definite changes in the vital manifestations. "If we call a leaf-rudiment at any stage 'indifferent,' it really means nothing but a denial of the causal connexion of evolutionary phenomena." These changes in the conditions of life can be produced experimentally, and extremely important experiences in regard to the organic construction of plants can be gained thereby. Space forbids our going closely into Goebel's experimental studies in "*Organographie*," as he calls the study of the relation between an organic form and function. He has hereby performed a great service in furthering the investigation of plant evolution and the mechanical process in evolution in general.

Even apart from Sachs's school there have been many important students of plant biology who have produced valuable results; a couple of them may be mentioned here. JULIUS WIESNER (1838–1916), professor at Vienna, has carried out useful experiments in the sphere of technical botany, especially in regard to the effect of light upon the vegetable world, which he studied in different localities and under different experimental conditions. HANS KARL ALBERT WINKLER (born in 1877), professor at Hamburg, brought to light the curious graft-hybrid phenomena: as a result of grafting related plant-species their tissues can be made to grow through one another in different ways.

In zoology the experimental method has made slow but sure progress. Strictly speaking, its development has gone hand in hand with the new conception of the fundamental problems of biology that has gradually usurped the place of the old phylogenetical idea, dating from the zenith of Darwinism. It may be said that the new principle had already been enunciated by Kleinenberg in his above-quoted saying that an organ's form depends upon its function and not upon its origin. Almost at the same time as this utterance was made, AUGUST RAUBER (1842–1917), professor of anatomy at Dorpat, sought to discover the conditions and laws governing the first construction of form in the vertebrate embryo. In opposition to the contemporary belief in the independence of the individual cells he maintains that the whole governs the parts and not vice versa; the egg-cell determines the directions of cleavage by its division of matter, growth is the primary and cleavage the secondary. Division into many cells facilitates the metabolism of substance and renders possible greater strength in the organism and, by specializing the elements, a far more extensive division of labour, but even when this division of labour is at its maximum, the organism remains a whole, the parts of which are developed under the influence of the whole. Through his efforts to find out in detail the conditions governing the various phases of development, Rauber became, along with His, Goette, and Kleinenberg, a precursor of the later school of evolutional physiology.

As its founder is named by universal accord WILHELM ROUX (1850–1924). He was born at Jena, where his father was a fencing-master. He studied first under Haeckel and then, at Strassburg, under Goette, and also in Berlin. He became professor, first at Innsbruck, then at Halle, where he worked during the period 1895–1921. He laboured with never-failing energy and powers of endurance, in speeches and in writing, as a teacher and an agitator, on behalf of the method of research and the line of investigation that he originated. As a research-worker he has already been outdistanced by younger minds, but he will always be regarded as a pioneer. However, the same line of thought has been followed from other quarters as well — as, for instance, by the disciples of the above-mentioned plant-physiologists, Sachs and Pfeffer, whose ideas were really in many respects in accord with those of zoological evolutional mechanics.

Roux was a pupil of both Haeckel and Goette; his works, in fact, bear traces of the influence of both, not only in his early days, but even at a far later age: even into the nineties phylogeny still represented the aim of his research work, and the struggle for existence and selection appear to him the most vital forces of life. But he would achieve this aim, not like Haeckel through a mere comparison between more primitive and more developed forms, but through investigating the mechanical process in ontogenetical evolution, such as through the program that Goette had in mind. As a matter of fact, Roux also has points of contact with Weismann, whose theory of the continuity of the germinal plasm he embraces; he consequently rejects the theory of the heredity of acquired characters. He resembles Weismann, too, in the fact that he likes discussing hypothetical entities of life, whereof he enumerates a whole series, which, however, it is not worth while dilating upon here. For these very reasons he was on bad terms with O. Hertwig, who, as we have seen, entertained quite different views, just as, on the other hand, he was at variance with Haeckel and other original Darwinists on account of his mechanical theory of evolution.

Creation of evolutional mechanics

THE theoretical speculations upon descent are, however, a less essential side of Roux's research work. He will mostly be remembered as the creator of experimental embryology; whether the theories that he based upon his experiments are eventually accepted or rejected, there is no doubt at any rate about the fact that he created a special method of research, which has proved productive and was largely employed by his contemporaries. But he also exercised considerable influence upon the theoretical conception of biology itself; he has directed research to a series of problems which many, following his precedent, have taken up for treatment and which have largely guided modern biological research. He himself defined as the aim of the new science that he desired to found the elucidation of the "true causes of formation" to which all living creatures and every single individual must attribute their origin — a subject in which the earlier "descriptive natural science," to his mind, failed to show any interest. These causes of "form-building forces," whereby the individual organism receives step by step the form that characterizes it, must, in his opinion, be studied primarily by way of experiment; if a process of development is altered by different kinds of interference, it is possible by combining the results to discover the cause of the process; thus is created a "causal-analytical form of research," such as chemistry and physics had already realized in many instances.

Natural selection within the organism

As previously mentioned, Roux began his activities with a theory of functional adaptation produced by means of natural selection within the organism. This theory he sought to apply to the organs of various vertebrate animals; he measured a large number of muscles in man and tried to determine to what extent their dimensions are dependent upon one another; he studied the caudal fin of the dolphin from a mechanical point of view and sought to determine the lines and curves in which these tissues are arranged in order mechanically to sustain the function of the whole. After a short time, however, he went over entirely to embryology, and in this field propounded the question to what extent and how far back in evolution certain organs and tissues are predestined to assume their prospective form and function, and whether this predestination can be affected by different influences. This problem, in fact, is the old antithesis of preformation and epigenesis formulated in a different way, and the discussion of the problem has to a great extent been carried on in the new form brought about by the employment of the methods of experimental embryology. Roux himself, starting from the theory of the continuity of the germinal plasm, which was influenced by Weismann and Sachs, maintained the idea of a very early determination of the various parts of the embryo; in his view, the first cleavage plane of the germinal egg establishes the median plane in the individual, the cleavage furrow is determined by the line of penetration of the sperm, and the front part of the future animal is already determined before the fertilization of the egg through the amassing of cytoplasm in that quarter. Each successive cleavage delimits a prospective part of the embryo; because in karyokinesis the substance becomes not equally apportioned to the daughter-nuclei, these latter acquire different values, so that, as he says, the whole process of development becomes a piece of mosaic work. In proof of his theory Roux carried out a series of experiments, which excited universal admiration at the time; he treated a newly-fertilized frog's egg in such a way that with a heated needle he burnt away one of the two first-formed blastomeres; then there developed at first a half-embryo, which afterwards regenerated in the usual manner of the Amphibia. Roux performed many other experiments with the same purpose in view, and though his technique was simple as compared with that which his successors afterwards elaborated, it was nevertheless he who first systematized this kind of research work.

Roux, however, was at once subjected to sharp criticism; O. Hertwig in particular, who from the very beginning had been an opponent of the Weismann heredity-theory and its founders, at once attacked the "mosaic" theory, maintaining that the different parts of the egg are by no means predetermined, but that, on the contrary, the egg's mass is equipotential-isotopic, as he calls it. Moreover, Hertwig, who invariably opposed narrow mechanistic tendencies in biology, strongly objected to the actual term "developmental mechanics," which appeared to him to imply an inadmissible schematizing of the phenomena of life. And, finally, he considered Roux's experiments to be inaccurate; he imitated them and found that the halved egg gave rise, not to a half-embryo, but to a whole embryo of small size. Hertwig found immediate support in DRIESCH, an observer who will be mentioned further in another connexion. He halved newly-segmented eggs of the sea-urchin and obtained from each half one larva half the size of the normal. On the basis of these and other experiments he propounded a special theory of evolution, according to which each part of the egg has, on the one hand, a "prospective value," and, on the other hand, a "prospective potentiality"; the meaning of these terms will best

be explained by an example: the half-egg of the sea-urchin has the prospective value of forming half a larva, but the potentiality to form a whole
larva; the former is thus the determination of the part that holds good in
normal cases, the latter the power of every part to compensate for another
if necessary. On the other hand, Roux found support in the American WILSON.
He isolated cleavage cells from the egg of a mollusc and found that they
became what they would have become in their normal connexion, and nothing more. And later Boveri discovered that the sea-urchin's eggs investigated
by Driesch actually possess a differentiation that from the beginning determines the succeeding developmental orientation. The fact is, the disputants
have gradually had to reconcile their views; Roux had to abandon his theory
of the different-shaped cleavage of the nuclei, while Driesch had to give up
his theory of the absolutely uniform character of the sea-urchin's egg. On
the whole, these experimental discoveries were too generalized; what was
discovered in the case of one animal's egg was applied without question to
the entire animal kingdom, whereas in actual fact a vast number of different
conditions prevail. In the main, however, it seems as if subsequent research
has obtained results that would indicate generally that a very early specialization and localization of the rudiments take place in the embryo. The
investigations that have been carried out especially by HANS SPEMANN, one
of the foremost champions of experimental morphology at the present time,
give some evidence of this. Born in 1869 at Stuttgart, he studied at Heidelberg, Munich, and Würzburg, and has been a professor at Rostock and
at the Kaiser Wilhelm Institute in Berlin; he is now working at Freiburg.
An extremely clever experimenter, he has specially taken upon himself to
prove the determination of different sections of the embryo by transferring
parts of the body of a batrachian embryo from one place to another and
then studying the successive development. As a general result he records that
the main organs of the amphibian embryo are definitely determined during
the process of gastrula formation; at an earlier period, in the blastula stage,
normally formed twins can be produced by means of dividing off with thread;
at the beginning of the gastrula stage the forward end can still be doubled
by means of binding, but after that the possibility of regulation decreases,
and disappears altogether when the gastrula is completely formed. We must
here pass over the details of the minutely planned and carefully carried-out
experiments by which these statements are proved, as also the particulars
of the various attempts he made to get mechanical influences to operate —
pressure, binding up, and such means, whereby different observers have
sought to trace out the forces operating inside the egg and to discover the
details of which they are composed.

Besides these purely mechanical experiments others have, of course, been
carried out, such as those in which eggs and embryos have been subjected

to the influence of electricity, light, heat, chemical compounds; of these the last in particular have produced results of great interest. Among such may be quoted the experiments of CURT HERBST (born in 1866), Bütschli's successor at Heidelberg. He placed sea-urchins' eggs in various saline solutions; in lime-free sea-water the eggs disintegrate into their first cleavage cells, which give rise to dwarf larvæ; the addition of lithium salt produces quite abnormally developed larvæ; treatment with sulphate-free sea-water also causes peculiar malformations in various parts of the egg. Undoubtedly the most remarkable of these chemical experiments in evolution are, however, those concerned with the initial development of the egg. In this field JACQUES LOEB not only has taken the initiative, but has also proved a leader. Born in Germany of Jewish parents in 1859, he studied medicine in his native country, becoming a doctor and assistant lecturer in physiology at Würzburg, but he soon migrated to America, where he held a number of professorships, the last one being at the Rockefeller Institute in New York.[1] Already, in the eighties, R. Hertwig had observed that unfertilized sea-urchins' eggs, if treated with a weak solution of strychnine, surround themselves with a membrane similar to that which appears after fertilization, before the cleavage begins. Similar observations were made later by MORGAN, the well-known student of heredity. Loeb took up this problem for systematic revision and achieved results that at once attracted great attention and in some directions produced very far-reaching conclusions and awakened high expectations. After performing a series of experiments he worked out a method of bringing the sea-urchin's eggs to the larval stage without fertilization. This method is somewhat complicated; first of all, the eggs are subjected to the influence of a weak organic acid, which induces the formation of membrane, after which they are placed for a carefully fixed period in sea-water, the salinity of which has been increased to one and one-half times the normal and which besides has been mixed with soda, and finally in normal sea-water, in which the development takes place. Several other simpler methods have also produced results, but they have been few and indefinite. The above method, however, subject to careful regulation, works safely, although, contrary to Loeb's statements, the larvæ thus formed are by no means invariably quite typical. On account of this, Loeb has tried to ascertain the chemical compounds that induce development and has come to the conclusion that the membrane formation is caused by the fat-dissolving capacity of the influencing acid, and that the spermatozoon, which upon penetrating the egg has the same effect, must produce a similar substance. The "hypertonic" sea-water then used "corrects" this effect and thereby contributes to the actual segmentation. Thus the multiplication of

---

[1] Loeb died in 1924.

cells would be reduced to a relatively simple sequence of chemical processes. There can be no doubt, however, that Loeb has simplified overmuch. Other investigators have, in fact, produced the same developmental phenomena by entirely different methods. Among those who have worked in this field may be mentioned, apart from Loeb's own pupils, YVES DELAGE (1854–1920), professor at Paris, who worked out his own method of developing the sea-urchin's egg, and A. BATAILLON, professor at Dijon, who acquired a name especially as a result of his experiments with the frog's egg; by simply pricking the egg with a needle dipped in serum, he caused the frog's eggs to develop into larvæ — a method that "activates" the egg in an entirely different way from Loeb's. Here we obviously have a metabolistic process within the very mass of the egg, set free by mechanical irritation — as a matter of fact, the dose of serum has latterly proved to be superfluous — although the adipose splitting can in this case also be established; it is here induced by the egg's own vital manifestations and not by any solvent introduced from outside. Loeb's theories, to which we shall revert, are governed entirely by his lack of interest in morphological phenomena and by his consequent passion for schematizing the complex vital process. Strictly speaking, most other evolutional physiologists of the new direction, from however mechanical a point of view they may otherwise have regarded evolution, have nevertheless divided its mechanical phenomena into those having a purely external origin and those that result from specific internal causes. Among the former are counted the universally observable influences of heat, electricity, chemical reagents; among the latter, the various organisms' peculiar ways of reacting to them, as well as to purely mechanical interferences, with their normal existence. In the course of studying these heterogeneous phenomena of reaction many of the experimental observers of our own time have produced and are still producing a great number of detailed results of immense interest, achieved by the employment of exquisitely delicate methods. For the details of this field of inquiry — still by no means exhausted — we must refer the reader to technical literature on the subject; a number of general points of view that have arisen in the course of the work carried out in connexion with these subjects will be discussed later on. We shall instead pass on to another form of experimental research — a field that is without doubt of the greatest interest to humanity at the present time.

2. Experimental Heredity-research

Earlier ideas on heredity

"INHERITANCE" and "heredity" are terms that originally belonged to the judiciary and have been borrowed from it to acquire a natural meaning.

Just as the right of a child to take over his parents' property is called inheritance, so the word is used to denote the fact, which has been known of old, that children resemble their parents in body and soul; facial features and figure are said to "be hereditary" from father and mother to son and daughter. It can be no matter for surprise that from the beginning the meaning of heredity in the biological sense has been considered to be this: the direct transmission of qualities from parents to children — that, in fact, has been the idea up to recent times and it has not been until the details of this transmission came to be studied that a deeper insight has been gained into the true facts, and an entirely new conception has taken the place of the old theory of transmission. It is through this research work that the problems of evolution have for the first time been dealt with on an entirely exact basis; the same mathematical exactness that formerly only experimental physiology was capable of achieving now characterizes the methods and results of heredity research.

Exact heredity-research has received contributions from various quarters. Investigators with a Darwinistic training have made weighty contributions to it, but besides these, others — and, in fact, the most valuable of all — have come from circles that have had nothing whatever to do with Darwinism. In a previous section have been mentioned the investigations into the hybridization of plants that were carried out in the eighteenth century by Koelreuter. His experiments were taken up by many other students, among the most highly reputed of whom may be named KARL FRIEDRICH GÄRTNER (1772–1850), a medical practitioner by profession, whose elaborate experiments with plant hybrids brought him a great reputation and were especially taken advantage of by Darwin. The experiments carried out by the Frenchman LOUIS LEVÊQUE DE VILMORIN (1816–60) were of a different type. De Vilmorin belonged to a family that for generations had carried on trade in grain and seed-cultivation; he himself was particularly interested in sugar-beet, which he cultivated with a view to increasing its percentage of sugar. In this he started from the principle that the offspring of each individual should always be kept separate; he collected the seeds of beets with a high sugar-content and sowed them separately, with the result that he obtained cultures of a very valuable quality; in doing so he discovered that individuals that look alike might have entirely different characters, and he thus came to hold the view that the power of inheriting characters might in itself vary and give rise to heterogeneous offspring. Through these results de Vilmorin became a pioneer of modern heredity-research, and his theses, based, as they are, upon exact observations, possess quite a different value from his contemporaries' "philosophical" speculations on heredity, numerous traces of which are to be found, *inter alia*, in belles-lettres of a naturalistic type.

The problem of heredity was dealt with from an entirely different point of view by FRANCIS GALTON (1822–1911). He was a cousin of Darwin's and his life reminds one of the latter's. The son of wealthy parents, he studied medicine, but never graduated, for as soon as he had inherited his father's fortune, he went on voyages, first to Eygpt and then to south-west Africa, where he made valuable geographical discoveries. After his return to England he applied himself for some years to meteorology, but eventually devoted his life entirely to heredity research. He began by seeking to prove experimentally Darwin's pangenesis theory, which, it will be remembered, assumes that particles from all the organs of the body are transmitted through the sexual products from generation to generation and bring about the offspring's resemblance to the parents. Galton injected blood from a foreign rabbit into a pair of grey ones, in the hope that their progeny would thereby become dappled, which would have been a proof of the pangenesis theory; but his expectations were not fulfilled, the young of the grey rabbits turning out grey. In virtue of this experience Galton produced a heredity theory of his own: all the organic units existing in a fertilized egg he terms "stirp," and he believes that the majority of these are used for purposes of organic structure, but that a number of them are left over, and these give rise to the sexual cells, the qualities of which are thus not influenced by the conditions of life of the individual. He gave expression to this opposition to the doctrine of the heredity of acquired characters as early as in 1875 — that is, before Weismann — but after that he passed on to quite different developmental problems. From the very beginning the evolution of man had been a subject of special interest to him; he is extremely fond of demonstrating physiological phenomena with the aid of social and political comparisons. The fact that brothers and sisters are often so unlike one another is explained by a reference to an electoral body, where a very slight difference of opinion can often give rise to an entirely different result when it comes to the vote; in the same way, a slight change in the hereditary substance can produce a complete change in the appearance of the individual.

Galton's statistical method

IN a later work, *Natural Inheritance*, Galton has presented the statistical heredity-theory that has made his name famous; his stirp theory is here abandoned and the transmission of acquired characters is no longer denied so absolutely as it had been before. Instead, basing his arguments on an exhaustive research-material, he tries to find out the variational direction in a large number of human qualities. The most highly valued of his investigations have been those into the question of height; the height of a number of parents was compared with that of their grown-up children arranged in series from the shortest to the tallest; there is thus obtained an average height, which is possessed by the majority, while the less numerous extremes of abnormally short and tall persons group themselves in either direction from the middle. Now, Galton finds as a result of his heredity research a variation towards the mean value, the children of the tallest groups of parents having become tall, but not so tall as the parents; those of the shortest parents, on the other hand, having become short, but not quite so short as the parents themselves. From this Galton infers that there is a heredity variability in a certain direction, which natural selection, of course, influences. This result, seeing that it first came to light when Darwinism was at the height of its popularity, was bound to win many supporters, as also did Galton's principle that it is necessary to deal statistically with as large a number of cases as possible, seeing that law-bound necessity in isolated cases is effaced by incidental circumstances. This principle was bound to attract a generation that preferred to regard humanity collectively and placed but little value on what was purely individual. Later, however, it has been found that this collectivism actually constituted Galton's most serious weakness; it is practically an impossibility to draw conclusions regarding the individual case from statistical mass-calculations, and if it is attempted, it leads to absurd results. On the other hand, Galton's service lies in the fact that he introduced exact measurements and mathematical calculations into the theory of evolution; his method of expressing the details of development graphically by means of curves has since been applied with great success by students who were able to isolate well-defined phenomena and to follow them up through different generations. For Galton's chief weakness was really this, that he believed that he could deal with practically anything statistically; he never realized that an object which is to be examined must first have its true essence determined. Thus, in his above-mentioned work he tried to determine statistically the laws governing "marriage selection" — *inter alia*, whether persons of different dispositions feel attracted to their likes or vice versa. From one of his tables it appears that 46% of married men are ill-tempered; of these, again, 22% have had good and 24% bad-tempered wives.[2] Statistics of this sort seem far more suited to a comic paper. The fact that Galton seriously tried to solve such a problem testifies to his extraordinarily dilettante mind, which cannot be excused by the fact that even at a later period an occasional student of heredity has sought to ascertain the existence and transmission of equally vague and indefinite human qualities. As a matter of fact, Galton applied his method not only to human beings; he also experimented in horticulture, dealing with the results thus obtained by the same statistical method. He bequeathed his fortune to an institute for heredity research in London, which afterwards worked in accordance with the principles that he had laid down. Galton had human welfare very much at heart; he wanted

---

[2] See "Appendix D" in that work.

to create a better human race and desired that all research work should serve that object; he gave to the science that he placed highest of all the name of "eugenics," a name that has become universally accepted.

A new phase in the history of heredity research was introduced by HUGO DE VRIES. Born at Haarlem in 1848, he studied at Würzburg under Sachs, held various posts in Germany, and finally became professor at Amsterdam. He applied himself first to the study of plant physiology and published valuable results of his investigations into the pressure conditions in plant-cells. At the same time he speculated over Darwinistic problems. He, too, produced a theory of life-entities, which he called "pangens," by which he meant those qualities in the organism which are capable of independent variation, and each of which must, in his view, be represented by one material entity. Like so many other biologists of the younger generation, however, he entertained doubts as to the ability of the traditional Darwinism to solve the problem of evolution. He was especially preoccupied with the undeniable fact that the species in nature remain constant and that the slight transitions whereby, according to the old theory, one species is converted into another can never be observed; the species is a self-contained entity, and yet the conversion of species must have taken place in the course of the ages. Kölliker's previously mentioned theory of sudden changes of species seemed to him to offer the possibility of adjusting this inconsistency, nor, indeed, had Darwin himself denied the existence of sudden "single variations." It was only a question of obtaining actual proof of the existence of such changes. Eventually de Vries believed that he had found it in *Œnothera lamarckiana* (evening primrose), a plant introduced from America, which has spread over various European countries. This plant grew in masses in a meadow in the neighbourhood of Amsterdam and exhibited, besides a number of typical forms, some few with an entirely divergent appearance. A number of specimens having been transferred to a garden and there allowed to multiply by self-fertilization, it was discovered that in the course of a few years there developed out of seed of the old species not only forms similar to it, but also isolated specimens with well-marked new species-characters, which were retained for the purpose of further cultivation: among them a dwarf form, a markedly latifoliate form, and some others. Here, then, we get, according to de Vries, a species that suddenly "exploded," as he expresses it, and gave rise to a number of new species, each with definite characteristic features. This case shows, according to him, how the species in general have arisen; the species, he says, are no arbitrary groups, but completely independent entities, delimited in time and space, which originate through old species' suddenly disintegrating into a number of new forms; of these some are capable of life and survive unchanged until the next mutation, while others cannot sustain the struggle for existence and succumb to natural selection. All fresh characters have thus been formed as a result of mutations; between the mutations a species survives with its characters unchanged; the slight variations that occur daily in the life of the species have no effect on evolution, because they are not hereditary, and the recombinations of characters that arise through the crossing of different forms have no new significance. "As many steps as an organism has made from the beginning, so many mutation periods must have occurred." These mutation periods, de Vries believes, must have arisen at a far more rapid pace during previous geological periods than they do nowadays, and he is thus able to explain on the basis of this theory the origin of the forms of life without the assumption of those infinite spaces in time that the old Darwinism required at its disposal.

When it first appeared, de Vries's theory naturally met with violent opposition on the part of loyal Darwinists. It was certainly admitted that in all essentials he really accepted the point of view of the old Darwinism, in that he maintained the theory of natural selection as the principle governing life, but the fact that he denied the heredity of the slight variations and their importance for selection and maintained the immutability of species in the normal existence between mutations was far too much at variance with old traditional ideas to be acceptable. Every possible effort was made to get away from the facts that he had adduced and the conclusions that he drew therefrom. As a matter of fact, these certainly have been open to objection. His cultural experiments with Œnothera have failed to withstand the criticism of later years. Johannsen has objected that the material with which the experiment was carried out was casually selected and was not kept as pure as it should have been, and finally a Swedish naturalist, HERI-BERT-NILSSON, carried out the entire experiment over again and came to the conclusion that the new generations of Œnothera lamarckiana only show fresh combinations of characters that already existed in the main species. De Vries, who was one of those who rediscovered Mendel's law of cleavage, has, in fact, denied the validity of that law as regards mutations, such as those of Œnothera, but this has been found to be a mistake. Consequently his theory of the formation of new species of that plant collapsed. His service to science, on the other hand, lies in the fact that he revealed the phenomenon of mutation, for that this phenomenon exists has since been proved over and over again. Moreover, on the basis of his "pangen" theory he insisted upon the necessity of analysing with regard to their elementary units the hereditary qualities that characterize the species. "It is not a question," he says, "of the origin of species, but of the development of the species-characters." Through these assertions he became a pioneer of modern heredity-research. "His mutation theory," Johannsen declares, "has represented the principal milestone in the transition from the old ideas to the modern conception of heredity and will always, therefore, retain its historical significance."

WILHELM LUDVIG JOHANNSEN was born in 1857 at Copenhagen, studied there and in Germany, and eventually became a professor, first at the Institute of Agriculture and afterwards at the University in his native country. Being a pupil of Pfeffer, he first of all went in for experiments in plant physiology, making a number of interesting observations in this sphere in connexion with the effect of ether on the metabolism and growth of plants. Soon, however, he devoted himself exclusively to experimental heredity-research and has gradually become one of the leading authorities in that field. Originally a supporter of Galton's statistical method, he quickly realized its deficiencies: that by working with mixed material it reached conclusions utterly at variance with the true facts. Starting from de Vries's insistence upon the necessity of investigating the units of hereditary characters, he began to follow with minute care the phenomena of heredity in generations of plants. He purchased a quantity of beans, weighed them, and then cultivated the seeds of every bean separately; he thereupon found that within a succession of individuals thus produced — a "pure line," as he called it — there exists a certain type of hereditary units, which remains unchanged throughout; if the plants are starved, both they and their seeds become small; if they are manured, they grow strong, but this has no effect upon the hereditary character; whether one sows small or large seeds of the same pure line, one obtains under the same external conditions the same plant-type. There is thus within one and the same pure line a certain hereditary type — a "genotype," as it is called — that is unalterably the same, whether or not the vital conditions alter the external form of the actual individual — that is, the phenomenon-type, or "phenotype," as Johannsen called it. Those characters which form the genotype are thus the only ones that are really hereditary, whereas the phenomenon-type produced by environment has nothing to do with heredity; stunted growth in generations of plants growing on poor soil is an external character, a "false heredity." There is therefore no possibility of acquired characters' being inherited, nor is there within the pure lines any chance of variation of the kind assumed by Darwinism. Those characters that go to make up the hereditary disposition Johannsen terms hereditary factors, hereditary units, or genetic elements; in a pure line, then, the genetic elements are the same: its individuals are homozygotes ("zygote" denotes the fusion-product of the male and female sexual cells), while the offspring in life-forms that multiply by the pairing of different individuals is a heterozygote, as it represents a fusion of the parents' various inheritable factors. A heterozygous individual therefore always has a hybrid nature, and special methods are necessary for the elucidation of its qualities. But at this point we come to the subject of hybrid research, which possesses a history of its own.

JOHANN MENDEL was born in 1822 of peasant parents at Heinzendorf, a German colony in the midst of the Slav population of Austrian Silesia. Having shown remarkable intelligence at an early age, he was sent to a grammar-school, and, probably with a view to obtaining better opportunities for devoting himself to study, he entered an Augustine monastery at Brünn in the district of Moravia. As a monk he adopted, after the Catholic custom, a new Christian name, GREGOR, by which he became known to posterity. He was sent at the expense of the monastery to Vienna, where he studied for three years, devoting himself especially to mathematics and natural science; upon returning home he became a schoolmaster and in his leisure hours cultivated plants in the cloister garden for scientific purposes. He published the account of his results in the little-known "treatises" that were brought out by the natural-science society at Brünn. In 1868 he was appointed head of the monastery, or prelate, as it was called. This appointment, however, actually proved his undoing. Four years later the then liberal parliament in Austria sought to reduce the country's financial distress by, *inter alia*, taxing the monasteries. The monks, like all the reactionary parties in general in the country, considered that the tax menaced the monasteries' ancient privileges and set themselves up in opposition to it. Eventually, however, the measure was carried through in several instances, but the one who refused to give in was Mendel; for twelve years he held out, defying penalties and warrants of distraint, but finally he broke down completely under the struggle, contracting a sickness that resulted in his death in 1884. Thus fell one of the pioneers of modern biology as a champion of Catholic clericalism — in its way an irony of fate.

Mendel's experiments with peas

MENDEL's fame, which was late in coming, rests simply and solely upon two short essays in the above-mentioned journal. They are, however, the fruit of many years' work and testify to a keen observation of nature and a thorough grounding in mathematical thought, which do not often go together; Darwin, for instance, had a genius for observation, but the summary accounts of his observations are vague and obscure; Galton was a mathematician, but he worked mostly upon material obtained second-hand as the result of inquiry, so that it was not truly accurate. Mendel applied himself to the study of the phenomena of heredity in garden plants; he selected, to start with, certain easily observable characters — the colour of the flowers, the shape of the seeds, the structure of the position of the blooms — and he studied their modifications in different generations. He crossed peas with white and red flowers; the hybrids then proved to be red throughout; when, again, these hybrids were allowed to fertilize themselves, the succeeding generations turned out to be coloured in a peculiar way: for every three red individuals there was one white. These white, if self-fertilized, invariably produced white offspring, one-third of the red remained similarly constant, while the remaining red flowers repeated the above-mentioned colour ratio.

Dominant and recessive characters

THE explanation that Mendel gave of this strange phenomenon was as ingenious as the observation itself; the fact that the flowers of the first hybrid generation are red and acquire no intermediate colour between red and white he accounts for by the red colour's being dominant over the white, which latter character he calls recessive. In the succeeding generation both dominant and recessive characters again appear; transitional characters cannot be observed. From this he concluded that in the sexual cells, or gametes, there exists no fusion of characters; in the hybrid red-white exist potentialities for red and white side by side in the male and the female cells; when these are united, the fusion must thus take place in accordance with one out of four possibilities: red-red, red-white, white-red, white-white. This explains the proportion between the offspring's qualities; the two single-coloured combinations no longer vary, but remain constant, while those with the double characters are capable of repeating the same four possibilities so long as they exist. The same system of law-bound heredity was shown by all the characters that Mendel investigated in various plant species. He afterwards took up experiments with the crossing of bees, which, as is well known, produce many different racial types, but, disappointed over the lack of encouragement that he received as a result of his investigations into plants, he did not publish the results of this subsequent research-work, and they have now been lost to us. During his own lifetime Mendel's achievements attracted no attention whatever; Nägeli, as we have seen, found them irreconcilable with his own theories, and the other botanists displayed utter indifference. It was not until the turn of the century that Mendel's remarkable results were rediscovered in connexion with the hybrid research-work that was then being carried out. Three observers — DE VRIES, CORRENS, and TSCHERMAK — simultaneously pointed out the agreement between Mendel's observations and their own results. Thenceforward Mendel's name has been one of the best-known in biology; even among the general public his fame has in more recent times competed with that of Darwin himself. Much surprise has been expressed over the fact that Mendel's brilliant observations did not attract greater attention, and the blame has been laid upon the unknown journal in which they were published. One might with greater justification ask oneself whether any of the more important publications of the time would have undertaken to print results of research so utterly at variance with the prevailing conception of biology. We have only to remember that Mendel denies variability in those characters that he observed, whereas all the biologists were just at the time seeking after variations as material in proof of natural selection; and then come these assertions as to absolutely constant or constantly divisible characters from the pen of a monk in a monastery! It would certainly have been a miracle if they had found support from the generation that had been brought up on Haeckel's *Natural History of Creation*.

Universality of the Mendelian laws

NEVERTHELESS, the Mendelian principle of cleavage now forms the basis of all hybrid research. All characters that it has been possible to observe in living beings have the quality of "mendelizing"; de Vries's statement that mutations cannot be subject to this law has proved to be incorrect, and the exceptions that have since been observed have actually been explained in accordance with the same principle. Mendelian research has been carried on to an ever-increasing extent year by year, both by theoretical observers and by practical breeders. And Mendelism has stood the test in regard to the improvement of seeds and domestic animals no less than in the theoretical field; it is only by its aid that the practical improvement of breeds has been successfully based on exact principles instead of on mere chance. In Scandinavia Mendelism has won many adherents; Johannsen has contributed greatly to its advancement, and in Sweden H. NILSSON-EHLE especially has applied it to practical purposes with universally acknowledged success; his work for the improvement of seed cultures, carried on at Svallöv, has been done in accordance with its principles and has received widespread recognition. And both in Europe and America there are a great many Mendelian students — to name only a few, W. BATESON and R. C. PUNNETT, in England, ERWIN BAUR and CARL CORRENS, in Germany, the previously mentioned A. LANG in Switzerland, L. CUÉNOT in France, T. H. MORGAN in America. As a result of the investigations of these and many others more and more light has been thrown upon the whole complex and manifold profusion of varied hereditary factors and their mutual relation. In this field, indeed, there must be a vast amount of research material; it only remains to select certain characters of importance in one respect or another and follow them up. This has in fact been done, and numerous Mendelian students have taken up various subjects for research at which they have worked with an ever-increasing tendency to specialize. Of these subjects we shall examine one somewhat closely — namely, that which has been made possible through the application of the methods of modern cell-research to the problem of heredity.

The Morgan school

THE home of this cytological heredity-research has mainly been America. The experimental biologist EDMUND BEECHER WILSON (born in 1856, professor at Columbia University), to whom we have previously referred, had already made valuable studies of the influence of the reproductive chromosomes upon heredity, but it was really a scientist who had been trained in the same school — namely, the above-mentioned THOMAS HUNT MORGAN (born 1866, likewise a professor at Columbia University) — who discovered both the aim of this research work and the means for carrying it out, thereby providing the study of heredity with a wealth of material by way of detailed discoveries of far-reaching theoretical application, such as had never been found elsewhere. His subject for investigation has been a small parasite fruit-fly, *Drosophila melanogaster*, of which it has been said that it has apparently been created by God solely as an object of heredity research. It reproduces itself with incredible rapidity and profusion — in heat an individual requires only twelve days to develop from the egg to sexual maturity — it is extraordinarily hardy and can stand every possible kind of experimental treatment; its cells contain only four pairs of chromosomes, all of a different size and easily recognizable; and, finally, it has given rise under laboratory conditions to a profusion of mutations, the factors of which have been found to be constant and well adapted to Mendelian investigations. Morgan and his numerous pupils have examined millions of these creatures, in the course of which he has built up a methodology of his own and a terminology, which, owing to its refined subtlety, is extremely difficult for anyone but a specialist to comprehend. Among the results of this research work we note, first of all, a number of fresh principles, as hard and fast as Mendel's original ones. As had already been assumed, the hereditary factors are localized in the chromosomes, and it has been discovered that the factors in the same chromosome are not free, but invariably follow one another upon cleavage: they are "linked," as it is called. Thus, all the factors in this animal are grouped into four linkage-systems, one for each pair of chromosomes. Further, after a series of ingenious experiments it has been possible largely to determine the position of the factors in each chromosome, and, finally, in certain cases to establish the absence of parts of chromosomes or entire chromosomes as being the cause of various external modifications — that is, a set-back for the theory of the absolute constancy of the chromosomes in the same species. Besides this, it has to a large extent been made possible to ascertain the part played by the previously mentioned sex-chromosome in the determination of sex and in heredity — a problem which, as a matter of fact, observers not belonging to Morgan's school have also studied. In this province likely explanations have been found for a number of hitherto incomprehensible phenomena of heredity; among those that are generally known may be mentioned the inheritability of certain diseases, such as hæmophilia and colour-blindness in man, and, further, a number of cases of heredity in the sphere of mental diseases. And, finally, mention should be made of the valuable studies in regard to the relation of the chromosomes in species hybrids; in this sphere may be mentioned, of Scandinavian students,

H. O. G. ROSENBERG, of Stockholm, who has investigated the hybrids of Drosera, H. FEDERLEY, of Helsingfors, who established the fact that in the butterfly hybrid the number of chromosomes is equal to the total of the father's and mother's combined, and O. L. MOHR, of Oslo, who has carried out independent investigations on the reproductive chromosomes and has besides done valuable work in subjects dealt with by the Morgan school.

Heredity has been the most popular field of research of the age; it has succeeded Darwinism in the way it has taken hold of the public mind and has nowadays to serve as an explanation for anything that presents any difficulty in the various spheres of life. Just as formerly it was natural selection, so now it is the mixing of breeds that has to bear the blame for every kind of circumstance and disparity even in human community life, in which political and social prejudices take good care that problems are at least not treated impartially in the scientific sense. As a result of exact heredity-research the theory of evolution has itself been directed along other lines; phylogenetical speculations have for the most part been abandoned — at least for the time being. Natural selection is certainly retained in principle by some students of heredity — by Baur, for instance — but it is really of no practical importance; the phenomenon cannot be observed and it is therefore not possible to fit it into a subject of research that is based on exact observations. And while the old Darwinism operated with outward resemblance as a positive proof of common origin, heredity research has established the fact that resemblance and affinity are not analogous terms, thus undermining the very foundations of phylogeny. Generally speaking, heredity research goes to work in a more limited sphere; for the very reason that it has become an exact science it has not been able to follow the old Darwinism in its speculative ranging, but whatever may have been lost in the way of the general conception of life has undoubtedly been won in the way of concentration on facts and reliable results.

There are still one or two other subjects for experimental research which must be briefly dealt with in this chapter, and we shall now pass on to these.

3. Biochemistry

Application of modern chemistry to biology

THE science of chemistry as applied to biology has always afforded it valuable assistance in the search for an explanation of vital phenomena. In modern times, as is well known, chemistry has made splendid progress, and every new step has at once had its influence on our knowledge of living organisms. At the present day biochemistry is a line of research that, in point of the

THOMAS HENRY HUXLEY

From a portrait by the Hon. John Collier, reproduced in *Some Apostles of Physiology* by William Stirling, Waterlow and Sons, Limited, London

OSCAR HERTWIG

value of its results, competes well with experimental morphology and heredity research. Unfortunately, these results are far less accessible for the purposes of popular presentation than any other of the advanced spheres of
biology; in fact, biochemistry requires a very special technical training for
both its students and its critics. Nevertheless, space may be found here for
a few brief indications as to the most important progress that has been made
in its various special provinces in order to complete the picture of the general progress made by experimental biology in modern times.

That branch of chemical research which has had the greatest success
in our own day and has excited the keenest interest is undoubtedly physical
chemistry; each stage of its progress has at once been applicable to the living substance. Biology has actually led the way in certain physico-chemical
discoveries, as in the question of osmotic pressure, in which the results of
Pfeffer's and de Vries's research-work formed the foundations on which
scientists have subsequently built further. On the other hand, the modern
theory of solutions, as created, among others, by van t'Hoff, Arrhenius,
and Nernst, has contributed towards explaining a great many biological
phenomena; as an instance may be mentioned the above-described parthenogenetical phenomena in eggs that have been subjected to hypertonic saline
solutions; further may be quoted the part played by hydrogen ions in the
metabolism of the fluids of the body: they play a decisive part especially in
producing respiratory irritation, while other ion-combinations have been
found to be necessary for growth in individuals and organs.

Colloid chemistry

Of special significance for forming a conception of the nature of protoplasm
has been modern colloid chemistry; it has founded a province of its own,
with its own methods, which have made it possible to study far more closely
than before the most minute structural details in the category of elements
in which living substance is included. Thanks to these accurate observations
and experiments, the granular and vacuolized structure of plasm has been
given a far more natural explanation than that once given by Bütschli in
his froth theory. It has in many instances been possible to compare the mutual interpenetration of the various structures even with physico-chemical
metabolistic phenomena occurring in inanimate colloid substance, while,
on the other hand, the old dispute as to the solid or fluid nature of plasm
has lost all point; the intrinsic character of and changes in the colloids are
investigated on entirely different principles and have given rise to problems
utterly different from this old question of the state of aggregation. Instead,
extensive investigations have been made into the question of the penetra-
bility of cells and tissues by solutions of various kinds; in this sphere
especially Charles Ernest Overton (born in England in 1865, and after
studying in Germany appointed professor in Sweden) propounded a theory,

which has been both contradicted and supported by others, that the cellular membranes consist of a peculiar category of elements called lipoids; elements dissolved in these permeate the cell-walls, and vice versa.

Ferment chemistry

WITH colloid chemistry the modern chemistry of ferments is closely allied. In regard to ferments it has been known for a hundred years that through their presence in extremely small quantities they are capable of causing various chemical changes in large masses of the substances on which they are acting. In modern times this influence they exert has been compared with the catalytic effect of certain inorganic substances (first pointed out by Berzelius), as, for instance, the part played by acid in the production of ether out of alcohol. Of fundamental importance is the discovery that not only are phenomena of disintegration induced by ferments, but also synthetic processes, as, for example, the production of starch in the leaves of plants through the chlorophyll granules under the influence of sunlight. The ferment syntheses of fats, albuminous substances, and other products of vegetable and animal life, which have been the objects of special study, are in themselves of immense interest, but they cannot be discussed in detail here; nor can we describe the extremely subtle investigations that have been carried out in connexion with the co-operation of various ferments within the same vital unit.

In connexion with fermentation research mention should be made of the process of internal secretion, our knowledge of which has increased more and more in recent times. BERNARD, whom we mentioned earlier, may claim to have been its discoverer. As we have seen, it was he who established the fact that the liver produces substances which are directly carried away by the blood. CHARLES EDOUARD BROWN-SÉQUARD (1817–94) succeeded to Bernard's professorship after the latter's death. The son of an American father and a French mother, Brown-Séquard studied at Paris and afterwards worked in England and America, but later on returned to France. He became famous for his valuable investigations in the sphere of neuro-physiology, but more especially for his experiments, which he published in his old age, with the injection of extract of genital glands, whereby he demonstrated that these glands contain a special secretion inducing sexual desire. The experiments in themselves were somewhat clumsy and were published with a good deal of advertisement, especially as regards the power of rejuvenating the individual, which was promised as a result of them; all the same, it cannot be denied that through them attention was drawn to a fact that has since been investigated by others with greater thoroughness than before. The interstitial gland of the genital organs has been anatomically investigated by the two collaborators ANCEL and BOUIN, of Nancy; its physiological function and, in general, the influence of the genital organs upon the vital manifestations of the body have been studied, *inter alia*, by J. MEISENHEIMER, of Leipzig, who experimented on the larvæ of butterflies, and by EUGEN STEIN-ACH, of Vienna, who studied vertebrates from the same point of view. Stein-ach has succeeded by means of operations in influencing the sexual character; the genital glands of male and female animals have been interchanged, with the result that both the outward appearance and the sexual behaviour of the animals have been correspondingly altered. Otherwise, Steinach is best known for his having resumed rejuvenation experiments similar to those of Brown-Séquard, but more carefully carried out both in theory and in their details. These experiments have won him a fame that, owing to the nature of the problem, has become associated with much of the glamour that surrounded his predecessor; moreover, they appear already to have exhibited defects that have rendered them impossible of realization in practice.

Internal secretion and rejuvenation experiments

IN the mean time our knowledge of internal secretion in other spheres has increased with great rapidity. So-called endocrine glands have been discov-ered in large numbers; among them may be mentioned the suprarenal cap-sules, hypophysis, the thymus and the thyroid gland; and our knowledge of their functions has at the same time been extended. But other organs have also become known as producers of internal secretions, or hormones, as they are also called; the small intestine produces such a secretion, which is termed "secretin" and which, when conveyed through the blood to the pancreas, induces secretion in that gland; another similar substance is produced in con-nexion with the pregnancy of female animals and causes the segregation of milk, and a third is the product of a special cell-category in the pancreas and has a definite influence upon the metabolism of the body, in that its absence induces diabetes. On the whole, many of these internal secretions have become known only by indirect means, through the diseases that arise if the organs which produce them are injured or removed.

Serology

IN modern times serology forms a separate field of research, embracing one of the most important chapters in practical medicine. Pasteur should be named as its founder, while later BEHRING, EHRLICH, and their pupils have particularly distinguished themselves in that subject. It has been established that the danger of the disease-producing bacteria lies in the fact that in the course of their multiplication in the body they produce special isolatable chemical compounds having a specific poisonous effect, which have been given the name of toxins; the body reacts against these by forming similarly specific elements, antitoxins, which counteract them. To isolate these latter and to use them as a counter-poison has in many cases proved the only way for medical science to cure infectious diseases. By injecting the bacterial poi-son into experimental animals, these latter have been allowed to produce the antitoxin, their blood-serum afterwards being used as a counter-poison against the disease. The actual elements — both toxins and antitoxins — that operate in this process always exist in very small quantities, and for that reason and owing to their complicated composition it has not been possible to analyse them; their quantitative effect upon one another has nevertheless been investigated by SVANTE ARRHENIUS and others.

In connexion with this research work the nature of the blood in general has been the object of very detailed investigations, often with quite remarkable success. Among the best-known of these is the precipitin reaction, which was discovered by PAUL UHLENHUTH (born 1870, Behring's successor at Marburg). If we take serum from an animal — for instance, a dog — and inject it into a rabbit, we obtain from the rabbit's blood within some days a serum that produces precipitation in the serum of the dog's blood. This reaction is specific and can therefore be utilized in medico-legal investigations in order to distinguish, for instance, human blood from animal blood. In this, however, similar forms act in an identical way; for example, human blood and the blood of anthropoid apes produce the same reaction. When this chemical resemblance between allied organisms was discovered, it excited great enthusiasm as a phylogenetical argument; closer consideration, however, at once makes it clear that this agreement of chemical composition demonstrates just as much or just as little as the morphological resemblance that can be demonstrated by the old comparative method; the fact that resemblance in bodily structure, food, and habit is accompanied by corresponding chemical agreement is essentially so obvious that the contrary would be more surprising.

We may here cite one more example of discoveries in this sphere. We have previously mentioned how the Russian naturalist ILJA METSCHNIKOFF (1845–1916), who after studying at German universities was a professor at the Pasteur Institute in Paris, produced the so-called phagocyte theory; he found, as indeed Haeckel had already observed, that the white blood-corpuscles absorb foreign substances into the body; in particular, he discovered that the leucocytes in this way free the body from bacteria that enter it, provided the latter are not too strong and do not get the upper hand. In more recent times it has been found that special substances are produced, which have been called "opsonins," which possess the ability to increase the leucocytes' power of killing the bacteria. These substances, however, are at present little known, but they are, of course, of the very greatest interest. They have been mentioned here as examples of the wide possibilities with which modern biochemistry has to reckon. The immense practical benefits that this research work has brought humanity can only be hinted at here; some of the theoretical speculations to which it has given rise will be dealt with in the following.

4. Animal Psychology

ANIMAL psychology would rightly have been made part of experimental biology if it had consistently remained on the basis of empirical research. This, however, has been far from being the case; on the contrary, ideas about the psychic life of the animals have varied infinitely up to the present day, from the standpoint of primitive man, who ascribes to the animals an intelligence of the same kind as his own, to Descartes's conception of animals as completely automatically operating mechanisms. The reason for this confusion is, of course, to be sought in the vague ideas of human psychology, which have varied according to the general point of view taken by the different schools of philosophy. It was not until the middle of last century that an exact psychological research began to appear, thanks to precursors like THEODOR FECHNER, WILHELM WUNDT, and their pupils. This empirical psychology treats of psychical phenomena just like any other material for observation; it is, as Höffding says, "no more bound to begin with an explanation of what the soul is than physics is bound to begin with an explanation of what matter is." Unfortunately, those biologists who have dealt with the phenomena of animal psychology have by no means always taken this warning to heart; not infrequently they have followed Haeckel's bad habit of involving themselves in speculations upon the soul as such without having the qualifications to give critical treatment to this complicated problem.

Self-observation the foundation of psychology

THE foundation of all empirical psychology is self-observation; on this point an animal psychologist of the old school, such as Romanes, is in agreement with a modern experimental psychologist of the type of ALFRED LEHMANN (1858–1921), a pupil of Wundt, professor at Copenhagen. Lehmann maintains that it is only in one's own consciousness that one can observe psychical states and functional manifestations; the assumption of psychical phenomena in other creatures depends on whether these latter are seen to act in given circumstances as man would do in the same circumstances. As regards one's fellow human beings, this conclusion can be confirmed by means of language, but such a mode of control is wanting in animals; as far as the vertebrate animals are concerned a good deal can be concluded from their bodily structure where it agrees with that of man, but even this check is lacking in the invertebrates. As a general principle of animal psychology there remains, then, according to Lehmann, the principle of ascertaining by experiment whether the animal can adapt itself to new and unexpected situations; whether it can by learning from experience modify its actions to suit the conditions. If this is done, then we have the right to assume the existence of an individual psychic life in the animal — to assume life-phenomena of a different kind from the instinctive adaptation to normal conditions of life which characterizes all animal life and which is based upon nervous reflexes or simply upon tropisms induced by chemical and physical reactions, such as are observed even in the most primitive of organisms. Such experiments with individual experiences as their object are extremely difficult to carry out, however, and still more difficult to interpret aright. On the one hand, animals should be brought into situations which prove that they can learn something new, but, on the other hand, they should not be faced with situations which involve violence to their true nature. The road to a true insight into the subject is thus both long and difficult, and this explains to some extent why so many contradictory statements and theories have arisen in this sphere, even among observers who have been comparatively successful in freeing themselves from preconceived opinions. The confusion has, of course, been increased by so many students and dilettanti, for "Darwinistic" purposes, *wanting* necessarily to find in the animals as great and as human-like an intelligence as possible. On the other hand, there have not been wanting in modern times zoologists who have seen in animals nothing but reflex mechanisms, and that, too, not merely as a result of their holding a conservative view of life, but quite as often owing to ultra-radical views — through endeavouring to restrict as far as possible the part played by the psychic life in nature. We shall here cite in brief a few examples of different views on these problems.

Insect psychology

ONE subject that has specially interested animal-psychologists from ancient times has been the psychology of the insects, particularly of the community-forming Hymenoptera. Here in earlier times the imagination and credulity have combined to celebrate veritable orgies of the fancy; one reads with un-feigned amazement of all that people even in the latter half of last century[3] imagined they could observe in the ant-communities. A much more sober atmosphere has latterly prevailed, and it has now begun to be realized that most of the actions of the ants must after all be due to inherited instinct. A very prominent observer in this sphere is the Jesuit father ERICH WASMANN (born 1859), who has especially elucidated a number of facts in regard to the ants' relations to many different kinds of parasites, which swarm in the ant-heaps and which are often carefully looked after. Wasmann has otherwise appeared as a keen opponent of Haeckelian monism and has elab-orated in opposition to it a history of creation approved by Roman Catholic authorities, in which Darwinism in most of its aspects has been ingeniously

---

[3] As an instance may be cited a story related by many authors — Romanes, for example — of an American species of ant, according to which those ants whose duty it is to guard the communities do so formed up in a regular square; within the square the workers carry out a great many equally intelligent and well-ordered movements.

introduced, though at the same time the old doctrine of creation has been retained. In regard to the psychic life of the animals, its intrinsic resemblance to human life is, of course, denied, and accordingly the reflexive and instinctive functions are sharply emphasized and the idea of individual consciousness rejected. Nevertheless, even the insects are allowed a certain power of methodical action. ALBRECHT BETHE comes to a far more radical conclusion, based on entirely opposite grounds; by means of extensive and in part highly ingenious experiments he endeavours to prove, and in many cases actually succeeds in doing so, that the actions of the ants are pure reflex-actions. On the other hand, the famous student of psychiatry and psychology AUGUSTE FOREL, of Geneva, holds another view; after a series of careful experiments he believes that he has been able to prove that insects can actually learn by experience and thus possess intelligence. Another distinguished observer of insect life who arrived at partially similar results was the well-known English naturalist, banker, and philanthropist JOHN LUBBOCK, LORD AVEBURY (1834-1913). A very important animal psychologist was GEORGE JOHN ROMANES (1848-94), professor at Oxford, an enthusiastic supporter of Darwin, whose theory he defended in a number of writings, in which he especially attacked Weismann for his denial of the heredity of acquired characters. In particular, he carried out experimental investigations into the psychic life of the higher animals, which he believed to be — in kind, if not also in degree — similar to that of human beings. Like Darwin, however, he has accepted without criticism stories and anecdotes derived from foreign sources, but his own observations are very keen. Animal psychology has been dealt with experimentally with special keenness in America; among its pioneers in that country may be mentioned R. M. YERKS, who has carried out a long series of experiments in order to try to discover the power of animals to learn by experience; he is especially well known for his "labyrinth," in which he placed animals, who then had to find their way out as best they could.

As a general result of the work of these investigators it may be mentioned that the vertebrate animals, at least the higher, can certainly acquire knowledge from experience; whether the higher invertebrates can also do so would seem to be more doubtful. But even experience can be exhibited in different ways; either the animal finds its way in every fresh case to a new experience, independent of what it has gone through before, or else it really possesses the power to retain its experiences in the memory and to profit by them in new situations. The former power is, of course, the more primitive, as it is also the more usual in the animal kingdom; the latter, the rational power of adaptation, has certainly been observed to exist only in a few of the very highly developed animals. Still more debatable are the cases in which the power of grasping relations of number and other abstract ideas has come into question. Isolated cases have been known of old of alleged purely human intelligence in domestic animals. Shakspere in one of his comedies alludes to a horse, on show at the time, that was able to count; in our own day one or two similar cases have certainly given rise to much discussion. Even professional biologists have believed, after investigation, in the famous horses of Elberfeld, which performed operations of counting in higher mathematics and other equally remarkable feats. On the other hand, the authenticity of these cases has been very keenly disputed — Alfred Lehmann describes them unreservedly as self-deception on the part of the spectators or conscious duplicity on the part of the exhibitor. At best, however, it could only be a matter of abnormal receptivity to training, which cannot be considered in any respect to characterize normal intelligence in the animals and thus cannot offer any guidance for judging the development of the intelligence in the animal kingdom, which, however, has been very generally asserted by those who have felt convinced of the reality of the feats exhibited. In cases like these keen criticism is more than ever essential; unfortunately many even distinguished biologists have shown themselves, owing to their preconceived ideas, far stronger in their beliefs than clear in their judgments, and this has been not least apparent in that much disputed chapter, the psychic life of the animals.

CHAPTER XVIII

MODERN THEORETICAL SPECULATIONS

1. Mechanism and Vitalism

IT WILL HAVE BEEN SEEN from the previous chapter that biology in our own day has distinguished itself far more through practical detailed research than through theoretical speculations. Actually no further generally accepted theory of life such as that offered by Darwinism has been discovered; instead, there has prevailed a restless search for fresh grounds on which a theory of life might be built up. Many have been the roads along which the search for the theoretical solution of the problem of life has been made since the turn of the century, and they have run in widely differing directions. On the one hand, we find repetitions in a newer form of the old materialistic and mechanistic theories of life from the days of Vogt and Haeckel; on the other hand, vitalistic ideas that have looked for support in Bichat, in Stahl, in Aristotle. The old antagonism, Christian conservatism versus Darwinistic radicalism, which half a century ago resulted in the formation of parties, has now been essentially adjusted, though in no wise everywhere eradicated; in this respect it must be acknowledged that the struggle in modern times is more definitely a matter of facts than it was previously, at least among students, but the differences of opinion have in many quarters certainly been sharp enough. It is possible to give here only a few examples of views taken from the rival camps, this as a final summary of the position of biology as it stands in the present generation; unfortunately it is more difficult than ever to draw conclusions from them as to the direction likely to be taken by evolution in the future.

MAX VERWORN (1862–1921) was born in Berlin, studied there and at Jena, and eventually became professor of physiology at Göttingen. He was a pupil of Haeckel and throughout his life retained both his admiration for his master and his association with Haeckel's fundamental ideas. To him the biogenetical principle as well as the theory of selection always was a proved fact; of the more recent contributions to the hypothesis of evolution he embraced de Vries's mutation theory, but he showed no interest in Mendelism. On these foundations, however, he built up a life theory of his own

603

that has found lively support in many quarters. He calls it "cellular physi-ology" and bases it, as its name implies, on the doctrine of the cell as the unit of life, which he maintains repeatedly, and with greater emphasis than anyone else has done, against all theories of smaller, independent entities, such as Altmann's granula. To him the body is exclusively a cell-state; the formation and co-operation of the organs have little interest for him and are dealt with only in passing; Roux's name is mentioned as little as Mendel's even in the most recent edition of his *Allgemeine Physiologie*. He chiefly oc-cupies himself with the living substance and its nature. And though he has removed from this field of research those hypothetical life-units of a mechan-ical kind that played such a large part in the theories of his predecessors and contemporaries, he has nevertheless reverted to the same unworkable imaginative system of thought, although on different grounds; that is to say, he has imagined a chemical unit of life of an albuminoid character, which he terms "biogen," whose chief characteristic apparently consists in an extreme chemical lability: a constant and simultaneous alternation of disintegration and reconstruction. In this chemical change consists, accord-ing to Verworn, the true essence of life; there is no difference between ani-mate and inanimate except that which is brought about by the extraordinary metabolistic possibilities of the biogen molecule. Verworn, it is true, admits that the biogen is really as hypothetical as plastidules or micellæ ever were, but he has the same weakness as other speculative biologists for allowing hypothesis, when it is once completed, to stand for fact. It is a more serious matter, however, that the biogen cannot be reconciled with modern biochem-istry; the latter's representatives have fairly unanimously condemned the whole hypothesis, maintaining that there exists no other albumen molecule than that which is already known to general chemistry. "There is no such thing as dead and living albumen any more than there is dead and living sugar or fat, and the reactional powers of the protoplasm depend upon the co-operation of its various component parts in definite proportions" (Höber). Verworn, however, paid no heed to these objections, especially as modern biochemistry did not interest him in the least; he makes no mention of the progress of colloid chemistry, but instead discusses the old problem as to whether plasm is solid or fluid.

Verworn's conditionism

On the whole, then, Verworn went his own way, heedless of the progress of contemporary science; he had made it a principle, one of his biographers relates, not to read too much, but to think for himself. He was, moreover, of an ardent disposition; ethical and social questions interested him keenly, and every new-year's eve he ceremoniously inaugurated the coming year's work by sitting down at his writing-desk at the stroke of twelve. Having such a temperament he was naturally inclined to indulge in philosophical speculations upon the main problems of life; he elaborated a theory of his own that he called "conditionism," which was to replace the old ideas of cause and effect by a conception of all phenomena's being due to a multiplicity of simultaneous conditions; what he was trying to get at was, of course, the old contrast of things and phenomena and thereby ultimately the contrast between the physical and the psychical, which he would replace by an all-comprehensive "psycho-monism." But Verworn was no clear-sighted thinker; conditionism is reminiscent of Mach's phenomenalism, which, however, is far better thought out, and psycho-monism merely leads to a great many far-fetched and unnatural attempts to get away from the actually existing and observed difference between animate and inanimate, in which the hypothetical "living" albumen must play the part of a universal remedy for all difficulties. Through his enthusiasm, his brilliant style, and his undeniably valuable contributions to the problem of the cell as a physiological unit of life, Verworn has at any rate been an important personality among the biologists of his age.

Loeb on the movements of animals

A MECHANISTIC explanation of nature on entirely different grounds has been produced by the experimentalist LOEB, who has been mentioned in the previous chapter. He was, as already pointed out, a pupil of Sachs, whose studies of the tropisms of plants became the basis of his entire conception of life. In his earlier years he himself carried out a number of valuable investigations on the subject of tropisms in the lower animals; since then he has made the above-described experiments on parthenogenesis in eggs caused by chemical reagents. The general theory of life that he set up is actually based on these two classes of experiments. He regards, as far as is possible, the movements of animals as tropisms caused by external influence; when an animal moves towards the light, there actually takes place through the effect of the light an oxidization of certain elements in the animal, and this causes the movement; other movements, again, are induced by chemical associations that arise directly in the innermost being of the animal, as, for instance, in the mating-flight of insects. On these facts he bases a "mechanistic conception of life," which, however, he hardly succeeds in formulating in a very convincing way. Indeed, he has no idea of a scientific student's duty of first thinking out his theories; when his theory suits one case, it is at once made to hold good for all cases without further investigation, and if it does not do so, then the inexpedient cases are simply passed over. He gives an account, for instance, of the phototropism of the Aphidæ, on which he carried out most ingenious experiments, and he traces the phenomenon to the said oxidizing process — the fact that this phenomenon upon repetition proceeds with greater rapidity "may be brought about by" the lactic acid produced by the muscles upon movement. Thus the phenomenon in question is ascribed to a cause that is stated to possess a general significance for all vital phenomena, but when shortly afterwards it is declared that the worker-ants do not exhibit any such tropism, no attempt is made to explain this notable exception. And the same lack of consecutive thought is displayed everywhere. The development of the egg is explained as being due to oxidization; that this development can be caused by such diverse external influences as, on the one hand, a solution of acid and, on the other, the prick of a needle certainly evokes some surprise, but no more than that the whole phenomenon is after all accounted for as being a physico-chemical process, it not being considered at all necessary to discuss the most remarkable feature of all — namely, the nature of the egg itself. Indeed, in the opinion of Loeb, there exist no structural conditions whatever; there is hardly any question of the organism's possessing a chemical composition of its own; all that takes place in the organism is the result of outside impulses, the result being that no discrimination whatever is made between one life-phenomenon and another, whether it is a question of sea-urchins, insects, or frogs. The goal to be attained is, as we have said, a mechanical explanation of life, but just because of this exclusive interest for external influences the explanation proves to be essentially negative — a denial of the existence of any operating forces other than the said external influences. When he comes to discuss more complicated problems, Loeb shows the most amazing lack of criticism; he gives an account of Mendelism and declares that the riddle of heredity is thereby solved, but a little further on he asserts the exact opposite, that any ossean can be crossed with any other osean;[1] he himself has kept such hybrids alive for a month. The explanation of what has taken place is manifestly a parthenogenetical development of the eggs through the "activating" influence of foreign sperm; but Loeb literally declares that it is possible to obtain a hybrid between a salmon and a flounder. With such facility in drawing conclusions and such irresponsibility as to their consequences there need be no limit to one's flights of imagination, and, moreover, one is free in the long run to dispense with all exchanges of views with scientists possessing a normal sense of responsibility. Loeb is without doubt a brilliant experimentalist, and as such he deserves mention among the pioneers, though among biological thinkers he can claim no place.

Vitalistic explanations of life

On the whole, the mechanistic speculations in the sphere of modern biology give a somewhat monotonous impression and it is therefore hardly worth while making the acquaintance of any more representatives of this line of thought. Those who are constantly making the assertion that there is no

---

[1] See *The Mechanistic Conception of Life*, p. 24: "It is possible to cross practically any marine teleost with any other."

essential difference between animate and inanimate very quickly lose all appreciation of what is truly characteristic in living matter and its metabolistic phenomena, which must otherwise be the chief interest even of those biologists who maintain the old assertion, already proclaimed by Kant, that only material phenomena can be subjects for natural-scientific treatment. It is not, then, the idea — in itself justifiable — of limiting discussion to the chemical and physical manifestations of the phenomena of life that constitutes the weakness of these mechanistic theories of life, but the stubborn insistence upon the rough comparisons between phenomena in animate and inanimate nature — comparisons, in fact, the weakness of which would undoubtedly be realized by their proponents if the latter were not really trying all the time to lay the foundations of some kind of general philosophical theory extending far beyond the bounds of natural science. When Verworn discusses and denies the possibility of the immortality of the soul, he is arguing, from the natural-scientific point of view, about nothing at all, for though, as we have seen, biology studies psychical phenomena, it does not imply that it has anything to do with the impossible problem of what the soul is or is not. And if we ask, quite apart from such metaphysical quibbles, whether all observable *material* processes in the living organism or its parts can be directly derived from known material processes in inanimate nature, the answer even today must still be in the negative; those who have attempted to do so have either reverted to gross schematism or else drawn a bill on the possible progress of tomorrow — an unworthy manner of wriggling out of the fact of the problem's insolubility. It may at once be assumed that the future will bring us nearer the heart of the problem; whether it will ever be entirely solved we know no more than the truth, laid down by Herbert Spencer and many others, that the capacity for knowledge is limited and the most general laws in existence must therefore remain unexplained.

Strictly speaking, the same causes have brought about the popularity of the mechanistic theories of life as those that at one time produced so many editions of Haeckel's *Natural History of Creation*. It is obvious, however, that a reaction against this conception of life was bound to set in, owing to the disappointment felt over its splendid but unfulfilled promises. And so we find, even before the turn of the century, vitalistic theories of life of various kinds being produced, supported by representatives of no small importance, as regards both their numbers and their attainments. And during the present century their number has still further increased; true, they can nowhere be said to have dominated the situation, but the part they have played has been quite an important one, many of them having had a perceptible influence even in circles in which vitalistic or spiritualistic ideas have otherwise never been very highly appreciated. They have for the most part come from the physiologists, while the morphologists, with far fewer exceptions, have adhered to the standpoint that they had maintained since the appearance of the descent theory. The views of some of the vitalists have arisen in connexion with certain religious or social principles, as in the case of the above-mentioned Jesuit priest Wasmann, who, owing to his ecclesiastical point of view, was bound to feel attracted to a theory of life that offers the possibility of a spiritualistic explanation of existence. So, too, the Lutheran conservative botanist and politician J. REINKE, mentioned in the preceding chapter as an opponent of Haeckel and known for his violent polemics against materialism. Greater impartiality has been shown by the physiologists GUSTAV BUNGE and R. NEUMEISTER, both of whom have interested themselves in the question of whether the phenomena of life can originate merely in physico-chemical causes. Bunge sustains his arguments especially upon the complicated structure and vital manifestations of the cell, which cannot be explained on a physical or chemical basis; moreover, he takes as his premisses that life as a whole cannot be studied except through self-observation and is therefore a psychical process. Neumeister, on the other hand, maintains against Haeckel and his supporters that the origin of life is a transcendent problem and that the psychical phenomena cannot be derived from the material. This criticism of contemporary materialism is in itself certainly justified, but it actually implies a negative attitude; to try to substitute for it an imaginary life-force is only to create a fresh complication of the problem; it militates against our striving after simplicity, as Henle once said. The most pronounced vitalists of our own day have been fully conscious of the fact, but they have deliberately exceeded the bounds of exact research and gone over into the sphere of abstract speculation.

The "autonomy of life-phenomena"

THE most interesting of these modern philosopher-scientists is undoubtedly the previously mentioned experimental biologist HANS DRIESCH, inasmuch as his history shows the natural development of the consistent vitalist from biologist to metaphysician. Born in 1867, the son of a wealthy merchant, he was allowed full liberty, regardless of prevalent trends of thought, to devote himself to science. His start as an experimental biologist has already been described above, as also how he interpreted his experiments in a markedly epigenetical way, thereby finding himself opposed to Roux and his preformation theory. Markedly antagonistic was also his attitude towards Darwinism, which when still a youth he declared to be a thing of the past. Owing to these two facts — the results of his experiments, and his anti-Darwinism — he adopted from the outset an aggressive attitude towards the older biological school, which accounts for his keen insight into its weaknesses. Moreover, his is a pronounced speculative nature, with wide philosophical interests and corresponding erudition, and he has been very deeply attracted by the Aristotelean abstract construction of life-phenomena.

On these grounds he has produced his arguments in proof of "the autonomy of life-phenomena"; he asserts that a living being forms a "harmoniously equipotential system," by which he means that a hydra or other primitive organism can be regenerated out of severed parts; this proves, to his mind, that the animal is not a machine, for a machine cannot be evolved out of its parts. This antithesis — machine and living being — he is constantly bringing forward as a proof of the impossibility of deriving the animate from the inanimate, and by drawing a comparison between them he comes to the same negative result as the vitalists referred to above, though on markedly abstract and schematic grounds.

The "entelechy"

He is not content with this, however, but seeks to discover what life really is. The answer is summed up in an expression borrowed from Aristotle: "an entelechy." By this word Aristotle meant the potentiality which is inherent in matter and which achieves reality to the extent of matter's development into an ever higher and higher form (see Part I, p. 36). Driesch has likewise a marked interest for form in living nature, which, he considers, lends itself far more readily to "philosophical analysis" than metabolism does. By entelechy, however, Driesch means something far more involved: it is supposed to mean "something that carries its purpose within itself." It is thus the functional adaptation of living beings that is here indicated, but in entering into a profound and far-reaching analysis of the idea Driesch becomes involved in a maze of abstract speculations, which become still more difficult to understand owing to the extremely complicated terminology he employs; really we have to go back to the heyday of Hegelian philosophy to find the counterpart, in point of difficulty of comprehension, of Driesch's definitions and characterization of the phenomena of life. This much, however, can be gathered from them, that as the ultimate proof of his vitalism he cites his own personal consciousness; it is thus, apparently, that we are to interpret his expression "phenomenological idealism," which, according to his conception, leads directly to vitalism, at any rate as far as his own body is concerned. He then draws the same conclusion in other living bodies. But this is certainly, if anything, pure metaphysics; it has nothing to do with biology; to give a detailed account of how the idea of the relation of entelechy to matter is further developed would in such circumstances be superfluous, all the more so as here the incomprehensibility of his language exceeds all bounds; sentences such as the chapter heading: "*Entelechie bezieht sich auf den Raum und gehört daher zur Natur, aber Entelechie ist nicht im Raum*," which is afterwards explained as follows: "*Sie wirkt nicht im Raum, sie wirkt in den Raum hinein*," do not make the reader much the wiser. The same is true of such a statement as that "*Materie ist nicht einmal in irgend einem Sinne die Grundlage des Lebens*." The new formula that Driesch gives in this connexion to the law of the conservation of energy is a matter for the physicists to consider. Driesch having thus already at an early stage taken up a position essentially on the other side of the boundary line between metaphysics and empirical research, he has ultimately adopted this step formally as well; he is now professor of philosophy at Leipzig and in that capacity has been engaged in speculations of the most abstract kind.

Another vitalist of whom a good deal has been heard is EMANUEL RÁDL. He was born in Bohemia in 1873, studied at Prague, and has been a lecturer in physiology there. His research work has concentrated partly on physiological subjects — he has dealt with the tropisms in the lower animals — and partly on the morphology of the brain. He is best known, however, for his *Geschichte der biologischen Theorien der Neuzeit*, a much read and widely quoted work, the first part of which has come out in two editions, which are essentially different from each other. To examine this work properly, however, we must have some knowledge of its author's biological standpoint, which is clearly apparent in his most important monograph, *Neue Lehre vom zentralen Nervensystem*. In its introduction the author shows that he holds particularly broad views, and, on the other hand, that he is fully convinced of the soundness and future value of his own ideas. The Darwinistic morphology is rejected as a soulless description and specification of different developmental forms one after another; he gives slightly more credit to experimental evolutional physiology on the lines of Roux, but the science that in Rádl's view has the greatest future is that of ideal morphology, of which he himself is an exponent. As a pioneer of this science he mentions Geoffroy Saint-Hilaire, and also, though in a less degree, Cuvier; its aim is stated to be to discover the ideas according to which the forms of living organisms are constructed. "Many ideas compete at the root of organic life for precedence, and an ideal structure forms the basis of every organism." This must be sought for by means of comparisons throughout the animal kingdom, for only thus can we gain any knowledge of the fundamental ideas of existence. This is, of course, simply the idealistic morphology of the beginning of the nineteenth century over again. When it comes to working out his idea in detail, however, we find only a collection of disjointed sentences taken from other authors, in conjunction with his own, not always very convincing, observations concerning the object of his investigation, the central nervous system. Apathy's fibrilla theory is thus maintained as against the doctrine of neurones; Rádl himself describes in word and illustration one category of fibrillæ, existing, according to him, throughout the animal kingdom, which are convoluted in a special way at the entrance to every ganglion and which he terms "cascade fibrillæ." They do not, however, give the impression of being very natural and would appear rather to have arisen through being cut obliquely or through the ordinary nerve-fibres' being wrongly fixed. The formal points of agreement in the nervous system, upon which ideal morphology is based, must, on the whole, seem rather uninteresting to a morphologist of the old school, for they prove nothing new; but Rádl imagines that his method has achieved extraordinary results. The theory of organic structure is to prove capable of building up a magnificent philosophy of its own, "such as speaks to us out of the Pythagoreans' theory of harmony, out of Plato's doctrine of ideas, and out of the romantic reveries of the obscure German natural philosophy." Rádl's conception of the world as a "*Schöpfung des sie betrachtenden Geistes*" is undeniably reminiscent of the latter; in fact, his expressed intention, underlying the entire work, is to excite surprise, "for it is through being surprised that man has now and always begun to philosophize." And the work certainly does evoke surprise, though not perhaps of the kind the author intended. It has manifestly not exercised any influence whatever upon the development of neurology, and would not have been worth while referring to had not the author's abovementioned historical work acquired such widespread fame.

Rádl's history of biological theories

THIS fame is based partly on Rádl's undeniable merits as a historian: a wide knowledge of literature, a lively style, and shrewd, often striking discernment (his account of the development of Darwinism in Germany in Part II is particularly animated and instructive); partly on his opposition to the original Darwinism, an opposition which came into force at a date when the old doctrine was certainly undermined, but nevertheless still officially accepted; and partly again on the numerous philosophical digressions, sometimes witty, but more often merely odd, which are found scattered throughout his history and which proved attractive to a generation that had wearied of the old phylogenetical speculations without having on that account acquired any other speculative foundation on which to build. Of the various parts of the book, the first edition of Part I is the one that contains most of the old biological ideas; in the foreword of Part II the author regrets the far too confident belief that he had earlier entertained in an objective science; and in the second edition of Part II, which was published last, Rádl declares that he intends to promulgate a "realistic cosmic view, such as finds its deepest expression in Dostoievsky's novels." The work expresses throughout, each part more extravagantly than the last, a purely panegyrical enthusiasm for Aristotle, who is declared to be the unattainable ideal as a natural philosopher, but at the same time a warm admiration for his opponent and very antithesis, Paracelsus; and, further, the book extols Stahl's vitalism and romantic and idealistic speculation in general, while it disparages exact research, particularly cytology (whose methods the author nevertheless himself employed, though not very skilfully), Darwinism, and exact heredity-research, all of which is described as materialism. The author's subjectivism culminates in the above-quoted saying that there is no such thing as objective science; all interest is centred upon the personalities figuring in the history of science. This may to a certain extent be justified, when it is a question of giving expression to the purely ideal strivings of humanity, but when it is a question of nature, our knowledge rests undeniably upon certain facts that reveal themselves equally to all. He who refuses to admit this had best turn his back upon exact natural science for ever. Rádl has in fact done so; he is now professor of natural philosophy at Prague.

Uexküll on the life-process

On the whole, little is to be gained from the biological point of view by becoming too deeply engrossed in the works of the modern vitalists. Some of them have gone in for speculations about the theory of knowledge, as, for instance, Jacob von Uexküll, who holds that only a part of the life-process is mechanically comprehensible, while that part of it which gives to the mechanical phenomena their "*Zielstrebigkeit*" is super-mechanical and must be referred to impulses produced by an organized natural force; mechanical biology is concerned with the fitting-in of every being into certain given conditions, which give to the organism its limitations; in the world of dew-worms there exist only dew-worm conditions, while man can observe only human things. To analyse the different conditions of life and to work out the laws governing this reciprocal action between individual and environment is, to his way of thinking, the aim of biology. Other vitalists have reverted to downright mysticism, and, finally, the neo-Lamarckian school previously referred to, which is represented by Pauly and his pupils, has tried to see in the phenomena of life, especially in evolution, expressions for consciously operating psychical forces in the living substance. What is common to all these different aims is the attempt to discover the difference between animate and inanimate matter and what it is that produces the peculiar character of the phenomena of life. For this purpose the methods of physics and chemistry have proved ineffective, with the result that other means have been sought to attain the end in view. We have already seen that these means have proved fruitless. In such circumstances obviously the wisest course would be: neither mechanism nor vitalism, but resignation in face of the inexplicable. But science has not yet struck into that path, nor is it likely to do so in the future. And fortunately too, we may well say, for had not humanity possessed a belief in the possibility of solving the insoluble riddles of life, there would never have been any science at all. Every delusion that has involved an honest striving after truth has at any rate contributed something to human knowledge, even if it is only negative, and the speculations that have just been described are in this respect by no means valueless.

ERNST HEINRICH HAECKEL

WILHELM LUDWIG JOHANNSEN

2. The Idea of Species and Some Problems in Connexion Therewith

In a popular account of the results of heredity research published some years ago occurs the sentence: "For the very reason of the great number of fresh facts that modern heredity-research has brought to light, chaos prevails at present in regard to the views on the formation of species." These words characterize both the situation as regards the problem of species at the present day and the causes that have brought it about. Modern heredity-research has completely dislocated the circles drawn by the old morphological classification. Linnæus's idea of species, it will be remembered, was essentially genetical; he counted as many species as had been created in the beginning, or, in later years, at any rate some species created in the dawn of time in respect of each genus, out of which the other species have since been evolved. This idea of species could very easily be reconciled with the idealistic idea of species which has existed since the days of Greek philosophy and which the biology of the romantic period preferred; in those days greater attention was paid to the idea expressed in the species form than to the question of origin. Darwinism brought the genetical idea of species once more into repute. To discover the origin of the different forms of life by a close comparison of their external and internal structure was, according to Gegenbaur, Haeckel, and their disciples, the end of biology; thus, a natural classification system was to be created with species based upon true relationship — species which, it is true, must be assumed to vary and overlap, but which in their typical forms could be determined and characterized. Nevertheless, this genetical idea of species rested upon an indispensable proviso — namely, that from resemblance one could positively conclude affinity: the greater and the more universal the resemblance, the closer the affinity. Every species, and even every variety, had a common origin, as proved by the mutual resemblance between its individuals. It is this foundation for the idea of species that modern heredity-research has undermined; it has clearly demonstrated that very close morphological resemblance can in certain cases be due to entirely different causes. It is not outward resemblance, but the concurrence of hereditary factors that proves true affinity; that is to say, it is not phenotypical, but genotypical resemblance that determines the affinity. But nowadays in all systematical works the species are described entirely according to phenotypes; genotypical agreement can be ascertained only by experimental means. And in practice, of course, this can take place only on a small scale; the plant-geographer who makes records of localities and distribution charts at home or abroad would be stranded if, every time he sees a form, he were compelled to carry out hybridizing experiments with it in order to establish its identity, and this applies also to the morphologist and the systematist. In these circumstances it seems to be absolutely necessary to decide exactly what the categories of the system are intended to denote. Unfortunately it has not been possible to obtain unanimity on this point; opinions have clashed and their advocates have been very little disposed to modify their views. We shall here cite one or two examples of these divergences of opinion.

Lotsy and the species question

THE Dutch student of the heredity problem J. P. LOTSY has taken up an uncompromisingly genetical attitude towards the idea of species. "What is a species?" he asks in one of his treatises. The answer is: A species in the Linnæan sense is no species, for it comprises a large number of different life-forms whose outward appearance bears a certain resemblance, but as to whose origin we can determine nothing. The Linnæan species are therefore nothing but products of the imagination, as also the Linnæan genera and other higher classified groups; consequently they should no longer be called species but linnæonts. Nor are the minor species into which some Linnæan species can be divided and which remain constant, true species; these forms, which were specially studied by the French botanist JORDAN in the beginning of the nineteenth century, should be named after him jordanonts, but they are not species, for their internal resemblance cannot be ascertained. On the other hand, a species is a summary of all the homozygous individuals having the same hereditary character. "Consequently, not even all the pure-lines in the Johanssenian sense are true species; they are so only if they are at the same time homozygotes. And in regard to organisms with sexless reproduction, we can never know whether they are species, for that can be discovered only through the analysis of hybridizing experiments." In these circumstances it becomes a matter for doubt whether any species exist at all in nature.

This conclusion of Lotsy's clearly proves that an insistence upon the genetical idea of species can only lead to sheer paradox; he admits himself that only linnæonts and jordanonts can possess any practical systematical significance. The whole of his reasoning is really a striking proof of the power of language over thought; he wants what is called species to be a genetical entity, and so he comes to a point where he does not know whether species in his sense of the word exist at all in nature. Other students of heredity have also taken warning from this result: thus, HERIBERT-NILSSON declares that the term "species" might well be used in the form in which Linnæus employed it; thereby, it is true, the idea of species becomes purely morphological — "The species of classification is a phylogenetical conglomerate," he says — but for the genetical entities we have of course the new nomenclature "genotype" and "pure-line." The necessity of thus dispensing with the genetical idea of species has, indeed, been realized by many others; ERNST LEHMANN, for instance, maintained in his controversy with Lotsy that the species are abstract and arbitrary ideas and he has quoted the statements of other observers in support of this view.

Mutations and hereditary factors

THAT the systematic species in the old sense of the term, comprising individuals having a certain mutual resemblance, must possess immense, even inevitable, significance from a purely practical point of view, will at once be realized; moreover, it has both morphological and physiological importance, in that resemblance of various kinds, both outward and inward, characterize the same type of organism, from the simplest characters referred to in the text-books to the precipitin reaction. But the species-characters — mutual resemblance — say, as is established, nothing about mutual affinity, and heredity research is the one branch of biology that cannot utilize the idea of species, except under experimental control. The old speculations upon species-phylogenesis, however, have not been given up by the representatives of modern heredity-research; on the contrary, there is current among them a veritable maze of ideas on the problem of origin. The problem has been further complicated by a divergence of views regarding the relation between mutations and Mendelian cleavage; while Heribert-Nilsson, for instance, has found that mutations, where they exist, only cause loss of hereditary factors, and on that account holds that the process of evolution can be conceived merely as a series of reductions in the original material of genes, the specialists in Drosophila have proved that there have existed mutations which have given rise to positively new rudiments. The conception of the development of life on the earth must of course be influenced by this problem. Heribert-Nilsson has resolutely followed up the consequences of his theory that only detrimental mutations exist and has declared that a theory of evolution is on the whole unthinkable, while other geneticists have deemed it unnecessary to take refuge in this desperate expedient. Baur in particular has endeavoured to create a theory of evolution by combining the results of research work on mutations and hybridization with Darwin's theory of selection; a similar view has been held by Morgan and his school. Other students of heredity, again, have sought to establish a mutational Lamarckism by assuming mutations induced by the influence of external conditions upon the germinal plasm. Earlier attempts to prove the possibility of transmitting to the offspring characters that have been experimentally imparted to the parents either have, as has been mentioned before, proved unsuccessful, or else could be given a different interpretation, as, for instance, Kammerer's and Tower's results referred to above. Recently, however, some new observations of this kind have been carried out, certain experimentally acquired characters in animals under investigation having been transmitted to the offspring along Mendelian lines; to this class belong, for example, the attempts of Little and Bagg[2] by passing Röntgen rays through female rats to induce defects in the eyes and other parts of the body of their offspring, and the experiments of Harrison[3] with melanism in butterflies induced by the introduction of metallic salts in the food. However, there are no doubt obstinate anti-Lamarckists who have explained or will eventually explain even these results in a different way. In this connexion may also be mentioned the recently published attempt of Professor Muller, of Austin, Texas, to produce mutations in Drosophila by means of extreme temperatures and Röntgen rays.

The whole of this problem of evolution is of course highly involved and its discussion must, as far as our own times are concerned, terminate in a number of unanswered questions. First of all, selection; that it does not operate in the form imagined by Darwin must certainly be taken as proved, but does it exist at all? It is obvious that by the influence of external conditions, especially such as interfere with sudden violence, a thinning-out of the species is possible. If, for instance, a quantity of seed from a southern climate is sown in a northern country, the delicate plants will die, whereas the hardy ones will live, but this selection is only a matter of relation to cold and proves nothing as to the quality of the individuals in other respects. But the competition between the individuals, in which Haeckel thought he saw true selection — does it exist at all, or is it only imaginary, as O. Hertwig affirmed? And outside influence — has it no effect whatsoever upon the germinal plasm and offspring of the individuals, or is there really any such influence in the form of some kind of mutational Lamarckism?

These and many other questions it is for the future to answer. We have now followed the history of biology up to our own times; our task is fulfilled.

---

[2] Little and Bagg, *Anat. Record*, Vol. XXIV, 1923.
[3] *Proceedings* of the Royal Society of London, ser. B, Vol. XCIX, 1926.

SOURCES AND LITERATURE

SOURCES

PART I

Chapters I–X

Ælianus, *Works*, translated by Jakobs. Stuttgart, 1839.

Aristotle, *Historia animalium*, ed. Aubert et Wimmer. Leipzig, 1868.

—, *De partibus animalium*, ed. Frantzius. Leipzig, 1853.

—, *De generatione animalium*, ed. Aubert et Wimmer. Leipzig, 1860.

—, *Physica*, ed. Prantl. Leipzig, 1854.

Diel, *Die Fragmente der Vorsokratiker*. Berlin, 1903.

Galeni opera, ed. Kühn. Leipzig, 1821.

—, *Œuvres choisis*, translated by Daremberg. Paris, 1854.

Lucretii de rerum natura, ed. Munro. Cambridge, 1893.

Plato, *Timæus*, selected dialogues, translated by Dalsjö. Stockholm, 1870. (Swedish)

Plinii naturalis historiæ, libri VIII–XI, ed. Junius. Leipzig, 1856.

Chapters XI–XIV

Aldrovandi, Ulysses, *Ornithologia, hoc est de avibus historiæ, libri XII*. Bologna, 1599.

Bacon, Francis, *Works*. London, 1847.

Belon, Pierre, *La Nature et diversité des Poissons*. Paris, 1555.

—, *L'Histoire des oyseaux*. Paris, 1555.

Bruno, Giordano, *Von der Ursache, dem Prinzip und dem Einen*, translated by Lasson. Berlin, 1873.

Cæsalpini, Andræe, *Quæstionum peripateticarum, libri V*, ed. Vignon, 1588.

Columbi, Realdi, *De re anatomica, libri XV*. Venice, 1559.

Fabricii ab Aquapendente, *Opera omnia anatomica et physiologica*. Leipzig, 1867.

Falloppii, Gabrielis, *Observationes anatomicæ*. Cologne, 1562.

Galilei, Galileo, *Opere, ed. nazionale*. Florence, 1890 ff.

Gesner, Conrad, *Historia animalium*. Zurich, 1551 ff.

Harvei, Guilielmi, *Exercitatio de motu cordis et sanguinis in animalibus*. Frankfurt, 1528.

—, *Exercitationes de generatione animalium*. London, 1551.

Rondeletii, Guilielmi, *De piscibus marinis, libri XC.* Lyons, 1554 ff.
Servetus, Michael, *Christianismi restitutio.* 1553 (ed. 1790).
Severinus, Marcus Aurelius, *Zootomia Democritea.* Nuremberg, 1645.
Vesalii, Andreæ, *De humani corporis fabrica, libri VII.* Basel, 1543.
—, Ibid., *Epitome.* Basel, 1543.

PART II

Chapters I–III

Descartes, René, *Œuvres philosophiques,* ed. Aimé Martin. Paris, 1882.
Helmont, J. B. van, *Ortus medicinæ.* Amsterdam, 1648.
Paracelsus, Theophrastus, *Bücher und Schriften,* ed. Huserus. Basel, 1589 ff.
Spinoza, Baruch, *Sämtliche Werke,* translated by Auerbach. Stuttgart, 1841.

Chapter IV

Aselli, Gaspare, *De lactibus sive lacteis venis.* Basel, 1628.
Bartholin, Thomas, *De lacteis thoracicis.* Copenhagen, 1652.
—, *Opuscula nova de lacteis.* Copenhagen, 1670.
Borelli, Alfonso, *De motu animalium,* ed. 2. Leyden, 1685.
Glisson, Francis, *Anatomia hepatis.* Amsterdam, 1659.
Graaf, Reinier de, *Opera omnia,* ed. 2. Amsterdam, 1705.
Grew, Nehemiah, *The Anatomy of Vegetables.* London, 1672.
Leeuwenhoek, Antony, *Opera omnia.* Leyden, 1722.
Malpighi, Marcello, *Opera omnia.* Leyden, 1687.
—, *Opera posthuma.* Amsterdam, 1698.
Pecquet, Jean, *Experimenta nova anatomica.* Amsterdam, 1661.
Perrault, Claude, *Essais de physique.* Paris, 1680.
Rudbeck, Olof, *De circulatione sanguinis.* Upsala, 1652.
—, *Nova exercitatio anatomica.* Upsala, 1653.
Ruysch, Frederik, *Opera omnia.* Amsterdam, 1721.
Steno, Nicolaus, *De musculis et glandulis.* Copenhagen, 1664.
—, *Elementorum myologiæ specimen.* Florence, 1667.
—, *De solido intra solidum naturaliter contento dissertationis prodromus.* Florence, 1669.
Swammerdam, Jan, *Bibel der Natur.* Leipzig, 1752.
Vieussens, Raymond, *Neurographia universalis.* Lyons, 1685.
Wharton, Thomas, *Adenographia.* London, 1656.
Willis, Thomas, *Cerebri anatome.* London, 1664.
—, *De anima brutorum.* London, 1672.

Chapter V

Boerhaave, Hermann, *Institutiones medicæ,* ed. 3. Leyden, 1720.
Hoffman, Friedrich, *Fundamenta medicinæ.* Halle, 1703.

Hoffman, Friedrich, *Medicina rationalis*. Halle, 1739.

Stahl, G. E., *Theoria medica vera*. Halle, 1737.

Swedenborg, E., *Œconomia regni animalis*. London, 1740.

Sydenham, Thomas, *Works*, ed. Greenhill. London, 1849.

—, *Regnum animale*. London, 1745.

Chapters VI, VII

Artedi, Peter, *Ichthyologia*. Leyden, 1738.

Bauhin, Caspar, *Prodromus theatri botanici*. Frankfurt, 1620.

—, *Pinax theatri botanici*. Frankfurt, 1623.

Brunfels, Otto, *Herbarum vivæ eicones*. Strassburg, 1530 ff.

Camerarius, R. J., *Das Geschlecht der Pflanzen*, translated by Möbius. Ostwald's *Klassiker der exakten Wissenschaften*. Leipzig, 1899.

Cesalpino, Andrea, *De Plantis*. Rome, 1603.

Fabricius, J. Ch., *Species insectorum*. Hamburg, 1781.

Fuchs, Leonhard, *Historia stirpium*. Basel, 1542.

Jung, Joach., *Isagoge phytoscopica*. Hamburg, 1678.

Linnæus, Carl von, *Amœnitates academicæ*, Vol. I–X. Leyden, 1749 ff.

—, *Classes plantarum*. Leyden, 1738.

—, *Fauna suecica*, ed. 2. Stockholm, 1671.

—, *Fundamenta botanica*. Leyden, 1736.

—, *Genera plantarum*, ed. 2. Leyden, 1742.

—, *Methodus plantarum*. Leyden, 1737.

—, Papers published by the Swedish Academy of Science. Upsala, 1907.

—, *Systema naturæ*, ed. 1, 2, 6, 10, 12. Leyden, 1735; Stockholm, 1740, 1748, 1758, 1766.

—, *Juvenile Works*, published by Ährling. Stockholm, 1888.

Ray, John, *Methodus plantarum*, ed. 1, 2. London, 1682, 1733.

—, *Historia plantarum*. London, 1686 ff.

—, *Synopsis animalium quadrupedum*. London, 1693.

Tournefort, J. P. de, *Institutiones rei herbariæ*, ed. 2. Paris, 1700.

Chapters VIII–XI

Albinus, *Icones ossium fœtus humani*. Leyden, 1737.

—, *Tabulæ sceleti et musculorum corporis humani*. Leyden, 1747.

Bonnet, Charles, *Contemplation de la nature*. Amsterdam, 1769.

—, *Insectologie*, ed. 2. Amsterdam, 1780.

—, *La Palingénésie philosophique*. Lyons, 1770.

Buffon, G. L. L. de, *Histoire naturelle*. Paris, 1749 ff.

—, *Œuvres complètes*. Paris, 1778 ff.

Camper, Petrus, *Kleinere Schriften*, Leipzig, 1788.

De Geer, Charles, *Mémoires pour servir à l'histoire des insectes*. Stockholm, 1752 ff.

Hales, Stephen, *Statical Essays*. London, 1738.

Haller, Albrecht von, *Anfangsgründe der Physiologie des Menschen*, translated by J. S. Haller. Berlin, 1759 ff.

—, *Bibliotheca anatomica*. Berne, 1774.

—, *Mémoires sur la nature*. Lausanne, 1746 ff.

—, *On Sensible and Irritable Parts of the Body*. Swedish Academy of Science, *Transactions*, Vol. XLV. Stockholm, 1753.

—, *Primæ lineæ physiologiæ*. Göttingen, 1747.

Hunter, John, *Works*. London, 1835.

Koelreuter, J. G., *Vorläufige Nachricht von einigen das Geschlecht der Pflanzen betreffenden Versuchen*. Ostwald's *Klassiker*, 1893.

La Mettrie, J. O. de, *Œuvres philosophiques*. Berlin, 1774.

Lieberkühn, J. N., *Dissertatio . . . de fabrica . . . villorum intestini*. Leyden, 1745.

Lyonet, Pierre, *Traité anatomique de la chenille*, etc. The Hague, 1746.

Pallas, P. S., *Elenchus zoophytorum*. The Hague, 1766.

—, *Miscellanea zoologica*. The Hague, 1766.

—, *Novæ species quadrupedum*, etc. Erlangen, 1778.

Réaumur, A. F. de, *Mémoires pour servir à l'histoire des insectes*. Paris, 1734 ff.

Rösel von Rosenhof, *Monatliche Insektenbelustigungen*. Nuremberg, 1753 ff.

Spallanzani, L., *Expériences pour servir à l'histoire de la génération*. Geneva, 1785.

—, *Programme . . . d'une ouvrage sur les réproductions animales*. Geneva, 1768.

Sprengel, Conrad, *Das entdeckte Geheimnis der Natur*. Ostwald's *Klassiker*. Leipzig, 1894.

Trembley, A., *Mémoires pour servir à l'histoire d'une polype d'eau douce*. Leyden, 1744.

Wolff, C. F., *Theoria generationis*, ed. 2. Halle, 1774.

Chapters XII–XIV

Agardh, C. A., *Text-book on Botany*. Malmö, 1829 ff. (Swedish)

Carus, C. G., *Grundzüge der vergleichenden Anatomie*. Dresden, 1828.

—, *Natur und Idee*. Vienna, 1861.

Darwin, Erasmus, *Zoonomie*, translated by Brandis. Hanover, 1795.

Goethe, J. W., *Sämtliche Werke*. Cotta, Stuttgart, 1851 ff.

Herder, J. G., *Sämtliche Werke*. Berlin, 1887 ff.

Hwasser, I., *Essays*. Upsala, 1839. (Swedish)

—, *Selected Essays*. Stockholm, 1868 ff. (Swedish)

Ingenhousz, J., *Experiments upon Vegetables*. London, 1779.

Kant, J., *Gesammelte Schriften*, Berlin, 1902.

Lavoisier, A. L., *Œuvres*. Paris, 1863 ff.

Nees von Esenbeck, C. G., *Handbuch der Botanik*. Nuremberg, 1820.

Oken, Lorenz, *Naturgeschichte für alle Stände*. Stuttgart, 1841 ff.

Oken, Lorenz, *Naturphilosophie*. Jena, 1809.

Priestley, J., *Experiments and Observations on Different Kinds of Air*. London, 1775.

Saussure, N. Th. de, *Recherches chimiques sur la végétation*. Paris, 1804.

Schelling, F. W. J., *Sämtliche Werke*. Stuttgart, 1856 ff.

PART III

Chapters I–III

Barthez, P. J., *Nouveaux Éléments de la science de l'homme*. Montpellier, 1778.

Bichat, F. M. X., *Anatomie générale*, ed. 2. Paris, 1818.

—, *Traité d'anatomie descriptive*. Paris, 1801.

—, *Traité des membranes*, ed. 2. Paris, 1802.

Blumenbach, J. F., *Collectionis suæ craniorum*, etc., decades VI. Göttingen, 1790.

—, *Handbuch der Naturgeschichte*. Göttingen, 1803.

—, *Über die natürlichen Verschiedenheiten im Menschengeschlechte*, translated by Gruber. Leipzig, 1798.

—, *Über den Bildungstrieb*. Göttingen, 1791.

Bordeu, H. de, *Recherches anatomiques sur la position des glandes*, ed. 2. Paris, 1799.

Cuvier, G., *Le Règne animal*. Paris, 1817 ff.

—, *Leçons d'anatomie comparée*. Paris, 1799 ff.

—, *Recherches sur les ossemens fossiles de quadrupèdes*. Paris, 1812 ff.

Gall, F. J., *Lehre über die Verrichtungen des Gehirns*. Dresden, 1805.

—, et Spurzheim, *Recherches sur le système nerveux*, etc. Paris, 1809.

—, *Sur l'origine des qualités morales*. Paris, 1822 ff.

Humboldt, A. von, *Ansichten der Natur*. Stuttgart, 1849.

—, *Kosmos*. Stuttgart, 1845 ff.

—, *Versuche über die gereizte Muskel- und Nervenfaser*. Berlin, 1797.

Lamarck, J. B. de, *Histoire naturelle des animaux sans vertèbres*. Paris, 1815 ff.

—, *Mémoires de physique et d'histoire naturelle*. Paris, 1797.

—, *Philosophie zoologique*, ed. 2. Paris, 1873.

—, *Recherches sur l'organisation des corps vivants*. Paris, 1802.

Reil, J. C., *Archiv für Physiologie*, 1796, Bd. I.

Sömmerring, S. T., *De corporis humani fabrica*. Frankfurt, 1794 ff.

—, *Über das Organ der Seele*. Königsberg, 1796.

Vicq d'Azyr, Félix, *Œuvres*. Paris, 1805 ff.

—, *Traité d'anatomie et de physiologie*. Paris, 1786.

Chapters IV–VI

Baer, K. E. von, *Über Entwicklungsgeschichte der Tiere*, Königsberg, 1828 ff.

—, *De ovi mammalium et hominis genesi*. Leipzig, 1827.

Bell, Ch., *Die menschliche Hand*, trans. by Hauff. Stuttgart, 1836.

—, *Idea of a New Anatomy of the Brain*, ed. Ebstein. *Klassiker de Medizin*, published by Sudhoff. Leipzig, 1911.

Bernard, Claude, *La Science experimentale*. Paris, 1878.

—, *Leçons de physiologie experimentale*. Paris, 1855.

—, *Leçons sur les phénomènes de la vie*. Paris, 1878.

Berzelius, J. J., *Lectures on Animal Chemistry*. Stockholm, 1806 ff. (Swedish, trans. into German by Wöhler)

Blainville, H. M. D. de, *Traité des animaux*. Paris, 1822 ff.

Fourcroy, A. F., *Philosophia chemica*, trans. by Sparrmann. Stockholm, 1795. (Swedish)

Magendie, F., *Leçons sur les phénomènes physiques de la vie*. Paris, 1836.

—, *Précis élémentaire de physiologie*, ed. 5. Brussels, 1838.

Meckel, J. F., *System der vergleichenden Anatomie*. Halle, 1821 ff.

Müller, Johannes, *Bildungsgeschichte der Genitalien*. Düsseldorf, 1830.

—, *Handbuch der Physiologie des Menschen*. Coblenz, 1837–40.

—, *Über den Bau*, etc., *des Amphioxus*. Berlin, 1844.

—, *Vergleichende Anatomie der Myxinoiden*. Brllin, 1835–40.

—, *Zur vergleichenden Physiologie des Gesichtssinnes*. Leipzig, 1826.

Pander, H. C., *Beiträge zur Entwicklungsgeschichte des Hühnchens im Eye*. Würzburg, 1817.

Purkinje, J. E., *Beobachtungen und Versuche zur Physiologie der Sinne*. Berlin, 1825.

—, *De phænomeno*, etc., *motus vibratorii continui*. Breslau, 1835.

—, *Symbolæ ad ovi avium historiam*. Breslau, 1825.

Rathke, H., *Beiträge zur vergleichenden Anatomie und Physiologie*. Danzig, 1842.

—, *Bemerkungen über den Bau des Amphioxus*, etc. Königsberg, 1841.

—, *Untersuchungen über den Kiemenapparat*. Riga, 1832.

Rudolphi, C. A., *Entozoorum synopsis*. Berlin, 1819.

—, *Grundriss der Physiologie*. Berlin, 1821 ff.

Scheele, C. W., *Letters and Annotations*, published by A. E. Nordenskiöld. Stockholm, 1892. (Partly in Swedish)

Chapter VII

Henle, J., *Allgemeine Anatomie*. Leipzig, 1841.

—, *Symbolæ ad anatomiam villorum intestinalium*. Berlin, 1837.

—, Essays in *Archiv für Anatomie und Physiologie*.

Kölliker, A., *Entwicklungsgeschichte der Cephalopoden*. Zurich, 1844.

—, *Entwicklungsgeschichte des Menschen und der höheren Tiere*. Leipzig, 1861.

—, *Handbuch der Gewebelehre des Menschen*. Leipzig, 1852.

—, Essays in *Zeitschrift für wissenschaftliche Zoologie* and in *Archiv für Anatomie und Physiologie*.

Leydig, F., *Lehrbuch der Histologie des Menschen und der Tiere*. Frankfurt, 1857.

—, Essays in *Archiv für Anatomie und Physiologie* and in *Zeitschrift für wissenschaftliche Zoologie*.

Mohl, H., *Grundzüge der Anatomie und Physiologie der vegetabilischen Zelle*. Braunschweig, 1851.

—, *Vermischte Schriften*. Tübingen, 1845.

—, Essays in *Botanische Zeitung*.

Nägeli, K., *Zur Entwicklungsgeschichte des Pollens*. Zurich, 1842.

Reichert, K. B., *Beiträge zur heutigen Entwicklungsgeschichte*. Berlin, 1843.

—, *Bemerkungen zur vergleichenden Naturforschung*, etc. Dorpat, 1845.

—, *Das Entwicklungsleben im Wirbeltierreich*. Berlin, 1840.

—, Essays in *Archiv für Anatomie und Physiologie*.

Remak, R., Essays in *Archiv für Anatomie*, etc.; in *Comptes rendus de l'Académie des Sciences; et al.*

Schleiden, M., *Grundzüge der wissenschaftlichen Botanik*. Leipzig, 1842.

—, Essays in *Archiv für Anatomie und Physiologie*.

Schwann, Th., *Mikroskopische Untersuchungen über die Übereinstimmung in der Struktur und dem Wachstum der Tiere und Pflanzen*. Berlin, 1839.

Virchow, R., *Die Cellularpathologie*. Berlin, 1858.

—, *Gesammelte Abhandlungen*. Frankfurt, 1856.

—, Essays in *Archiv für pathologische Anatomie*.

Chapter VIII

Brown, R., *Botanische Schriften*, ed. Nees von Esenbeck. Leipzig, 1825 ff.

Candolle, A. de, *Physiologie végétale*. Paris, 1832.

—, *Théorie élémentaire de la botanique*, ed. 2. Paris, 1918.

Du Bois-Reymond, E., *Reden*. Leipzig, 1886.

—, *Untersuchungen über tierische Elektrizität*. Berlin, 1848 ff.

Dujardin, F., *Histoire naturelle des infusoires*. Paris, 1841.

Ehrenberg, C. G., *Die Infusionstiere als vollkommene Organismen*. Leipzig, 1838.

Endlicher, S., *Enchiridion botanicum*. Leipzig, 1841.

—, *Genera plantarum*. Leipzig, 1836 ff.

Fries, E., *Lichenographia europæa reformata*. Lund, 1831.

—, *Systema mycologicum*. Lund and Greifswald, 1820 ff.

Hedwig, J., *Theoria generationis*, etc., *plantarum cryptogamicarum*, ed. 2. Leipzig, 1798.

Helmholtz, H., *Vorträge und Reden*, ed. 3. Braunschweig, 1903.

—, *Wissenschaftliche Abhandlungen*. Leipzig, 1882.

Jussieu, A. L. de, *Genera plantarum*. Paris, 1789.

Leuckart, R., *Über die Morphologie der Wirbellosen Tiere*. Braunschweig, 1848.

—, *Über den Polymorphismus der Individuen*. Giessen, 1851.

—, *Die menschlichen Parasiten*. Leipzig, 1863.

Leuckart, R., Essays in Müller's *Archiv.*

Ludwig, K. F. W., *Lehrbuch der Physiologie des Menschen.* Heildelberg, 1852.

Mayer, R., *Die Mechanik der Wärme* in *Gesammelte Abhandlungen,* published by Weyrauch. Stuttgart, 1893.

Müller, O. F., *Vermium terrestrium et fluviatilium historia.* Copenhagen, 1773.

—, *Animalcula infusoria.* Copenhagen, 1786.

Nordmann, A., *Micrographische Beiträge.* Berlin, 1832.

Owen, R., *On the Archetype and Homologies of the Vertebrate Skeleton.* London, 1848.

—, *Lectures on the Comparative Anatomy.* London, 1843 ff.

—, *On the Anatomy of Vertebrates.* London, 1866.

—, *On the Nature of Limbs.* London, 1849.

Pasteur, L., *Études sur le vin.* Paris, 1868.

—, *Études sur le vinaigre.* Paris, 1866.

—, Essays in *Annales de chim. et de phys.* and *Comptes rendus de l'Académie des Sciences.*

Siebold, Th. and Stannius, H., *Lehrbuch der vergleichenden Anatomie.* Berlin, 1848.

Siebold, *Über die Band- und Blasenwürmer.* Leipzig, 1854.

—, *Wahre Parthenogenesis.* Leipzig, 1856.

—, Essays in Müller's *Archiv* and *Zeitschrift für wissenschaftliche Zoologie.*

Stannius, *Das perpherische Nervensystem der Fische.* Rostock, 1849.

Chapter IX

Büchner, L., *Kraft und Stoff,* ed. 7. Leipzig, 1862.

Comte, A., *Cours de philosophie positive,* ed. 3. Paris, 1869.

Liebig, J., *Chemische Briefe.* Heidelberg, 1844.

—, *Die organische Chemie in ihrer Anwendung auf Agrikultur und Physiologie.* Braunschweig, 1840.

—, *Die organische Chemie in ihrer Anwendung auf Physiologie und Pathologie.* Braunschweig, 1842.

Mill, J. S., *System of Logic,* ed. 5. London, 1852.

Moleschott, J., *Kreislauf des Lebens.* Mainz, 1852.

Vogt, K., *Köhlerglaube und Wissenschaft,* ed. 4. Giessen, 1855.

Wagner, R., *Menschliches Gehirn als Seelenorgan.* Göttingen, 1862.

—, *Vortrag mit Diskussion im Bericht über die 31. Versammlung deutscher Naturforscher und Ärzte.* Göttingen, 1854.

Chapters X–XII

Agassiz, L., *An Essay on Classification.* London, 1859.

—, *Recherches sur les poissons fossiles.* Neuchâtel, 1833 ff.

—, Essays in *American Journal of Science and Arts.*

Darwin, Ch., *The Structure of Coral Reefs*. London, 1842.

—, *Geological Observations on Volcanic Islands*. London, 1844.

—, *A Monograph on Cirripedia*. London, 1851 ff.

—, *The Origin of Species*, ed. 1, 2, 5. London, 1859, 1860, 1872.

—, *Variations of Animals and Plants under Domestication*. London, 1868.

—, *The Descent of Man*. London, 1871.

—, *Expression of Emotions*. London, 1872.

—, *Movements and Habits of Climbing Plants*. London, 1875.

—, *Insectivorous Plants*. London, 1875.

—, *The Effects of Cross- and Self-fertilization*. London, 1876.

—, *On Earth-worms*. London, 1881.

—, *Life and Letters*, ed. Fr. Darwin. London, 1887 ff.

Gray, A., *Darwiniana*. New York, 1887 ff.

Hooker, J. D., *Flora Tasmaniæ*. London, 1860.

Huxley, T. H., *Evidence as to Man's Place in Nature*. London, 1863.

—, *Lectures on the Elements of Comparative Anatomy*. London, 1864.

—, *Lay Sermons, addresses and reviews*. London, 1871.

—, *Life and Letters*, ed. L. Huxley. London, 1900.

—, Essays in *Transactions* of Royal Society.

Lyell, Ch., *Elements of Geology*. London, 1832 ff.

—, *The Geological Evidences of the Antiquity of Man*. London, 1863.

Malthus, T., *Essay on Population*, ed. 5. London, 1817.

Quatrefages, A. de, *Charles Darwin et ses précurseurs français*. Paris, 1870.

Spencer, H., *Essays*. London, 1868.

—, *A System of Synthetic Philosophy*. London, 1870 ff.

Wallace, A. R., *Contributions to the Theory of Natural Selection*. London, 1870.

—, *Island Life*. London, 1880.

—, *Darwinism*. London, 1889.

Wigand, A., *Der Darwinismus*. Braunschweig, 1874 ff.

Chapters XIII, XIV

Fürbringer, M., *Untersuchungen zur Morphologie und Systematik der Vögel*. Amsterdam, 1888.

Gegenbaur, C., *Grundzüge der vergleichenden Anatomie*, ed. 1, 2. Leipzig, 1859, 1870.

—, *Untersuchungen zur vergleichenden Anatomie der Wirbeltiere*. Leipzig, 1864 ff.

—, *Grundriss der vergleichenden Anatomie*. Leipzig, 1874.

—, *Vergleichende Anatomie der Wirbeltiere*. Leipzig, 1898.

—, *Gesammelte Abhandlungen*. Leipzig, 1912.

Haeckel, E., *Die Radiolarien*. Berlin, 1862, 1888.

—, *Die Kalkschwämme*. Berlin, 1872.

—, *Das System der Medusen*. Jena, 1879.

Haeckel, E., *Generelle Morphologie*. Berlin, 1866.

—, *Natürliche Schöpfungsgeschichte*. Berlin, 1868.

—, *Anthropogenie*. Leipzig, 1874.

—, *Gesammelte Vorträge*. Bonn, 1878.

—, *Welträtsel*. Bonn, 1903.

—, *Fünfzig Jahre Stammesgeschichte*. Jena, 1916.

—, *Entwicklungsgeschichte einer Jugend*. Published by Schmidt, Leipzig, 1921.

—, *Italienfahrt*. Published by Schmidt, Leipzig, 1921.

Hubrecht, A., *Die Säugetierontogenese*. Jena, 1909.

Mach, E., *Analyse der Empfindungen*. Jena, 1906.

—, *Erkenntnis und Irrtum*. Leipzig, 1906.

Müller, F., *Für Darwin*. Leipzig, 1864.

Chapter XV

Altmann, R., *Die Elementarorganismen*. Leipzig, 1890.

Balfour, F., *Works*. London, 1884 ff.

Beneden, E. van, *Recherches sur la maturation de l'œuf*. Paris, 1883.

Boveri, Th., *Zellenstudien*. Jena, 1887 ff.

Braun, A., *Betrachtungen über die Verjüngung in der Natur*. Leipzig, 1851.

Buchner, E., Essays in *Berichten der Deutschen chemischen Gesellschaft*.

Bütschli, O., *Studien über die ersten Entwicklungsvorgänge der Eizelle*. Frankfurt a. M., 1876.

—, *Protozoa*. Bronn's *Klassen und Ordnungen*. Leipzig, 1889.

Cajal, S. Ramón, *Histologie du système nerveux*. Paris, 1909 ff.

Engelmann, Th., Essays in Pflüger's *Archiv für Physiologie*.

Engler, A., *Entwicklungsgeschichte der Pflanzenwelt*. Leipzig, 1879.

Flemming, W., *Zellsubstanz, Kern und Zellteilung*. Leipzig, 1882.

Fol, H., *Lehrbuch der vergleichenden mikroskopischen Anatomie*. Leipzig, 1896.

—, Essays in *Mémoires de la société de physique et d'histoire naturelle*. Geneva.

Goette, A., *Entwicklungsgeschichte der Unke*. Leipzig, 1874.

Hansen, E. C., Essays in *Mitteilungen des Carlsberg-Laboratoriums*. Copenhagen.

Heidenhain, M., *Plasma und Zelle*. Jena, 1907 ff.

Heidenhain, R., Essays in *Studien des Physiologischen Instituts zu Breslau*.

Hensen, V., *Die Planktonexpedition*. Kiel, 1890 ff.

Hertwig, O., *Allgemeine Biologie*, ed. 4. Jena, 1912.

—, *Das Werden der Organismen*. Jena, 1916.

—, *Zur Abwehr des ethischen*, etc., *Darwinismus*. Jena, 1918.

—, Essays in *Morphologische Jahrbücher*, *Jenaische Zeitschrift*, and *Zeitschrift für wissenschaftliche Mikroskopie*.

Hertwig, R., *Lehrbuch der Zoologie*. Jena, 1890.

—, Essays on the Protozoa in scientific journals.

Hertwig, O. and R., *Untersuchungen zur Morphologie der Zelle*, I–VI. Jena, 1884 ff.

His, W., *Unsere Körperform*. Leipzig, 1874.

Hofmeister, W., *Vergleichende Untersuchungen der Keimung höherer Kryptogamen* Leipzig, 1851.

Holmgren, E., *Text-book on Histology*. Stockholm, 1920. (Swedish)

Kleinenberg, N., *Hydra*. Leipzig, 1872.

—, Essays in the *Zeitschrift für wissenschaftliche Zoologie*, 1886.

Koch, R., *Gesammelte Werke*. Leipzig, 1912 ff.

Korschelt, E., and Heider, K., *Lehrbuch der vergleichenden Entwicklungsge- schichte der wirbellosen Tiere*. Jena, 1902.

Kowalewsky, A., Essays in the *Acta* of the Imperial Academy, St. Peters- burg.

Lang, A., *Lehrbuch der vergleichenden Anatomie*. Jena, 1890 ff.

Lankester, E. R., *Treatise on Zoology*. London, 1900.

Laveran, A., *Nature parasitaire des accidents de l'impaludisme*. Paris, 1881.

Möbius, K., *Fauna der Kieler Bucht*. Leipzig, 1865.

Nägeli, C., *Entstehung des Begriffes der naturhistorischen Art*. Munich, 1865.

—, *Mechanisch-physiologische Theorie der Abstammungslehre*. Munich, 1884.

—, Essays in *Zeitschrift für wissenschaftliche Botanik*.

Ranvier, L., *Leçons d'anatomie générale*. Paris, 1881.

Rollett, A., *Untersuchungen über die quergestreiften Muskelfasern*. Vienna, 1885.

Ross, R., *Researches on Malaria*. Stockholm, 1904.

Schaudinn, F., Essays in *Arbeiten aus dem Kaiserlichen Gesundheitsamte* and in *Sitzungsberichte der Königl. Akademie der Wissenschaften*. Berlin.

Schimper, A. F. W., *Pflanzengeographie auf physiologischer Grundlage*. Jena, 1898.

Schwendener, S., *Untersuchungen über den Flechtenthallus*. Leipzig, 1860.

—, *Das mechanische Prinzip im anatomischen Bau der Monocotyle*. Leipzig, 1874.

—, *Mechanische Theorie der Blattstellungen*. Leipzig, 1878.

Strasburger, E., *Zellbildung und Zellteilung*, ed. 1–3. Jena, 1875–80.

Wiedersheim, R., *Vergleichende Anatomie der Wirbeltiere*, ed. 6. Jena, 1906.

Chapters XVI–XVIII

Bataillon, A., Essays in *Archives de Zoologie expérimentale et générale*.

Bateson, W., *Mendel's Principles of Heredity*. Cambridge, 1909.

Baur, E., *Einführung in die experimentelle Vererbungslehre*, ed. 4. Berlin, 1919.

Bayliss, W., *Principles of General Physiology*, ed. 2. London, 1918.

Bethe, A., Essays in *Archiv für die gesamte Physiologie*, Vol. LXX.

Bunge, G., *Lehrbuch der physiologischen und pathologischen Chemie*, ed. 2. Leip- zig, 1889.

Driesch, H., *Analytische Theorie der organischen Entwicklung*. Leipzig, 1904.

—, *Philosophie des Organischen*. Leipzig, 1909.

—, Essays in *Morphologische Jahrbücher* and *Zeitschrift für wissenschaftliche Zoologie*.

Eimer, Th., *Die Entstehung der Arten*. Jena, 1888 ff.

Galton, F., *Natural Inheritance*. London, 1889.

—, Essays in the *Journal* of the Anthropological Institute. London, 1875.

Giard, A., *Controverses transformistes*. Paris, 1904.

Goebel, K., *Organographie der Pflanzen*. Jena, 1898.

—, *Einleitung in die experimentelle Morphologie der Pflanzen*. Leipzig, 1908.

Herbst, C., Essays in *Archiv für Entwicklungsmechanik*.

Heribert-Nilsson, N., *Experimentelle Studien über Variabilität*, etc. *in der Gattung Salix*. Lund, 1918.

Hertwig, O., *Zeit- und Streitfragen der Biologie*, I, II. Jena, 1894 ff.

—, *Der Kampf um die Kernfragen der Entwicklungs- und Vererbungslehre*. Jena, 1909.

Höber, R., *Physikalische Chemie der Zelle und der Gewebe*, ed. 4. Leipzig, 1914.

Johannsen, W. L., *Elemente der exakten Erblichkeitslehre*, ed. 2. Jena, 1913.

—, *Erblichkeit*. Copenhagen, 1917. (Danish)

—, *False Analogies*, Swedish by Larsson. Stockholm, 1917.

—, *Biology in the Nineteenth Century*. Copenhagen, 1919. (Danish)

Lehmann, A., *Grundzüge der Psychophysiologie*. Leipzig, 1912.

Loeb, J., *Über den chemischen Charakter des Befruchtungsvorganges*. Leipzig, 1908.

—, *The Mechanistic Conception of Life*. Chicago, 1912.

Lotsy, J., *Qu'est-ce qu'une espèce*. Amsterdam, 1916.

Lubbock, J., *On the Senses, Instincts and Intelligence of Animals*. London, 1888.

Mendel, G., *Versuche über Pflanzenhybriden*. Ostwald's *Klassiker der Naturwissenschaft*. Leipzig, 1913.

Morgan, T., *The Mechanism of Mendelian Heredity*. New York, 1915.

Neumeister, R., *Betrachtungen über das Wesen der Lebenserscheinungen*. Jena, 1903.

Nilsson-Ehle, H., Essays in the *Journal* of the Swedish Society of Seed-Culture and in the journal *Hereditas*. Lund.

Overton, E., "*Uber den Mechanismus der Resorption und Sekretion.*" Nagel's *Handbuch der Physiologie*. Braunschweig, 1907.

Pauly, *Darwinismus und Lamarckismus*. Munich, 1905.

Pfeffer, W., *Pflanzenphysiologie*, ed. 2. Leipzig, 1897 ff.

Plate, L., *Selektionsprinzip und Probleme der Artbildung*, ed. 4. Leipzig, 1913.

Punnett, R. C., *Mendelism*. London, 1911.

Rádl, E., *Geschichte der biologischen Theorien der Neuzeit*. Leipzig, 1905–13.

—, *Neue Lehre vom zentralen Nervensystem*. Leipzig, 1912.

Rauber, A., Essays in *Morphologisches Jahrbuch*, 1883.

Romanes, G., *Animal Intelligence*, ed. 3. London, 1888.

Roux, W., *Collected Essays*. Leipzig, 1895.

—, Essays in *Archiv für Entwicklungsmechanik*.

Sachs, J., *Geschichte der Botanik*. Munich, 1875.

—, *Vorlesungen über Pflanzenphysiologie*. Leipzig, 1882.

Semon, R., *Das Problem der Vererbung erworbener Eigenschaften.* Leipzig, 1912.

Spemann, H., Essays in *Archiv für Entwicklungsmechanik.*

Uhlenhuth, P., Essays in *Deutsche medizinische Wochenschrift.*

Verworn, M., *Allgemeine Physiologie*, ed. 5. Jena, 1909.

Vries, H. de, *Die Mutationstheorie.* Leipzig, 1901 ff.

—, Essays in *Berichten der deutschen botanischen Gesellschaft.*

Wasmann, E., *Die moderne Biologie*, ed. 3. Freiburg, 1906.

Weismann, A., *Aufsätze über Vererbung.* Jena, 1892.

—, *Das Keimplasma.* Jena, 1892.

—, *Über Germinalselektion.* Jena, 1896.

—, *Die Selektionstheorie.* Jena, 1909.

Wundt, W., *Vorlesungen über die Menschen- und Tierseele*, ed. 2. Hamburg, 1892.

LITERATURE

Boucke, *Goethes Weltanschauung.* Stuttgart, 1907.

Burckhardt, *Geschichte der Zoologie*, ed. 1, 2. Leipzig, 1907, 1921.

Carus, *Geschichte der Zoologie.* Munich, 1872.

Eckermann, J. P., *Gespräche mit Goethe.* Leipzig, 1836.

Fries, Th., *Linné.* Stockholm, 1903. (Swedish)

Gomperz, *Griechische Denker.* Leipzig, 1896.

Haeser, *Geschichte der Medizin.* Jena, 1875.

Höffding, *Geschichte der neueren Philosophie.* Leipzig, 1895.

—, *Modern Philosophers.* Copenhagen, 1904. (Danish)

Hulth, J. M., *Bibliographia linneana*, I. Upsala.

Kohlbrugge, J. H. T., *Historisch kritische Studien über Goethe als Naturforscher.* *Zoologische Annalen*, Vols. V, VI. Würzburg, 1913–14.

Kopp, H., *Geschichte der Chemie.* Braunschweig, 1843.

Lamm, M., *Swedenborg.* Stockholm, 1915. (Swedish)

Lange, *Geschichte des Materialismus*, ed. 1, 9. Leipzig, 1914.

May, W., *Haeckel.* Leipzig, 1909.

Paulsen, Fr., *Kant.* Stuttgart, 1899.

Rádl, *Geschichte der biologischen Theorien*, ed. 1, 2. Leipzig, 1907, 1909, 1913.

Roth, M., *Andreas Vesalius.* Berlin, 1892.

Sachs, J., *Geschichte der Botanik*, Munich, 1875.

Zeller, *Die Philosophie der Griechen.* Tübingen, 1879.

INDEX

i

A NOTE ON THE TYPE
IN WHICH THIS BOOK IS SET

This book is set on the Monotype in Garamont, a modern rendering of the type first cut in the sixteenth century by Claude Garamond (1510–1561). He was a pupil of Geofroy Tory and is believed to have based his letters on Venetian models, although he introduced a number of important differences. It is to him we owe the letter which we know as Old Style. He gave to his letters a certain elegance and a feeling of movement which won for their creator an immediate reputation and the patronage of the

French King, Francis I.

MANUFACTURED BY MONTAUK BOOKBINDING CORP.